Materials, Chemicals, and Energy from Forest Biomass

ACS SYMPOSIUM SERIES **954**

Materials, Chemicals, and Energy from Forest Biomass

Dimitris S. Argyropoulos, Editor
North Carolina State University

American Chemical Society, Washington, DC

Library of Congress Cataloging-in-Publication Data

Materials, chemicals, and energy from forest biomass / Dimitris S. Argyropoulos, editor.

p. cm.—(ACS symposium series ; 954)

Includes bibliographical references and index.

ISBN 978–0–8412–3981–4 (alk. paper)

1. Biomass energy. 2. Forest biomass.

I. Argyropoulos, Dimitris S., 1956–

TP339.M367   2006
620.1′.17—dc22   2006043053

The paper used in this publication meets the minimum requirements of American National Standard for Information Sciences—Permanence of Paper for Printed Library Materials, ANSI Z39.48–1984.

PRINTED IN THE UNITED STATES OF AMERICA

Foreword

The ACS Symposium Series was first published in 1974 to provide a mechanism for publishing symposia quickly in book form. The purpose of the series is to publish timely, comprehensive books developed from ACS sponsored symposia based on current scientific research. Occasionally, books are developed from symposia sponsored by other organizations when the topic is of keen interest to the chemistry audience.

Before agreeing to publish a book, the proposed table of contents is reviewed for appropriate and comprehensive coverage and for interest to the audience. Some papers may be excluded to better focus the book; others may be added to provide comprehensiveness. When appropriate, overview or introductory chapters are added. Drafts of chapters are peer-reviewed prior to final acceptance or rejection, and manuscripts are prepared in camera-ready format.

As a rule, only original research papers and original review papers are included in the volumes. Verbatim reproductions of previously published papers are not accepted.

ACS Books Department

Contents

Chemicals from Forest Biomass

Energy from Forest Biomass

Novel Analytical Methods for Structural Elucidation of Forest Biomass

Indexes

Preface

Without a doubt, our society's concerns over increasing fuel prices, green house gas emissions, and the associated global warming have created a tremendous interest in the science and technologies that promise the sustainable production of materials, chemicals, and energy from domestic resources. In this respect, lignocellulosic biomass has the unique ability to supply all of the above because carbohydrates and lignin are among the most abundant organic compounds on the planet, representing a vast amount of biomass (in the range of hundreds of billions of tons). It is interesting to note that only 3% of this vast resource is actually used by humans. Despite the fact that a significant amount of research has been carried out aimed at augmenting the industrial use of readily available carbohydrates as organic raw materials, the systematic exploitation of this vast resource is still in its infancy. The fundamentally different chemistries of hydrocarbons and of carbohydrates and lignins are perhaps pivotally important in imposing serious difficulties for their use as organic raw materials. Our fossil carbon-based economy relies on distinctly hydrophobic hydrocarbon molecules that are devoid of oxygen and functional groups. In contrast to hydrocarbons, carbohydrates are highly functionalized and hydrophilic molecules. Because environmental pressures are mounting and our dependence on fossil fuel continues to grow, any prevailing economic advantages for a petrochemical-based economy will likely fade away within the next 50 years. Our chemical and energy industry therefore needs to redevelop in a major way if it is to use lignocellulosic biomass as its feedstock. These considerations unambiguously dictate the need for practically oriented scientific research and development covering a wide range of applications for the production of *Materials, Chemicals, and Energy from Forest Biomass*.

The literature is abundant with research accounts aimed at offering an understanding to these complex chemical processes. To date no concerted effort has been made about bringing this knowledge together with the aim to connect the past with the future. Consequently, during the Pacifichem meeting of 2005 a Symposium was held in Hawaii, focused at bringing together the global expertise from academia, government, and industry with the aim to disseminate their latest findings and to exchange their ideas for the future in the realm of *Materials, Chemicals, and Energy from Forest Biomass*. The present book attempts to offer the reader a thorough view of the information presented at this meeting and beyond. Despite the fact that the material emerged from a symposium, it is not a collection of fragmented research findings in the form of conference proceedings. Most chapter contributors attempted to provide a good review of the literature, creating a sound foundation for the science to be subsequently developed. Furthermore, a collection of authoritative reviews is also provided at the onset of the book prior to embarking on specific topics.

In an effort to convey the material in a coherent fashion the 33 chapters of this book were divided into the following distinct sections that deal with:

- Critical Reviews
- Materials from Forest Biomass
- Chemicals from Forest Biomass
- Energy from Forest Biomass
- Novel Analytical Methods for the Structural Elucidation of Forest Biomass

The editor anticipates that this volume will provide a resource for new ideas, guidance, and a good embarkation point for any future endeavors in *Materials, Chemicals, and Energy from Forest Biomass*.

Acknowledgments

This compilation is a product of a concerted effort of numerous individuals to whom the editor expresses his appreciation. Primarily, thanks are due to the many scientists who have contributed their time and effort in documenting their research findings. Without their enthusiastic response and support this volume would not have been possible. No book of this nature would have been possible if it were not for the support of nearly 100 members of the international scientific community who acted as critical reviewers to the submitted chapters. Their invaluable critical and constructive comments ensured that each chapter was of the highest scientific standard, reflecting the state of the art, and was presented in the best possible way. Last but not least, the numerous editorial and administrative contributions of Ms. Eliza Root are gratefully acknowledged.

Dimitris S. Argyropoulos

Department of Forest Biomaterials Science and Engineering
North Carolina State University
Raleigh, NC 27695–8005

Materials, Chemicals, and Energy from Forest Biomass

Critical Reviews

Chapter 1

Chemicals, Materials, and Energy from Biomass: A Review

Lucian A. Lucia, Dimitris S. Argyropoulos, Lambrini Adamopoulos, and Armindo R. Gaspar

Department of Forest Biomaterials Science and Engineering, North Carolina State University, Raleigh, NC 27695–8005

There are approximately 89 million metric tonnes of organic chemicals and lubricants produced annually in the United States (1). The majority of these are fossil fuel-based materials that have the potential to become environmental pollutants during use and that carry end-of-life cycle concerns such as disposal, pollution, and degradation. As a result, the need to decrease pollution caused by petrochemical usage is currently impelling the development of green technologies. It is virtually inarguable that the dwindling hydrocarbon economy will eventually become unsustainable. The cost of crude oil continues to increase, while agricultural products see dramatic decreases in world market prices. These trends provide sufficient basis for renewed interest in the use of biomass as a feedstock and for the development of a lignocellulosic-based economy as the logical alternative to fossil fuel resources.

Petroleum: The Current Resource

Petroleum is known by a number of names: crude oil, naphtha, and "black gold." It is perhaps the most chemically versatile and important of the fossil fuels, the so-called "biomass of earlier eras," such as coal and natural gas, formed by a combination of geological, biological, and chemical factors over geological time scales (2). Petroleum is an indispensable resource, whose availability is integral to the function of our modern society. However, since the energy crises of the 1970s, the world realizes that petroleum is a finite resource and its availability is limited. Taking into account the production and existing reserves, Hubbert has predicted a maximum peak for the oil production (Figure 1a) (3) which ominously coincides with current increasing petroleum prices. The price of petroleum tends to fluctuate, but is very closely related to world events (Figure 1b) (4).

In the United States, 40% of its total energy consumption is petroleum-based. This makes it perhaps the world's most important commodity, and obtaining it has been a factor in several military conflicts. Petroleum is primarily used as fuel oil (90% by volume), such as motor gasoline and diesel fuel. Other uses include finished non-fuel products, like chemical solvents and lubricating oils. It is also used as a raw feedstock (naphtha and various refinery gases) for many sundry chemicals by the petrochemical industry. All the hydrocarbon fractions of petroleum are typically separated by their boiling point range (5).

Besides its intrinsic fuel value, it is also used to produce petrochemicals. In general, petrochemicals, chemicals that originate from gas or petroleum, can be divided into two product classes, from which other chemicals are derived. These chemicals are olefins and aromatics (2). Olefins are straight- or branched-chain unsaturated hydrocarbons. These include ethylene, propylene, and butadiene. Aromatics are cyclic, unsaturated hydrocarbons. The most important of these are benzene, toluene, p-xylene and o-xylene.

Presently, petroleum resources are used to manufacture billions of pounds of chemical products. In order to manufacture these industrial chemicals and materials, about one million barrels of oil are consumed annually. This corresponds to about 6.5 quadrillion BTU of petroleum equivalents.

Biomass: The Alternative Resource

The National Renewable Energy Laboratory defines biomass as organic matter available on a renewable basis (6). Biomass includes forest and mill residues, agricultural crops and wastes, wood and wood wastes, animal wastes, livestock operation residues, aquatic plants, fast-growing trees and plants, and municipal and industrial wastes. There has been much debate about the definition of biomass, however, in general, biomass can include anything that is not a fossil fuel that can be argued to be organic-based.

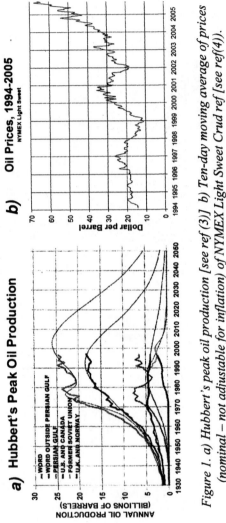

Figure 1. a) Hubbert's peak oil production [see ref (3)] b) Ten-day moving average of prices (nominal – not adjustable for inflation) of NYMEX Light Sweet Crud ref [see ref(4)).

The utilization of biomass as a raw material presents several advantages It provides a naturally abundant resource that is not only cheap but also sustainable. Also, the bioindustry presents an environmentally friendly alternative to the petroleum industry. The productive use of waste residues also allows for lower emissions to the atmosphere and could have ancillary benefits such as providing a new source of economic growth for rural communities. Presently, the available biomass resources could provide as much as 6-10 quadrillion BTU of feedstock energy. This corresponds to the energy required to manufacture over 300 billion pounds of organic chemicals (6).

Rationale for its Use

Presently, technologies that rely on the use of fossil fuels for energy and chemicals are predominant. These are directed to produce fuels, power, chemicals and materials. Increasing energy demands coupled with crude oil prices from 60-70 USD/barrel, accompanied by growing concerns about global change and environmental pollution allow biomass technologies the opportunity to become competitive. With the introduction of stricter emission laws, lower margins, current petroleum prices, offshoring, and declining profits, chemical industries must reinvent themselves. Consequently, there exists a real opportunity to discover new ways to produce novel and quality products, within the context of sustainability issues that are beginning to permeate recent industrial thinking. In this context, issues such as producing biomaterials from renewable natural resources are at the forefront. As fossil fuel feedstocks are irreversibly diminishing, environmental pressures are escalating and the availability of inexpensive crude oil is coming to an end (7,8,9). Thus, the evolution to renewable feedstocks for energy and chemicals seems inevitable (9, 10).

The combustion of fossil fuels contributes to carbon dioxide accumulation in the atmosphere. Its accumulation, along with deforestation activities, are the main culprits for radical climactic changes in the global environment (11). During the 1990s, greenhouse gases released to the atmosphere annually from the burning of fossil fuels was 6.3 Gtonnes in carbon units and an additional 1.6 Gtonnes from deforestation activities (12,13). It has also been calculated that the annual increase of CO_2 to the atmosphere is 3.3 Gtonnes as carbon (14). The remaining 4.6 Gtonnes is absorbed annually by the world's oceans and terrestrial vegetation (13). Figure 2 is a picture of the carbon cycle mass balance.

Traditional Uses

Bioproducts are industrial and consumer goods manufactured wholly or in part from renewable biomass. They are mainly created from primary resources,

Carbon Cycle

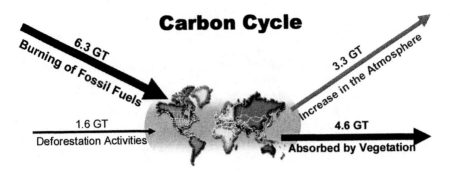

Figure 2. Mass balance of the carbon cycle during the 1990's per year
(12,13,14).

which are corn, wood, soybeans and plant oils. The tremendous diversity of the bioproducts, ranging from paper, to solvents, to pharmaceuticals makes them an integral part of our lives.

Variations in the definition of what should be considered a bioproduct are responsible for disparities in the estimates of annual usage of biomass in the manufacture of industrial and consumer products. The National Renewable Energy Laboratory estimated that about 21 billion pounds of biomass were used in 2001 (6). On the other hand, Energetics Inc.'s estimate was much more conservative, at approximately 12.4 billion pounds for 2001 (6).

The thousands of different industrial bioproducts produced today can be considered as stemming from sugar and starch, oils and lipids, gum, wood and finally cellulose (6, 15, 16, 17, 18). Sugar and starch bioproducts are derived from fermentation and thermochemical processes. The feedstocks include sugarcane, corn, potatoes, barley and sugar beets. These products include alcohols and acids. The annual production of industrial corn starch products is estimated at 6500 million pounds, which corresponds to an estimated value of 2200 million dollars. Industrial ethanol production is estimated at 3.41 billion gallons, an estimated value of 961 million dollars. Oil- and lipid- based bioproducts include fatty acids and oils derived from soybeans, rapeseed, or other oil seeds. There are 400 million pounds of glycerine, equivalent to 320 million dollars, produced annually. Over 1200 million pounds of soy oil, peppermint, spearmint and other plant oils produced annually have an estimated value of 550 million dollars. Gum and wood chemicals include tall oil, alkyd resins, rosins, pitch, fatty acids, turpentine, and other chemicals derived from trees. The annual production of these is approximately 3200 million pounds per year. The value of these is estimated at 890 million dollars. Cellulose derivatives, fibers and plastics include products derived from cellulose, including cellulose acetate (cellophane) and triacetate, cellulose nitrate, alkali cellulose, and regenerated cellulose. The primary sources of cellulose are bleached wood pulp and cotton linters. Cellulose derivatives have an annual production of 2140 million pounds. Their estimated value is 1400 million dollars.

Finally, a portion of the global and US economy is biobased and relies on carbohydrates. Conventional uses of biomass already account for over 400 billion dollars in products annually (6).

Future Uses

There exists a real opportunity to perform valuable, exploitable research, and discover new ways to produce novel and quality products, within the context of sustainability issues that permeate every recent industrial endeavor. Environmental sustainability can be understood as the long-term maintenance of valued environmental resources in an evolving human context. In this context, the production of biomaterials from renewable natural resources is at the forefront.

The use of biomass for energy, materials and chemicals parallels the concept of the biorefinery, as well as that of sustainability. A biorefinery is a facility that integrates biomass conversion processes and equipment to produce fuels, power, heat and high-value chemicals from biomass. Byproducts, residues and a portion of the produced fuels would be used to fuel the biorefinery itself. The biorefinery is analogous to a petroleum refinery, producing multiple fuels and products. A biorefinery might produce transportation fuels (low-value product) in high volumes, high-value chemicals in low volumes, while generating electricity and process heat for its own use and perhaps surplus for sale into the power grid. The high-value products enhance profitability, the high-volume fuel helps meet national energy needs, and the power production reduces costs and avoids greenhouse-gas emissions.

The pulp and paper industry has been practicing aspects of the "biorefinery" philosophy almost since its inception. Indeed, wood is converted into pulp for papermaking and black liquor is used in the recovery furnace for power and steam generation while the tall oil is sold for conversion into high-value chemicals.

For example, Archer Daniels Midland Co. is a chemical company that has an expanded biorefinery in Decatur, Illinois. Their large corn wet-milling plant produces industrial enzymes, lactic acid, citric acid, amino acids and ethanol. Electricity and steam are obtained through an on-site cogeneration system.

Another company that has received a lot of attention for its biorefinery is Cargill Dow (19, 20). Their polylactide facility in Blair, Nebraska, currently relies on corn grain for glucose, which it converts to lactic acid. The plant capacity is 300 million pounds, and demand for the polylactide product is strong. Cargill Dow projects a possible market of 8 billion pounds by 2020. Their polylactide is cost-competitive, high performing and requires 30-50% less fossil fuel for production than conventional polymers derived from petroleum. Finally, when it has reached the end of its life, it can be melted and reused or made into compost.

With the right technology, abundant biomass resources may be converted into valuable bioproducts and energy. What's more, the development of low-cost enzymes and recombinant technologies allows for the engineering of exemplary microbes. These should also be examined in conjunction with chemicals derived from biomass, as they are an essential part of the biorefinery. Advances in chemical, biochemical and separation technologies allow for the emergence of more and more avenues for bioproducts, and enhanced performance and cost-competitiveness will almost certainly lead to increased market share for biomass-based products.

Bio-Based Chemicals Versus Parallel Petrochemicals

Cellulosic biomass from plants is a raw source of sugars for industrial processes. The advantage of this biomaterial is that it will theoretically be less expensive than petroleum as a feedstock, it will not affect food supplies, and all chemicals derived from it will have a lower environmental impact than petrochemicals. Additionally, it is considered carbon dioxide neutral since burning it with coal in power plants doesn't add carbon to the environment beyond what was required for the plant to grow. Five percent of all global chemical sales relate to "green" products such as ethanol, pharmaceutical intermediates, citric acid, and amino acids. This market share may go as high as 20% by 2010 and may reach as high as 2/3 of the total global economy if low-cost enzymes and new recombinant technologies to make more efficient enzymes are available.

The Pacific Northwest National Laboratory took part in a collaborative study with the National Renewable Energy Laboratory in an effort to identify the top-tier building block chemicals for biorefineries (21). The top twelve chemicals (Figure 3) were selected based on their compatibility with existing petrochemical processing as well as their ease of synthesis. The compounds are: 1,4 diacids (succinic, fumaric and malic), 2,5 furandicarboxylic acid, 3-hydroxypropionic acid, aspartic acid, glucaric acid, glutamic acid, itaconic acid, levulinic acid, 3-hydroxybutyrolactone, glycerol, sorbitol and xylitol/arabinitol. These molecules have six to twelve carbon atoms and multiple functional groups. Examining some of these is worthwhile in the effort to show different synthesis pathways for the same product, as well as product derivatives.

Many other chemicals can also be made using biomass, for example, formaldehyde, formic acid, acetic acid, methanol, methane, ammonia, ethylene, carbon monoxide, hydrogen, phenol, etc.

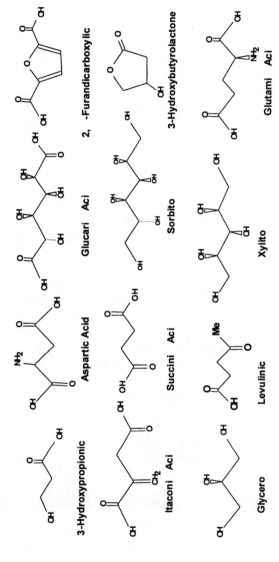

Figure 3. Chemical building blocks from biomass compatible with existing petrochemicals.

Different Synthesis Pathways for Succinic Acid

Succinic acid $(C_4H_6O_4)$ occurs frequently in nature in animal tissues, vegetables and fruits and in amber (22). Succinic acid and its salts may serve as building blocks for numerous chemical intermediates and end products (see Figure 4) in industries producing food and pharmaceutical products, surfactants, detergents, green solvents and biodegradable plastics. Reaction with glycols gives polyesters, and specifically, esters formed by reaction with monoalcohols are important plasticizers and lubricants (23).

The total market for succinic acid use is more than 400,000,000 USD/year (24). Currently, the food market is the only one using succinic acid produced by fermentation. For other uses, succinic acid is by a large extent produced petrochemically from butane through maleic anhydride. Succinic acid can be produced by hydrogenation of maleic acid, maleic anhydride, or fumaric acid with standard catalysts such as Raney nickel, Cu, NiO, or CuZnCr, Pd-Al$_2$O$_3$, Pd-CaCO$_3$ or Ni-diatomite (23).

Succinic acid is a common metabolite for plants and microorganisms. It is derived through the fermentation of glucose, and is a very green technology because it is CO_2-fixing. Succinate is formed from sugars or amino- acids by propionate-producing bacteria such as the genus *Propionibacterium*, gastrointestinal bacteria such as *Escherichia coli* and rumen bacteria such as *Ruminococcus flavefaciens*.

For the rumen ecosystem, succinate is an important precursor for propionate, which, after oxidation, provides energy and biosynthetic precursors to the animal. Rumen microorganisms are major cellulose-digesting anaerobes that produce acetic and succinic acids in the rumen. (25) *Actinobacillus succinogenes* 130 Z is a ruminal anaerobic bacterium that is capable of producing very high concentrations of succinate from many different substrates, such as glucose, maltose, mannose, sucrose, cellobiose and D-xylose.(26)

A. succinogenes uses the phosphoenolpyruviate (PEP) carboxykinase pathway to produce succinic acid (see Figure 5). Four key enzymes come into play: PEP carboxykinase, malate dehydrogenase, funarase and fumarate dehydrogenase. The pathway used for succinic acid production is regulated by CO_2 levels in which PEP carboxykinase fixes CO_2 and synthesizes oxaloacetate from phosphoenolpyruvate. At low CO_2 levels (10 mol CO_2/100 mol glucose), *A. succinogens* produces mostly ethanol. At high CO_2 levels (100 mol CO_2/100 mol glucose), succinate is the major product, with only traces amounts of lactic acid or ethanol produced.

In actuality, the CO_2 concentration regulates the levels of the key enzymes in the pathway, thus determining the majority product. At high CO_2 levels, phosphoenolpyruviate carboxykinase levels rise and alcohol dehydrogenase and lactate dehydrogenase are imperceptible. The CO_2 functions as an electron acceptor and PEP flows to succinate at high levels.

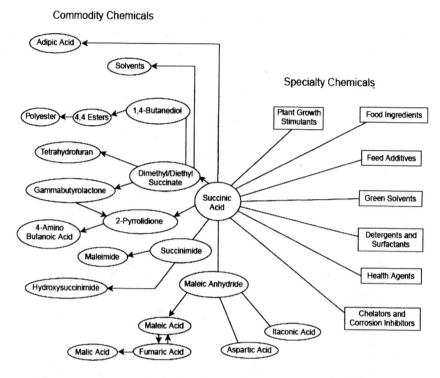

Figure 4. Map of routes to succinic acid- based products [see ref (24)].

Most fermentation organisms are not tolerant of acidic conditions and the fermentation is usually neutralized to obtain a salt of succinic acid. This must then be separated, recovered and re-acidified to form free succinic acid. Significant research in microbial production of succinic acid has resulted in the development of a strain of *E. coli*, AFP111, which shows greatly improved productivity, and fermentation using this microorganism has been successfully tested at a commercial scale (6). Moreover, the development of a two-stage desalting and water-splitting electrodialysis system has facilitated the separation, concentration, purification and acidification of the product (27).

In 1992, fermentation production costs for SA ranged from $1.50 to $2.00 per pound (6). Advances in fermentation and especially separation technology for the bio-based route have reduced the potential production costs to about $0.50 per pound (24). Further improvements in separation and fermentation technologies could drop the production costs even more, enabling commodity scale production. This would allow for the expansion of current markets and for the opportunities in new markets.

12

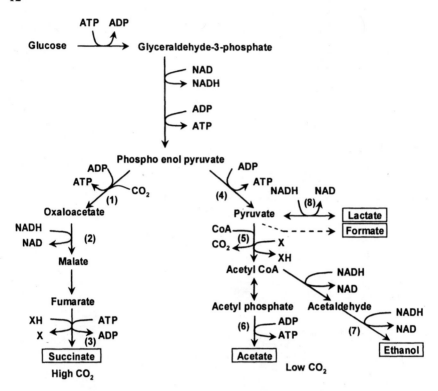

Figure 5. Proposed catabolic pathway for glucose fermentation in A.succiniciproducens and A. succinogens. Steps: 1 phosphoenolpyruviate carboxykinase, 2 malate dehydrogenase, 3 fumarate reductase, 4 pyruvate kinase, 5 pyruvate ferredoxin oxidoreductase, 6 acetate kinase, 7 alcohol dehydrogenase, 8 lactate dehydrogenase. [see ref (28)]

Different Synthesis Pathways for Itaconic Acid

Itaconic acid (2-methylenebutanedioic acid) is an unsaturated dicarbonic acid. It can be regarded as an α-substituted acrylic of methacrylic acid and may serve as a substitute for these petrochemical-based chemicals. Due to its two carboxyl groups, itaconic acid may be easily incorporated into polymers.

Currently, polymerized methyl, ethyl or vinyl esters of itaconic acid are used as plastics, adhesives and coatings. Also, it is used as a co-monomer in rubber-like resins and in the manufacture of emulsion paints, where it improves the adhesion of the polymer. Acrylic lattices are supplemented with itaconic acid and used as non-woven fabric binders. Furthermore, itaconic acid may have agricultural, pharmaceutical and medicinal applications. It is used as a hardening agent in organosiloxanes for use in contact lenses. In addition, mono and diesters of partly substituted itaconic acid possess analgesic properties while some display plant-growth related activities.

The first synthesis of itaconic acid was by pyrolysis of citric acid and anhydride hydrolysis (Baup 1837) (Figure 6).

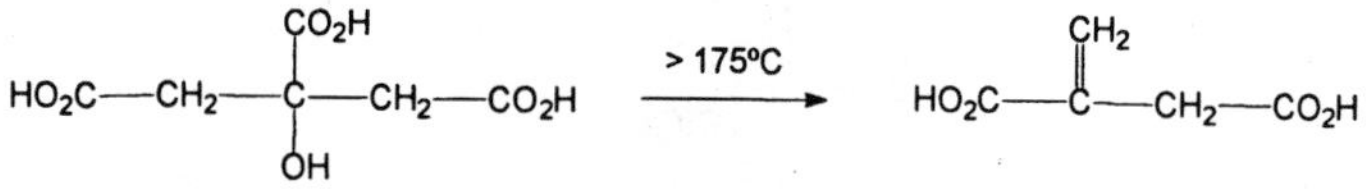

Figure 6. Thermal decomposition of citric acid(29).

Similarly, chemical synthesis is mainly carried out by dry distillation of citric acid and treatment of the anhydride with water. However, use of itaconic acid has been limited because the petroleum route is still expensive.

It cannot compete with the fermentation of carbohydrates by fungi as a source of itaconic acid. In fact, approximately 15,000 tons of itaconic acid are produced annually by fermentation. Production rates do not exceed $1g\ l^{-1}h^{-1}$ and production concentrations of $80g\ l^{-1}$ (*30*). This makes itaconic acid an expensive product, with limited usage.

The biosynthesis of itaconic acid most probably happens through glycolysis, followed by the tricarboxylic acid cycle. Citric acid and aconitic acid are intermediates. Enzymatic decarboxylation of the latter allows for the formation of itaconic acid (Figure 7).

The most frequently used commercial application of itaconic acid is the cultivation of *Aspergillus terreus* with molasses. Molasses products are less expensive than other kinds of carbohydrates, but the cost of itaconic acid is still high because molasses contains many impurities that are not consumed by microorganisms. Therefore, downstream processing and waste treatment become expensive. Thus, alternate carbon sources are being examined. The best yields of itaconic acid are achieved with glucose or sucrose, but these are expensive raw materials. Raw starch materials show particular potential as a substitute to molasses. Itaconic acid was produced by *A. terreus* TN-484 in a medium containing raw corn starch ($140g\ l^{-1}$) at pH 2.0, hydrolyzed with nitric acid at a concentration of more than $60g\ l^{-1}$ in a 2.5-L air lift bioreactor (*31*).

There are efforts are also being made in the selection of different microorganisms such as various *Ustalgo* and *Candida* species. Furthermore, mutant microorganisms are also being produced and used to increase yields and biomass immobilization is being considered.

Finally, fermentation conditions are important and must be optimized. The itaconic acid fermentation is most favorable under phosphate-limited growth conditions at sugar concentrations between 100 and $150g\ l^{-1}$. During fermentation, the pH drops to 2.0 and itaconic acid becomes the main product. The temperature is normally kept at 37°C, but certain mutated microorganisms

14

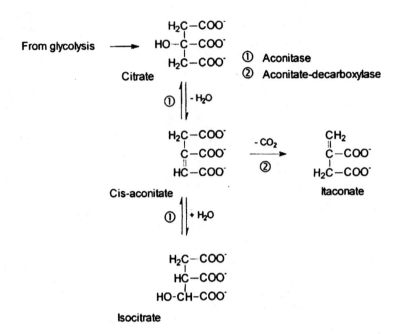

Figure 7. Biosynthesis of Itaconic Acid (30).

may perform better at higher temperatures. Also, aerobic conditions are required and culture medium components, such as Fe, Mn, Mg, Cu, Zn, P and N, must be carefully monitored.

The market for itaconic acid is still growing because of possibilities for substitution of acrylic and methacrylic acid in polymers. Moreover, its potential production from agricultural plant biomass makes it an attractive bio-friendly alternative. It also provides a source of inexpensive raw feedstock. Efforts in effectively utilizing such waste, as well as the introduction of new technologies, are required for reducing itaconic acid production costs and market expansion.

Synthetic Pathways for Levulinic Acid

Levulinic acid ($H_8C_5O_3$), also known as γ-ketovaleric acid, β-acetylpropionic acid, 4-oxopentanoic acid, is a short chain fatty acid with a ketone carbonyl group and an acidic carboxyl group. These two highly reactive functional groups make it a highly versatile chemical capable of serving as a building block for other chemicals (Figure 8). Levulinic acid is the starting compound for many heterocycles. It can be used as a raw material for resins, plasticizers, textiles, animal feed, coatings and antifreeze. It is an auxiliary in electroplating and levulinic acid esters are used in cosmetics.

Industrially, levulinic acid is prepared from wood processing and agricultural wastes. It basically hails from the transformation of hexose sugars in acidic media. The hexoses are obtained by the hydrolysis of cellulose at atmospheric pressure with strong acids such as HCl and H_2SO_4 at 100°C. The hydrolysate is then heated to 110°C with 20% HCl and kept at this temperature for 1 day. Free halogens, as well as transition metals and anion-exchange resins, accelerate the process.

Initially, the intermediate compound 5-hydroxymethyl furfural is formed from the hexoses. This happens through a series of reactions (Figure 9). The first step is the enolization of D-glucose, D-mannose or D-fructose in acidic media to give the ene-diol. Next, this compound is dehydrated and the enol form of 3-deoxyhexosulose. This substance then produces 3,4-dideoxyglycosulosene-3, which is converted to the dienediol which results in 5-hydroxymethylfurfural (*34*).

5-Hydroxymethylfurfural can then be converted to levulinic acid by the addition of a molecule of water to the C2-C3 bond of the furan ring. The ring opens and an unstable tricarbonyl intermediate is formed, which is finally decomposed to formic and levulinic acid (Figure 10. 5-Hydroxymethylfurfural to Levulinic Acid (34).). Levulinic acid is isolated from the mixture with a yield of approximately 40% with respect to the hexose content (*34*). Furthermore, for each levulinic acid molecule generated, a formic acid molecule is also created and insoluble residues, including humans are produced.

Furthermore, levulinic acid can be synthesized starting from furfuryl alcohol (Figure 11). Heating furfuryl alcohol in aqueous organic acids in the presence of hydrochloric acid allows an 80% yield of levulinic acid. The yield can be increased to approximately 90% when the reaction is performed in boiling ethyl methyl ketone.

Moreover, the oxidation of 5-methylfurfural with 28% H_2O_2 at 60°C in the presence of HCOOH produces levulinic acid as well as 4-oxopent-2-enoic acids (Figure 12).

Levulinic acid can also be formed in high yields from the reaction of 4-(diphenylmethylsilyl)butyrolactone with MeMgI.

Levulinic acid can also be prepared from biomass, such as rice straw, paper and cotton (*32*). In this case, biomass is reacted at 40-240 °C for 1 to 96 hours in the presence of 5-90% sulfuric acid. Another substrate which can be used for levulinic acid production is sorghum grain. It has low cash value when sold as a feed grain, and is a major source of carbohydrates. Pentosans, starch and cellulose make up 80-85% of the sorghum grain. Flour made from this grain, at a 10% loading, blended with 8% sulfuric acid gave a yield of approximately 33% levulinic acid at 200 °C (*33*).

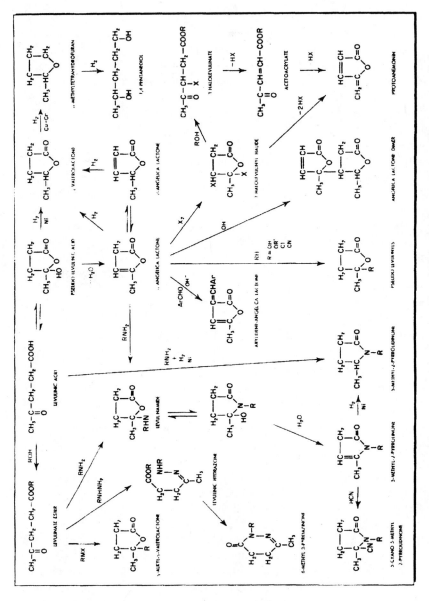

Figure 8. Principal Reactions of Levulinic Acid and Relationship among its Derivatives (34).

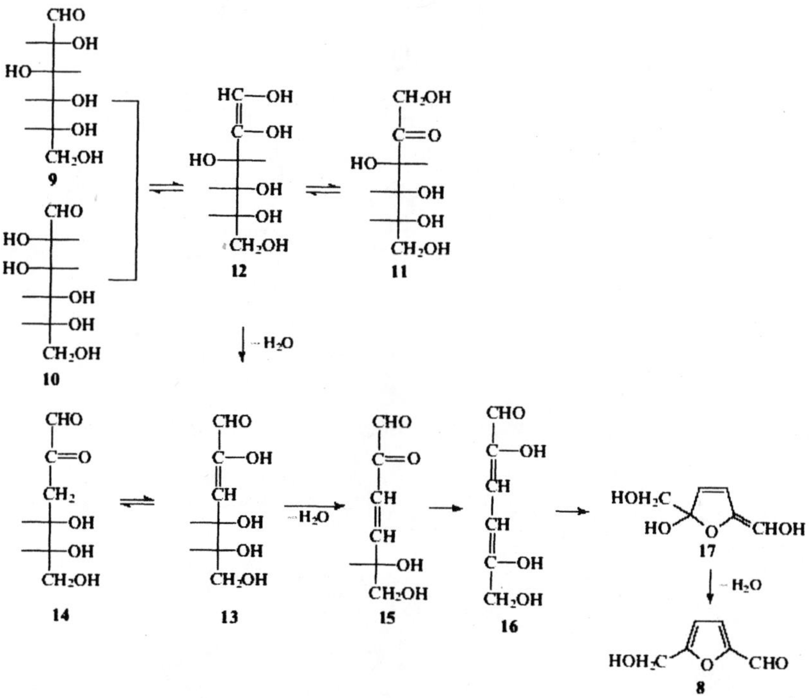

Figure 9. 5-Hydroxymethylfurfural (8) obtained from D-glucose (9), D-mannose (10) or D-fructose (11) (34).

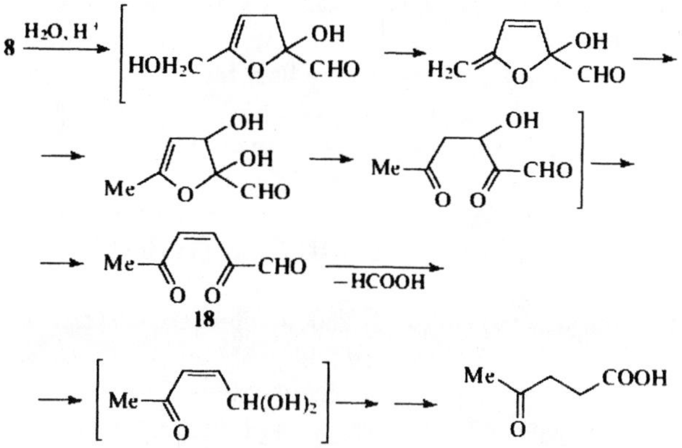

Figure 10. 5-Hydroxymethylfurfural to Levulinic Acid (34).

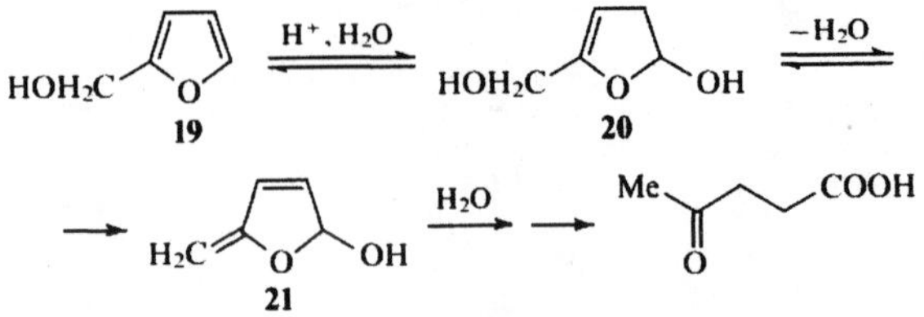

Figure 11. Levulinic Acid from Furfuryl Alcohol (19) (34).

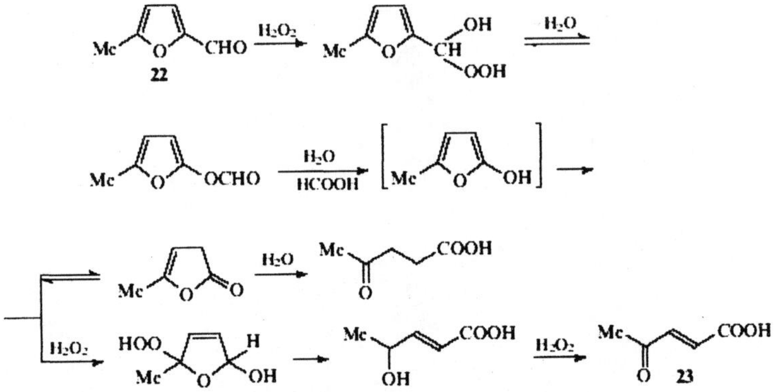

Figure 12. Oxidation of 5-methylfurfural (22) to levulinic acid (34).

Finally, a two-step dilute mineral acid hydrolysis is used, the Biofine process (35), to break down biomass containing lignocellulose into intermediate chemicals. It was developed by BioMetics Inc., with funding from the Department of Energy.

This high temperature, dilute acid hydrolysis process fractionates lignocellulosic biomass into platform chemicals. The primary "fractions" of the process are levulinic acid, formic acid, furfural and a bone dry char. The char residue contains approximately 11,000 BTU per pound and is a suitable feedstock for gasification.

Biofine has constructed and operated a small commercial demonstration unit in New York State for four years and has just completed the first large scale commercial facility in Italy. This large scale facility will produce levulinic acid for monomer use, oxygenated diesel esters and formic acid-based solvents. The char residue will be converted to synthesis gas which will be burned in a gas engine to generate electricity. Approximately 0.5 pounds of levulinic acid is produced per pound of cellulose processed (6). Currently, levulinic acid has a worldwide market of about one million pounds per year at a price of $4-6/lb (36). Large-scale commercialization of the Biofine process could produce levulinic acid for as little as $0.32/lb (36), spurring increased demand.

Parallel with the process development program, Biofine has an ongoing program aimed at expanding commercial markets for the chemical derivatives of levulinic acid in cooperation with New York State, Rensselaer Polytechnic Institute (NY), National Renewable Energy Laboratory (CO), Pacific Northwest National Laboratory (WA) and major chemical companies.

This effort has allowed for the development of a one step catalytic process facilitating the conversion of levulinic acid to MTHF approved for "P-series" alternative fuel formulations (37). Also, it has provided a low cost route to the herbicide and plant growth regulator δ-aminolevulinic acid. New polycarbonate resin formulations using diphenolic acid are being examined and a new oxygenated diesel additive (ethyl levulinate) which reduces particulate matter emission has been developed.

The Biofine process allows for the manufacture of many high-value chemical and fuel product from cheap lignocellulosic biomass. This process is fast and continuous and is capable of utilizing a wide range of biomass feedstocks. Moreover, the gasification of the char covers the plant's entire energy needs, making it self-sufficient. Finally, the process can be retrofitted into existing plants.

Levulinic acid is a compound that has many industrial applications. Its efficient preparation from biomass has many avenues. Its efficient extraction from decomposition products of natural materials would prove to be worthwhile for both industry and the environment.

Sample Reactions for High Value Chemicals

As opposed to many synthetic chemicals, forest-based and agricultural biomaterials are readily renewable, inexpensive, and environmentally benign. Despite a significant amount of recent research for augmenting the industrial use of readily available carbohydrates as raw materials (38,39,40,41,42,43) (e.g., car moldings, cosmetics, food and additives), the systematic exploitation of this vast resource as already indicated is still in its infancy. Since pressures are mounting for clean air while the availability of cheap and abundant fossil fuel becomes increasingly questionable as seen in the current climate, it is likely that any economic advantages for a petrochemical-based economy will literally evaporate within the next fifty years.

For example, DuPont has determined that once the price point at which an extractive from plants, tulipalin (A–α-methylenebutyrolactone) shown in Figure 13, becomes competitive with its commercial analogue, methyl methacrylate (MM), the natural material will be used commercially for polymer applications

(*44*). MM is a feedstock that is used for polymerization to manufacture a number of plastics, moldings, and related materials. Its natural analogue, tulapilin A, produces a polymer whose durability and refractive index rival that of MM (*44*). Remarkably, DuPont has also found that limonene, a citrus tree extractive, mimics the chemistry of 4-vinyl-1-cyclohexene in that it can be used as an anchoring point for the introduction of alkoxylsilanes for exterior coatings (*44*).

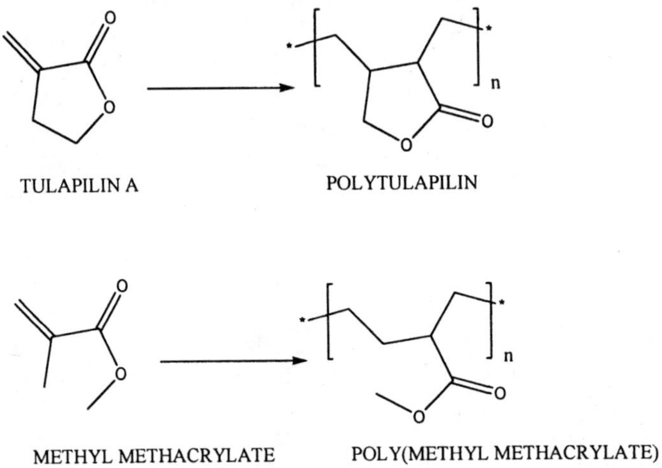

TULAPILIN A   POLYTULAPILIN

METHYL METHACRYLATE   POLY(METHYL METHACRYLATE)

Figure 13. Shown above in the top structure is the natural product extract, tulapilin A (A–α-methylenebutyrolactone), found in tulips, and below it is its synthetic analogue, methyl methacrylate. Notice the strong chemical structural similarity in both the monomer and polymer.

An early example of a bio-based "green" solvent is a new class of chiral ionic liquids that can be derived from α-pinene, an extractive from pine trees, as shown in Figure 14. Ionic liquids are salts that are liquid at or near room temperature. As such, they are composed of an anion and a cation, like any salt, but they do not have the high melting points typical of such species. Since the combination of organic cations and anions is virtually limitless, several classes of ionic liquids have actually been reported with more and more such systems discovered continuously.

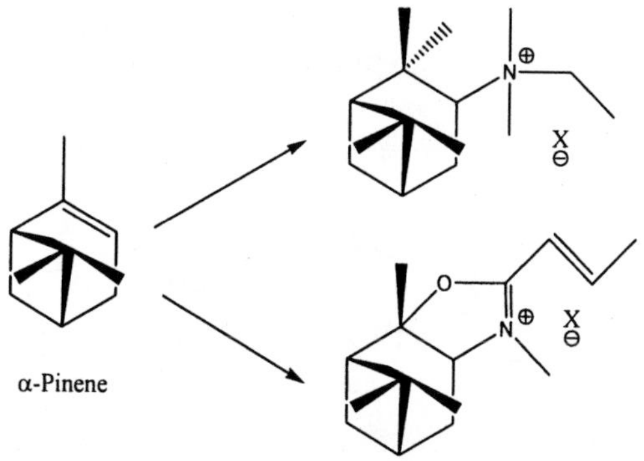

$$X = BF_4, PF_6, CF_3CO_2, CF_3SO_3, (CF_3SO_3)_2N$$

Figure 14. Structure of α-pinene, a natural extractive from pine trees, which can be converted to two distinct chiral ionic liquids.

Related Research at North Carolina State University (NCSU): The Concept of the Biorefinery

The Department of Wood and Paper Science at NCSU is currently witnessing a renaissance in the concentration area of forest products science and engineering. They are beginning to explore the possibilities of using the forest for more than just pulp, paper, and tall oil extractives. One of the unifying themes in all of their biomaterial efforts is the redefinition of the pulp mill as a "biorefinery," i.e., a conceptualized site that provides bio-based resources and energy that are outside the scope of traditional functions. Figure 15 illustrates this concept using a flow diagram. The pulp and paper mill participates in the biorefinery concept by providing raw feedstock (byproducts such as hemicellulose, extractives, etc.) to chemically convert into a variety of fine chemicals, polymers, and other materials.

The Department of Wood and Paper Science is realizing that a strong academic component must be included in any future efforts directed toward the establishment of a "green" economy. A "green" economy is based on the efficient utilization of renewable resources such as wood and agricultural products and byproducts. For example, the mill economy on a typical Southern softwood kraft mill is approximately 55% self-sustaining through the utilization of all of the components from the softwood and then maintaining softwood plantations for future furnish. In addition, green science in forest biomaterials is highlighted by efficient utilization of a renewable feedstock, i.e., trees can be replenished. In addition, green science will prevent waste, maximize resources,

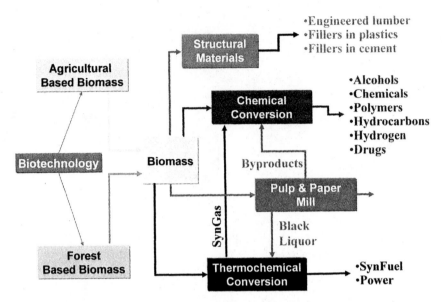

Figure 15. A flow diagram is shown above that represents the "biorefinery" concept in pulp, paper, agricultural residues, and related biomass.

enhance safety, and be rationally conceived for recyclability or efficient disposal of by-products, and thus have low environmental impact. At NCSU, a strong driver to support the realization of an economy based on green science has been to establish a "forest biomaterials" program that encompasses work from a number of disciplines to address the needs of the U.S. in a bio-based economy. Most agree that a revolution in the manner in which we develop products from petrochemicals and use energy from petroleum must take place because of scarcity considerations and environmental sustainability. Thus, researchers at NCSU are working together to assemble practical research projects and cooperation efforts that will eventually provide a critical mass for industrial exploration and implementation.

One such effort is the development of a center at NCSU that will tackle a broad agenda of associated forest chemical, material, and energy research and educational considerations for a future bio-based economy. The Center for Forest Biomaterial Research and Education will be a locus for many of the current efforts that are being explored. Some of the research endeavors undertaken or in progress in the department that contribute to the center are shown below:

- Modification of Cellulosic Structures and Properties by Interfacial Interactions with Colloids and Polyelectrolytes

- Value Prior to Pulping & Characterization of Extracted Hemicellulose Oligomers from Southern Loblolly Pine
- Commodity Chemicals from Biomass using Ionic Liquids as Processing Media
- Novel Utilization of Starch Microcellular Foam Materials as Opacifying Agents
- Fatty Acid Esters as Agents for Enhanced Food Safety in Paper Packaging
- Immobilized, Green Catalysis of Cellulose Hydrolysis Products to Higher value Chemicals
- Conversion of Lignin in Supercritical $CO2$ into Commodity Phenolics & Value-added Chemicals
- Agricultural Biomaterials
- Lignins in the Formulation of Heavy Crude Oil and Bitumen Emulsions
- Nanoscale Sensing of Lignin and Lignin nano Fibers via Electrospinning
- Biorefining for Fuel and Paper
- Green Production of Fermentable Sugars for Ethanol from Biomass using Autohydrolysis followed by Oxygen Delignification
- Value-Added Lignin Products from Bioethanol Process
- Nanodevises via Cellulose Nanocrystal Manipulation
- Smart Polymers and Hydrogels based on Hemicelluloses
- High Performance Fiber Structural Composites

Energy Derived from Biomass

Since antiquity, biomass has been used by humanity as a source of energy, i.e., mainly for heating. Nowadays, new developing energy technologies are broadening its use well beyond simple combustion for heating. The conversion processes used on biomass feedstock allow the generation of steam and electric power, as well as, a wide range of energy products such as ethanol, biodiesel, fuel gas, and chemical intermediates and products. The conversion of biomass feedstock into energy products are well established technologies, with enhanced improvements in the last decades. This subject was reviewed by Paisley (*45*).

Steam and Electric Power

Biomass has been used for simple heating by direct combustion, which by nature is a very low efficient operation (approx. 7% in an open fireplace and approx. 15% in a fireplace with convective tubes) (46). Efficiency improvements are realized by larger scale combustion processes in boilers or furnaces, thus improving heat recovery and therefore overall efficiency.

Boiler applications further provide the opportunity to generate electricity power from the incoming biomass. The biomass is burned to generate steam, which is then used to turn a turbine for the generation of electric power. The majority of the biomass-based power plants are found in the pulp and paper industry. Woody residues from the wood raw material and the black liquor, as a by-product from the pulping process, are consumed in order to generate the necessary power for the plant operations. When black liquor is burned for energy recovery, the pulping chemicals are recovered and recycled, thus further increasing energy efficiency and reducing the costs.

Biomass based Combined Heat and Power (CHP) systems provide the primary energy for large segments of the population in many European countries, mainly Scandinavian and northern countries. These systems, typically combust biomass to produce high pressure steam for further electric power generation. Lower pressure steam is then extracted from the turbine system and used for district heating.

Ethanol

Ethanol production is historically known for the manufacture of alcoholic beverages such as wines and beers. It is produced through natural fermentation of the starch and sugars present in different forms of biomass by biological organisms. Nowadays, grains of wheat or corn (biomass materials) are typically the primary source for the production of ethanol as fuel, which can be blended with other fuels, as gasoline (47,48). Other sources for ethanol production, which are also in use are straw, sugarcane and sugar beet, and wood (49).

Large-scale production of ethanol makes use of the fermentation process as well. In a first stage, the biomass material is milled through a dry or wet milling mode. The fermentation takes place then by mixing the resulting flour (biomass polysaccharides) in water with yeast. The mixture is heated at approximately 30°C to initiate the fermentation reactions, under an anaerobic environment and take approximately 1–2 days for completion. After fermentation, the ethanol is filtered and then distilled to increase its concentration from 10 to 95%. The concentrated ethanol is then blended with petroleum-based fuels to attain the desired concentration.

Ethanol can also be produced from cellulosic biomass such as agricultural residues, forestry residues, waste paper, yard wastes, portions of municipal residues and industrial residues (50). The long-chain polysaccharide materials are initially treated with acid or enzymes to cleave the glycosidic bonds. The acid recovery system can be a complex process. Research continues to reduce the acid levels and maintain high conversion levels. The resulting small fractions can then readily be fermented to ethanol. In parallel, small fractions of biomass can be used to grow fungi and other organisms to produce the enzymes (cellulase), which hydrolyze cellulose in pretreated biomass to glucose. Figure 16 presents a scheme for the production of ethanol from cellulosic biomass (*50*).

U.S. ethanol production has grown significantly in recent years. In 2004, a record 3.41 billion gallons of ethanol were produced,, an incredible 109 percent from 2000 (*51*). But ethanol plants do not make just ethanol. Depending on the type of facility, a number of other coproducts result from ethanol production, adding even more value to corn and to the economy. Dry mill facilities also produce distillers dried grain with solubles (DDGS) and carbon dioxide. Ethanol wet mills can also produce corn gluten meal, corn gluten feed, sweeteners and corn oil. These coproducts and byproducts from the cellulosic biomass ethanol production can be used to produce value-added chemicals, contributing to the economy of fuel production from biomass. On the other hand, ethanol can also be used as a chemical intermediate in the production of other organic chemicals.

Oxygenated transportation fuels, as the ethanol-gasoline blends, can benefit environmental conditions. Mixtures of up to 10% ethanol referred to as (E10) can be utilized in most gasoline designed engines with no modification. More concentrated blends, namely, the E85 and E95 blends (85 and 95% ethanol, respectively) require specifically designed engines, referred to as "flexible fuel" engines to perform properly. These flexible fuel engines can run on either gasoline or the high concentration ethanol blends. Automobile manufacturers are producing more vehicles that can use these higher ethanol blends each year, thus increasing demand. The spark ignitions, Otto-cycle engines used in today's automobiles, even in their early stages of development, were designed to operate with ethanol-containing fuels. Henry Ford designed the early Model T to use ethanol as a major fuel source. Such plans were changed subsequently when storage and transportation difficulties combined with high corn prices at the time caused the supply of ethanol-containing blends to be reduced.

Biodiesel

Biodiesel is a fuel that can be made from renewable resources such as vegetable oils or animal fats. It is usually produced from soy, palm or canola oil. Other possible resources are recycled fryer oils. Through a transesterification process, the oils are converted into biodiesel, which can be used, as pure fuel or blended at any level with petro-diesel, by diesel engines. The biodiesel is designated B100 and it meets the requirements of ASTM (American Society for Testing & Materials) D 6751.

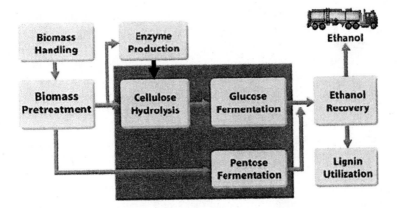

Figure 16. Basic steps in production of ethanol from cellulosic biomass.

Biodiesel comprises a mix of mono-alkyl esters of long chain fatty acids. A lipid transesterification production process is used to convert the base oil to the desired biodiesel. The most common form uses methanol to produce methyl esters, though ethanol can also be used to produce an ethyl ester biodiesel. In the production of biodiesel, the triglicerides in the oils and fats are reacted with methanol (or ethanol) to make methyl esters (or ethyl esters) and glycerol, as a byproduct. The process uses a catalyst, typically NaOH or KOH to enhance the reaction rates. Some oils may have to undergo some pretreatment before reaction with methanol, to avoid the formation of high concentrations of free fatty acids. The reactions take place at low temperatures (~65°C) and at modest pressures (2 atm). The biodiesel is further purified by washing and evaporation to remove any remaining methanol. The oil (87%), alcohol (9%) and catalyst (1%) are the inputs in the production of biodiesel (86%), the main output. Also resulting as byproducts output are glycerine (9%), alcohol (4%) and fertilizer (1%). There is nothing is wasted in biodiesel production (*52*).

Biodiesel can be a direct substitute for petro-diesel, either as neat fuel (B100 or BD 100) or as an oxygenate additive at any concentration blended with diesel. Typically, the most utilized biodiesel is the B20 blend (20% of biodiesel and 80% of petro-diesel). Pure biodiesel or blends can be used in any compression ignition (diesel) engines. The B20 blend earns credits for alternative fuel under the US Energy Policy Act of 1992. (*53*). In 2004, almost 30 million gallons of commercially produced biodiesel were sold in the U.S., an increase from less than 0.5 million gallons in 1999 (*54*). Due to increasing pollution control requirements and tax relief, the U.S. market is expected to grow to 1 or 2 billion gallons by 2010.

In the past, Rudolf Diesel had successively presented his engine powered by peanut oil (biofuel) at the World Fair in Paris, France in 1900, where he received the "Grand Prix" (highest prize). He believed that the utilization of a biomass fuel was the real future of this engine. In a prophetic 1912 speech, Rudolf Diesel said, "the use of vegetable oils for engine fuels may seem insignificant today, but such oils may become, in the course of time, as important as petroleum and the coal-tar products of the present time." (55).

Conclusions

In retrospect, the petroleum age has served the growth of civilization quite well. However, it has also become an albatross to further growth given its intrinsic deficiencies for environmental compatibility, sustainability, and micro- and macroeconomics. As Rudolph Diesel so astutely intimated, we are now at the nexus of a major crossroads in the future of humanity with respect to the production of chemicals, materials, and energy. The only real solution which is viable and necessary is the efficient and intelligent use of biomass as derived from plant and animal residues. Capturing the chemical and material value of bioderived succinic, itaconic, and levulinic acids as indicated in this review will provide the world with a vast window of freedom from the economic and geopolitical uncertainties associated with petroleum dependency. In addition, tapping into bio-based fuel sources such as ethanol and diesel expands that horizon considerably by supporting environmental stewardship by their renewable nature and carbon neutrality.

References

1. Carole, T.M.; Pellegrino, J.; Paster, M.D. Opportunities in the Industrial Biobased Products Industry, Applied Biochemistry and Biotechnology, Vol.113-116, 2004, pp 871-885.
2. Boepple, J.T. Petrochemicals, Feedstocks, Kirk-Othmer Encyclopedia of Chemical Technology, 2005.
3. Campbell, C.J. "The Imminent Peak of World Oil Production" 1999.
4. Petroleum in Wikipedia, the free encyclopedia (2006). Data from webpage: http://en.wikipedia.org/wiki/Image:Oil_Prices_Medium_Term.png
5. Speight, J.G. Handbook of Petroleum Analysis, John Wiley & Sons, 2001.
6. Paster, M.; Pellegrino, J.L.; Carole, T.M. Industrial Bioproducts: Today and Tomorrow, July 2003.
7. Campbell, C.J.; Laherrère, J.H. "The End of Cheap Oil," Sci. Am. 1998, March, pp 60-65.

8. Klass, D. H. "Fossil Fuel Reserves and Depletion," *Biomass for Renewable Energy, Fuels and Chemicals*, Acad. Press, San Diego, 1998, pp 10 – 19.

9. Okkerse, C.; van Bekkum, H, "Fossil to Green," Green Chem. (1999) pp 107 – 114.

10. National Research Council, USA. Biobased Industrial Products: Priorities for Research and Commercialization, Natl. Acad. Press, Washington, 2000.

11. Biomass Energy Research Association, internet site, www.beral.org.

12. FAO (2001) Global Forest Resources Assessment 2000. Summary Report. Food and Agriculture Organisation website (www.fao.org/docrep/meeting/003/x9835e/x9835e00.htm).

13. Matthews, R.; Robertson, K. "Answers to ten frequently answered questions about bioenergy, carbon sinks, and their role in global climate change," IEA Bioenergy Task 38, internet site, www.joanneum.at/iea-bioenergy-task38/publications/faq/, 2001.

14. IPCC (2000a) Land use, land-use change and forestry. Watson, R. T.; Noble, I. R.; Bolin, B.; Ravindranath, N. H.; Verardo, D. J. and Dokken, D. J., Eds.; Special Report of the Intergovernmental Panel on Climate Change. Cambridge: Cambridge University Press.

15. "Vertec Biosolvents on Verge of Breaking Out with New Replacement Applications," Chemical Marketing Reporter, September 2, 2002.

16. *1999 Annual Survey of Manufactures*, U.S. Department of Commerce, Bureau of the Census.

17. Chemical Profiles, Chem Expo, service of Chemical Marketing Reporting, 2002, www.chemexpo.com.

18. Biomass Use for Power, Fuels and Products: Current Use and Trends. April 2002. Energetics, Inc., Columbia, MD.

19. Verespej, M.A., Winning Technologies: Polylactide Polymers, Industry Week, Dec. 2000.

20. http://www.natureworksllc.com/corporate/news.asp

21. Ritter, S.K. **Biomass or Bust**, C&EN, May 31st, 2004.

22. Kirk- Othmer Encyclopedia of Chemical Technology, Fourth Edition, Volume 22, **Succinic Acid and Succinic Anhydride**, 1997, pp 1074-1088.

23. Ullmann's Encyclopedia of Industrial Chemistry, Sixth Edition, Volume 10, **Dicarboxylic Acids**, 2003, pp 525-526.

24. Zeikus, J. G., Jain, M. K., Elankovan, P., **Biotechnology of succinic acid production and markets for derived industrial products**, Journal of Applied Microbiology and Biotechnology, Vol. 52, 1999, pp 545-552.

25. Weimer, P. J., Effects of dilution rate and pH on the ruminal cellulolytic bacterium *Fibrobacter succinogenes* S85 in cellulose-fed continuous culture. Arch Microbiol , Vol.160, 1993, pp 288-294.

26. Guettler M.V., Rumler D., Jain, M.K., Actinobacillus succinogenes sp., a novel succinic acid producing strain from the producing strain from the bovine rumen, Int J Syst Bacteriol, Vol.49, 1999, pp 207-216.

27. Oak Ridge National Laboratory, Argonne National Laboratory, Pacific Northwest NationalLaboratory, National Renewable Energy Laboratory, and Applied Carbochemicals, Inc., "Production of Chemical Derivatives from Renewables: CRADA Final Report for CRADA ORNL 96-0407," November 1999.

28. Samuelov, N.S., Lamed, R., Lowe, S., Zeikus, J.G., Influence of CO_2-HCO_3^- levels and pH on growth, succinate production, and enzyme activities of Anaerobiospirillum succiniciproducens, Applied Environmental Microbiology, Vol.57, 1991, pp 3013-3019.

29. Kirk- Othmer Encyclopedia of Chemical Technology, Fourth Edition, Volume 6, pp 353-359.

30. Wilke, T.; Vorlop, K.-D. **Biotechnological production of itaconic acid,** Applied Microbiology and Biotechnology, Vol.56, 2001, pp 289-295.

31. Yahiro, K.; Shibata, S.; Jia, S-R; Park, Y.; and Okabe, M.; **Efficient Itaconic Acid Production from Raw Corn Starch,** Journal of Fermentation and Bioengineering, Vol.84, No.4, 1997, pp 375-377.

32. Farone, W.A., Cuzens, J., **Method for the production of levulinic acid and its derivatives.** US Patent 6,054,611, 2000.

33. Fang, Q.; Hanna, M.A. **Experimental studies for levulinic acid production from whole kernel grain sorghum,** Bioresource Technology, Vol.81, 2002, pp 187-192.

34. Timokhin, B.V.; Baransky V.A.; Eliseeva, G.D. Levulinic Acid in Organic Synthesis, Russian Chemical Reviews, Volume 68, No1, pp 73-84, 1999.

35. Anastas, P.; Kirchchoff, M.; Williamson, T. Green Chemistry, October 1999, The Royal Society of Chemistry.

36. **http://www.epa.gov/greenchemistry/sba99.html**

37. **http://www.iags.org/biofine.htm**

38. *Carbohydrates as Organic Raw Materials*; Lichtenthaler, F.W., Ed.; VCH Publ., Weinheim/New York, 1991.

39. *Carbohydrates as Organic Raw Materials II*; Descotes, G., Ed.; VCH Publ., Weinheim/New York, 1993.

40. *Carbohydrates as Organic Raw Materials III;* van Bekkum, H.; Roper, H. Voragen, A.G.J., Eds.; VCH Publ., Weinheim/New York, 1996.

41. *Perspektiven Nachwachsender Rohstofe in der Chemie*; Eierdanz, H., Ed.; VCH Publ., Weinheim/New York, 1996.

42. Lichtenthaler, F.W.; Mondel, S. Perspectives in the Use of Low Molecular Weight Carbohydrates as Organic Raw Material. *Pure Appl. Chem.* **1997,** *69,* pp 1853 – 1866.

43. Hugill, A. Introductory Dedicational Metaphor, in Sugar and all That. In *A History of Tate and Lyle*, Gentry Books, London, 1978.

44. Mullin, R. Sustainable Specialties. *Chemical & Engineering News* 2004, *82*, pp 29-37.

45. M.A. Paisley, "Biomass Energy", *Kirk-Othmer Encyclopedia of Chemical Technology*, John Wiley & Sons, Inc., 2003.

46. Houck, J. E.; Tiegs, P. E. *Residential Wood Combustion Review Volume 1, Technical Report* U. S. Environmental Protection Agency, Office of Research and Development, 1998.

47. General Biomass Corporation, "Bioenergy and Biofuels," corporate internet site, www.generalbiomass.com.

48. United States Department of Energy, Office of Transportation Technologies, "History of Biofuels", 2002.

49. Department for Transport, UK, "International resource costs for biodiesel and bioethanol", internet site, www.dft.gov.uk/stellent/groups/dft_roads/documents/page/dft_roads_02405 4.hcsp, 2005.

50. Renewable Fuels Association, Resource Center, "How Ethanol is Made", internet site, http://www.ethanolrfa.org/resource/made/, 2005.

51. National Corn Growers Association, "Ethanol–America's Clean Renewable Fuel", internet site, http://www.ncga.com/ethanol/main/production.htm, 2005.; Renewable Fuels Association, Media Center, "Ethanol Industry Outlook 2005", internet site, http://www.ethanolrfa.org/objects/pdf/outlook/outlook_2005.pdf, 2005.

52. National Biodiesel Board, "Biodiesel – The official website", internet site, http://www.biodiesel.org/, 2005.

53. http://www.eere.energy.gov/biomass/renewable_diesel.html

54. Soy Stats[TM], "A Reference Guide To Important Soybean Facts & Figures", U.S Biodiesel Consumption, internet site, http://www.soystats.com/2005/Default-frames.htm, 2005.

55. "Biodiesel", *Wikipedia, the free encyclopedia*, internet site, http://en.wikipedia.org/wiki/Biodiesel, 2005.

Chapter 2

Lignocellulosic Biorefineries: Reality, Hype, or Something in Between?

Stephen S. Kelley

Department of Forest Biomaterials Science and Engineering, North Carolina State University, Raleigh, NC 27695–8005

With growing concerns over increasing fuel prices, green house gas emissions and the national security issues surrounding reliance on imported oil, there is an increasing interest in technologies that allow for sustainable production of energy and fuels from domestic resources. Among the various renewable energy options, lignocellulosic biomass is unique in its ability to produce liquid transportation fuels, which can be integrated into the current fuel infrastructure. There are a wide variety of technologies that can be used to convert lignocellulosic biomass into liquid fuels, including fermentation of sugars and production of liquid fuels from biomass derived syngas. There are also two industries with well-developed biomass conversion infrastructure, the corn ethanol industry, and the pulp and paper industry, that could be used as a launch point for developing lignocellulosic biorefineries. These industries offer opportunities for 1) producing monomeric sugars from biomass residues that can be fermented into ethanol, and 2) additional opportunities for converting the lignin-rich fermentation residues and low quality biomass residues to syngas that can be used to produce liquid fuels. The pulp and paper industry offers opportunities for 3) extracting fermentable sugars prior to pulping and converting these sugars to ethanol, and 4) gasifying the biomass or lignin-rich spent pulping liquors for the production of liquid fuels. Finally, there are attractive technologies being developed for the direct conversion of biomass-derived oils into transportation fuels in petroleum refineries. In all of these technology areas there have been significant advances, but there is still a need for additional technology development to overcome economic challenges. There is also a need for a better understanding of the energy and environmental benefits of these lignocellulosic biorefineries. This paper will highlight the technical challenges to be overcome, and potential for producing fuel products from lignocellulosic biomass.

Introduction: What is a Biorefinery?

With all the current interest in biorefineries, it is useful to step back and consider the fact that some biorefineries have been successfully operating for almost 130 years. The first biorefineries were operated by the pulp and paper industry and more recently, a different type of biorefinery has been developed by the agricultural industry. One of the first chemical pulp mills was opened in 1874 in Europe and was based on the sulfite process (1). This was rapidly followed by the first kraft pulp mill in 1879. This same basic kraft pulping technology dominates the industry today. A modern pulp and paper mill is a biorefinery that buys clean wood chips, manufactures a variety of pulp and paper products, and generates process heat and power from the residues.

Nearly 100 years ago, the Ford Model T ran on ethanol, gasoline or on a combination of the two fuels. The agricultural industry currently operates corn wet mills and dry mills, which convert corn grain into a suite of products including food and feed products, corn oil, high fructose corn syrup and ethanol.

The current discussions on biorefineries are focused on improving the value of its product streams. The key challenges for the next generation of biorefineries involve 1) production of fuels and chemicals from non-food (lignocellulosic) biomass and 2) production of higher-value products from the non-fermentable components of the biomass.

While increasing the value of the products from the biorefinery another aspect of the discussion is to match the products with the cost and characteristics of the feedstocks. There are several examples of feedstocks prices shown in Table 1.

Table 1. Cost of different biomass feedstocks

	$/lb (dry basis)	Common Units	Cost of Sugar in the biomass ($/lb)
Feedstock Costs			
Wood Residues	0.015-0.02	30-40 $/ton	0.021-0.028
Corn Stover (target)	0.025-0.03	50-60 $/ton	0.036-0.043
Corn Grain	0.035-0.04	2.00-2.25 $/bushel	0.043-0.050
Pulp Wood Chips	0.04-0.05	60-100 $/ton	0.043-0.071

The data in Table 1 shows that sugars from corn grain are more expensive than sugars from wood residues or corn stover. But the corn grain sugars are primarily glucose derived from starch, and more easily accessed and processed than those in wood or corn stover. This data also shows that wood chips are more valuable as a feedstock for production of pulp and paper than as a feedstock for production of ethanol. Thus, wood chips with the proper fiber characteristics should be used for production of paper or wood products, while lower quality wood resources should be used for production of fuels. Table 1 also shows that all four feedstocks contain about 25-35% non-fermentable materials. In the case of wood, these non-fermentable materials include lignin and extractives, while in the case of corn stover, these non-fermentable materials include lignin, protein and ash. Corn grain also contains about 20% non-fermentable materials in the form of oil, protein and ash.

It is less obvious from the data in Table 1, but wood and corn stover contain several types of sugars, hexoses and pentoses, while corn grain contains primarily a single hexose sugar, glucose (2). There are also differences in the relative amounts of hexose and pentose sugars between hardwoods and softwoods, which can have a significant impact on the technology used for conversion of these raw materials into products. If lignocellulosic feedstocks are going to be competitive with corn grain as a feedstock for production of ethanol, all the hexose and pentose sugars must be converted to ethanol.

The non-fermentable residues from corn grain can be used as animal feed, although the market for this product will become saturated as ethanol production rates increase. The non-fermentable residues from lignocellulosic biomass could be used as process heat and power, or as a feedstock for other products. The market for process heat and power is not likely to be affected by very high levels of ethanol production, but process heat and power is a lower value product. Thus, there is the classic economic trade-off between the size of the market and the value of the products.

Table 2 shows the value of several products that can be obtained from biomass sugars. It is useful to note that ethanol at $2.00/gal has about the same value as paper at $600/ton. This table also highlights the importance of moving from the commodity markets, e.g., animal feed or fuels, to value-added chemicals or plastics.

Table 2. Relative value of products from biomass sugars (3)

Product Values	$/lb (dry basis)	Common Units
Distiller Dry Grains	0.04-0.06	80-120 $/ton
Wood Pulp	0.25-0.40	500-800 $/ton
Ethanol	0.30-0.35	1.97-2.30 $/gal
Propylene glycol	0.55-0.65	1,100-1,300 $/ton
Cellulose esters	1.00-1.50	2,000-3,000 $/ton

Thus, the current discussion about the forest biorefinery is essentially a discussion on how to increase the overall value of the products that can be derived from biomass. For example, can the carbon that is currently used to produce process heat and power be used to produce chemicals or fuels? Can a fraction of the sugars in woody biomass be converted to specialty chemicals or fuel ethanol and the rest used for pulp and paper? Can some of the lower quality feedstock, which is not useful for production of pulp or wood products, be used to produce commodity fuels? The technology needed to answer these questions is the driver for the commercial development of the next generation of biorefineries.

Can Lignocellulosic Biomass Have an Impact?

There are two key questions regarding the national impact of using lignocellulosic biomass for large-scale production of fuels or chemicals. The first question is the net energy balance for ethanol from biomass. The second question is whether there is enough biomass available, on a sustainable basis, to have an impact at the national scale.

This first question of the energy balance has been recently addressed by a very complete study that concluded fuel ethanol from corn has a favorable non-renewable energy balance (4). Specifically, this study reviewed several prior studies and concluded that there is more energy contained in a gallon of fuel ethanol than non-renewable energy required to make the ethanol from corn grain. This study also concludes that less than 20% of the non-renewable energy used in the production of fuel ethanol comes from oil, thus, fuel ethanol has a significant benefit in reducing the national security threats caused by a heavy reliance on imported oil. In addition, this work concludes that the production and use of fuel ethanol has a favorable impact on green house gas emissions.

More importantly, this study reinforces the conclusions of prior work on the very positive energy balance and green house gas benefits from production of ethanol in an integrated lignocellulose biorefinery. The production of ethanol from biomass also has the potential for carbon sequestration, although the main green house gas emission benefits will come from displacement of fossil fuel energy sources (5). Thus, while there may continue to be a vocal minority arguing against the benefits of ethanol as a fuel source (6,7), the overwhelming majority of the current literature agrees that there are energy and environmental benefits (4,5).

The question of the sustainable availabilityof biomass has also been clearly and positively answered. A 2005 study by US Department of Energy (DOE) and US Department of Agriculture (USDA) concluded that there are more than 1.3 billion dry tons of forest and agriculture biomass available, on a sustainable basis, for production of fuels and energy (8). This 1.3 billion dry tons of sustainable biomass has the energy equivalence of 3.5 billion barrels of oil, which is about 75% greater than our current domestic oil production of 2.0 billion barrels of oil. The next part of this question is the effectiveness of current or proposed technology that can be used to convert this biomass into energy or fuels. A preliminary study conducted by the National Renewable Energy Laboratory (9) concluded that using technology, which has been demonstrated in the laboratory or at the pilot plant scale but not commercially deployed, this 1.3 billion tons of biomass could produce 1.9 billion barrels of oil equivalents. This is essentially equivalent to the current annual US petroleum production of 2.0 billion barrels.

This projected 1.9 billion barrels of oil equivalents from sustainable biomass comes from three sources: ethanol from grain, ethanol from lignocellulosic sugars and thermochemical processes for production of fuels from non-fermentable biomass. Specifically, this study projects production of ethanol from grains at a very high level, 20 billion gallons or 0.3 billion barrels of oil equivalent; production of ethanol from lignocellulose sugars for 1.1 billion barrels of oil equivalents; and the production of fuels from lignin or low quality biomass using thermochemical conversion technology for 0.5 billion barrels of oil equivalents. While there is a large number of assumptions embodied in this preliminary study, it highlights the potential impact of biomass. In addition, it clearly establishes that there is enough biomass to make a difference at the national scale.

With positive answers to the questions on the national benefits of using biomass as a feedstock, the focus then falls on the technologies that can be used to produce a higher value suite of fuels and chemicals and from sustainable lignocellulosic biomass.

It is also important to note that there are many other sources of sugars and biomass derived carbon, e.g., food processing wastes, and crop oils and animal fats. However, the national impact of producing fuels or chemicals from these resources is relatively small. The limited national energy impact obtained from the utilization of these other biomass resources should not diminish the economic, rural development benefits of producing fuels or chemicals from these resources.

Issues for Producing Fuels and Energy in a Biorefinery

Prior to delving into a discussion on the merits and liabilities of alternative routes for the conversion of lignocellulosic biomass to fuels and energy, it is useful to have a discussion about the relationship between the composition of the biomass and potential products.

As indicated in Table 1, lignocellulosic biomass is purchased by the ton. It is also known that lignocellulosic biomass contains between 40-45% oxygen (2). But if we are selling fuels or chemicals that are primarily or exclusively carbon, then it is very likely that we are not obtaining significant value from the oxygen in the original biomass. A second consideration is that the common fermentation processes used for the production of fuel ethanol can only utilize the sugars in lignocellulosic biomass. Therefore, processes that use lignin and extractives to produce a value-added product are needed. A related consideration is the maximum theoretical yield of ethanol from six-carbon sugar such as glucose is 67%, while the actual practical yield may be closer to 60%. Thus, after accounting for lignin and extractives, and the co-production of carbon dioxide via fermentation, the overall yield of ethanol from lignocellulosic biomass may be close to 50%. The fermentation of lignocellulosic biomass sugars becomes even more challenging if hardwoods are used as the feedstock since commercially viable processes for fermentation of five-carbon sugars (pentoses) are still being developed. On the plus side, ethanol production requires process heat and power for fermentation and distillation, and most of this process heat can be generated from the non-fermentable lignin and extractives.

The inherent limitations on fermentation processes for conversion of lignocellulosic biomass to ethanol is one reason that some have argued for thermal processes that can use all of the biomass. Thermochemical processes, such as gasification can make all of the lignocellulosic biomass into synthesis gas (carbon monoxide and hydrogen) essentially independent of the biomass source. There are also well-known, commercial processes for production of syngas to chemicals and fuel products such as methanol, dimethyl ether, Fischer-Tropsch liquids, or acetic anhydride. However, all of these commercial processes have some significant differences from the proposed biomass gasification scenarios, including the scale of the gasification process, gas clean-up issues and the absence of oxygen in the current gasification feedstocks (10).

It is important to remember that the process yield is only one aspect of an economical process. The yield of ethanol from wood may be modest relative to the yield of ethanol from corn grain, but wood is a less expensive feedstock and the residual wood components can be used for relatively high quality process heat. Thus, ethanol from wood may be a more commercially attractive process than ethanol from corn grain, and it is clear that there is enough lignocellulosic biomass to have a national impact.

Technology Options for Producing Fuels and Energy in a Biorefinery

Total Hydrolysis of Lignocellulosic Biomass

Total hydrolysis of biomass for the production of fuel ethanol has received a great deal of the attention over the past few years. The DOE Office of the Biomass Program has focused on the total hydrolysis of biomass, especially corn stover. This process involves six major process steps, 1) feedstock collection, 2) feedstock pretreatment, 3) hydrolysis of cellulose, 4) fermentation of mixed sugars, 5) separation and distillation of ethanol, and 6) recovery of value or energy from residues (11). Each of these process steps has technical challenges but feedstock pretreatment, hydrolysis of cellulose and fermentation of mixed sugars presents the greatest challenges for the total hydrolysis process.

Biomass pretreatment is a challenging process that may include several different processes (12). The common goal for all pretreatment processes is to increase the accessibility of the cellulose in biomass to further processing. In addition, many pretreatment processes separate some or most of the non-cellulose sugars and lignin from the cellulose. The separation of the more soluble sugars, especially five carbon sugars such as xylose, offers the potential for improvements in the downstream fermentation processes. The major concern with the different pretreatment processes is minimizing subsequent losses of the soluble sugars to furans and other degradation products. The simplest pretreatments include hot water wash or steam explosion pretreatment. In both cases the time and temperature will vary depending on the source of the biomass, generally the temperature is between 150°C and 200°C, while the pretreatment times may vary between one and 30 minutes. Some of the processes such as ammonia fiber explosion or concentrated acid pretreatments require complex recycling and recovery schemes or relatively expensive materials of construction, respectively (12). There is a second suite of pretreatment technologies that resemble chemical pulping; oxidative delignification, organosolv pulping, alkaline extractions, and biopulping (13). These technologies work well to produce high quality cellulose, but are relatively expensive.

Cellulose hydrolysis can be accomplished through either chemical or enzymatic means. While the chemical hydrolysis is well known and simple, it also processes large amounts of degradation products, reducing the overall yield of fermentable sugars and producing many products that inhibit the fermentation. Enzymatic processes are much more attractive for their long-term sustainability and their potential for maximizing the production of ethanol (14). The biggest limitation of enzymatic hydrolysis is the overall cost of the enzymatic systems. These systems include many different specific enzymes that produce cellulose fragments (glucose dimers), which then hydrolyze the dimers to fermentable glucose. The overall cost of these enzymes has been significantly reduced in the past 3 years due to large development projects led by Novozymes and Genencor (15). However, the cost of cellulose hydrolysis enzymes are still 5-7 times higher than comparable starch hydrolysis enzymes.

The final major technical challenge for the total hydrolysis process is the conversion of the complex mixture of 4 or 5 different sugars to ethanol. Three of the sugars are hexoses (glucose, mannose and galactose) which can be relatively easily converted to ethanol, while two of the sugars (xylose and arabanose) are pentoses which are much more difficult to ferment under commercial conditions. The fermentation technology for pentose sugars, especially xylose, has been significantly improved by several research groups (16-18), but these improved systems all have limitations that must be overcome prior to commercial deployment.

In addition to these three major technical hurdles there are improvements needed in the areas of biomass collection and adding value to the fermentation residues. In the feedstock collection arena there are many questions about the impacts of collection and storage on the overall quality of the biomass. This is especially true for agricultural residues that can be stored in the open for 6-9 months after harvest. In contrast, the pulp and paper industry has shown that wood feedstocks can essentially be harvested year-round and stored for months without any negative impact on the quality of the sugars. The production of process energy from the residual lignin and unfermented products is relatively simple. However, this is the lowest value product and does not take advantage of the potential value of the aromatic structure of the lignin. There have been several promising schemes for conversion of lignin into products including fuel additives (19) and adhesives (20) proposed. However, these processes have not progressed past the laboratory scale.

Gasification of Lignocellulosic Biomass

The gasification of biomass and fossil based carbon has been extensively studied and well developed for some applications. According to the DOE, in 2004 there were 117 plants with 385 gasifiers operating around the world. Essentially all of these gasifiers were operating on fossil feedstocks including coal, oil, natural gas and petcoke. Large-scale gasification plants making liquid fuels and chemicals include Sasol's (South Africa) complex for making liquid fuels from coal, Eastman Chemical Company's (USA) coal to acetic anhydride process and a variety of plants in the Mideast and Southeast Asia converting syngas to liquid distillates.

There are a number of pilot and production scale biomass gasification systems that have been built and operated around the world. The characteristics and performance of many of these systems have been recently reviewed by Babu (21). Many of the more successful systems have been built in Northern Europe. For example, a circulating fluid bed system in Lahti, Finland built in 1988 was operated for more than 30,000 hours with 97% online availability. A second system in Varamo, Sweden based on a pressurized gasification unit was operated for more than 8,500 hours. This system was close-coupled to a gas turbine. This system has been re-engineered as an oxygen blown, pressurized gasification unit that can be used for production of fuels or chemicals. Very recently, a collaboration between Choren, and DaimlerChrysler AG and Volkswagen AG focused on the production of Fischer-Tropsch liquids (FTL) was announced. The collaboration has targeted the production of a very low sulfur SunDiesel from biomass (22)

Gasification has the significant advantage of converting both the carbohydrate and lignin components of biomass into syngas. This contrasts fermentation, which can only convert sugars into a final product, and only at a 67% yield based on carbon. Gasification is also relatively insensitive to the quality of the biomass feedstock and offers a number of options for heat and power integration.

However, production of syngas from biomass has several fundamental differences from fossil fuel derived syngas (10). These differences include differences in the feed handling, gas clean up and conditioning, the nature of the ash and the potential for very large-scale plants.

While gasification can utilize a wide variety of biomass feedstocks, especially those that are not suitable for fermentation, there is one downside for the use of biomass for direct production of hydrocarbons. The biomass is 40-45% oxygen and the gasification process requires that the oxygen in the biomass feedstock ultimately is converted to carbon dioxide or water, materials without significant value. Since the biomass is purchased on a weight basis, the conversion of the oxygen to carbon dioxide or water puts additional pressure on the overall economics of the process. Production of methanol or dimethyl ether, ethanol or mixed alcohols is a more attractive option for the more efficient utilization of the oxygen in the biomass feedstock.

The fibrous nature of biomass, especially agricultural feedstocks, makes handling and feeding into any process a challenge. The viscoelastic properties of biomass that make it attractive for many materials applications make it difficult to grind and pulverize. Thermal pretreatments or torrefaction can be used to increase the energy density of biomass and reduce the energy needed to grind and pulverize the biomass. This is especially important for high-pressure gasification, where relatively expensive lock hopper systems may be required.

A second key technical issue is the clean up and conditioning of syngas (10, 23, 24). To be suitable for production of liquid fuels, either FTL or alcohols, the syngas must be clean. This involves the removal of tars, alkaline earth elements, ammonia, chlorides, sulfur compounds and particulates. Many of these gas clean-up technologies are used in the petrochemical industry at hundreds of locations.

Sulfur and ammonia removal can be accomplished using physical and chemical absorption technologies, including proprietary processes such as LO-CAT, Selexol and Rectisol. Since biomass syngas is relatively low in sulfur, a physical absorption onto iron or zinc beds may be an attractive technology (23, 25). There is a critical interaction between the gas clean-up technology and the catalysts used for the fuel synthesis. For example, alkaline-doped molybdenum sulfide catalysts are particularly attractive since they are more resistant to sulfur poisoning and have high activity for the water-gas shift reaction (23, 26). These features allow for potential cost savings on the up-stream sulfur removal process, although the sulfur would have to be removed from the final product. Cyclones and high-temperature filters can both be used to remove particulates and alkaline earth elements.

For biomass gasification the key technology hurdle in the destruction of the tar compounds. Tars result from the incomplete conversion of the biomass to syngas, due to the relative low temperature of the processes. A number of technologies are being evaluated for the removal of tars, including Ni-based catalysts (27).

While biomass gasification faces several challenges, most involve the demonstration of more economical systems, and unlike the total hydrolysis of biomass to make fermentable sugars, most of the technology has been demonstrated at commercial scale. More importantly, biomass gasification is much less sensitive to the quality, i.e. species, seasonal variations, storage conditions, etc., of the biomass feedstock.

Integrated Total Hydrolysis and Gasification of Lignocellulosic Biomass

One alternative to the narrow focus of converting all the biomass into fermentable sugars has recently been proposed by researchers at NREL (28). This process is intended to address the issue of low quality or off-spec feedstocks, increase the value of the products derived from the lignin rich residues, and increase the overall efficiency of the biorefinery. This approach integrates the total hydrolysis and gasification processes previously outlined. This integrated approach can utilize low quality or off-spec feedstocks, and the lignin-rich fermentation residues to produce syngas, which is used to provide heat and power for the fermentation process, while also producing ethanol and higher-value alcohols. One advantage of the heat integration is the potential for using low syngas conversion to produce a liquid product since some of the syngas will be required for process heat (29).

The production of mixed alcohols from syngas has been extensively studied at both the laboratory and commercial pilot scale (26). In general, the work has focused on the development of catalysts that maximize the production of higher alcohols and minimize the production of methanol. Alkaline-doped molybdenum sulfide catalysts are particularly attractive since they are more resistant to sulfur poisoning and have high activity for the water-gas shift reaction. These features allow for potential cost savings in the sulfur removal process. The integration of ethanol from the fermentation of sugars and mixed alcohols from syngas will allow synergies in the product recovery, storage, distribution and marketing.

The integration of gasification with the fermentation process also offers potential for using lignin for the production of a higher value product. The integrated plant will require a significant amount of process heat for fermentation and distillation, and this process heat can come from the syngas exiting the fuel synthesis reactor. Typically, the fuel synthesis requires very high conversion of the syngas to the liquid product, but in this case the need for process heat will allow for lower conversion or fewer recycles.

While an integrated biorefinery combining total hydrolysis and gasification technology allows for a number of synergies and can more easily handle wide variations in the quality of the feedstocks, it also includes the highest level of technological risk and high capital costs. All of the technical risks associated with the total hydrolysis processes, e.g., pretreatment, enzyme hydrolysis and fermentation of mixed sugars, and all of the technical risks associated with biomass gasification, e.g., syngas clean-up and conditioning, and scale-down of gas-to-liquid synthesis processes are still present in this technology. However, the integrated biorefinery does offer the greatest potential for total revenues.

The Forest Biorefinery

The forest biorefinery draws on several technical aspects of the total hydrolysis concept and the biomass gasification processes previously outlined. The forest biorefinery also has several unique attributes that could serve to lower the capital costs and reduce the technical risks of a stand alone, green field plant.

A forest biorefinery is a generic term for the conversion of wood to value-added products, although two options have attracted most of the attention. The first option is the extraction of hemicellulose sugars that are then converted to ethanol (30). This option has also been termed *value prior to pulping*. These hemicellulose sugars are then used for production of ethanol. The extracted wood chips continue through the pulping process. This option only makes commercial sense if the quality and yield of pulp and paper products is maintained at the current levels. Although pre-extraction of the hemicellulose sugars may also allow for improvements in the pulping and bleaching process. A major challenge for this option is the need to demonstrate production of high quality pulp from the extracted wood chips. However, the pulp and paper industry is highly skilled at modifying and optimizing pulping and bleaching processes with "new" wood feedstocks.

The *value prior to pulping* option is very attractive since there is no need to develop improved processes for the total hydrolysis of the cellulose or for utilization of lignin. The cellulose is being sold as paper, while the lignin is being combusted in the recovery boiler. While this is a low value use of the lignin, it has little if any technical risk. Thus, it appears that the greatest technical challenge is the production of ethanol from the extracted sugars. Depending on the wood species and the extraction process, it is likely that the extracted solution will contain a significant portion of the wood extractives, and in the case of hardwoods, a significant amount of acetic acid. Both of these components must be removed prior to fermentation of the sugars. In addition, any oligomeric sugars must be converted to monomeric sugars using either acid or enzymatic catalyst. The fermentation of the mixed sugars also holds the same challenges discussed for the total hydrolysis concept. If the pulpwood feedstock is softwood, then the hemicellulose sugars are largely hexoses (glucose, mannose, and galactose) that can be fermented to ethanol using yeasts that are similar to the current systems. However, if the pulpwood feedstock is hardwood, then the hemicellulose sugars will contain hexoses and xylose, a pentose. As previously mentioned, the pentoses are more difficult to ferment and the technology will have to be developed and demonstrated at the commercial scale (16-18). If it is commercially attractive, this process may also allow for the recovery of wood extractives and acetic acid. There are relatively well-developed options for steam stripping or extraction that can be used to recover the wood extractives and acetic acid. The main advantage of the *value prior to pulping* option is that it allows for the rapid implementation of ethanol production, without the technical risks and capital costs associated with the total hydrolysis of cellulose.

A second option for the forest biorefinery is the gasification of biomass or spent pulping liquors. These options have been extensively studied and a number demonstration units have been constructed (29). These demonstration units include four gasifiers specifically designed to operate on spent pulping liquors. These four demonstration units are at a Weyerhaeuser mill in New Bern, North Carolina; a Georgia-Pacific mill in Big Island, VA; a Norampac mill in Ontario, Canada; and a unit built by a consortium lead by Chemrec in Pitea, Sweden. The main driver for the development of high-efficiency gasification systems for spent pulping liquors is the onsite demand for heat and power. A very detailed study of a spent pulping liquor gasifier integrated into a pulp mill showed economic, environmental and national security benefits from the larger scale deployment of the technology (31). Specifically, this work showed the reference mill could become a net producer of electricity, significantly lower the production of CO_2, NO_x and SO_2, and produce more pulp from less wood. However, all of these systems require significant improvements in the durability of the high temperature refractory materials used in the gasifiers. In spite of a significant investment and substantial progress, problems with these refractory materials still limit the on-stream time to the 80-85% range, well below what is needed for a commercial system. Also, the modifications to the chemical recovery system that is required for gasification of kraft pulping liquors are still not well developed.

Integration of Bio-oil into a Petroleum Refinery

The concept of introducing bio-oil into petroleum refining has very recently attracted some attention (32). The general concept is to produce a liquid bio-oil or bio-crude which can then be introduced into a petroleum refinery. The petroleum industry is very familiar with the conversion of low quality crude oils into fuel products.

Fast pyrolysis technology can be used to produce bio-oil at 60-65% yield along with some water, non-condensable gases and ash (33). There are several practical systems that can used to produce the bio-oil and this approach is less sensitive to the quality of the biomass feedstock than the fermentation technology. In fact, the fast pyrolysis process is most appropriate for biomass like bark or other sources with a high lignin content. If needed, the bio-oil can be stabilized, shipped and stored prior to being introduced into the refinery. This approach makes more sense if the biomass, the bio-oil plant and petroleum refinery are all located in a reasonable proximity to one another.

Bio-oil could be introduced at a number of different locations within a refinery. These locations include: at the fluid catalytic cracker (FCC) for production of a "green" gasoline; at the petro-FCC for production of "green" ethylene and propylene; at the diesel hydrotreater for production of a "green" diesel; or at the hydrotreater for production of "green" paraffins. In each case the bio-oil could be co-fed directly into the unit with the petroleum feedstock or it could be processed in a similar unit operation to generate a product that could be blended into the product downstream of the unit. If the bio-oil and petroleum were processed in parallel, then the catalysts and operating conditions could be optimized for each feedstock and the product quality controlled to a fine degree. Based on a combination of experimental work and process modeling, this approach shows economic promise even at 30 $/bbl.

This concept has some of the same advantages as the forest biorefinery concept. It uses in-place capital, produces a product that the industry knows how to distribute and market, and it gives the petroleum companies control over the entire process that is used for production of a "green" fuel. The major limitation is that the process "discards" the oxygen in the biomass feedstock. So from the point of view of the total utilization of biomass, this is a less attractive technology. However, LCA of the processes suggests that there is a net benefit for the CO_2 emission profile (32).

Conclusions

There is enough biomass available, on a sustainable basis, in the United States to make the large scale production of biomass derived fuels, heat and power, a goal worthy of significant government and industrial support. Using technology that has been demonstrated at the pilot scale, the 1.3 billion tons identified by USDA and DOE has the potential to produce more liquid fuels than all of the current domestic oil production.

However, there are significant technical and financial hurdles that must be overcome to make most of these technologies commercially attractive based on the historic price of crude oil.

In the case of the total hydrolysis of biomass to sugars and the subsequent production of ethanol, these hurdles include pretreatment of the biomass to make the cellulose and hemicellulose polymers more accessible; improved enzyme systems for hydrolysis of the polysaccharides to simple sugars; and the development of fermentation systems that can effectively convert mixed sugars to ethanol. The question of biomass quality and consistency is also an unknown.

Biomass gasification is very attractive as a way to minimize questions regarding the quality and consistency of the biomass feedstocks. However, there is a need to develop better systems for syngas clean up and conditioning, and to develop fuel synthesis technology that can accomodate the relatively small scale anticipated for biomass gasifiers.

The national energy impact of either hemicellulose extraction or gasification of spent pulping liquor in the forest biorefinery is modest, however, technologies could be quite valuable to the economic future of the industry. The hemicellulose extraction technology must demonstrate the ability of make high quality pulp and paper from the extracted wood chips, and the fermentability of the mixed sugars must be demonstrated. The gasification of the spent pulping liquor has the same challenges of biomass gasification and the added technical challenges for improved refractory materials and complete integration of the gasifier into the kraft chemical recovery system.

Finally, the recent developments of bio-oil as a feedstock into a petroleum refinery is very attractive since there is relatively little technical risk. While new catalysts for upgrading the bio-oil need to be developed, this approach significantly leverages the tremendous infrastructure, skills and resources of the petroleum industry. The major limitation of this approach is the incomplete utilization of the oxygen in the biomass feedstock.

It is undeniably clear that with continued technology development and demonstrations, high oil prices and global security concerns, several of these technologies will be tested in the marketplace in the foreseeable future.

References

1. Gellerstedt, G., Pulping Chemistry, In *Wood and Cellulose Chemistry*, 2nd edition; Hon, D.N.S.; Shiraishi, N., Eds.; Marcel Dekker: New York, NY, 2001.
2. Brown, R.C. In *Biorenewable Resources, Engineering New Products from Agriculture*; Iowa State Press, Ames, IA, 2003.

3. *Chemical Economics Handbook*, SRI Consulting, Menlo Park, CA.
4. Farrell, A.E.; Plevin, R.J.; Turner, B.T.; Jones, A.D.; O'Hare, M.; Kammen, D.M. *Science* **2006**, *311*, pp 506-508.
5. Lemus, R.; Lai, R. *Critical Reviews in Plant Science*, 2005, 24:1-21
6. Pimentel, D.; Patzek, T. W. *Natural Resources Research* **2005**, *14(1)*, pp 65-74.
7. Pimentel, D. *Natural Resources Research* **2003**, *12(2)*, pp 127-134.
8. Perlack, R.D.; Wright, L.L.; Turhollow, A.; Graham, R.; Stokes, R.D.; Erbach, D. Biomass as a feedstock for a Bioenergy and bioproducts industry: The technical feasibility of a billion-ton annual supply, ORNL/TM-2006/66, Oak Ridge National Laboratory, Oak Ridge TN, 2005.
9. Wallace, R. The Integrated Biorefinery: How Do We Get There From Here, presented at the 1st International Biorefinery Workshop U.S. DOE and EU, July, 20 and 21, 2005, Washington, DC.
   www.biorefineryworkshop.com/abstracts/analysis_systems_wallace.doc
10. Spath, P.L.; Dayton, D.C. Preliminary Screening – Technical and Economic Assessment of Synthesis Gas to Fuels and Chemicals with Emphasis on the Potential for Biomass-Derived Syngas, NREL/TP-510-34929, National Renewable Energy Laboratory, Golden, CO, 2003.
11. Schell, D.J.; Riley, C.J.; Dowe, N.; Farmer, J.; Ibsen, K.N.; Ruth, M.F.; Toon, S. T.; Lumpkin. R.E. *Bioresource Technology* **2004**, *91*, pp 179-188.
12. Sun, Y.; Cheng, J. *Bioresource Technology* **2002**, *83*, pp 1-11.
13. Bozell, J.J.; Black, S.K.; Myers, M.D. International Symposium on Wood and Pulping Chemistry, 8th, Helsinki, June 6-9, 1995, 1, pp 697-704.
14. Nieves, R.A.; Ehrman, C.I.; Adney, W.S.; Elander, R.T.; Himmel, M.E. *World Journal of Microbiology and Biotechnology*, **1998**, *14*, pp 301-304.
15. McMillan, J. Technology Advances in Sugars Platform Biorefining R&D, presented at the 1st International Biorefinery Workshop U.S. DOE and EU, July, 20 and 21, 2005, Washington, DC.
    http://www.biorefineryworkshop.com/presentations/McMillan.pdf
16. Zaldivar, J.; Martinez, A.; Ingram, L.O. *Biotech. Bioeng.* **2000**, *68*, pp 524-530.
17. Sedlak, M.; Ho, N. W. Y. *Appl. Biochem. Biotech.* **2004**, *113-116*, pp 403-405.
18. Jeffries, T.W.; Jin, Y-S. U.S. Pat. Appl. Publ., US 2004142456, 2004.
19. Johnson, D.K.; Chornet, E.; Zmierczak, W.; Shabtai, J. Preprints of Symposia - American Chemical Society, Division of Fuel Chemistry **2002**, *47(1)*, pp 380-383.
20. Alma, M.H.; Kelley, S.S.; *J. Mat. Sci. Lett.* **2000**, *19*, pp 1517-1520.

21. Babu, S.P.; Bain, R.L.; Craig, K.R.; VTT Symposium, Power Production from Biomass II with Special Emphasis on Gasification and Pyrolysis R&DD **1996**, pp 55-66.
22. Seyfried, F. Renewable Fuels for Advanced Powertrains, RENEW, presented at the 1st International Biorefinery Workshop U.S. DOE and EU, July, 20 and 21, 2005, Washington, DC. http://www.biorefineryworkshop.com/presentations/Seyfried.pdf
23. Dayton, D. A Review of the Literature on Catalytic Biomass Tar Destruction (2002) NREL/TP-510-32815.
24. El-Rub, Z.A.; Bramer, E.A.; Brem, G. *Ind. Eng. Chem. Res.* **2004**, *43*, pp 6911-6919.
25. Jin, Y.; Yu, Q.; Chang, S-G. *Envir. Prog.* **1997**, *16*, pp 1-8.
26. Li, D.; Yang, C.; Li, W.; Sun, Y.; Zhong, B. *Topics in Cat.* **2005**, *32(3-4)*, pp 233-239.
27. Bain, R.L.; Dayton, D. C.; Carpenter, D.L.; Czernik, S.R.; Feik, C.J.; French, R.J.; Magrini-Bair, K.A.; Phillips, S.D. *Ind. Eng. Chem. Res.* **2005**, *44*, pp 7945-7956.
28. Jechura, J., 2006, personal communications.
29. Van der Drift, A.; Boerrigter, H. Synthesis Gas from Biomass for Fuels and Chemicals, 2006, ECN-C-06-001.
30. Closset, G.; Raymond, D.; Thorpe, B. The Integrated Forest Biorefinery, a Preliminary and Compelling Business Case, presented at the 1st International Biorefinery Workshop U.S. DOE and EU, July, 20 and 21, 2005, Washington, DC. http://www.biorefineryworkshop.com/abstracts/existing_raymond.doc.
31. Larson, E.D.; Consonni, S.; Katofsky, R.E.; A Cost-Benefit Assessment of Biomass Gasification Power Generation in the Pulp and Paper Industry, 2003, DOE Final Report.
32. Holmgren, J.; Czernik, S.; Bain, R.; Kalnes, T.; Markowiak, D.; Marker, T.; McCall, M.; Petri, J. Opportunities for Biorenewables in Petroleum Refineries, 2005, DOE Contract, DE FG36-05GO-15085.
33. Oasmaa, A.; Czernik, S. *Energy and Fuels* **1999**, *13*, pp 914-921.

Chapter 3

Recent Contributions to the Realm of Polymers from Renewable Resources

Alessandro Gandini[1] and Mohamed Naceur Belgacem[2]

[1]CICECO and Chemistry Department, University of Aveiro,
3810–193 Aveiro, Portugal
[2]Laboratoire de Génie des Procédés Papetiers, UMR 5518, École Française
de Papeterie et des Industries Graphiques (INPG), BP65, 38402 Saint
Martin d'Hères, France

In this chapter recent investigations on novel materials derived from renewable resources are summarized, with particular emphasis on polymers based on lignins, starch, cork components, furans and vegetable oils.

Polymer science and technology was born in the late nineteenth century, when two natural polymers were chemically modified to prepare novel materials. Both of these are major commodities today, namely cellulose esters and vulcanized rubber. The use of renewable resources for the manufacture of macromolecular materials gradually lost ground during the twentieth century, in favor of synthetic polymers derived from fossil-based resources. However, the arrival of the third millennium coincided with its vigorous revitalization, due to ecological and quality considerations. This renewed activity for both basic and applied research extends over a wide range of topics, (*1-6*).

The Oxypropylation of Natural Polymers and the Exploitation of the Ensuing Polyols

The oxypropylation of cellulose and starch has attracted significant attention. Wu and Glasser (*7*) reported the oxypropylation of kraft and sulfite lignins and showed that the conversion of these solids required elevated temperatures, pressures and reaction times. Recently, this reaction was applied to industrial by-products such as sugar beet pulp (*8-10*), kraft, organosolv, and soda lignins (*11,12*) as well as cork powder (*13-15*). The aim of this work was to convert rather intractable solid residues into viscous liquid polyols to be used as multifunctional macromonomers, e.g. in polyurethane (**PU**) syntheses. The process introduces oligo(propylene oxide) (**OPO**) grafts from the OH groups of the natural macromolecules, as depicted in Scheme 1. However, this reaction is always accompanied by some homopolymerization of propylene oxide (**PO**), which gives **OPO** macrodiols, i.e. difunctional co-macromonomers.

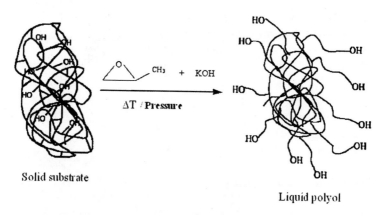

Scheme 1. A pictorial view of the oxypropylation
of OH-bearing macromolecules

These polyols were thoroughly characterized in terms of **OPO** content (extraction with n-hexane), structure (FTIR, ^{1}H-NMR), Mn (VPO), OH index, and viscosity. This systematic study showed that the chain extension of different solids with **PO** occurred smoothly and efficiently yielding polyols with functional characteristics similar to industrial materials used in PU manufacture. Indeed, for some substrates, after the reaction temperature had reached its peak value, the time needed for the total consumption of **PO** was only about 30 min., which is particularly interesting. This is because even the industrial oxypropylation of sugars like sorbitol, the precursor which is used in most PU systems, takes longer.

Polyurethanes from Oxypropylated Natural Polymers

Rigid PU foams (RPU) were prepared from a selection of the polyols described above (*8,11,12,15*). The formulations studied contained the following:

Polyol, ca. 10% w/w of glycerol; an industrial methylene diphenyl isocyanate with an NCO functionality of 2.7 (added to give a [NCO]/[OH] molar ratio of 1.15); a silicone-based surfactant; a blowing agent (here commercial 141b or pentane at 20% w/w with respect to the total weight of polyol was used); a catalyst combination (0.8% w/w, with respect to the total weight of polyol, of equal amounts of cyclohexyldimethyl amine and Niax A-1 (Witco Chemicals); and a small amount of water as initial blowing agent (about 3% w/w with respect to the total weight of the polyol). The water was used to induce the reaction of the excess of NCO functions to creating carbon dioxide. Three commercial polyols were also used to prepare reference RPU foams.

The data collected from the various RPU's prepared in this study included density, thermal conductivity, glass transition temperature, cell morphology and dimensional stability, prior to and after accelerated and natural ageing. The data indicated that several foams prepared from these oxypropylated natural substrates possessed (i) insulating properties, (ii) dimensional stability and (iii) resistance to natural and accelerated ageing, very similar to those prepared with industrial polyols. Moreover, measurements of the time evolution of the gas composition inside the foam cells (8,11,12,15) showed that the blowing agents with powerful insulating properties remained inside the cells, which explains the negligible effect of time on the foams' thermal conductivity.

The best performance was obtained with polyols from organosolv and soda lignins, as well as sugar beet pulp. Thus, thousands of tons of agriculture and forestry industrial residues constitute a potential source of useful materials, prepared through a straightforward and economic oxypropylation process, followed by the use of standard RPU technology. This is a promising alternative to their present use in combustion for energy recovery. Obviously, such oxypropylated polyols should also be examined as precursors to other polymers, e.g. polyesters and polyethers. In a separate investigation the oxypropylation of chitosan aimed at preparing film-forming polymer electrolytes (16) from the ensuing polyols was also examined.

The Chemical Modification of Wood

Wood preservation and the improvement of some of its properties are long-standing issues (17) which are being tackled increasingly by specific chemical modifications, rather than through the physical addition of various (and often toxic) additives. Some of the approaches used to modify cellulose fibers were extended to wood in order to enhance its density and hydrophobic character by generating covalently-linked macromolecules in its inner structure. In addition, the aim was at altering its resistance to fungal attack by the chemical condensation of appropriate biocides within its macromolecular constituents (18).

The strategy selected for this chemical modification consisted in using the diisocyanate **I** and the dianhydride **II** as "carriers" of the two OH-bearing biocides **III** and **IV** through the preparation of the reactive intermediates **V-VII**, as shown in Scheme 2.

Two different materials were used as substrates to be modified and tested in terms of antifungal efficiency, namely: (i) ground samples or small wood blocks of *Pinus sylvestris* and (ii) sheets from Whatman filter paper n° 5 as reference. They were treated with **V-VII** in media of varying polarity, viz. dimethyl formamide, pyridine, methylene chloride and tetrahydrofuran. All treated samples were first submitted to a Soxhlet extraction with the appropriate solvent in order to remove the excess reagent that had not been chemically bound to the lignocellulosic substrate. After drying, they were characterized by weight gain, FTIR spectroscopy (multiple reflections on the block surface at different depths and transmission on wood powder mixed into KBr pellets), elemental analysis and biological tests.

The extent of chemical incorporation of the three reagents, through the condensation of their NCO or anhydride functions with the OH moieties of the wood macromolecules or the paper cellulose, was always satisfactory, but much higher when polar media and appropriate catalysts were used.

Ten representative fungi were selected among the 1,413 available at the mycology collection of the Grenoble University Pharmacy. The presence of the structures associated with both **III** and **IV** chemically bound within the wood and paper morphology retarded or inhibited the fungal multiplication. This demonstrates that the selected approach is promising as an alternative to the adsorption of toxic additives.

Vegetable Oils as Precursors to Photosensitive Polymers

UV-initiated photocuring processes of inks and varnishes represent a fast-growing research and technology field due to their advantages associated mostly with their high reaction rates and the elimination of volatile solvents. Recent efforts were initiated, aimed at using renewable resources as starting materials. More specifically, this work aims at modifying vegetable oils, imparting in them reactivity toward free-radical or cationic cross-linking (*19,20*) reactions. Different common oils were examined, but only soybean oil (**SO**) will be discussed in this chapter.

The Epoxidized **SO** (**ESO**) used had ~5 oxirane moieties per triglyceride molecule. This was treated with an excess of acrylic acid (**AA**) to introduce, after optimization, an equivalent number of acrylic functions in their place. The structure of one of these highly acrylated **ESO** molecules (**AESO**) is shown below:

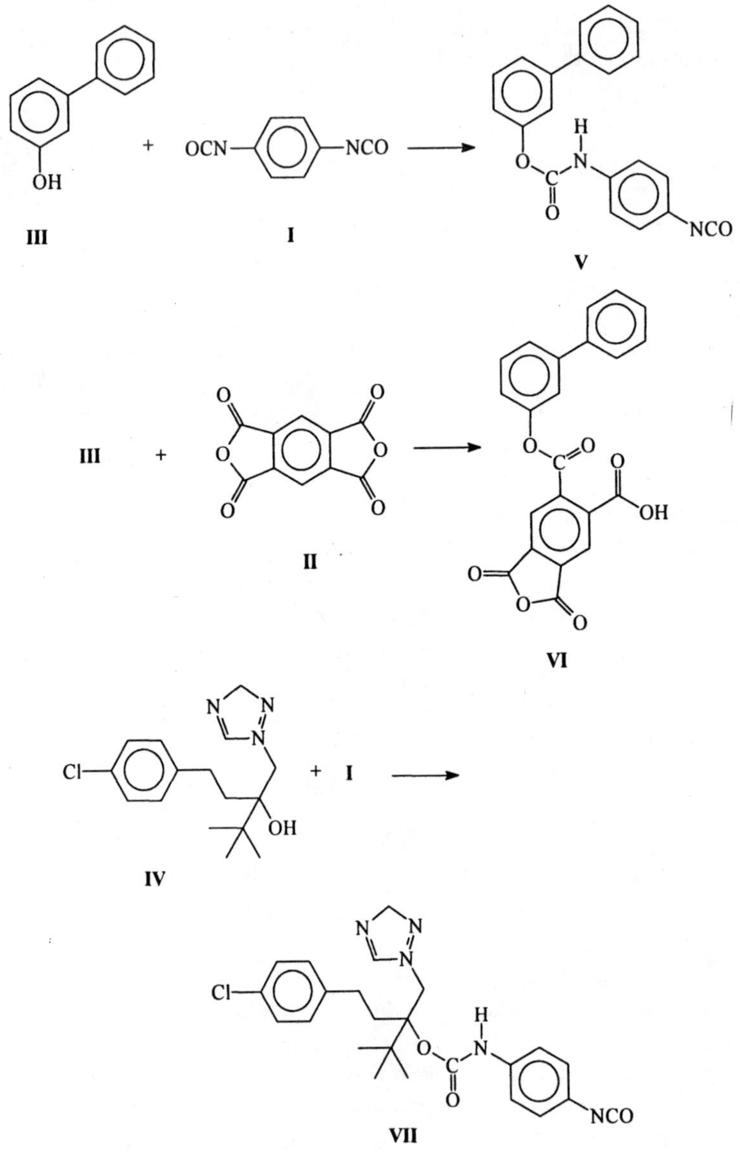

Scheme 2. Synthesis of the bioactive reagents to be condensed
to the wood macromolecules

Their bulk photopolymerization, as thin films, was conducted using standard irradiation equipment equipped with a filtered medium-pressure Hg lamp ($\lambda > 280$ nm) and a conveyor belt. Three Ciba-Geigy free radical photoinitiators, Darocure 1173 (**VIII**), Irgacure 651 (**IX**) and Irgacure 184 (**X**), were tested at 2-4% concentration with respect to the **AESO**. Under these conditions, near-complete conversion (% of acrylic polymerization) was reached within 5 to 10 seconds.

| **VIII** | **IX** | **X** |

FTIR spectroscopy showed the rapid decrease of the acrylic C=C band at 1630 cm^{-1} and the concomitant amalgamation of the two C=O ester peaks, at 1740 (trygliceride moiety) and 1728 cm^{-1} (acrylic group), respectively, into a single band at 1740 cm^{-1}, due to the progressive loss of conjugation, associated with the polymerization.

The ensuing cured materials were submitted to a Soxhlet extraction with methylene chloride to determine the extent of cross-linking within the polymer. Table 1 summarizes these results. As the reactivity increased, the gel point was attained more rapidly but at a lower gel fraction, thus slowing down the polymerization of the sol fraction due reduced macromolecular mobility.

Table 1. AESO photopolymerizations

n° acrylic/ triglyceride	Photoinitiator, %			Conversion, %	Gel content, %
	VIII	IX	X		
4.6	2	-	-	97.1	76.1
3.9	2	-	-	98.2	94.2
2	2	-	-	94.7	94.1
4.6	4	-	-	97.1	81.6
4.6	-	2	-	92.9	80.2
4.6	-	-	2	90.6	80.7

Novel Materials Bearing Furan Moieties

Photocrosslinkable Polymers

Photosensitive furan moieties (Fu) were introduced into a polymer structure either as pendant groups (21), or as an integral part of the main chain (22). The purpose of these syntheses was to prepare novel negative resists for printing plates or microcircuits. Previous work on model compounds had shown that the Fu-CH=CH-Fu chromophore readily dimerizes through a [π2+π2] molecular coupling between a ground state and an excited moiety. This prompted us to attach such chromophores to macromolecular structures to render them photocrosslinkable.

In the initial efforts (21), the unsaturated furan moieties (readily obtained from 5-methylfurfural) were condensed as photosensitive side groups (5 to 20%) onto poly(vinyl alcohol). Irradiation of films of the ensuing polymers induced their fast crosslinking by interchain photochemical coupling. These reactions were followed by UV spectroscopy, which showed the progressive disruption of the conjugation between the two heterocycles through the corresponding decrease in absorption at 330nm (the peak associated with the Fu-C=C-Fu structure). Scheme 3 illustrates this approach, which was in fact successfully applied to both furan- and thiophene-based chromophores.

In a similar manner, but with the aim of preparing photosensitive polymer electrolytes, the same chromophore was attached to chitosan macromolecules (another polymer from a renewable animal resource, chitin) together with polyether chains and the resulting doubly grafted structure, mixed with LiClO$_4$, was irradiated as a thin film to give a crosslinked conducting material (22).

The second strategy (*23*) called upon the insertion of a similar chromophore in the linear backbone of a copolyester prepared by copolytransesterification of an α-ω-aliphatic hydroxyester with the furan homologue 2-hydroxymethyl-5-furanacrylic acid ethyl ester (**HMFAE**), readily obtained from hydroxymethylfuraldehyde. The former comonomer was chosen to impart flexibility (low Tg) to the ensuing copolymer films, whereas **HMFAE**, introduced in amounts of 4% to 10% of comonomer units, insured the possibility of photocrosslinking. The photochemistry of model compounds bearing the basic structural features of **HMFAE** was first studied (*24*) to verify that the same coupling mechanism operated.

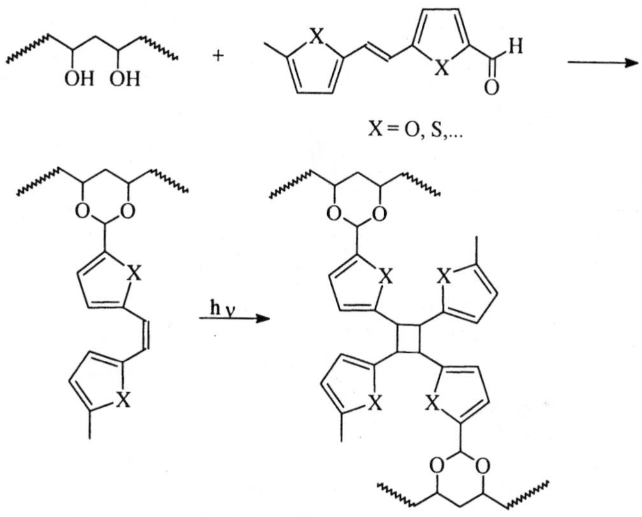

Scheme 3. Grafting the furan chromophore onto poly(vinyl alcohol) and photochemical cross-linking of the ensuing polymer

The UV irradiation of the ensuing copolyester films resulted in their progressive insolubilization through crosslinking. The photochemical intermolecular coupling was monitored by FTIR spectroscopy by following the decrease in the C=C peaks. Scheme 4 illustrates this mechanism.

Polyamides and Homologues

Following an extensive study of the synthesis of furan polyesters and polyamides by different techniques, the optimization of the interfacial

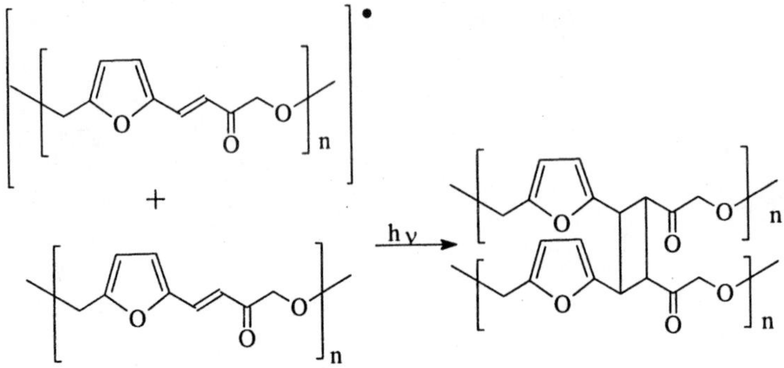

Scheme 4. Photocross-linking of the furan unsaturated polyester

polymerization led to the recent investigation of furan-aromatic polyamides
(*25*). Regular polymers of high molecular weight were obtained, with Tg values
around 100°C, melting temperatures of the crystalline fraction of 220-236°C and
onset of thermal decomposition at ~400°C. This suggests the possibility of melt
processing (compared with the difficulties associated with all aromatic
structures), while at the same time ensuring good thermal stability.

In an analogous piece of research, the synthesis and characterization of a
novel class of polyamide-imides incorporating furan moieties was undertaken
(*26*) in order to prepare processable polyimide materials. Scheme 5 illustrates
the synthetic procedure adopted to prepare these polymers, which were
characterized spectroscopically and in terms of molecular weight, thermal
properties (DSC and TGA) and solubility in a variety of polar media. The results
showed Tg values around 250°C, thermal stability up to 400°C and good
solubility in several protic (DMA, DMF, m-cresol) and aprotic (NMP, DMSO)
solvents. This work confirmed that these materials could be processed by both
melt and solution techniques.

Finally, the study of the reaction of furan diamines with aliphatic
diisocyanates provided the first example of polyureas bearing this heterocycle
(*27*) and the possibility of converting them into the corresponding
poly(parabanic acid)s, following the sequence of reactions shown in Scheme 6.

As with most investigations of new polymers, model reactions were first
conducted and optimized. The polymerizations and subsequent cyclization
reactions gave the expected macromolecules in good yields and reasonably high
molecular weights, all of which were soluble in polar solvents. The properties of
these materials are currently under examination.

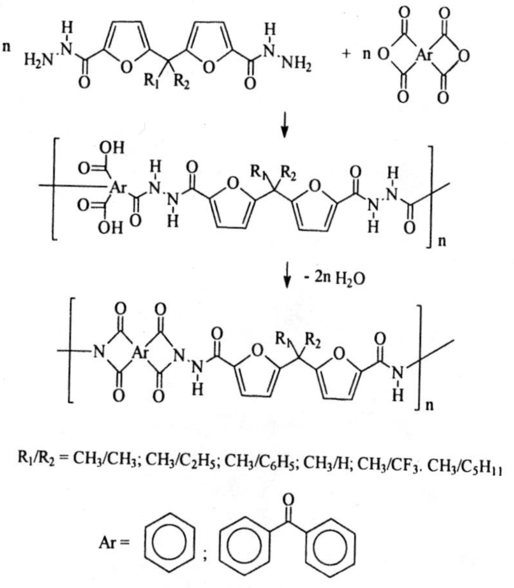
- 2n H_2O
$R_1/R_2 = CH_3/CH_3; CH_3/C_2H_5; CH_3/C_6H_5; CH_3/H; CH_3/CF_3. CH_3/C_5H_{11}$
Ar =

Scheme 5. Synthesis of furan poly(amide-imide)s

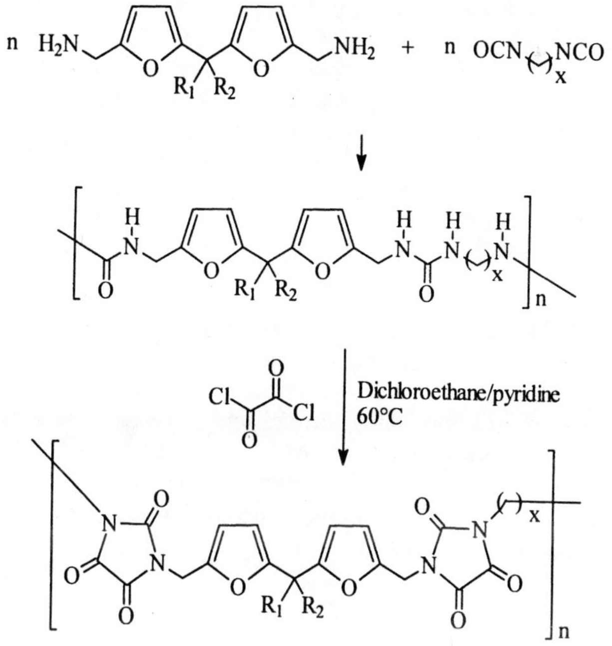
Dichloroethane/pyridine
60°C

Scheme 6. Synthesis of furan polyureas and poly(parabanicacid)s

Thermoplastic Starch

The use of starch as a source of polymeric materials has stimulated much research, particularly in recent years (28), because this renewable resource is ubiquitous, cheap and biodegradable. The three major aspects associated with the processing and usage of thermoplastic starches (TPS) are (i) the reduction of their hydrophilic character, (ii) the search for good plasticizers and (iii) the study of blends and composites.

Hydrophobization

TPS based materials are intrinsically prone to moisture uptake, which induces dimensional instability and loss of mechanical properties. The chemical modification of each starch macromolecule, aimed at reducing hydrophilicity, does not appear to be a reasonable economic solution when compared to a surface chemical treatment. Various reagents were therefore examined using the latter approach, which called upon a simple dipping and heating treatment of the TPS samples (29). Although excellent results were obtained with phenyl isocyanate solutions, the obvious practical drawbacks associated with this reagent, suggested its replacement with a commercial blocked oligomeric isocyanate (Desmodur AP Stabil from Bayer), i.e. a non-volatile, non-toxic water-insensitive compound which only becomes active by generating aromatic NCO functions upon heating. The success of the treatment was indicated by the very drastic reduction of the polar component of the surface energy and a corresponding increase of the contact angle with water (which reached ~100°). Other reagents, viz. stearoyl chloride and a styrene-glycidyl methacrylate copolymer, also produced a very significant surface hydrophobization.

New Plasticizers

The typical plasticizer used in TPS is glycerol, but ethylene glycol, urea and some sugars have also been utilized. In order to establish a rational structure/property relationship, a large series of alcohols, diols and polyols, including macromolecular structures, were examined as potential plasticizers by mixing them vigorously with starch in variable proportions at 150°C (30). Some additives did not mix, others gave intractable materials, some produced a stiffening effect (antiplasticization) and some produced the desired softening. The structural requirements associated with a good plasticizing effect were found to be:

$$HO \left[\begin{array}{c} R \\ \diagdown \\ O \end{array} \right]_n \right]_m H$$

with R = H or CH$_3$, n = 1 or 2 and m = 1, 2 or 3. In other words, the "density" of OH groups must be high, while keeping the molecular weight relatively low.

This systematic study provided a reliable criterion of anticipating the possible plasticizing role of starch OH-bearing additives.

Blends with Natural Rubber

The idea of preparing novel materials based on two natural polymers is obviously attractive. The combination of glycerol TPS with the latex from *Hevea Brasiliensis* was an interesting example. The aqueous component, as extracted from the tree, was an excellent mixing medium with starch and glycerol (*31*). It turned out that the choice was appropriate because the proteins and lipids present in the latex played a useful role in promoting good interfacial adhesion between the rubber particles (2 to 8 μm in diameter) and the TPS matrix. This was demonstrated to be the case with scanning electron microscopy and by the mechanical properties of the ensuing blends. However, phase separation occurred above 7% to 12% of added rubber, depending on the amount of glycerol present. This study requires further work to optimize the nature and the quantity of plasticizer.

References

1. Heinze, T.; Ficher, K. *Cellulose and Cellulose Derivatives*, Wiley-VCH, Weinheim, 2005.
2. Hu, T.Q., ed. *Chemical Modification, Properties and Usage of Lignin*, Kluwer Academic, New York, 2002.
3. Tomasik, P.; Yuriev, V.P.; Bertoft, E., eds. *Starch: Progress in Structural Studies, Modifications and Applications*, Cracow, 2004. Tomasik, P.; Schilling, C.H. *Adv. Carbohydr. Chem. Biochem.* **2004**, *59*, p 175.
4. Hatakeyama, T.; Hatakeyama, H. *Thermal Properties of Green Polymers and Biocomposites*, Kluwer Academic, Dordrecht, 2004.
5. Dufresne, A., ed. *Comp. Interface* (*special issue*) **2005**, *12*, n° 1,2.
6. Gandini, A.; Belgacem, M.N. *J. Polym. Environm.* **2002**, *10*, 105 and refs. therein.

60

7. Wu, L. C.-F.; Glasser, W. G. *J. Appl. Polym. Sci.* **1984**, *29*, p 1111.
8. Pavier, C.; Gandini, A. *Carbohydrate Polym.* **2000**, *42*, 13; Pavier C., PhD thesis of National Polytechnic Institute of Grenoble, 1998.
9. Pavier, C.; Gandini, A. *Indust. Crops Prod.* **2000**, *12*, p 1.
10. Pavier, C.; Gandini, A. *European Polym. J.* **2000**, *36*, p 1653.
11. Gandini, A.; Belgacem, M.N.; Guo, Z. X.; Montanari, S. *"Lignins as macromonomers for polyesters and polyurethanes"* in ref. 2, p 57.
12. Nadji, H.; Bruzzèse, C.; Belgacem, M.N.; Benaboura, A.; Gandini, A. *Macromol. Mater. Eng.* **2005**, *290*, p 1009.
13. Evtiouguina, M.; Barros, A.M.; Cruz-Pinto, J.J.; Pascoal Neto, C.; Belgacem, M.N.; Pavier, C.; Gandini, A. *Biores. Technol.* **2000**, *73*, p 187.
14. Evtiouguina, M.; Gandini, A.; Pascoal Neto, C.; Belgacem, M.N. *Polymer Intern.* **2001**, *50*, p 1150.
15. Evtiouguina, M.; Barros, A.M.; Cruz-Pinto, J.J.; Pascoal Neto, C.; Belgacem, M.N.; Gandini, A. *Biomacomolecules* **2002**, *3*, p 57.
16. Velazquez-Morales, P.; Le Nest, J.F.; Gandini, A. in *Advances in Chitin Science*, Vol. 2, Jacques André Publ., Lyon, 1996, p. 236.
17. *Wood Deterioration and Preservation: Advances in Our Changing World*; Goodell, B.; Nicholas, D. D.; Schultz, T. P., editors; ACS Symposium Series 845; American Chemical Society: Washington, D.C., 2003.
18. Bach, S.; Gandini, A.; Belgacem. M.N. submitted to *Polym. Degr. Stab.* Bach, S. *Doctorate Thesis*, Grenoble National Polytechnic Institute, 2000.
19. Pelletier, H.; Gandini A. *Eur. Lipid Sci. Technol.* in press; Pelletier, H. *Doctorate Thesis*, Grenoble National Polytechnic Institute, 2005.
20. Pelletier, H.; Belgacem, M.N.; Gandini, A. *J Appl. Polym. Sci.,* **2006**, *99*, p 3218.
21. Waig Fang, S.; Timpe, H.J.; Gandini, A. *Polymer* **2002**, *43*, p 3505.
22. Hariri, S.; Le Nest, J.F.; Gandini, A. *Polymer* **2003**, *44*, p 7565.
23. Lasseuguette E.; Gandini A.; Belgacem, M.N.; Timpe H.J. *Polymer* **2005**, *46*, p 5476.
24. Lasseuguette, E.; Gandini, A.; Timpe, H.J. *J. Photochem. Photobiol. A: Chem.* **2005**, *174*, p 222.
25. Abid, S.; El Gharbi, R.; Gandini, A. *Polymer* **2004**, *45*, p 5793.
26. Abid, S.; El Gharbi, R.; Gandini, A. *Polymer* **2004**, *45*, p 6469.
27. Abid, S.; Matoussi, S.; El Gharbi, R.; Gandini, A. submitted .
28. Stepto, R.T.F. *Macromol. Symp.* **2003**, *201*, p 203.
29. Carvalho, A.J.F.; Curvelo, A.A.S.; Gandini, A. *Ind. Crops Prod.* **2005**, *21*, p 331.
30. Da Róz, A.; Carvalho, A.J.F.; Gandini, A.; Curvelo, A.A.S. *Carbohydrate Polym.*, **2006**, *63*, p 417.
31. Carvalho, A.J.F.; Job, A.E. Alves, N.; Curvelo, A.A.S.; Gandini, A. *Carbohydrate Polym.*, **2003**, *53*, p 95.

Chapter 4

Transforming Academic Curricula: From Pulp and Paper to Biobased Products

Shri Ramaswamy, Ulrike Tschirner, and Yi-ru Chen

Department of Bio-based Products, University of Minnesota, St. Paul, MN 55108

This chapter describes the ongoing efforts at the University of Minnesota to transform the academic curricula to meet changes anticipated within the forest products and pulp and paper industries. The new program places emphasis on the efficient utilization of renewable bio-resources including forestry, agricultural residue and other biomass. The department-wide transformation of teaching and learning practices has been undertaken. through the reformation of degree programs, the development of curriculum and the improvement in instructional strategies. The specific objectives are organized around key thematic areas and include developing and implementing a new, transdisciplinary, inter-collegiate, bio-based products engineering and marketing and management degree program. The broader impacts of the ongoing effort include the transformation of an academic curriculum in the emerging field of bio-based products, training well-prepared technical and business professionals for the future "bio-economy."

Impetus for Change

The U.S. is approaching a bio-based revolution that will fundamentally change the way it produces and consumes energy, materials and products (*1*). As clearly envisioned in the "Technology Roadmap" of the U.S. D.O.E. and the U.S.D.A. and in the "Bio-Vision" of the Biomass Technical Advisory Committee, the fraction of materials and products from renewable bio-resources are estimated to grow significantly, if not exponentially (*2*). While wood products and pulp and paper are estimated to comprise the majority of this growth (greater than 90%), industrial and consumer products from renewable resources comprise the remaining 10%. This substantiates that emerging bio-based products are poised to grow significantly as projected in the reports from USDOE and USDA (*2, 3*) (Figure 1). Similarly, wood, wood waste, black liquor and biomass will continue to increase their contribution to the U.S. bioenergy consumption, including liquid transportation fuels (Figure 2). The importance of renewable resources in the 21st century is also highlighted in the report by the Bio-based Industrial Products Committee of the National Research Council (NRC). The report concludes that bio-based products have the potential to improve the sustainability of natural resources, environmental quality, and national security while being economically competitive as well (*4*). Increasing cost and dependence on foreign fossil fuels is continuing to motivate the development of innovative technologies for sustainable use of our domestic renewable resources for energy, chemicals, and materials beyond the traditional realm of the production of food and feed from agriculture and wood products, and the production of paper from forest resources. Decreasing the dependence on foreign oil and developing newer technology for increasing sustainable use of domestic agriculture, forest resources and other biomass for energy and products is the only means of ensuring U.S.'s future energy security.

Recognizng the importance of renewable resources and the bio-economy, the University of Minnesota established several initiatives to promote renewable energy and bio-based products, including the Initiative for Renewable Energy and the Environment (IREE) and the Presidential Initiative on Renewable Energy and the Environment (PIERE), in addition to the "biocatalysis" initiative. Additionally, there were many factors specific to conventional bio-based products industry, i.e. the pulp and paper and forest products sector, that provided the impetus for change including: the North American pulp and paper industry has not earned its cost of capital in more than a decade; changing fortunes within the nation's pulp and paper industry, and a corresponding drop of more than 50 % in undergraduate pulp and paper program enrollment nationally; growing wood products manufacturing capacity outside of North America, and increasing competition from Asia and elsewhere in domestic and global wood

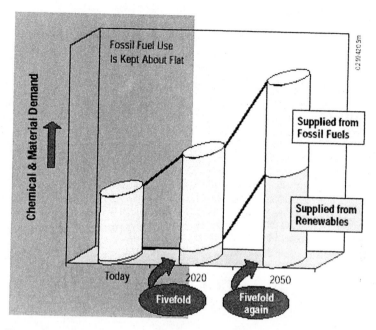

Figure 1. Projected growth of materials and chemicals from renewable resources (Reproduced from reference 3.)

products markets; ongoing consolidation in the domestic building products distribution industry which translates into steady erosion of employment and career track opportunities for marketing-oriented forest products graduates; rising concern about the future supply of fossil fuels and fossil-based chemical feedstocks; growing societal interest in bio-resources as a source of fuels and chemicals; realization that research and development efforts and capital investment in the bio-products arena are not occurring within the nation's forest products industry, but rather in the chemical, pharmaceutical, and agri-business sectors; increasing emphasis within the University of Minnesota on pursuit of excellence across all programs; intensifiedpressure within the College of Natural Resources for increases in student numbers and tuition income.

Feedback from the pulp and paper industry indicated that employment opportunities for new graduates nationwide were unlikely to return to levels from the 1980s and 1990s. In response to questions about the desired academic backgrounds of new hires, employers indicated only a slight preference for graduates of pulp and paper programs over pure chemical engineering graduates, The greatest preference was for chemical engineering graduates who have

Table 1. Role of Wood as a Resource in Biomass Energy Consumption

Exhibit 2
Recent Trends in U.S. Biomass Energy Consumption
1990 - 2001 (Quadrillion Btu)

Sector	Source	1990	2001	% Change '90-'01
Residential	Wood [a]	0.581	0.433	-25%
Commercial	Wood [a]	0.037	0.052	40%
Industrial	Wood [b]	1.254	1.702	36%
Transportation	Waste [c]	0.271	0.287	6%
Transportation	Alcohol Fuels [d]	0.063	0.147	133%
Electric Utilities	Wood [b]	0.008	0.006	-25%
Electric Utilities	Waste [c]	0.013	0.013	—
Non-Utility Power Producers [e]	Wood [a]	0.308	0.379	23%
Non-Utility Power Producers [e]	Waste [c]	0.124	0.324	161%
Total		**2.659**	**3.343**	**25%**

SOURCE: U.S. Department of Energy, Energy Information Administration, *Monthly Energy Review*, July 2002 (Tables 10.2, 10.3a, 10.3b); Data from reference 2.

NOTES:

[a] Wood only

[b] Wood, wood waste, black liquor, red liquor, spent sulfite liquor, wood sludge, peat, railroad ties, and utility poles

[c] Municipal solid waste, landfill gas, methane, digester gas, liquid acetonitrile waste, tall oil, waste alcohol, medical waste, paper pellets, sludge waste, solid byproducts, tires, agricultural byproducts, closed loop biomass, fish oil, and straw

[d] Ethanol blended into gasoline

[e] Includes the portion of ono-utiltiy power producers' use of renewable energy to produce electricity; excludes the portion used to produce useful thermal output, which is included in "industrial"

completed four to six courses in pulp and paper as part of their degree programs as a specialization or as an area of emphasis. This response suggested that there was considerable latitude in redesigning academic programs to serve a broader clientele (5).

It has been noted in a recent paper that "Converting a pulp and paper mill into an integrated forest bio-refinery, producing fuels, power, chemicals and fiber-based products from renewable resources, affords an exciting range of technical challenges and opportunities while offering the real prospect of significant improvement in industry profitability and societal acceptance." (6)

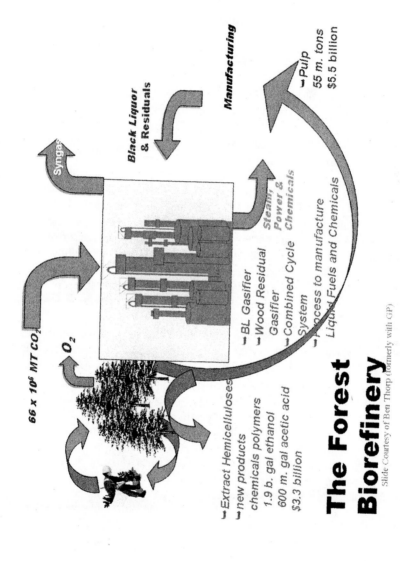

The Forest Biorefinery

Slide Courtesy of Ben Thorp (formerly with GP)

66 x 10⁶ MT CO₂

O₂

Syngas

Black Liquor & Residuals

Manufacturing

Pulp
55 m. tons
$5.5 billion

Steam, Power & Chemicals

↳ BL Gasifier
↳ Wood Residual Gasifier
↳ Combined Cycle System
↳ Process to manufacture Liquid Fuels and Chemicals

↳ Extract Hemicelluloses
↳ new products
chemicals polymers
1.9 b. gal ethanol
600 m. gal acetic acid
$3.3 billion

Figure 2. Conceptual schematic of a forest biorefinery (Source: Reference 20)

The concept of a Forest Biorefinery is schematically depicted in Figure 2. This statement describes almost precisely the prevailing view within the University of Minnesota Bio-based Products faculty.

Program and Curricula

General Program and Curriculum Development

In step with the technology roadmap and Minnesota's and the University of Minnesota's long term visions to meet the workforce and educational needs of the conventional and emerging U.S. bio-based industry, the faculty and staff in the Department of Bio-based Products (formerly the Department of Wood and Paper Science) undertook an innovative and major departmental level reform with the introduction of an inter-disciplinary undergraduate curriculum in Bio-based Products Engineering (BPE) in Fall 2004. The name of the curriculum refers to non-food and feed, industrial and consumer products including materials, chemicals and energy derived from renewable bio-resources . The term 'bio-based products' as defined by Farm Security and Rural Investment Act (FSRIA), means a product determined by the U.S. Secretary of Agriculture to be a commercial or industrial product (other than food or feed), that is composed in whole or in significant part, of biological products or renewable domestic agricultural materials (including plant, animal, and marine materials) or forestry materials.

The inter-disciplinary Bachelor of Bio-based Products Engineering degree (BPE) is a joint program between the College of Natural Resources and the Institute of Technology, the science and engineering college at the University of Minnesota. This program is the first such collaborative program between the two colleges. A formal agreement has been entered into between the two colleges that allows bio-based products engineering students to enter the University through either the College of Natural Resources or the Institute of Technology and to earn their degrees from the Institute of Technology. This arrangement not only offers new opportunities for students, but it also directly involves the Institute of Technology in student recruiting for the bio-based products curriculum as one of 19 science and engineering majors the University has to offer. There are efforts underway which implement an integrated, trans-disciplinary approach, by leveraging the full breadth and expertise of a variety of disciplines. We will be seeking ABET accreditation of BPE curriculum in Fall 2007 under the general criteria. The faculty and staff are very committed to this major departmental reform embracing a broad scope of materials, chemicals and energy from renewable bio-resources in their teaching, research and outreach.

Program Reformation Embracing a Broader Scope of Bio-based Products

Traditional agricultural education has been "production-oriented" with the emphasis solely on food and feed supplies (7). Likewise, conventional forest products curricula focused on the manufacturing aspects of wood and fiber products. In general, the manufacturing and process industries are going through a monumental change from a traditional process, manufacturing and operations emphasis to a more globally competitive, product and customer-focused emphasis (8).

In order to prepare the future professionals for the bio-economy, traditional agriculture and forest products academic curricula have to embrace an unprecedented change by completely revamping and reshaping its curricula (9). Thus an interdisciplinary approach, similar to what is becoming increasingly common practice in research, is essential in academic curricula of the future. Indeed the need for a trans-disciplinary approach for broadly trained professionals for the bio-economy with a true convergence of traditional agriculture, forestry and petrochemical disciplines and emerging industrial biotechnology has been clearly envisioned in the Technology Roadmap (3).

Other disciplines involved in the process include departments of biosystems and agricultural engineering, biotechnology institutes, colleges of biological sciences, departments of chemistry, chemical engineering and materials science, mechanical engineering and biomedical engineering (all in the Institute of Technology). Industry stakeholders include members of our industry advisory council with representatives from traditional bio-based products industry including forest products, pulp and paper and allied industries and emerging bio-based products industry. Based on the above extensive discussions and consultations, and recognizing the need for change, faculty arrived at a consensus to broaden the scope of our programs and curricula. The Pulp and Paper Industry Advisory Council at their annual meeting in Fall 2003 voted unanimously to support curricular and programmatic changes.

The programmatic essentials of the new curriculum are outlined in Table 2. While the overall make-up of the curriculum is similar to the previous pulp and paper curriculum, virtually all of the courses have been changed in some way and the changes in many courses are substantial. Examples of the course changes and their comparison to the previous pulp and paper curriculum is described in the section below. Several completely new courses, such as molecular and cellular processes, bioprocess engineering, bio-based products engineering, and bio-based composites engineering, have been incorporated as part of the curriculum as well.

Multi-disciplinary, Integrated Bio-based Products Engineering Curriculum and ABET Accreditation in 2007

The Engineer of 2020 lists the following attributes of the successful engineer in 2020 (*10*) helping to provide guidelines for future engineering curricula: possess strong analytical skills; exhibit practical ingenuity; posses creativity; good communication skills with multiple stakeholders; business and management skills; understand the principles of leadership; high ethical standards and a strong sense of professionalism; dynamic/agile/resilient/flexible; lifelong learners.

Engineers must now complement technical competence with the skills in communication, teamwork, understanding of the non-technical aspects that affect engineering decisions, and a commitment to lifelong learning. Traditional engineering undergraduate programs are not capable of handling such demands in liberal education components or radically different curriculum delivery methods (*11*). The traditional lecture-based learning and competitive reward structure have to be changed to prepare engineers for the 21st century.

Based on discussions among faculty, consultations with other departments and the industry advisory council, and following the ABET guidelines, rigorous Program Educational Objectives (PEO's) and Program Outcomes (PO's) for the new curriculum were formulated.

Close cooperation among the previously separate disciplines is essential for better preparing future engineers to tackle the problems that have no obvious disciplinary boundaries. The bio-based products engineering curriculum (4-year plan of courses shown in Table 1) spans across multiple colleges and disciplines with required course work and faculty participation from the Department Biosystems and Agricultural Engineering (BAE), College of Agricultural, Food and Environmental Sciences; Biotechnology Institute, College of Biological Sciences and the Institute of Technology (U of M's College of Engineering); and the newly formed Department of Bio-based Products, College of Natural Resources. Basic sciences and some of the basic engineering courses are offered by faculty from the Institute of Technology. Basic biological systems and bioprocess engineering courses are offered by faculty in BAE and the BioTechnology Institute. Basic science and engineering of bio-based products are offered by faculty in the Bio-based Products department. Capstone process and product design is co-taught by faculty in College of Natural Resource and the Institute of Technology.

For example, one semester long course on wood chemistry has been changed to embrace a broader scope of chemistry of plant materials. In addition to carbohydrates, lignin and extractives, the broadened scope of topics includes proteins, alkaloids, chemicals and other products from biomass. The pulp and paper unit operations course has been changed to a 4 credit bio-based products engineering course which includes basic unit operations relevant to the conventional bio-based products industry as well as the emerging bio-based products industry. The unit operations considered include, in addition to filtration, drying, evaporation, combustion, and dewatering, membrane separation, distillation and extraction, gasification, etc. In each of the unit operations in addition to drawing examples from pulp and paper, examples from emerging bio-based products industry such as bio-ethanol, bio-diesel, bio-based polymers are also used. Basics of bioprocesses such as fermentation, bioreactors, cell growth kinetics, recovery and purification are considered in a separate class on bioprocess engineering. We have taken a more basic, fundamental approach to the undergraduate curriculum providing the students with the basic fundamental principles and some flavor for applications but more importantly an ability to carry on life learning for new scientific principles and applications technology. For example, the traditional wet end chemistry course taught in many pulp and paper curriculums has been changed to a more fundamental surface and colloid science course covering the aspects of not only wet end chemistry but also coatings.

Integration of Design – Process and Product Design

Given the changes in business and industry, it is unlikely that many of the future engineering graduates are expected to design large-scale continuous plants. Rather, they need to be prepared for more diverse and flexible roles in industry. This is also true for Bio-based Products Engineering graduates. To realize the BPE program educational objectives, and in response to the significant changes in emerging bio-based products industries, a new two-course sequence, Process and Product Design, was created. The new curriculum provides the capstone design experience in the Bio-based Product Engineering curriculum incorporating novel, multi-disciplinary, cross-functional process and product design approach to engineering design. Integrated design concepts are also planned to be incorporated across the entire curriculum. Course development will be guided by the Backward Design Process as outlined by Bransford, Vye and Bateman (*12*) in *The Knowledge Economy and Post-Secondary Education* and by Wiggins and McTighe (*13*) in *Understanding by Design*. Process and Product Design involves making decisions on what to make and how to make it. A four-step procedure set forth by Moggridge and Cussler (*14, 15*) for chemical product design will be followed and include the following topics in sequence: identifying customer need, generating ideas, selecting a design, and manufacturing the products.

Table 2. Four Year Bio-based Products Engineering Curriculum

BIO-BASED PRODUCTS ENGINEERING

Freshman Year

Fall Semester	crs
BP 1001 BP Orientation	1
Chem 1021 Chem Princ I [1011 or 1015 or passing plcmnt]	4
Math 1271 or 1371 Calculus I [1371 Fall only][1151or1155orplcmnt or	4
Phys 1301 Intro Phys I [&Math 1271 or &Math 1371]	4
Rhet 1101	4
	17

Spring Semester	crs
BP 1002 Wood & Fiber Sci	3
BP 1003 Wood & Fiber Sci Lab	1
Chem 1022 Chem Princ II [Chem 1021]	4
Math 1272 or 1372 Calculus II [Math 1271 or 1371]	4
UD Phys 1302 Intro Phys II [&Math 1272 or &1372, Phys 1301]	4
	16

*You should try to take English composition in the freshman year.

Sophomore Year

Fall Semester	crs
Biol 1009 [high school chem]	4
Chem 2301 Organic Chem I [Chem 1022]	3
UD Math 2263 or 2374 Multivar Calc [Math 1272 or 1372]	4
BP 3001 Stat., Mech., Struct Des [Gen physics, algebra, trig]	4
Lib Ed	3
	18

Spring Semester	crs
UD Chem 2302 Organic Chem II [Chem 2301]	4
Chem 3501 Physical Chemistry I [Chem 1022, Math 2263 or 2374, Phys 1302]	3
UD Math 2243 or 2373, Lin Alg/Diff Eqs [Math 1272 or 1372]	3
ChEn 4001 Material & Energy Balances [Chem 2302 or &2302, Chem 3501 or &3501, Math 2373/2374 or &2373/2374, Phys 1302]	4
Lib Ed	3
	16

Junior Year

Fall Semester	crs
BP 4001 Chem of Plant Mat'ls [1002 or WPS1301,chem2302,UD or #]	4
BP 4303 BP Mat'ls Sci** [3001 or WPS 4301, ME3321, UD or #]	3
CE 3502 Fluid Mechs [IT, AEM 3031, Math 2243, or BP major]	4
ChEn 4101 Chem Eng Thermodynamics [ChEn 4001,Chem3501,UD or dept con]	4
ChEn 3701 Intro Biomolecular Eng. Chem 2302 or conc, Math 2373,ChenN4001	3
	18

Spring Semester	crs
BAE 4013 Transp in Biol Sys [BAE 3013 or &3013 or ChEn 3701, CE3502, ME3331 or ChEn 4101]	4
BAE 4713 Bio Process Engrg [UD IT, BAE 4013 or &4013]	3
BP 4301 Surf. &Col. Sci in BP Mfg [Chem 3501, ME3321, UD or #]	3
BP 4402 BP Eng Lab I [BP 1002 or WPS1301, BAE4713, UD or #]	1
Lib Ed	3
	14

Senior Year

Fall Semester	crs
BP 4401 BP Prod Engrg [BAE 3013, BAE 4013 or #]	4
BP 4403 BP Prod Engrg Lab II [BP1002 or WPS1301, BP4402, BAE 4713, o...	1
BP 4405 Proc Control & Simul [BP4401 or &4401, BP4402 or &4402, BP440...	3
BP 4501 Proc and Prod Des I [BAE3013, BAE4013, UD BP, or #]	4
Lib Ed	3
	16

Spring Semester	crs
BP 4302 Organ Impcting BP [BP 1002 or WPS 1301 or #]	3
BP 4404 Bio-based Comp Engrg [...]	3
BP 4502/W Proc and Prod Des II [UD, BP 4501]	3
BP 4504/W Bio-based Prod Dev Mgmt [UD]	3
Lib Ed	3
	15

Total 127

In addition, students will be involved in a professional work environment which requires oral and written communication among team members and between the team and engineering professionals. The students will be given opportunities to discuss the development of their design, its limitation and other issues in both weekly meetings and project presentations. The ethical, safety and environmental consideration and recommendations are required in their final report. These measures are designed on the basis of ABET criteria.

Pedagogy

In addition to embracing a broader focus, as part of the new Bio-based Products Engineering curriculum, the plan is to implement the learning from recent advances in the pedagogy of engineering education, thus contributing to its own advancement. During the recent past there have been numerous discussions about effective engineering education (16) and the consensus is that the goals in teaching college level science and engineering include helping students to develop problem-solving and communication skills, the ability to be a team member and the desire for lifelong learning. These goals clearly go beyond the traditional discipline-oriented knowledge acquisition. Young engineers today need to be equipped with a broader background in related subject areas and be able to apply this knowledge in a complex problem environment (17). Another well-accepted fact is that learning is best facilitated by giving students the opportunity for active learning and hands-on practice rather than by simply presenting the information (18, 19). The days of the "sage on the stage" clearly are over. Effective instructors see themselves in the role of coaching and encouraging students to achieve the required skills and providing constructive feedback. Cooperative learning is one example of pedagogies of engagement that has been shown to be effective in achieving a broad range of student learning outcomes (20).

Summary and Path Forward

The forest products and pulp and paper industry has a long, successful history of making industrial and consumer products from wood, the primary renewable resource. This industry has been under severe global competition from low cost manufacturers, and instead of being focused on a narrow range of commodity products based on cellulose fiber, the forest products and paper industry has an opportunity to play a vital role in the 21st century bio-economy. With the advent of industrial biotechnology, this industry can be at the forefront of developing a multitude of materials, chemicals, fuels and energy for the bio-economy. The success of the bio-economy can only be realized if a trans-disciplinary approach within traditional forestry/agriculture, and the petrochemical and industrial biotechnology industries is pursued.

Recognizing the opportunities in conventional and emerging bio-based products and the role a traditional department such as wood and paper science can play, faculty at the University of Minnesota have undertaken a major initiative developing a bio-based products program and curricula which embraces a broader scope of bio-based products including wood products, paper and other bio-based products. The new curricula in bio-based products engineering and marketing and management are inter-disciplinary, borrowing on the wide-ranging expertise in a multitude of relevant disciplines throughout the campus. The broader impacts of the ongoing effort include transformation of an academic curriculum in the emerging field of bio-based products training; well-prepared technical and business professionals for the future "bio-economy", thus ensuring U.S. energy independence and economic security; environmental sustainability; and economic growth.

The implementation of the curricula is in its second year. Already, positive response in terms of increased student interest is apparent. Given the magnitude of the curricular change, it will be a few years before faculty feel more comfortable embracing a broader bio-based products focus in their classes. Already, a broadened scope of sustainable renewable bio-resources utilization and bio-based products has gained increased attention and visibility within the campus and elsewhere, attracting additional opportunities for research. The future success of this effort, as with any curriculum, will certainly depend on career job opportunities in the conventional wood products and pulp and paper industry, as well as job opportunities in the emerging bio-based products industry.

It is believed that the curricular and programmatic changes underway in collaboration with other departments and colleges across the campus will advance the University of Minnesota's efforts to provide vitally important education, research and training for the bio-based industry and bring the University of Minnesota to the leading edge of the bio-economy.

Acknowledgements

The on-going efforts at the University of Minnesota presented here is a collaborative effort with participation and commitment from all faculty in the

Department of Bio-based Products including Dr. J. Bowyer, Ms. M. Cheple, Dr. D. Grimsrud, Mr. P. Huelman, Mr. H. Petersen, Dr. E. Schmidt, Dr. S. Sarkanen, Dr. R. Seavey, Dr. S. Severtson, Dr. T. Smith, Dr. S. Suh, Dr. U. Tschirner, Dr. W. Tze and Dr. K. Yin. I would also like to acknowledge the contributions and participation from other faculty across the campus including Dr. E. Cussler, Dr. R. Morey, Dr. K. Smith, Dr. M. von Keitz. Funding from the U.S. Department of Agriculture (USDA) Higher Education Challenge Grant 2005-38411-15870 for the development and implementation of the bio-based products engineering curriculum is greatly appreciated. Also, a matching grant from the University of Minnesota's Initiative for Renewable Energy and the Environment is also greatly appreciated.

References

1. EESI. "Elements of a bio-based economy: bio-based energy, fuels and products" 2003.
   http://www.eesi.org/briefings/2003/EnergyandClimate/5.20.03%20Biomass/5.20.03%20Biomass.htm
2. Biomass R& D Act. Vision for Bioenergy and BioProducts in the United States Report by the Biomass Technical Advisory Committee, established by Biomass Research & Development Act of 2000, 2000. http://www.bioproducts-bioenergy.gov/about/bio_act.asp
3. Renewable Vision. "US Department of Energy Technology Roadmap for Plant/Crop based Renewable Resources 2020", Renewable Vision 2020 (1999) http://www.newuses.org/pdf/ technology_roadmap.pdf
4. NRC. Bio-based Industrial Products: Research and Commercialization Priorities (2000) Committee on Bio-based Industrial Products, National Research Council, National Academy Press, Washington DC, p 162. http://www.nap.edu/books/0309053927/html/
5. Personal communication, Industry Advisory Council, 2003.
6. Cullinan, H. Where are our universities headed? *Solutions* **2004**, *87(12)*.
7. Myers, B.; Dyer, J.E. "Making science applicable- the need for a modern agricultural education curriculum," *Agricultural Education Magazine* **2002**, *74(5)*, p 24.
8. Cussler, E. L., and J. Wei, "Chemical Product Engineering," *AIChE J.* **2003**, *49*, pp 1072-1075.
9. Singh, S.P.; Ekanem, E.; Wakefield, Jr., T.; Comer, S. "Emerging importance of bio-based products and bio-energy in the U.S. economy: information dissemination and training of students," *International Food and Agribusiness Management Review* **2003**, *5(3)* http://www.ifama.org/nonmember/OpenIFAMR/Articles/v5i3/singh.PDF

10. NAE (National Academy of Engineering). 2004. The Engineer of 2020: Vision of Engineering in the New Century. Washington, DC: National Academies Press.

11. Lang, J.D.; Cruse, S.; McVey, F.D.; McMasters, J. "Industry expectations of new engineers: a survey to assist curriculum designers," *J. Eng. Ed.* **1999**, *88*, p 43.

12. Bransford, J.; Vye, N.; Bateman, H. "Creating high-quality learning environments: Guidelines from research on how people learn." In *National Research Council. The Knowledge Economy and Post-Secondary Education: Report of a Workshop. Committee on the Impact of the Changing Economy on the Education System*, 2002.

13. Wiggins, G.; McTighe, J. "Understanding by Design," Alexandria, VA: Association for Supervision and Curriculum Development, 1998.

14. Moggridge, G. D.; Cussler, E.L. "Chemical Product Design," *Chemical Engineering* (centennial issue) **2002**, *109*, p 133.

15. Moggridge, G. D. and Cussler, E.L. "Chemical Product Design," In *Kirk-Othmer Encyclopedia Chem. Tech.* (in press; available online, 2003).

16. Rugarcia, A.; Felder, R.M.; Woods, D.R.; Stice, J.E. "The Future of Engineering Education I. A vision for a new century," *Chem. Eng. Ed.*, **2000**, 34 (1), p 16.

17. Meier, M.A.; Hungerbuehler, K.; Alean-Kirkpatrick, P. "Integrated products and process development using case-studies: a challenge to the education of science and engineering undergraduates," *Chimia* **1997**, *51(5)*, p 171.

18. Woods, D.R.; Hyrymak, A.N.; Marshall, R.R.; Wood, P.E.; Crowe, C.M.; Hoffman, T.W.; Wright, J.D.; Taylor, P.A.; Woodhouse, K.A. and Bouchard, C.G.K. "Developing Problem Solving Skills: The McMaster Problem Solving program", *J. Eng. Ed.* **1997**, *86 (2)*, p 75.

19. Bandura, A. "Self-Efficacy Mechanism in Human Agency", *Amer. Psychologist* **1982**, 37, p 122.

20. Smith, K.A., Sheppard, S. D., Johnson, D.W., & Johnson, R.T. "Pedagogies of engagement: Classroom-based practices". *Journal of Engineering Education Special Issue on the State of the Art and Practice of Engineering Education Research* **2005**, *94 (1)*, pp 87-102.

21. Thorp, B.A. "Business Model for North American Mill Survival" *Pulp and Paper*, November **2005**.

Materials from Forest Biomass

Chapter 5

Composite Materials from Forest Biomass: A Review of Current Practices, Science, and Technology

Roger M. Rowell

FS, Forest Products Laboratory, U.S. Department of Agriculture, and University of Wisconsin, Madison, WI 53706

Renewable and sustainable composite materials can be produced using forest biomass if we maintain healthy forests. Small diameter trees and other forest biomass can be processed in the forest into small solid wood pieces, sliced veneers, strands, flakes, chips, particles and fiber that can be used to make construction composite products such as glued-laminated lumber, plywood, structural composite lumber, flakeboard, particleboard, fiberboard, and molded materials. It can also be combined with other resources to make new materials taking advantage of the unique properties of each resource. Non-construction composites can also be made such as geotextiles, filters, sorbents, packaging and nano-materials. The adhesive used in the various products depends on the intended application.

Introduction

We have used wood, in its natural form, for many applications since the beginning of the human race. The earliest humans used wood to make shelters, cook food, construct tools, and make weapons. There are human marks on a climbing pole that were made over 300,000 years ago. We have found wood in the Egyptian pyramids, Chinese temples and tombs and ancient ships that attest to the use of wood by past societies. Collectively, society learned very early the great advantages of using a resource that was widely distributed, multifunctional, strong, easy to work, aesthetic, sustainable, and renewable.

In 1900, almost half of the materials used in the United States, on a per weight basis, came from renewable resources. By 1995, materials from renewable resources had decreased to less than 8%.

The United States ranks third in the world in the volume of standing forest biomass and is the largest producer of industrial timber products. These products contribute over $240 billion to the gross national product and employ over 1.1 million people.

Forest biomass is defined as all photosynthetic mass that is produced within a forest. This includes all trees, large and small, and underbrush. There are 749 million acres of forest lands in the United States. Of this, 319 million acres are held by the public (local, state and federal) and 430 million acres are privately owned. Of the private sector, 363 million acres are non-industrial. These forests produce approximately 368 million dry tons of biomass per year. Biomass from agricultural sources is approximately 998 million dry tons per year making a total of 1.366 billion dry tons of biomass produced ever year that could potentially be used to produce composite materials (*1*). The total global inventory of all biomass is over 4 billion metric tons of which wood represents 1.75 billion, straw, 1.15 billion, stalks, 0.97 billion, and 0.13 billion for all other (grass, leaf, core, bagasse, etc). This chapter will concentrate only on the biomass from US forests that could be used to make composite materials.

Wood is not renewable! It comes from trees and trees are renewable. So any discussion of renewable and sustainable development for wood, must concentrate on sustainable forests. Sustainable forests are healthy forests. And, healthy forests have many demands put on them including: recreation, biodiversity, source of clean water, a buffer against storms, a vast carbon storage, animal habitat, aesthetics, as well as a source of products such as lumber, fuels, composites, pulp and paper, and chemicals. Healthy forests are also critical in converting carbon dioxide to oxygen. Because of this, a discussion of composite materials from forest biomass must start with a consideration for managing and maintaining healthy forests.

Healthy Forests

There will be increased competition for land use as the world population increases. In 1830, the world population was 1 billion. Over the next hundred years, it doubled to 2 billion. At the present rate of growth, the population increases by 1 billion people every eleven years. Composite materials from forest biomass provide an opportunity to fill a growing need for materials, however, there will be a greater and greater need for food and feed. Forest lands are being converted to agricultural lands and to industrial and residential lands.

Healthy forests play an important role in providing clean air and water as well as to regulate the climate. About 80% of the fresh water in the United States originates on the 650 million acres of forestlands that cover about 1/3 of the Nation's land area.

In order to insure a continuous supply of forest biomass, management of the forest land should be under a proactive system of land management whose goal is both sustainable forestry and the promotion of healthy ecosystems. Ecosystem management is not a euphemism for preservation, which might imply benign neglect. Sustainable forests denotes a balance between conservation and utilization of forest lands to serve both social and economic needs, from local, national and global vantage points. Sustainable forestry does not represent exploitation but rather is aimed toward meeting all the needs of the present generation without compromising the ability of future generations to meet their needs. It encompasses, in the present case, a continuous production of biomass for composites but also considerations of multi-land use, and the protection, restoration, and conservation of the total ecosystem.

Many of our forests, especially forests in the Western part of the US, are over crowded with small diameter timber. This, so called "fire-prone forest biomass" is responsible for the devastating forest fires we have had recently. With a healthy forest where, at least, 60-75% of the ground is covered with vegetation, only about 2% or less of rainfall becomes surface runoff and erosion is low (less than 0.05 tons loss per acre). After a forest fire, burned land usually sorbs water more slowly than unburned land.

So, it is important to forest health for this fuel load to be removed. But, the argument is always that it is too expensive to remove and there is no demand for such a large amount of biomass. The question might be that while it is too expensive to remove this biomass, it is more costly not to remove it. In a study of the cost to fight forest fires, Lynch has shown that is more expensive to fight a fire (reactive) than to remove the forest biomass (proactive) (2). Looking at the cost to fight fires in Colorado over a seven year period, he concludes that it costs approximately $600 per acre to fight a fire and that does not include costs such as restoration of watersheds, equipment loss, loss of productivity, and loss of life.

Based on the results of the Lynch study, one strategy for producing composites from forest biomass is to go into an overgrown area and harvest all small diameter and other fuel load material and convert it to composites. This biomass with be crooked, have a large percentage of bark, diverse mixed species, and hard to handle but it can processed in the woods into transportable furnish. Mara has suggested that this material can be converted to a mixture of fibers that can be used to produce high valued products (3). The total removed biomass can also be converted to chips, flakes or particles that can be more easily transported to a mill for processing into composites. There will be a high percentage of bark and many mixed species in the furnish that will require a higher level of adhesive in the composites. Fractionation of various sizes and types of elements can also be done in the forest. The objective of this approach of producing composites from biomass is not based on the recoverable cost from the composites, but to clean the forest of the fuel load.

The in forest processing can be done using the biomass as fuel to run small-scale power units (4, 5).The units are portable and can be used in remote locations to process the small diameter forest biomass into a usable, transportable furnish.

Composites from Forest Biomass

A composite can be defined as two or more elements held together by a matrix. The matrix can be physical or chemical. Chemical matrixes include anything from hydrogen bonding to covalent bonding using resins and adhesives. Physical matrixes include anything from physical entanglement and non-woven technologies to porous containers (bags).

Solid wood is a three dimensional composite composed of cellulose, hemicelluloses and lignin polymers with smaller amounts of inorganics and extractives held together in a lignin matrix. Solid wood can be broken down into smaller and smaller elements, i.e. veneers, flakes, strands, chips, flakes, particles, fiber and cellulose. As the element size becomes smaller, it is possible to either remove defects (knots, cracks, checks, etc.) or redistribute them to reduce their effect on product properties and uniformity increases. Composites made from these smaller elements become more like a true material, i.e. consistent, uniform, continuous, predictable, and reproducible. Size reduction, however, requires energy and may increase water consumption.

Composite materials from forest biomass can be classified by several different systems: density (for example, medium density fiberboards), application (for example, insulation board), raw material form (for example, particleboard) and process type (for example, dry process fiberboard). The breakdown of forest biomass can include large timbers, dimensional lumber, very thick laminates for glued-laminated beams, thin veneers for plywood, strands for strandboard, flakes for flakeboard, chips for chipboard, particles for particleboard, and fibers for fiberboard. However, since the bulk of this chapter deals with the utilization of small diameter biomass, only composites made using small solid wood members, strands, chips or flakes and fiber will be covered.

Two major types of composites will be reviewed in this chapter: construction composites used for building and non-construction composites.

Construction Composites

The largest single use for forest-based composites is for residential construction. We have used wood for light wood-framed housing for over 100 years. Over 95% of the approximate 1.5 million homes built in the United States each year are framed using wood. Many wood-based composites are also used in this application and are also used for non-residential construction, especially exterior wall paneling.

Table 1 shows the number and size of the average American home from 1970 to 2000 and the projection of growth in both number and size for 2020 and 2050. The number of homes has increased with the increase in population and the size of the average home has also increased. It is projected that the average size of a home will continue to increase but this will depend on income, energy costs and raw material availability.

Table 1. Sizes of Single Family Houses in the US Past and Future Projections

Year	Households (Millions)	Average size (Square Meters)
1970	63.9	140
1980	79.1	162
1990	93.2	193
2000	102	204
2020	138.5	210
2050	178.1	240

Glued-Laminated Lumber

From the large diameter trees and to some extent smaller diameter trees as well, structural glued-laminated beams (glulam) can be made using thick, wide wood members that are used as structural elements in large open buildings.

Glued-laminated lumber could be produced from small diameter trees using small pieces finger-jointed together. Glulam is a structural product that consists of two or more layers of lumber glued together with the grain all going parallel to the length. The laminated beam is usually fitted into a steel plate that joins the beam to the ground. It can be formed straight or curved depending on the desired application. Typically the laminates are 25 to 50 mm in thickness. Douglas fir, southern pine, hem-fir and spruce are common wood species used in the United States (6). Lower quality wood can be used, thinner lumber can be dried much faster than large thick beams and a variety of curved shapes can be produced. Solid wood and glulam have a specific gravity of 0.4 to 0.8.

Plywood

The modern plywood industry began around 1910 but the furniture industry had used veneers over solid wood for several hundred years before this. Plywood is made using thin veneers which are glued together and are used as a structural underlayment in floors and roofs and in furniture manufacturing. The veneers are produced from large diameter tree bolts by peeling but they can also be produced by slicing small diameter trees bolts.

Structural Composite Lumber

Structural composite lumber (SCL) is manufactured by laminating strips of veneers or strands of wood glued parallel to the length. There are several types of SCL: oriented strand lumber (OSL), parallel strand lumber (PSL) and laminated veneer lumber (LVL). Laminated strand lumber (LSL), oriented strand board (OSB) and OSL are produced using different lengths and sizes of strands. LSL uses strands that are about 0.3 m in length while OSB is produced from shorter strands. OSB is usually made of three to five layers of strands that are oriented at right angles to each other and the product is used for exterior sheathing. Table 2 shows some of the mechanical properties of sheathing grade OSB. PSL is made from strands that are 3 mm thick, approximately 20 mm wide and 100 to 300 mm in length. Usually Douglas-fir, southern pines, western hemlock and yellow-poplar are used but other species are also used. LVL is produced from veneers that are approximately 2.5 to 3.2 mm thick and of varying lengths. The major adhesives used to produce SCL products are phenol-formaldehyde or isocyanates. All of these SCL products are used as replacements for solid wood and have a specific gravity of 0.5 to 0.8.

Table 2. Properties of Sheathing Grade OSB (Wood Handbook 1999)

Property	Value	ASTM test method
Modulus of rupture	20.7 – 27.6 MPa	D3043-87
Modulus of elasticity	4.8 – 8.3 GPa	
Tensile strength	6.9 – 10.3 MPa	D3500-90
Compression strength	10.3 – 17.2 MPa	D3501-76

ASTM D304-87, Standard methods for testing structural panels in flexure. ASTM D3500-90, Standard test methods for structural panels in tension. ASTM D3501-76, Standard method of testing plywood in compression.

By combining several elements, composite structural beams can be produced. An I beam can be made of curved plywood sides and laminated plywood top and bottom. They can also be made using a flakeboard center with plywood top and bottom or plywood, hardboard, flakeboard and oriented strandboard. Prefabricated I-beams are used by builders because they are light weight, have increased dimensional stability, uniform, easy to use and meet codes and standards. In most, if not all cases, performance requirements of these composites are spelled out in codes and/or in specifications set forth by local or national organizations.

Waferboard and Flakeboard

Waferboard is a structural panel used in exterior applications bonded with a phenolic adhesive. Large thin wafers or smaller flakes can be produced by several methods and used to produce a composite board. Wafers are almost as wide as they are long while flakes are much longer than they are wide. Wafers are also thicker than flakes. A waferizer slices the wood into wafers typically 38 mm wide by 76 to 150 mm long and 7 mm thick with a specific gravity between 0.6 and 0.8.

The first waferboard plant was opened in 1963 by Macmillan Bloedel in Saskatchewan, Canada. Aspen was the raw material and the wafers were randomly oriented. In the late 1980's, most wafers were oriented resulting in oriented waferboard (OWB) that is stronger and stiffer than the randomly oriented board. The orientation distribution may be tailored to the application. OWB and OSB compete with plywood in applications such as single layer flooring, sheathing, and underlayment in lightweight structures, however, OSB has largely replaced OWB in most places in the US. Pines, firs and spruce are usually used as well as aspen.

The flakeboard industries started in the early 1960's. These are made using an exterior grade adhesive and are used as the structural skin over wall and floor joists. The specific gravity of flakeboard is usually between 0.6 and 0.8 and made using a waterproof adhesive such as phenol formaldehyde or an isocyanate (6).

Particleboard

The particleboard industry started in the 1940's out of a need to use large quantities of waste products such as sawdust, planer shavings and other mill residues. Particles of various sizes are formed into an air or mechanically formed mats and glued together to produce a randomly oriented flat panel. Almost all particleboards are produced by a dry process. Particleboard is usually made in three layers with the faces made using fine particles and a core of coarser material. Most applications of particleboard are for interior use and are bonded using a urea-formaldehyde adhesive. Paraffin or microcrystalline wax emulsion is usually added to improve short-term moisture resistance. Phenol-formaldehyde, melamine-formaldehyde and isocyanates are rarely used but are when increased moisture resistance is required. The resin content ranges from 4 to 10% but is usually made using 6 to 9%. The resin content of the two faces is usually slightly higher than the core. Table 3 shows some properties of different grades of particleboard.

Particleboard is often used as a core material for veneers and laminates. These are often used in counter tops, shelving, doors, room dividers, built-ins and furniture. Particleboard generally does not warp in use. It is available in several thicknesses from 6 to 32 mm in sheets of 120 by 240 cm.

Fiberboard

Wood can be broken down into fiber bundles and single fibers by grinding or refining. In the grinding process, the wood is mechanically broken down into fibers. In the refining process, wood chips are broken down into fibers between one or two rotating plates in a wet environment to fibers. If the refining is done at high temperatures, the fibers tend to slip apart due to the softening of the lignin matrix between the fibers, and the fibers have a lignin rich surface. If the refining is done at lower temperatures, the fibers tend to break apart and the surface is rich in carbohydrate polymers. Fiberboards can be formed using a wet forming or a dry forming process. In a wet forming process, water is used to distribute the fibers into a mat and then pressed into a board. In many cases an

Table 3. Properties of Particleboards

Grade*	Modulus of rupture (MPa)	Modulus of elasticity (MPa)	Internal bond MPa)
H-1	16.5	2400	0.90
H-2	20.5	2400	0.90
H-3	23.5	2750	1.00
M-1	11.0	1725	0.40
M-2	14.5	2225	0.45
M-3	16.5	2750	0.55
LD-1	3.0	550	0.10
LD-2	5.0	1025	0.15

H = density greater than 800 kg/m^3, M = density 640 – 800 kg/m^3, LD = density less than 640 kg/m^3. (National Plywood Association 1993 Particleboard, ANSI A208.1-1993. Gaithersburg, MD.)

adhesive is not used and the lignin in the fibers serves as the adhesive. In the dry forming process, fibers from the refiner go through a dryer and blowline where the adhesive is applied and then formed into a web which is pressed into a board.

Low density fiberboards (LDF) have a specific gravity of between 0.15 and 0.45 and are used for insulation and for light weight cores for furniture. They are usually produced using a dry process using a ground wood fiber. Medium density fiberboard (MDF) has a specific gravity of between 0.6 and 0.8 and is mainly used as a core for furniture and is usually made using a dry process. High density fiberboard (HDF), sometimes called hardboard, has a specific gravity of between 0.85 and 1.2 and is used as an overlay on workbenches, floors and for siding. It is usually made using a dry process but is also made using a wet fiber process. The hardboard industry started around 1950 and is produced both with wax (tempered) and without wax and sizing agents. The wax is added to give the board water resistance.

Urea-formaldehyde resin is usually used for interior applications and phenol-formaldehyde for exterior applications. Table 4 gives the mechanical properties of LDF, MDF and HDF. Standard (without wax) and tempered (with wax) hardboards come in many different thicknesses from 2.1 to 9.5 mm and have required standard minimum average modulus of rupture of 31.0 MPa for standard and 41.4 MPa for tempered. Standard hardboard must have a required standard minimum average tensile strength parallel to the surface of 15.2 MPa and 20.7 MPa for tempered. The required standard minimum average tensile strength perpendicular to the surface is 0.62 MPa for standard and 0.90 MPa for tempered.

Table 4. Properties of Fiberboards

Product	Thickness (mm)	Modulus of Rupture (MPa)	Modulus of Elasticity (MPa)	Internal bond (MPa)
Internal MDF				
HDF	<21	34.5	3,450	0.75
MDF	>21	24.0	2400	0.60
LDF	21	14.0	1400	0.30
Exterior				
MDF	21	34.5	3,450	0.90

National Plywood Association 1994. Medium density fiberboard, ANSI A208.2-1994. Gaithersburg, MD.

Molded Products

The present wood-based composite industry mainly produces two-dimensional (flat) sheet products. In some cases, these flat sheets are cut into pieces and glued/fastened together to make shaped products such as drawers, boxes, and packaging. Flat sheet fiber composite products are made by making a gravity formed mat of fibers with an adhesive, and then, pressing. If the final shape can be produced during the pressing step, then the secondary manufacturing profits can be realized by the primary board producer. Instead of making low cost flat sheet type composites, it is possible to make complex shaped composites directly using the long fibers alone or combinations of long and short fibers.

Wood fiber come in two lengths: short (softwoods) and shorter (hardwoods). So wood fiber is limited to short fiber applications unless it is combined with longer agricultural fibers for applications in a wider array of products. In this technology, fiber mats are made by combining long bast or leaf fibers from such plants as kenaf, jute, cotton, sisal, agave, etc. with wood fiber and then formed into flexible fiber mats. These can be made by physical entanglement (carding), nonwoven needling, or thermoplastic fiber melt matrix technologies. In carding, the fibers are combed, mixed and physically entangled into a felted mat. These are usually of high density but can be made at almost any density. A needle punched mat is produced in a machine which passes a randomly formed machine made web through a needle board that produces a mat in which the fibers are mechanically entangled. The density of this type of mat can be controlled by the amount of fiber going through the needle board or by overlapping needled mats to give the desired density. In the thermoplastic fiber matrix, the bio-based fibers are held in the mat using a thermally softened thermoplastic fiber such as polypropylene or polyethylene.

During the mat formation step, an adhesive is added by dipping or spraying the fiber before mat formation or added as a powder during mat formation. The mat is then shaped and densified by a thermoforming step. The final desired shape is determined by the mold in the hot press. Within certain limits, any size, shape, thickness, and density is possible.

Combinations with Other Resources

It is possible to make completely new types of composites by combining different resources. It is possible to combine, blend, or alloy woof fiber with other materials such as glass, metals, plastics, and synthetics to produce new classes of materials. The objective will be to combine two or more materials in such a way that a synergism between the components results in a new material that is much better than the individual components.

Wood-fiber/glass fiber composites can be made using the glass as a surface material or combined as a fiber with bio-based fiber. Composites of this type can have a very high stiffness to weight ratio. Wood fiber can also be used in place of glass fiber in resin injection molding (RIM) or used to replace, or in combination with, glass fiber in resin transfer molding (RTM) technologies. Problems of dimensional stability and compatibility with the resin must be addressed but this could also lead to new markets for property-enhanced forest biomass composites.

Metal films can be overlayed onto smooth, fiber composite surfaces or applied through cold plasma technology to produce durable coatings. Such products could be used in exterior construction to replace all aluminum or vinyl siding, markets where wood-based resources have lost market share.

Metal fibers can also be combined with fiber in a matrix configuration in the same way metal fibers are added to rubber to produce wear-resistant aircraft tires. A metal matrix offers excellent temperature resistance and improved strength properties, and the ductility of the metal lends toughness to the resulting composite. Application for metal matrix composites could be in the cooler parts of the skin of ultra-high-speed aircraft. Technology also exists for making molded products using perforated metal plates embedded in a phenolic-coated fiber mat, which is then pressed into various shaped sections.

Wood fiber can also be combined in an inorganic matrix. Such composites are dimensionally and thermally stable, and they can be used as substitutes for asbestos composites. The inorganic phase can be gypsum, magnesia-cement, or Portland-cement. Inorganic bonded fiber composites can also be made with variable densities that can be used for structural applications.

One of the biggest new areas of research incorporating wood flour or wood fiber in a non-wood matrix is in combination with thermoplastics (7). Since prices for plastics have risen sharply over the past few years, adding wood flour or fiber to plastics provides a cost reduction to the plastic industry (and in some cases increases performance as well) but to the forest-based industry, this represents an increased value for the wood component.

Blending of the plastics with the wood fiber may require compatibilization to improve dispersion, flow and mechanical properties of the composite. Extrusion of wood-filled plastics for the automotive industry is well known and has been used for more than twenty years. Typical blending involves the plastic-filler/reinforcement to be shear mixed at temperatures above the softening point of the plastics. The heated mixture is then typically extruded into "small rods", which are then cut into short lengths to produce a conventional pellet. The pellets can then be used in typical injection or compression molding techniques. To reduce the cost of this blending process, direct injection molding of wood-fiber/plastics can be done. The direct injection molding process probably has limitations on the amount of filler/fiber that can be used in the composite, and is also likely to be limited to particulate or shorter fiber. The chemical characteristics of the surface and bulk of the woof fibers are also important in the blending with plastics. The ability of the matrix of the lignocellulosic (hemicellulose and lignin) to soften in the presence of moisture at plastic processing temperature may give these materials unique characteristics to develop novel processing techniques. The performance of injection molded cellulose fiber thermoplastics can be significantly improved by the use of microfibrils (8).

The primary advantages of using wood-based fibers as fillers/reinforcements in plastics are: low densities; non-abrasive, high filling levels possibly resulting in high stiffness properties; high specific properties; easily recyclable; biodegradable; the fibers will not fracture when processing over sharp curvatures unlike brittle fibers; wide variety of fibers available throughout the world; generate rural jobs; increase non-food agricultural/farm-based economy; low energy consumption; and low cost.

Non-Construction Composites

Geotextiles

Medium- to high-density fiber mats, described previously, can be used in several ways other than for molded composites. One is for use as a geotextile.

Geotextiles derive their name from the two words geo and textile and, therefore, mean the use of fabrics in association with the earth.

Geotextiles have a large variety of uses. They can be used for mulch around newly planted seedlings. The mats provide the benefits of natural mulch; in addition, controlled-release fertilizers, repellents, insecticides, and herbicides can be added to the mats as needed. Medium density fiber mats can also be used to replace dirt or sod for grass seeding around new homesites or along highway embankments. Grass or other type of seed can be incorporated in the fiber mat. Fiber mats promote seed germination and good moisture retention. Low and medium density fiber mats can be used for soil stabilization around new or existing construction sites. Without root stabilization, steep slopes lead to erosion and loss of top soil.

Medium and high density fiber mats can also be used below ground in road and other types of construction as a natural separator of different materials in the layering of the back fill. It is important to restrain slippage and mixing of the different layers by placing separators between the various layers.

It is estimated that the cost of controlling erosion in the United States is in excess of $55 million dollars per year. This is a large potential market for forest resource composites.

Filters and Sorbents

Filter systems are presently used to clean our water but new innovation in filtration technology is needed to remove contaminants from water. The development of filters to clean our water supply is big business. It is estimated that global spending on filtration (including dust collectors, air filtration, liquid cartridges, membranes and liquid macro-filtration) will increase from $17 billion in 1998 to $75 billion by 2020. The fastest-growing non-industrial application area for filter media is for the generation of clean water.

Medium to high density mats can be used as filtering aids to remove particulates out of drinking water, waste water and solvents. Wood fibers have also been shown to sorb oil from water. While not as good as other agricultural fibers such as kenaf, wood fiber can remove oil from both fresh and sea water (9).

Bark has been found to be very effective in removing heavy metal ions from contaminated water. Table 5 shows the effectiveness of several barks in removing copper from solution. Since the size of the bark particle affects its sorption ability, the bark is hammer-milled into small particles and then placed in bags that are put into a filter box where water passes through.

Medium and high density fiber mats or fiber filler containers can be used for air filters. The density of the mats can be varied, depending on the size and quantity of material being filtered and the volume of air required to pass through

Table 5. Effectiveness of Several Barks in Removing Copper from Solution

Bark	Average percent copper removed
Eastern black walnut	88.1
Osage orange	78.1
Ponderosa pine	72.4
American sycamore	63.8
Eastern red cedar	61.7
Burr oak	61.5
Utah juniper	51.7
Eastern cottonwood	48.2
Green ash	46.8
Silver maple	44.9

the filter per unit of time. Air filters can be made to remove particulates and/or can be impregnated or reacted with various chemicals as an air freshener or cleanser.

Packaging

Medium and high density wood-based fiber composites can be used for small containers, for example, in the food industry and for large sea-going containers for commodity goods. These composites can be shaped to suit the product by using the molding technology described previously or made into low cost, flat sheets and made into containers.

Wood-based composites can and have been used for pallets where cost and weight are critical factors. Moldability has been a key factor in the development of the flake-, particle- and fiber-based pallets.

Wood-based fiber composites can also be used in returnable containers where the product is reused several times. These containers can range from simple crease-fold types to more solid, even nestable types. Wood fiber mats can be overlayed with thermoplastic films such as polyethylene or polypropylene to be used to package such products as concrete, foods, chemicals, and fertilizer. Corrosive chemicals require the plastic film to make them more water-resistant and reduce degradation of the wood fiber.

Nano-composites

The smaller the wood element is, the better the homogeneity and the smaller the variability in properties. The smallest load-bearing element in wood is the cellulose fibril. Wood derives its strength and stiffness from cellulose and its respective matrix. Lignin provides some moisture resistance and along with the hemicelluloses, bonds cellulose fibrils together, but the intrinsic mechanical performance comes from the cellulose fibrils. They are strongly aligned at an angle close to the axial direction of the wood cells. The Young's modulus of the cellulose unit cell is 134 GPa, a value similar to that of Kevlar[c] fibers. The reason for this high value is that the cellulose chains are densely packed and as they are deformed, strong covalent bonds of high energy take the load (*10*).

Cellulose microfibrils are 30 nm in diameter and a few microns in length and can be disintegrated from wood pulp fibers using a mechanical milling procedure. It is also possible to take pulp fibers, subject them to acid hydrolysis and then disintegration will take place in conventional mixing equipment used in plastics processing. As with other wood composites, mechanisms of failure such as microcracking and debonding, are delayed to higher stresses and strains. The reason is that small-scale reinforcements also cause damage at a very small scale, which is less detrimental to material performance. Cellulose microfibrils also show interesting properties as a material of their own. If a water suspension of cellulose microfibrils is dried, it forms a hard and tough material, similar to ivory in character. The reason is millions of strong hydrogen bonds forming between the cellulose entities.

The ultimate wood composite would have a high content of cellulose microfibrils oriented in the main direction of loading. Strength and stiffness would be competing with high performance composites used in the aerospace industry. Lightweight sandwich structures could be produced by introducing porosity, for instance through foam cores where the material primarily consists of cellulose.

Yano et al (*11*) has shown that wood veneer layers with significant parts of lignins and hemicelluloses removed and impregnated with a phenol-formaldehyde resin, and then compressed and cured, resulted in a wood composite with a Young's modulus of 40 GPa. This is more than twice the value of any commercial wood material or wood composite.

The most recent research in micro-crystalline cellulose has shown that it can be used to make clear flexible films that are very strong and tough and may be the future television and computer screens. (*12*).

Adhesives

The adhesives that are now used in the wood industry for composites are based on the performance requirements of the composites. That is, phenol-formaldehyde, resorcinol-formaldehyde, phenol-resorcinol-formaldehyde, emulsion polymer-isocyanate, melamine-formaldehyde, and melamine-urea-formaldehyde are all used for exterior grade composites where they must withstand long term water-soaking and drying cycles.

Melaimine-urea-formaldehyde, isocyanates and epoxy resins are used when short-term water resistance is required. Urea-formaldehyde and casein adhesives are used for interior composite products.

Research is underway to develop new adhesive systems that are not petroleum-based, that are based on using resources, and that are based on a better understand of the mechanisms of adhesion (13). These mechanisms require a much better understanding of interface and interphase relationships between the forest resource elements and the matrix. Some of the newest research in this area involves the use of enzymes, surface activation, biotechnology, chemical modification, and cold plasma (14,15).

Conclusions

The future of forest resource-based composite materials will be very exciting and dynamic. It will be driven by traditions, trends, costs, performance, availability of resources, and legislation. Of these, the most critical issue is cost, however, concepts of sustainable development are changing the profit structure focused not only on the profit margin, but developing sustainable industry. There are several competing ideologies today that are driving public opinion. On the one hand, there is a growing need to create jobs, expand recreational opportunities, and improve the standard of living. On the other hand, there are concerns about energy consumption, an expanding world population, maintaining wilderness areas, cleaning up the environment, and the conservation of our natural resources. There is no "right" or "simple" answer.

There is no question that renewable, recyclable and sustainable resources will play a major role in future world developments. Forest resources and composite materials will be part of this dynamic future. We must not allow ourselves to be locked into a mental framework tied to past technologies or close our eyes to exciting new possibilities.

Existing forest resource-based composite markets will continue to grow, but the real excitement will come in the new potential markets we have either lost to other resources or in totally new markets. We are no longer limited by shape, density, size, or texture. We are only limited by our imagination and our knowledge and understanding of how to achieve the highest level of performance from these composites.

References

1. Perlack, R.D., Wright, L.L., Turhollow, A., Graham, R.L., Stokes, B., Erbach, D.C. Biomass as feedstock for a bioenergy and bioproducts

92

industry: The technical feasibility of a billion-ton annual supply. USDA, DOE Report, **April 2005**, ORNL/TM-2005/66, OakRidge National Laboratory, Oak Ridge, TN.

2. Lynch, D.L. What do forest fires really cost. *J. Forestry*, **Sept 2004**, pp 42-49.

3. Marra, A.A. High value products from low value wood. *For Prod J*, **Feb 2003**, pp 6-13.

4. *TechLine*, Wood biomass for energy. USDA, FS, Forest Products Laboratory, S&PF, TMU, WOE-1, **April 2004**.

5. TechLine, Biomass for small-scale heat and power. USDA, FS, Forest Products Laboratory, S&PF, TMU WOE-2, **April 2004**.

6. *Wood Handbook*. Wood as an engineering material. USDA, Forest Service, Forest Products Laboratory, General Technical Report FPL-GTR-113, **1999**.

7. Caulfield, D.F., Clemons, C., Jacobson, R.E., and R.M. Rowell, R.M. Wood-Thermoplastic Composites. In: Handbook of wood chemistry and wood composites, R.M. Rowell, ed, Taylor and Francis, Boca Raton, FL, 2005, Chapter 13, pp 365-380.

8. Boldizar A, Klason C, Kubat J, Näslund P and Saha P. *Intern J Polymeric Mater*,**1987**, (11): 229.

9. Rowell, R.M., Han, J.S., and Byrd, V.Y. Fiber Webs. In: *Handbook of wood chemistry and wood composites*, R.M. Rowell, ed, Taylor and Francis, Boca Raton, FL, **2005**, Chapter 12, pp 349-364.

10. Bergland, L. and Rowell, R.M. Wood Composites. In: *Handbook of wood chemistry and wood composites*, R.M. Rowell, ed, Taylor and Francis, Boca Raton, FL, **2005**, Chapter 10, pp 279-302.

11. Yano H, Hirose A, Collins PJ and Yazaki Y. *J Mat Sci Lett* **2001**, (20):1125.

12. Yano, H., Sugiyama, J., Nakagaito, A.N., Nogi, M., Matsuura, T., Hikita, M. And Handa, K. Optically transparent composites reinforced with networks of bacterial nanofibers. *Adv. Mater.* **2005**, Vol 17 (2), pp 153-155.

13. Frihart, C.R. Wood adhesion and adhesives. In: *Handbook of wood chemistry and wood composites*, R.M. Rowell, ed, Taylor and Francis, Boca Raton, FL, **2005**, Chapter 9, pp 215-278.

14. Rowell, R.M. Cell Wall Chemical Modification. In: *Handbook of wood chemistry and wood composites*, R.M. Rowell, ed, Taylor and Francis, Boca Raton, FL, **2005**, Chapter 14, pp 381-420.

15. Denes, F.S., Cruz-Barba, L.E., and Manolache, S. Plasma treatment of wood. In: *Handbook of wood chemistry and wood composites*, R.M. Rowell, ed, Taylor and Francis, Boca Raton, FL, **2005**, Chapter 16, pp 447-475.

Chapter 6

Surface and In-Depth Modification of Cellulose Fibers

Alessandro Gandini[1] and Mohamed Naceur Belgacem[2]

[1]CICECO and Chemistry Department, University of Aveiro,
3810–193 Aveiro, Portugal
[2]Laboratoire de Génie des Procédés Papetiers, UMR 5518, École Française de Papeterie et des Industries Graphiques (INPG), BP65, 38402 Saint Martin d'Hères, France

This chapter is divided into three parts, according to the application envisaged for the modified fibers, namely (i) as reinforcing elements in macromolecular composite materials; (ii) in wood densification and protection; and (iii) for trapping organic pollutants. The major emphasis in the first part is devoted to the controversial aspects related to the interactions between cellulose and siloxanes. The second part illustrates our approach through the use of bifunctional coupling agents and the subsequent grafting of the densifying polymer. The third part shows how admicelles or aliphatic brushes, built around the fibers, play a useful role in capturing organic impurities from aqueous media.

Cellulose is the most abundant natural polymer on earth, found mostly in vegetal biomass, where it is produced by photosynthesis in the form of semicrystalline fibers. The exploitation of this ubiquitous and inexpensive *renewable* resource remains a very important issue today, not only for the production of high volume commodities like textiles and paper, but also for novel value-added materials. Such materials may take advantage of the specific properties associated with the cellulose fibers (low density, good mechanical performances, biodegradability, etc.) and the versatility toward partial or total chemical modification.

The syntheses of cellulose esters, ethers and other derivatives, some of which are well-established industrial processes, involve the destruction of the fibrous morphology and its transformation into thermoplastic materials soluble in organic solvents (*1-3*). Recently, similar chemical processes were applied to cellulose, but limiting their impact to the macromolecular layers which constitute the fiber surface. The main purpose of this novel approach is to conserve the fiber integrity, and thus its mechanical properties, but to modify some specific features associated with its surface. One major driving force for the vast amount of research conducted on this topic (*4-8*) is the growing interest in composite materials in which a polymeric matrix is reinforced by cellulose fibers, replacing glass counterparts. The driving forces behind this are: (i) ecological considerations including the renewable character of cellulose and its biodegradability, (ii) economic factors, including, low cost and abundant fiber supply, lower density (which implies lower fuel consumption when the composites are transported) and lower abrasion resistance during processing; (iii) satisfactory mechanical performance of the composites. The fiber surface treatment fulfills two roles, *viz.* the improvement of the interfacial adhesion with the matrix and the reduction of moisture uptake by the fibers.

On another front, the *in situ* surface chemical modification of lignocellulosic fibers of wood (*9*) can bring about notable improvements related to hydrophobization and resistance to biocides. Due to the demonstrated efficiency of cellulose as the stationary phase in gas-liquid chromatography, we recently extended this feature by using appropriately modified cellulose substrates to trap organic pollutants from contaminated waters (*10-12*).

Modification of Cellulose Fibers for Use as Reinforcing Elements in Macromolecular Matrices

Siloxane coupling agents are commercial compounds commonly used to modify the surface of glass fibers by reacting with their OH groups directly or after (partial) hydrolysis, as shown in Scheme 1.

Since cellulose fibers also possess an OH-rich surface, it was thought that these reagents could be suitable for similar modifications. However, a fundamental study aimed at establishing the reactions occurring between different silane coupling agents and the surface of cellulose (*13*) showed that reaction 1 (Scheme 1) did *not* take place with six silane coupling agents, namely vinyltrimetoxy-(**VS**), γ-methacryloxypropyltrimetoxy- (**MPS**), cyanoethyltrimetoxy- (**CES**), γ-aminopropyltrimethoxy- (**APS**), octyltriethoxy- (**OS**) and glycidilpropyltrimethoxy-silane (**GPS**). In fact, these reagents were found to condense with cellulose OH groups *only* if they had previously been hydrolyzed, i.e. only through reaction 2 (Scheme 1). This finding was attributed to the more nucleophilic character of the cellulosic OH's, compared with their counterparts present in glass. In other words, the siloxane moiety was not cleaved if the OH group attached to the solid substrate was not sufficiently acidic, whereas the same substrate underwent condensation reactions between its OH groups and silanol moieties (*13*). This was confirmed by the observation that lignin reacts with the unhydrolyzed siloxanes, through its acidic phenolic OH groups (*13*).

In a subsequent investigation, we studied the hydrolysis kinetics **MPS**, **APS** and γ-diethylenetriaminopropyl trimethoxy silane (**TAS**) (*14*).

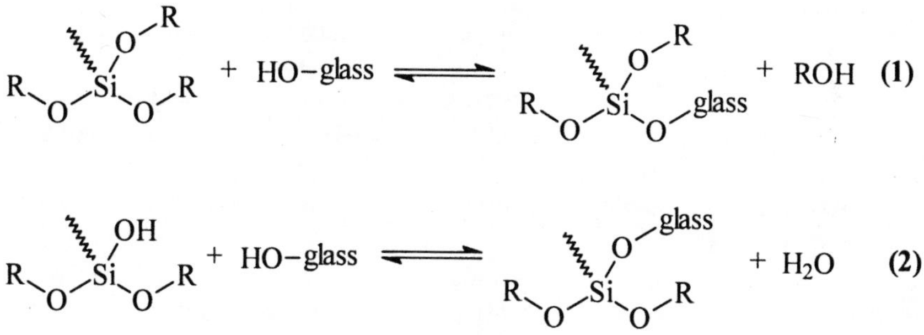

Scheme 1.Condensation reactions between pristine and hydrolyzed siloxanes and glass surface OH groups

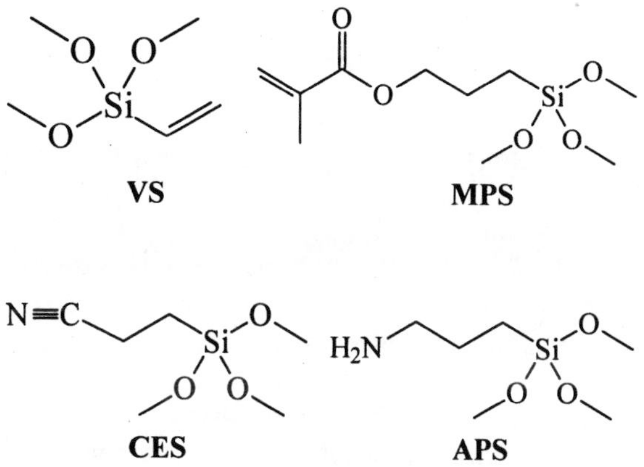

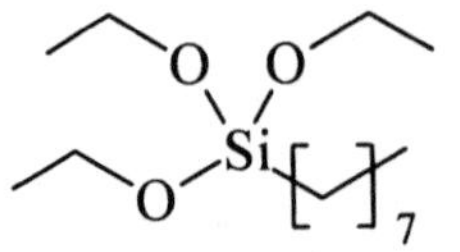

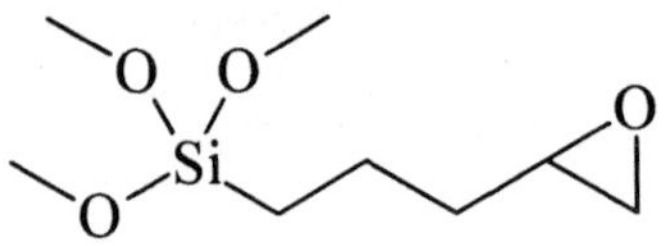

OS   GPS

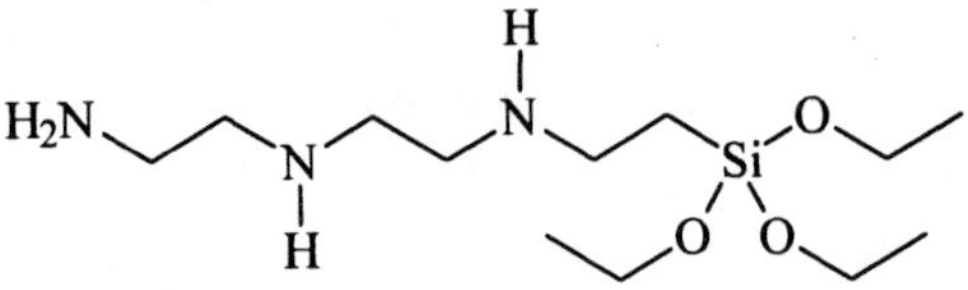

TAS

The rate of these reactions, carried out in an 80/20 (w/w) ethanol(or methanol)/water solution, increased in the order **MPS<APS<TAS**, as shown in Figure 1, which represents the conversion of siloxane groups into silanol moieties of the three siloxanes tested and that corresponding to triethyl amine-catalyzed (TEA) **MPS**.

The same reaction kinetics were followed *in situ* by ^{1}H-, ^{13}C- and ^{29}Si-NMR spectroscopies under the same conditions of temperature, water concentration and solvent composition. First order kinetics were found to fit all the data (Equation 1) related to the removal of the alkoxy groups and gave straight lines up to conversions exceeding 95%, as shown in Figure 2. These data offered apparent first order rate constants $k_{H\,a}$,calculated from the slope of each straight line, viz. 0.0085, 0.378, 0.76 and 5.04 min^{-1}, for **MPS**, **MPS-TEA**, **APS** and **TAS**, respectively.

$$-\frac{d[S]}{dt} = k_h[S][H_2O] \Rightarrow$$
$$\ln\frac{[S]}{[S]_0} = -k_h[H_2O]t = -k_H.t$$

(1)

It is worth noting that with **MPS**, the presence of TEA greatly enhanced the hydrolysis and the subsequent OH-OH mutual condensation (oligomerization) rates, but the concentration of partially-hydrolyzed species did not exceed the maximum obtained in the absence of the catalyst.

The rate of hydrolysis was determined by measuring the changes in the peak intensity of the **CH$_2$** of the ethanol released by **APS** and of the **CH$_3$** of the methanol liberated by **MPS** and **TAS**, using as reference the corresponding areas of the initial siloxanes. The kinetic plots deduced from the ^{1}H- and the ^{13}C-NMR

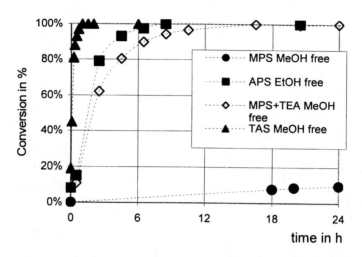

*Figure 1. Reaction rate of alcohol release for **MPS**, TEA-catalyzed **MPS**, **APS** and **TAS**. (Reproduced from reference 14. Copyright 2005 Elsevier.)*

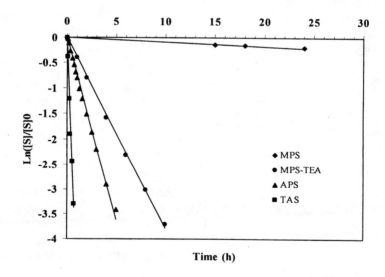

*Figure 2. First-order plots for **MPS**, TEA-catalyzed **MPS**, **APS** and **TAS** hydrolysis. (Reproduced from reference 14. Copyright 2005 Elsevier.)*

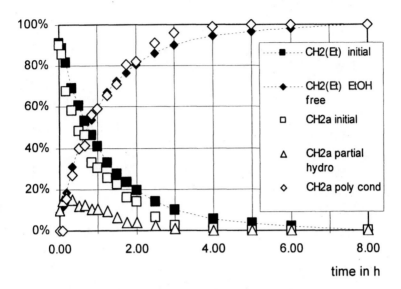

*Figure 3. APS hydrolysis kinetics, as deduced from the ¹H-(filled symbols) and
¹³C-NMR (empty symbols) spectra. (Reproduced from reference 14. Copyright
2005 Elsevier.)*

spectra showed a good overlapping of the curves, indicating that ¹³C-NMR spectroscopy was a reliable technique for the quantitative study of the hydrolysis of **MPS** and **APS** (see Figure 3). The transient partially-hydrolyzed **APS** was clearly detected only in the corresponding ¹³C-NMR spectra.

²⁹Si-NMR spectroscopy provided additional insights into the hydrolysis and the condensation reactions of silicon alkoxides and gave novel information about the structure of the oligomeric species. The chemical shift of a silicon atom is determined by the number of siloxane bridges attached to it. The observations deduced from this study corroborated the ¹H-, ¹³C-NMR data in accordance with the decrease in the initial siloxane units. Subsequently, the amount of hydrolyzed species increased up to a maximum value before starting to decrease. Intermediate species with increasing numbers of Si-O-Si bridges were also detected. These moieties apparently started to form only after the maximum in the hydrolyzed units had been reached. Finally, an increase in the concentration of the most condensed silane structures $(Si-O-Si)_3$ took place, to the detriment of other less condensed intermediates. Interestingly, **APS** and **MPS** gave only soluble products, while colloidal particles precipitated in the medium when **TAS** was hydrolyzed.

Given the lack of reactivity of unhydrolyzed siloxanes toward cellulose OH groups, subsequent studies aimed at grafting these entities onto cellulose fiber surfaces were conducted using prehydrolyzed reagents. The contact angle measurements on additive-free tracing papers before and after treatment with partially-hydrolyzed hexadecyltrimethoxy silane (**HDS**) showed that the polar component of the surface energy decreased drastically from 20 mJ/m^2 to practically zero (*15*). This significant amount of hydrophobization was also attained by applying a double grafting strategy (*8*) consisting of an initial reaction between cellulose and a hydrolyzed difunctional siloxane, namely **MPS**, γ-mercaptopropyltrimethoxy silane (**MRPS**) or **APS**, followed by copolymerization of the functionalized surface with the appropriate co-monomers, viz. styrene (**ST**) or methyl methacrylate (**MMA**) for **MPS**-treated fibers and an epoxy-based resin in the case of **APS**-treated materials (*15*).

HDS   **MRPS**

APS-, MPS- and MRPS-treated cellulose fibers from the annual plant *alfa esparto* were incorporated into either an unsaturated polyester (for **MPS** and **MRPS**) or an epoxy resin (for **APS**) matrix with a loading of 40% w/w (*16*). In all instances the reactive moieties of the respective silanes, namely, the primary amino groups, the acrylic moieties and the SH groups reacted with the complementary functions borne by the matrices, namely the C=C unsaturation or the terminal OH or COOH of the polyester and the oxirane moieties of the epoxy resin. These coupling agents were found to be efficient in improving the mechanical properties of the ensuing composites, since they increased the flexural strength of the materials by up to 40% (*16*). However, the immersion of the composites in water induced a drastic loss of their mechanical properties, indicating that the grafted Si-O-C moieties had been (partly) hydrolyzed. The observation of the fractured surfaces by scanning electron microscopy confirmed that the quality of the interface had deteriorated, in agreement with the dynamic mechanical data (*16*). The composites were also studied in terms of water uptake and showed that the siloxane treatments were not very efficient as a water barrier for cellulose.

Another investigation, based on the same double coupling strategy, called upon the use of two other coupling agents, viz. 3-isopropenyl-α,α'-dimethylbenzyl isocyanate (**TMI**) and 2-isocyanatoethyl methacrylate (**IEM**) (*17*) and showed that up to 13% of polystyrene could be copolymerized with both **TMI**- and **IEM**-treated fibers. After enzymatic hydrolysis of the cellulose substrate, the residual synthetic macromolecules were characterized and found to be polystyrene with molecular weights of 8100 and 19,000 for M_n and M_w, respectively. Fragmentation tests proved that the quality of the interface was improved substantially when using this double grafting strategy (*17*).

A further family of reagents recently tested were fatty acid chlorides which were reacted with microcrystalline cellulose and fibers from sugar cane and bleached kraft pulps from *Eucalyptus globolus* (*18,19*). The effect of the solvent (toluene or DMF), reaction time and fatty acids chain length (C6, C12, C16, C18 and C22) on the extent of esterification and on the properties of the ensuing modified celluloses was evaluated. This study showed that the fiber surface had become totally hydrophobic and the polar contribution to its energy had been minimized, in tune with the presence of grafted brush-like aliphatic chains. Interestingly, when the modification was carried out using a swelling medium and allowed to reach high degrees of substitution (DS), these in-depth modified materials could be hot-pressed into films (*19*). These films were transparent when long aliphatic chains had been appended to the fibers and opaque (heterogeneous) with shorter brush grafts. This observation was corroborated by measurements of the loss of cellulose crystallinity, which was higher relative to the higher the DS and the longer the appended aliphatic chains.

TMI   **IEM**

The possibility of using cellulose fibers as a *single* source of composite materials by their *partial* conversion into a thermoplastic polymer through in-depth modifications was further investigated using oxypropylation as the source of partial softening (*20*).

Modification for Wood Densification

The densification of wood was achieved through the double grafting strategy (*8*), using swelling media to enhance the grafting efficiency and to extend it to the bulk of pine wood. In the initial step, **TMI, IEM**, methacrylic anhydride (**MA**) and glycidyl methacrylate (**GM**) were used. In the second step, i.e. the modified fibers-matrix copolymerization, **ST** and **MMA** were polymerized *in situ*. Different experimental conditions (temperature, solvent, catalyst, reaction time, pressure) were examined and the optimized systems produced up to 75% and more than 100% weight gains after the first and the second grafting steps, respectively. The pore size of the tested samples shifted from about 5 μm for untreated samples to 2.6 μm after the **TMI** treatment and 0.07 μm after densification with poly **ST** (*21,22*). The contact angles of a drop of water deposited at the surface of the wood samples, before and after densification, increased from about 40° to more than 110° for the most efficient modifications. Finally, the anti-shrinkage behavior was also substantially improved, since it shifted from about 35% to practically 90%, for **TMI-ST**-densified wood (21,22).

MA   **GM**

This study showed that various reactive groups can be grafted within the wood morphology, providing new possibilities in the field of the chemical modification of wood. The partial filling of the wood cavities with a polymer bound to its structure promoted both its densification and its stabilization towards water uptake (21,22).

Modification for Trapping Organic Pollutants

This work started a few years ago with a study of the adsorption of cationic surfactants onto the surface of cellulose fibers in an aqueous medium and the subsequent inclusion of **ST**, **MMA** and 2-ethylhexyl acrylate (**EHA**) in the ensuing admicelles (23). The adsorbed monomers were then polymerized in order to generate a macromolecular sleeve around the fibers, which therefore became hydrophobic, as shown by the decrease of the polar component of their surface energy from 20 to less 4 and 1.2 mJ/m^2, for poly**ST**- and poly**EHA**-treated cellulose, respectively (23).

Further experimentation was pursued using various cationic surfactants and more negatively-charged cellulose substrates, in order to boost the formation of admicelles and consequently enhance their adsorption capacity (10,11). Thus, the primary alcohol groups of the cellulose glycosyl units were oxidized to carboxylic groups in order to obtain substrates with different surface charges, viz. 39, 150, 300 and 600 µmol/g. The cationic surfactants used: octadecyltrimethyl ammonium bromide (**C18**), hexadecylpyridinium chloride (**C16**), tetradecyltrimethyl ammonium bromide (**C14**) and dodecylpyridinium chloride (**C12**) were commercial analytical-grade products. Their critical micelle concentrations at 25°C (CMC) were 2.1 10^{-4}, 7.5 10^{-4}, 3.4 10^{-3}, and 1.2 10^{-2} M, respectively. The amount of surfactant adsorbed onto the fiber surface, ζ-potential and the amount of surfactant counter ions (Cl$^-$) released in the solution were measured. From this study it was established that the adsorption of the surfactant onto the cellulose surface increased with increasing surface charge density (Figure 4) and hydrophobic aliphatic chain length of the surfactant (Figure 5).

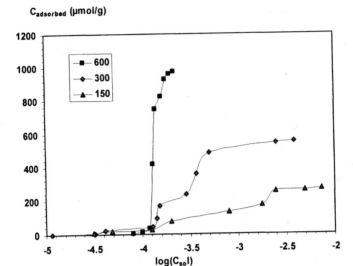

Figure 4. Adsorption isotherm of **C16** onto the oxidized cellulose surface at pH of 7.5-8, in the absence of any added salt. The figures in the inset box refer to the fibers' charge density expressed in μmol of COO⁻ per gram of cellulose (Reproduced from reference 11. Copyright 2005 American Chemical Society.)

These surface-modified fibers were used to adsolubilize organic water-soluble compounds. These compounds are summarized in Table 1, together with their maximum adsolubilized concentration and the corresponding adsolubilization constants (partition coefficient, K_{ads}, and maximum adsorption concentration of the adsolubilized compound, $C_{max\ adsol}$). The first parameter was determined from the mole fraction of the solute in the adsorbed surfactant layer (X_{adsol}) and the solute concentration in the aqueous phase (C_{aq}):

$$K_{ads} = \frac{X_{adsol}}{C_{aq}} = \frac{C_{adsol}}{C_{adso}^{surf} \cdot C_{aq}}$$ (2)

Plotting X_{adsol} vs C_{aq} gave a straight line for the all water-soluble organic molecules studied and the slope was the partition coefficient according to Equation 2. The $C_{max\ adsol}$ was deduced from the adsolubilization isotherms of these molecules on the modified surfaces, i.e. when C_{adsol} reached a maximum value (10). These studies showed that different water-soluble organic pollutants could be successfully squeezed into the admicelles by adsolubilization.

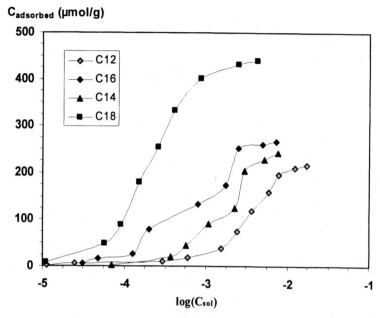

Figure 5. Adsorption isotherm of the four cationic surfactants onto the cellulose surface with 150 μmol/g charge density at pH of 6.5-7, in the absence of any added salt. (Reproduced from reference 11. Copyright 2005 American Chemical Society.)

Table 1 also contains the values of K_{ads} and $C_{max\ adsol}$ relative to 2-naphtol after its adsolubilization on the cellulose substrate modified with cationic surfactants of different lengths. These data show clearly that the longer the hydrophobic tail of the surfactant, the better is the adsolubilization capacity of the modified surface.

The availability and low cost of cellulose fibers makes the present approach promising in applications related to the removal of organic pollutants and toxic substances in waste waters. However, the possible desorption of the surfactant molecules from the cellulose surface should be impeded. For this reason, further studies were done in which the cellulose surface was chemically grafted with two fatty acid chlorides (C12 and C16), in order to avoid desorption when recovering the trapped organic substances (*12*). The partition coefficients and the maximum adsorbed amounts of water-soluble organic molecules on the C12-treated surface are given in Table 1. This work showed that when *chemically-bound* surfactant-like structures are appended to cellulose fibers, the substrates obtained can be recycled without losing their adsorption capacities. Their good performance is attributed to the particularly porous structure of the cellulose fibers, which is preserved after grafting (*12*).

Table 1. Adsolubilization of Organic Compounds on Cellulose Fibers Bearing Surface Admicelles or Grafted Hydrophobic Chains

Organic solute	Cn (adsorbed) (ref.10)		C12 (chemically bound) (ref. 12)	
	$10^{-3}K_{adso}$	$C_{max.adsol}$ $(\mu mol/l)$	$10^{-3}K_{adsol}$	$C_{max.adsol}$ $(\mu mol/l)$
Benzene	8.9	390	-	-
Naphthalene	6.8	300	-	-
Chlorobenzene	6.5	310	10	190
TCB	5.0	190	29	470
Nitrobenzene	1.9	200	16	150
2-Naphthol	2.0	290	14	320
Quinoline	1.1	200	35	330
2-Chlorophenol	1.0	170	-	-
DCB	-	-	9.1	250

Surfactant chain length (with 2-naphthol)	$10^{-3}K_{adso}$	$C_{max.adsol}$ $(\mu mol/l)$
C18	1.9	300
C16	1.3	180
C14	0.9	140
C12	0.7	120

TCB: 1,3,5-Trichlorobenzene; DCB: 1,4-Dichlorobenzene.

Conclusions

The numerous and interesting approaches to the modification of cellulose fibers published from a growing number of laboratories, testify to the vitality of this field. The results of the work described in this chapter show that (i) the use of siloxanes can be optimized thanks to a better knowledge of their reactivity toward the cellulose hydroxy groups. Consequently, continuous covalent links can be produced between the fibers and different polymeric matrices; (ii) wood can be densified by a process involving reagents capable of establishing chemical connections between the wood macromolecules and the densifying polymer produced *in situ*; and (iii) cellulose fibers, surrounded by admicellar sleeves of cationic surfactants or grafted with long aliphatic chains, are excellent traps for organic pollutants. (*24*)

References

1. Heinze, T.; Ficher, K. *Cellulose and Cellulose Derivatives*, Wiley-VCH, Weinheim, 2005.
2. Klemm, D.; Philipp, B.; Heinze, T.; Heinze, U., Wagenknecht, W. In *Comprehensive Cellulose Chemistry*; Wiley-VCH: Weinheim, 2001.
3. Klemm, D.; Heublein, B.; Fink, H.-P.; Bohn A. *Angew. Chem. Int. Ed.* 2005, *44*, pp 3358-3393.
4. Mohanty, A. K.; Misra, M.; Hinrichsen, G. *Macromolecular Mater. Eng.* 2000, 276/277, pp 1-24.
5. Eichhorn, S. J.; Baillie, C. A.; Zafeiropoulos, N.; Mwaikambo, L. Y.; Ansell, M. P.; Dufresne, A.; Entwistle, K. M.; Herrera-Franco, P. J.; Escamilla, G. C.; Groom, L.; Hughes, M.; Hill, C.; Rials, T. G.; Wild, P. M. *J. Materials Sci.* 2001, *36*, pp 2107-2131.
6. Gassan, J.; Bledzki, A. K. *Prog. Polym. Sci.* 1999, *24*, pp 221-271.
7. Lu, J. Z.; Wu, Q. McNabb, H. S. Jr. *Wood Fibre Sci.* 2000, *32*, pp 88-104.
8. Belgacem, M. N.; Gandini, A. *Comp. Interface* 2005, *12*, 41-75.
9. Fabbri, P.; Castellano, M.; Champon, G., Belgacem, M.N.; Gandini, A. *Polym. Intern.* 2004, *53*, pp 7-11.
10. Aloulou, F.; Boufi, S.; Belgacem, M. N.; Gandini A., *Colloid Polym. Sci.* 2004, *283*, pp 344-350.
11. Alila, S.; Boufi, S; Belgacem, M. N.; Beneventi, D. *Langmuir* 2005, *21*, pp 8106-8113.
12. Boufi, S.; Belgacem, M. N. *Cellulose*, in press.
13. Castellano, M.; Gandini, A.; Fabbri, P.; Belgacem M. N. *J. Colloid Interface Sci.* 2004, *273*, pp 505-511.
14. Brochier-Salon, M. C.; Abdelmouleh, M.; Boufi, S.; Belgacem, M. N.; Gandini, A. *J. Colloid Interface Sci.* 2005, *289*, pp 249-261.
15. Abdelmouleh, M.; Boufi, S.; Belgacem, M. N.; Duarte, A. P.; Ben Salah, A.; Gandini, A. *Intern. J. Adhesion Adhesives* 2004, *24*, pp 43-54.
16. Abdelmouleh M., Boufi S., Belgacem M. N., Dufresne A., Gandini A. *J. Appl. Polym. Sci.* 2005, *98*, pp 974-984.
17. Botaro, V. R.; Gandini, A; Belgacem, M. N. *J. Thermoplastic Comp. Mater.* 2005, *18*, pp 107-117.
18. Pasquini, D.; Belgacem, M. N.; Gandini, A.; Curvelo, A. A. S. *J. Colloid Interface Sci.,* 2006, *295*, pp 79-83.
19. Freire, C.S.R.; Silvestre, A.J.D.; Pascoal Neto, C.; Belgacem, M.N.; Gandini, A. *J. Appl. Polym. Sci.,* 2006, *100*, pp 1093-1102.
20. Gandini, A.; Curvelo, A.A.S.; Pasquini, D.; de Menezes A.J. *Polymer* 2005, *46*, pp 10611-10613.

21. Bach, S.; Belgacem, M. N.; Gandini, A. *Holzforschung* **2005**, *59*, pp 389-396.
22. Gandini, A; Belgacem, M. N. *Macro*mol. *Symp.* **2005**, *221*, pp 257-270.
23. Boufi, S.; Gandini, A. Cellulose **2001**, *8*, pp 303-312.
24. Gandini, A.; Belgacem, M.N. *J. Polym. Environm.* **2002**, *10*, pp 105-115 and refs. therein.

Chapter 7

Opportunities for Hardwood Hemicellulose in Biodegradable Polymer Blends

Arthur J. Stipanovic[1], Jennifer S. Haghpanah[2], Thomas E. Amidon[3], Gary M. Scott[3], Vincent Barber[3], and Kunal Mishra[3]

[1]Faculty of Chemistry, [2]Faculty of Chemistry–NSF Research Experience for Undergraduates, and [3]Faculty of Paper Science and Engineering, State University of New York–College of Environmental Science and Forestry (SUNY-ESF), Syracuse, NY 13210

Hardwoods indigenous to the northeastern U.S., including birch, beech, maple and short-rotation woody crops such as willow, are relatively rich in the hemicellulose xylan. In this study, a combined biodelignification and hot water extraction procedure was employed to isolate polymeric xylan in its native acetylated state. To enhance the mechanical properties of xylan as a biodegradable material, solutions of xylan were mixed with solutions of commercially available cellulose esters followed by casting into solid films. In this fashion, it was possible to prepare acetylated xylan / cellulose triacetate "blends" with mechanical properties comparable to the cellulose triacetate itself up to 25 wt% xylan. Plasticizers were effective in increasing the strain to break for these materials but lowered the modulus at 1% strain.

As the world's petroleum supplies continue to dwindle, renewal biomass will become an increasing important feedstock for the manufacture of fuels, chemicals and advanced materials. A recent report issued jointly by the U.S. Departments of Agriculture and Energy, estimated that over one billion tons per year of agricultural and forest biomass will become available with the ultimate potential of providing one-third of the current U.S. consumption of transportation fuels (*1*). Since the cost of shipping and storing biomass is a critical factor in the economics of biobased industries, it is envisioned that "biorefineries" will be regional facilities that use locally harvested feedstocks (*2*). For the northeastern region of the U.S., forest hardwoods including birch, beech, maple and short-rotation woody crops such willow (*5*), are abundant sources of biomass. From a compositional perspective, hardwoods typically contain 40-55% cellulose, 20-35% hemicellulose and 18-30% lignin on a dry wt% basis (*3,4*). By comparison, softwoods contain a higher fraction of lignin (up to 35%) and lower level of hemicellulose. In addition, softwood hemicellulose contains a significant fraction of 2-3 different polymers, while hardwood hemicellulose tends to be predominately a single polymer, 4-O-methylglucuronoxylan (xylan). The β-(1-4)-linked D-xylopyranosyl (xylose) backbone structure of hardwood xylan, shown in Figure 1, also contains about one 4-O-methyl-α-D-glucuronic acid residue and 7 acetyl ester substituents per 10 xylose monomers (*6*).

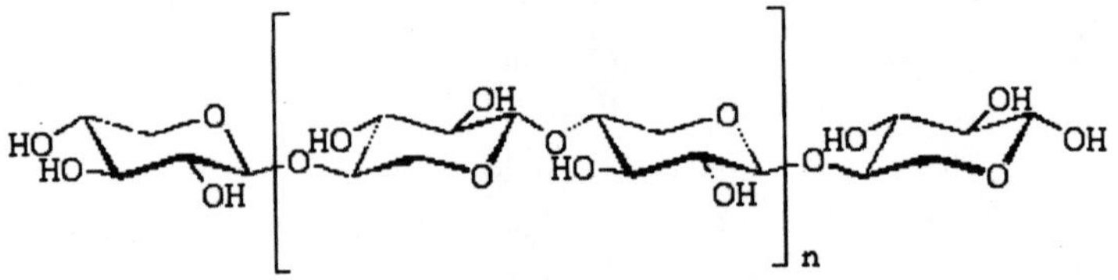

Figure 1: Idealized backbone structure of xylan (pyranose ring hydrogen atoms omitted for clarity).

Although the molecular weight of xylan is the subject of considerable debate because this polymer is typically associated with lignin (*7*), it degrades during chemical delignification processes (*8*), and it self-assembles in solution into aggregate structures (*6*), it is widely held that the degree of polymerization (DP) of xylan is in the range of 100-200 (*6,8*). Comparatively, native bacterial and plant celluloses were found to have DP's from 400-10,000 (*9,10*).

Based on the relative abundance of hardwoods in the northeastern region of the U.S., efforts are being made at our institution and at others to fully utilize the hemicellulose fraction in fuels (xylose fermentation to ethanol), chemicals (microbial synthesis of propanediols), and biodegradable thermoplastic polymers (xylose biosynthesis into polyhydroxyalkanotes). In this study, we exploited a sequential biodelignification / hot water extraction process reported previously (*11*) to recover xylan from northeastern hardwoods (paper birch, sugar maple) in an effort to characterize the material properties of films cast from these extracts. It was envisioned that these materials would be relatively inexpensive compared with other biodegradable polymers but due to the relatively low DP of xylan, we expected that the mechanical strength of these films would be limited. As a result, the objective of this work was to blend xylan with other commercially available polysaccharides and polyesters to determine if these "hybrid" materials could be useful in packaging and other disposal consumer items.

Experimental Methods

The hardwood biodelignification process and hot-water hemicellulose extraction procedures used in this study to isolate xylan were reported previously (*11*). Typically, for sugar maple chips, 60% of the original xylan present in the original ground wood can be recovered by this treatment. Solid xylan was recovered by precipitation in acetone / ethanol. A water-soluble birch xylan sample in dry powdered form, originally isolated in the laboratory of Professor T.E. Timell (Professor Emeritus, - SUNY-ESF, circa 1960's) by dilute KOH extraction was also studied (*12*). ^{1}H-NMR results on this sample indicated a degree of acetylation of 1.06 acetate groups per xylose residue. Xylan valerate was synthesized in the Timell laboratory (*6*).

Films of individual polymers and blends were obtained by solvent evaporation (vacuum oven at 105°C) after mixing polymer solutions prepared in either dimethyformamide (DMF) or dimethylsulfoxide (DMSO) at 4 wt%. Glass transition temperatures (T_g) for polymer films were recorded using a TA Instruments 2920 Modulated Differential Scanning Calorimetry (DSC) apparatus in open Al pans under dry nitrogen by cycling from room temperature to 190°C, cooling to -40°C and then back to 200°C. Only T_g values from the second heating stage were reported. Tensile strength measurements were performed in a force ramp mode using a TA Instruments 2980 Dynamic Mechanical Analyzer (DMA) where deformation (% strain) was monitored as the applied force was increased from 0-14N on films of dimensions 0.05mm thick, 6mm width and 10mm length (strain direction). Typically, measurements were made 2-3 times and the values reported were an average of the replicates.

The relative molecular weights of xylan extracts were characterized by Size Exclusion Chromatography (SEC) on a Waters Breeze system with RI detector using three Styragel HR columns (HR3, 4 and 6) with DMSO as a mobile phase. The molecular weight calculation was calibrated with pullulan standards (Showa Denko–Shodex) ranging in molecular weight from $6x10^3$ to $8x10^5$.

Cellulose triacetate (CTA; 40% acetyl, DS = 2.5) and cellulose acetate butyrate (CAB; 51% butyrate and 4% acetate) are products of the Eastman Chemical Company. CTA and CAB molecular weight values are reported by

Eastman to be in the range of 50,000-80,000 (M_w) and 57,000 (M_n), respectively. The molecular weight of the xylan samples isolated in this study and characterized by SEC will be discussed below. Polyhydroxybutyrate-co-valerate (PHB-HV) was obtained from the Aldrich Chemical Company (30% valerate content.)

Results and Discussion

Impact of Biodelignification on Xylan Extraction

Although biodelignification and its potential benefits in papermaking have been explored for the past several decades (13-27), the impact of this process on hemicellulose extraction has received less attention. Our laboratory has previously shown that the application of a white-rot fungus (*Ceriporiopsis subvermispora*) pretreatment significantly increases the amount of hemicellulose that can be extracted from hardwood by water at 140-150°C, as shown in Figure 2, for a fast-growing willow species (*Salix sp.* ; *11, 28*) when comparisons are made to untreated control samples at comparable extraction circulation times (2 hours).

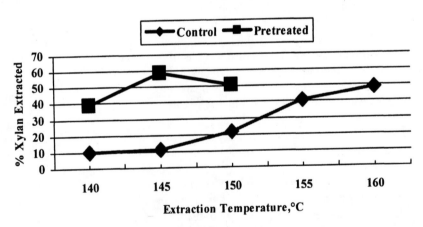

Figure 2. Impact of biodelignification on xylan extraction efficiency in water – Reproduced from Reference 11. Copyright 2005 American Chemical Society.

Over the period of several weeks, *C. subvermispora* is capable of reducing the lignin content of woody biomass by 5-10% while spreading its hyphae throughout the cellular ultrastructure of the wood creating permeable channels that facilitate hemicellulose removal by water (*29*). For example, if the original lignin content of woody biomass is 25%, this value is reduced by 1.3-2.5% after biodelignification. This organism exhibits essentially no cellulase or hemicellulase activity (28) that could degrade the polymeric components of wood. SEC results show that extracts obtained at higher temperatures contain a high molecular weight peak due to polymeric xylan and a low molecular peak due to xylo-oligomers while material isolated at lower temperatures is predominately polymeric as discussed below. As a result, biodelignification creates specific opportunities for xylan utilization in polymeric form.

Size Exclusion Chromatography of Extracted Hardwood Xylan

Samples of xylan hemicellulose extracted from biodelignified hardwood were subjected to SEC in DMSO on a column set calibrated with pullulan standards. This bacterial polysaccharide consists of maltotriose units linked through an α-(1-6)-linkage which renders this molecule very flexible. As a result, in comparison to cellulose standards of known molecular weight, the hydrodynamic volume of pullulan is 1.8-2.6 times lower than cellulose at comparable molecular weight and conditions of solvent and temperature (30). As a result, we expected that the SEC-based hemicellulose molecular weight values reported in this study would be higher than those based on direct measurements since xylan is less flexible than pullulan and hydrodynamically larger at any molecular weight. This is a common limitation of SEC.

SEC results have shown that un-fractionated hemicellulose isolated by hot-water extraction from biodelignified wood typically exhibits a high molecular weight component (Peak 1) and a lower molecular weight fraction (Peak 2) containing xylo-oligomers (molecular weight 10^2-10^3). The results shown in Table I highlight that degradation of xylan to oligomeric species is strongly a function of extraction temperature when compared at comparable extraction times as evidenced by the relative decrease in magnitude of Peak 1. This is attributable to the autohydrolysis effect induced by acetyl groups that are cleaved from the xylan chain in water at elevated temperature. Based on the high molecular weight and low oligomer content of sample VB31, it was selected for additional study of its film-forming properties as discussed below. The anomalously high M_w observed for the birch xylan sample could be the result of aggregation in DMSO although this solution was clear and readily passed through a 0.45 micron filter. Such aggregation has been previously reported for hardwood xylans (6).

Thermal Properties of Xylan Films and Blends

Solid films cast from DMSO or DMF solutions of maple xylan (MX; sample VB31) were observed to be very brittle and not capable of withstanding

Table I. SEC of Hardwood Xylan Samples

Sample (1)	Extraction Time (min)	M_w (x10^3)	% High Mat Fraction – Peak 1
140°C Isolate	60	217	99
140°C Isolate	180	190	77
180°C Isolate	180	158	72
190°C Isolate	180	134	53
200°C Isolate	180	160	46
VB31 (2) – 160°C	120	252	100
Birch Xylan	-	509	100

NOTES: (1) Sugar maple extraction after biodelignification unless indicated, (2) Fractionated in ethanol / acetone to selectively precipitate high molecular weight species.

mechanical deformation (bending, cutting, twisting). This is consistent with recent work by Grondahl et. al. on aspen wood xylan which demonstrated that the films were brittle and flakey upon drying (31). However, upon humidification and the addition of a plasticizer (xylitol or sorbitol), films with improved mechanical properties were obtained. In our study, emphasis was directed to the solution blending of xylan with other polymers to improve mechanical properties rather than through the incorporation of plasticizers which can diffuse out of the resulting polymer film over time.

The T_g of solid polymers is the temperature at which the mechanical properties of the material change from glassy and brittle to rubbery and flexible due to the onset of long range, cooperative molecular segmental motion. For most underivatized, unplasticized polysaccharides, T_g is significantly above room temperature (usually >150°C) and the polymer is glassy under normal application conditions. Differential Scanning Calorimetry was employed to determine the T_g of naturally acetylated xylan extracts and blends of xylan with cellulose triacetate (CTA), cellulose acetate butyrate (CAB) and PHB-HV as summarized in Table II. A DSC thermogram where heat flow is monitored with increasing temperature is provided in Figure 3 for CTA, birch xylan and a 50:50 wt:wt blend of CTA and birch xylan.

Table II illustrates that CTA, birch xylan and maple xylan each exhibit a single T_g although differences are observed between the two xylan sources. Irvin has shown (32) that xylan from the hardwood *Eucalyptus regnans* exhibited T_g values in the range of 160-180°C for dry samples while moisture served as a plasticizer lowering T_g to almost 45°C at 20% water content. It is possible that the T_g differences seen for the maple and birch xylan samples arise from variations in acetyl content, associated lignin or small differences in moisture content that might occur between sample preparation and DSC analysis.

Table II. Thermal Properties of Xylan Films

Polymer	Solvent	T_g (°C) by DSC
1. Cellulose Triacetate (CTA)	DMF	195
2. Birch Xylan (Acetate)	DMF	151 (Weak)
Blend of 1 and 2	Film (1)	164 (Weak)
3. Maple Xylan (Acetate)	DMSO	128 (Weak)
4. CTA	DMF	195
Blend of 3 and 4	Film (1)	163 (Broad)
5. PHB-HV Bacterial Polyester	DMF	-7
6. Birch Xylan (Acetate)	DMF	151
Blend of 5 and 6	Film (1)	2 Tg's: -4, 139

(1) A solid film cast from the solvent shown

When two solid polymers are mixed, a true "thermodynamic" blend is reflected in a single T_g instead of two T_g's associated with each individual component (33). In general, the free energy of mixing for two polymers ($\Delta G_{mix} = \Delta H_{mix} - T \Delta S_{mix}$) is negative only if the enthalpy of mixing (ΔH_{mix}) is negative because the entropy of mixing (ΔS_{mix}) for two high molecular weight species is usually not positive (34). In the case of xylan blends with cellulose esters, a negative ΔH_{mix} could result from hydrogen bonding or the association of hydrophobic ester side chains.

DSC results shown in Figure 3 (summarized in Table II) provide some evidence that CTA and acetylated birch xylan form a homogenous blend since a single T_g is observed although this thermal feature is relatively weak and broad. Clear evidence is provided that xylan and a thermoplastic bacterial polyester are not thermodynamically compatible since two T_g's are observed (Table II).

An alternative approach was also attempted to obtain "compatible" blends in which ester derivatives of cellulose and xylan were mixed in solution followed by film casting. In this case, commercially available CAB was blended with a hardwood xylan valerate (XV) previously prepared at our institution (6). In general, long hydrocarbon side chains on polymers reduce T_g and enhance film extensibility. DSC results, shown in Table III, illustrate that lower T_g's were generally observed for CAB and XV compared to CTA and naturally acetylated xylans. However, a blend of these derivatives exhibited two T_g's indicating that true thermodynamic mixing had not occurred. A blend of CTA with XV also displayed two T_g values characteristic of a two domain, non-homogenous solid-state morphology. It is interesting to note that the two T_g's observed for the XV

114

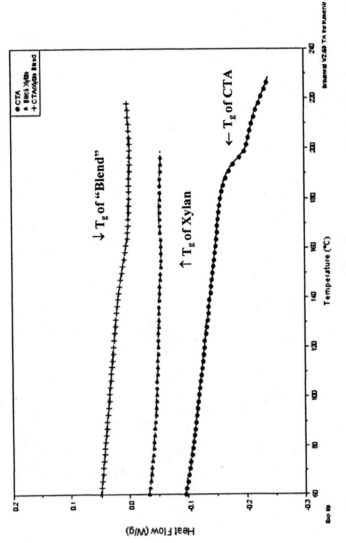

Figure 3. DSC of CTA, birch xylan, and a 50:50 wt:wt blend of CTA and birch xylan cast from DMF.

Table III. Thermal Properties of Polysaccharide Ester Films

Polymer	Solvent	T_g (°C) by DSC
1. Cellulose Acetate Butyrate (CAB)	THF	105
2. Xylan Valerate (XV)	THF	10-20
Blend of 1 and 2	Film	2 Tg's: 8, 105
3. PHB-HV	DMF	-7
Blend of 2 and 3	Film	2 Tg's: -22, 1
4. CTA + XV	Film	2 Tg's: -4, 195

Table IV. Tensile Properties of Xylan Blends

Sample	Stress at 1% Strain (MPa, 40°C)	Strain at Break (%)
Cellulose Triacetate (CTA)	23.2	1.8
Maple Xylan (MX)	Brittle, Friable	-
90:10 CTA:MX	23.8	2.2
75:25 CTA:MX	21.0	1.2
65:35 CTA:MX	15.9	1.3
50:50 CTA:MX	2.9	1.6
CAB	7.6	1.7
XV	5.2	3.5
50:50 CAB:XV	4.7	5.3

blend with PHB-HV are observed at lower temperatures than the "parent" polymers. This could be a reflection of some domain compatibility or a cooperative "plasticizing" effect induced by the short valerate side chains present on each polymer.

Mechanical Properties of Xylan Films and Blends

From an applications perspective, the tensile properties of xylan films and blends are critically important in defining the commercial viability of these potentially biodegradable materials. Typically, unplasticized or un-blended hardwood xylan cast into films is extremely brittle and cannot withstand any mechanical deformation while CTA, widely used in film, fiber and coating applications, exhibits a relatively high modulus but modest % strain at break when uniaxially deformed. Results provided in Table IV illustrate that blending maple xylan films with CTA, at concentrations of xylan ranging from 10-25%, can provide tensile properties very similar to CTA alone. At higher concentrations of xylan (35-50%) a significant deterioration is observed in stress at 1% strain while % strain at break remains comparable. (Note: Young's Modulus at low deformation = Stress at 1% Strain / 0.01). Table IV also contains results for the "higher" esters CAB and XV. These materials show a lower modulus at 1% strain compared to CTA but XV appears to exhibit an improved strain to break value. The 50:50 mixture of CTA and XV showed very good elongational properties.

The xylan film tensile properties reported above can be compared to recent work by the Gatenholm group who has demonstrated that the open chain form sugars xylitol and sorbitol are effective plasticizers for aspen wood (hardwood) xylan providing tensile stress values at 1% strain in the range 5-30 MPa (30). Further, at high levels of plasticizer (35-50 wt%), up to 12% strain to break was observed by these authors. As a result of these findings, we evaluated the effect of triethylcitrate (TEC), shown to be good plasticizer for CTA, for CTA / xylan blends (35). Data are given in Table V.

Table V. Effect of Plasticizer on the Mechanical Properties of Xylan Blends

Sample	Stress at 1% Strain (MPa, 40°C)	Strain at Break (%)
Cellulose Triacetate (CTA)	23.2	1.8
CTA + 20% TEC	14.2	4.1
CTA / MX Blends + 20% TEC	-	-
90:10 CTA:MX	18.3	4.8
75:25 CTA:MX	16.9	1.4

For CTA alone, TEC has the effect of decreasing modulus compared to unplasticized CTA while significantly enhancing the degree to which the film can be stretched before breaking. Similar results were observed for a 90:10 blend of CTA:MX. For both plasticized and unplasticized CTA:MX, the 75:25 blend showed good tensile stress at 1% strain but a low strain at break in both cases.

Conclusions

As biorefineries evolve as regional sites for the production of fuels, chemicals and polymeric materials, locally available biomass feedstocks will be exploited as sources cellulose, hemicellulose and lignin. In the northeastern U.S., forest and plantation grown hardwoods represent an attractive and abundant source of biomass, comparatively rich in the hemicellulose xylan. Prior work at our institution has shown that biodelignification is an "enabling" technology which significantly increases the yield of polymeric xylan potentially extracted from hardwoods. Despite its relatively low molecular weight and brittle solid-state morphology, this study has shown that blends of xylan with other biobased polymers can provide potentially useful film materials. Specific results are summarized below:

- Biodelignification followed by hot water xylan extraction at 140°C can yield up to 60% of the hemicellulose originally present in the biomass feedstock largely in polymeric form as evidenced by SEC.

- In the absence of plasticizers, hardwood xylan forms very brittle films with essentially no capacity for deformation under load.

- Mixtures of xylan with CTA may form polymeric blends based on DSC results that show a single broad T_g.

- These CTA blends show good mechanical properties at 10-25% xylan in CTA, above which modulus and extensibility deteriorate.

- The addition of triethylcitrate plasticizer reduces the modulus of CTA and xylan blend films but enhances the % strain to break.

Future studies will be focused on: (1) the characterization of xylan molecular weight after biodelignification and hot water extraction, (2) quantification of residual lignin and its effect on film properties, (3) preparation and characterization of synthetically acetylated xylan blends with CTA with and without TEC plasticizer, (4) evaluation of a broader spectrum of commercially available cellulose derivatives as blend components with xylan and its derivatives and, (5) the thermal characteristics of xylan blends leading to a more fundamental understanding of mixing thermodynamics establishing a more rational approach for the selection or synthesis of other polymers for optimized blending compatibility with xylan.

118

Acknowledgements

Support of this research was provided by the USDA McIntire-Stennis Cooperative Forestry Research Program and a US Department of Interior – US Forest Service grant to SUNY-ESF. Ms. Haghpanah was supported by a National Science Foundation – Research Experience for Undergraduates (REU) program grant during the summer of 2005. Consultation and facilities were also provided by the SUNY Cellulose Research Institute and the Empire State Paper Research Institute (ESPRI), both located at SUNY-ESF.

References

1. A *Billion-Ton Feedstock Supply for a Bioenergy and Bioproducts Industry - Technical Feasibility of Annually Supplying 1 Billion Dry Tons of Biomass*, A joint study sponsored by the US Dept. of Energy and US Dept. of Agriculture, 2005.
2. *Biobased Industrial Products – Priorities for Research and Commercialization*, National Research Council, National Academy Press, Washington, DC, 2000; p 103.
3. Sjostrom, E. *Wood Chemistry – Fundamentals and Applications*, 2nd Edition, Academic Press, 1993.
4. Stenius, P. *Forest Products Chemistry*, TAPPI Press, 2000.
5. Heller, M.C.; Keoleian, G.A.; Volk, T.A. Life Cycle Assessment of a Willow Bioenergy Cropping System, *Biomass and Bioenergy* **2003**, *25*, pp 147-165.
6. Koshijima, T.; Timell, T.E.; Zinbo, M.; The Number-Average Molecular Weight of Native Hardwood Xylans. *J. Polymer Sci - Part C* **1965**, *11*, pp 265-279.
7. Glasser, W.G.; Kaar, W.E.; Jain, R.K.; Sealey, J.E. Isolation Options for Non-Cellulosic Heteropolysaccharides, *Cellulose* **2000**, *7*, pp 299-317.
8. Teleman, A.; Tenkanen, M.; Jacobs, A. and Dahlman, O. Characterization of O-acetyl-(4-O-methylglucurono)xylan isolated from birch and beech, *Carbohydr. Res.* **2002**, *337*, pp 373-377.
9. Srisodsuk, M.; Kleman-Leyer, K.; Keranen, S.; Kirk, K and Teeri, T.T. Modes of action on cotton and bacterial cellusloe of a homologous endogluxcanase-exoglucanase pair from *T. Reesei. Eur. J. Biochem.* **1998**, *251*, pp 885-892.
10. Pettersen, R.C. The Chemical Composition of Wood, *The Chemistry of Solid Wood*, R. Rowell, Ed., Advances in Chemistry 207, American Chemical Society **1984**, p 60.
11. Stipanovic, A.; Amidon, T.E.; Scott, G.M.; Barber, V.; and Blowers, M.K. Hemicellulose From Biodelignified Wood: A Feedstock for Renewable Materials and Chemicals. *Feedstocks for the Future*, ACS Symposium Series 921, Joseph J. Bozell Ed., Oxford University Press, **2005**.

119

12. Timell, T.E. "Wood Hemicelluloses I", *Adv. Carbo. Chem.* **1964**, *19*, p 247.
13. Lawson, Jr., L.R.; Still, C.N. The biological decomposition of lignin--literature survey, *Tappi* 40:56A, **1957**.
14. Kawase, K. Chemical components of wood decayed under natural conditions and their properties, *J. Fac. Agri. Hokkaido Univ.* **1962**, *52*, p 186.
15. Reis, C.J. and Libby, C.E. An experimental study of the effect of *Fonnes pini* (Thure) Lloyd on the pulping qualities of pond pine *Pinus serotina* (Michx) cooked by the sulfate process, *Tappi J.* **1960**, *43*, p 489.
16. Ander, P. and Eriksson, K.-E. Mekanisk massa fran forrotad flis--en inledande undersokning, *Svensk Papperstidning* **1975**, 18, p 647.
17. Johnsrud, S.C. and Eriksson, K.E., Cross-Breeding of Selected and Mutated Strains of *Phanerochaete chrysporium* K-3: New Cellulase-Deficient Strains with Increasde Ability to Degrade Lignin, *Appl. Micro. and Biotech.* **1985**, *21(5)*, pp 320-327.
18. Eriksson, K.-E.; Johnsrud, S.C. ; Vallander, L. Degradation of lignin and lignin model compounds by various mutants of the white-rot fungus *Sporotrichun pulverulentum, Arch. Microbiol.* **1983**, 135, 161.
19. Eriksson, K.-E. "Biotechnology in the pulp and paper industry," *Wood Sci. Technol.* **1990**, *24*, p 79.
20. Samuelsson, L.; Mjober, P.J.; Harler, N.; Vallander, L.and. Eriksson, K.-E. Influence of fungal treatment on the strength versus energy relationship in mechanical pulping, *Svensk Papperstidning* **1980**, *8*, p 221.
21. Eriksson, K.-E. and Vallander, L. Properties of pulps from thermomechanical pulping of chips pretreated with fungi, *Svensk Papperstid.* **1982**, *85*, R33.
22. Pearce, M.H.; Dunlop, R.W.; Falk, C.J. and Norman, K. Screening lignin degrading fungi for biomechanical pulping of eucalypt wood chips, *Proc. 49th Appita Annual General Conf.,* Australia, p 347, 1995.
23. Kirk, T.K.; Koning, Jr., J.W.; Burgess, R.R., et al. Biopulping: A glimpse of the future, *Res. Rep.FPL-RP-523*, Madison, WI., 1993.
24. Kirk, T.K., Akhtar, M. and Blanchette, R.A. Biopulping: seven years of consortia research, *Proc. 1994 TAPPI Biological Sciences Symposium*, Tappi, Atlanta, p 57, 1994.
25. Akhtar, M.; Blanchette, R.A.; Myers, G.C.; Kirk, T.K. An overview of biomechanical pulping research, *Environmentally Friendly Technologies for the Pulp and Paper Industry*, R.A. Young and M. Akhtar, Eds, John Wiley & Sons, New York, p 309, 1998.
26. Messner, K., Koller, K.; Wall, M.B.; Akhtar, M.; Scott, G.M. Fungal pretreatment of wood chips for chemical pulping, *Environmentally Friendly Technologies for the Pulp and Paper Industry*, R.A. Young and M. Akhtar, editors, John Wiley & Sons, New York, p 385, 1998.
27. Scott, G.M., M. Akhtar, M.J. Lentz, and R.E. Swaney, Engineering, scale-up, and economic aspects of fungal pretreatment for wood chips, *Environmentally Friendly Technologies for the Pulp and Paper Industry*, R.A. Young and M. Akhtar, editors, John Wiley & Sons, New York, p 341, 1998.

28. Blowers, M.K. *Xylan extraction from short rotation willow biomass*, M.S. Thesis, State University of New York- College of Environmental Science and Forestry, Syracuse, NY, 2003.
29. Bartholomew, J. Identification and isolation of lignolytic enzymes of Phlebia subserialis and an analysis of white-rot fungi on Picea abies for Mechanical Pulp, M.S. Thesis, State University of New York- College of Environmental Science and Forestry, Syracuse, NY, 2003.
30. Chen, Y. *The Effect of Cellulose Crystal Structure and Solid-State Morphology on the Activity of Cellulases*, M.S. Thesis, State University of New York- College of Environmental Science and Forestry, Syracuse, NY, July 2005.
31. Grondahl, M., Eriksson, L. and Gatenholm, P. Material Properties of Plasticized Hardwood Xylans for Potential Application as Oxygen Barrier Films. *Biomacromolecules* **2004**, 5, pp 1528-1535.
32. Irvine, G.M. The Glass Transitions of Lignin and Hemicellulose and Their Measurement by Differential Thermal Analysis. *TAPPI Journal* **1984**, May, pp 118-121.
33. Sperling, L.H. *Introduction to Physical Polymer Science*, 4th Edition, Wiley Interscience, 2006, p 694.
34. Fried, J.R. *Polymer Science and Technology*, 2nd Edition, Prentice Hall Publishers, Chapter 7, 2003.
35. Park, H-M., Liang, X., Mohanty. A.K., Mishra, M. and Drzal, L.T. Effect of compatibilizer on nanostructure of the biodegradable cellulose acetate / organoclay nanocomposites. *Macromolecules* **2004**, 37, pp 9076-9082.

Chapter 8

Cellulose–Silica Hybrid Materials Obtained by Heteropolyacid Catalyzed Sol–Gel Synthesis

S. Sequeira, D. V. Evtuguin, and I. Portugal

CICECO and Chemistry Department, University of Aveiro,
3810–193 Aveiro, Portugal

Organic-inorganic hybrids based on cellulose fibers from *Eucalyptus globulus* kraft pulp were prepared at room temperature by a sol-gel method using TEOS as the silica precursor and $H_3PW_{12}O_{40}$ as the catalyst. These cellulose/silica hybrids (CSHs) were chemically and structurally characterized. The absence of unconverted TEOS and the formation of a silica network composed essentially by Q^3 and Q^4 structures were confirmed by FTIR, solid-state ^{13}C and ^{29}Si NMR. Image analysis (SEM and AFM) revealed that the silica had deposited on the fibers essentially in the form of a thin film or mesoparticles, which bridged the fibers. The silica network improved hydrophobicity, dimensional and thermal stability and bending strength of the starting fibrous material. The potential applications of these CSHs are discussed.

Hybrids, or composites, are composed by at least two different materials belonging to the same (e.g. two metals) or different classes (e.g., a metal and a ceramic). In particular, organic-inorganic hybrids (OIHs) comprise typically an organic polymer host matrix and embedded inorganic formulations. The organic polymer may be used as a pre-formed material or synthesized *in situ* during the hybrid preparation. OIHs are a relatively new type of materials with interesting mechanical, optical, electrical, and thermal properties. The inorganic phase usually confers thermal and mechanical resistance to the hybrid, whereas the organic phase governs the hybrid's low density, toughness, and flexibility *(1)*. The properties of hybrid materials depend not only on the properties of the starting materials and the preparation methods used, but also on the composite's phase morphology and interfacial properties.

A significant proportion of recently reported OIHs are synthesized using a sol-gel process, which is a rather flexible and versatile technique *(1)*. This allows the incorporation of the inorganic component in thermally unstable organic materials, since the synthesis occurs under mild conditions (room temperature and atmospheric pressure). In a sol-gel process, the *in situ* generated inorganic particles are evenly dispersed at a nanometer scale in the polymeric host matrix and are bound to it through hydrogen or covalent bonds thus forming a network. A brief outline of the fundamentals of the sol-gel syntheses of OIHs is given below.

Sol-Gel Synthesis of OIHs

In a typical sol-gel synthesis of OIHs, a metal alkoxide $M(OR)_n$ (where M is Si, Ti, Al, etc. and R is CH_3, C_2H_5, C_3H_7, etc.) is mixed with water, an alcohol (or another co-solvent) and a small amount of acid or base as catalyst, in the presence of an organic polymer. Hydrolysis and polycondensation reactions occur simultaneously and a metal oxide three-dimensional network with -O-M-O-M- linkages is formed on the polymer surface *(2)*. Silicon alkoxides are most frequently used for these systems and are discussed here as an example. The basic reactions involved in the formation of a silica network, using tetraethyl orthosilicate (TEOS) as chemical precursor, are (where $y=n+1$):

$$\text{Et}-\text{O}-\underset{\underset{\text{OEt}}{|}}{\overset{\overset{\text{OEt}}{|}}{\text{Si}}}-\text{OEt} \; + \; H_2O \; \rightleftharpoons \; \text{H}-\text{O}-\underset{\underset{\text{OEt}}{|}}{\overset{\overset{\text{OEt}}{|}}{\text{Si}}}-\text{OEt} \; + \; \text{EtOH} \qquad (1)$$

$$y \; \text{H}-\text{O}-\underset{\underset{\text{OEt}}{|}}{\overset{\overset{\text{OEt}}{|}}{\text{Si}}}-\text{OEt} \; \rightleftharpoons \; \text{Et}-\text{O}-\underset{\underset{\text{OEt}}{|}}{\overset{\overset{\text{OEt}}{|}}{\text{Si}}}-\left[\text{O}-\underset{\underset{\text{OEt}}{|}}{\overset{\overset{\text{OEt}}{|}}{\text{Si}}}\right]_n\!\!-\text{OH} \; + \; n \; \text{EtOH} \qquad (2)$$

The primary hydrolysis of TEOS by water replaces ethoxy moieties with hydroxyl groups (Eq. 1), thereby rendering the silane active for low temperature polymerization (Eq. 2). The hydrolysis reaction is partially reversible and, in the presence of an alcohol as the co-solvent (for example, ethanol), re-etherification takes place *(2)*. The condensation reaction eliminates either water (if two $\equiv$Si-OH moieties react) or alcohol (if $\equiv$Si-OH and $\equiv$Si-OEt react) to produce -Si-O-Si- linkages (Eq. 2). Partial alcoholysis (or hydrolysis) of the -Si-O-Si- bonds constitutes the reverse reactions. TEOS hydrolysis and condensation reactions are promoted by either acids or bases. However, the hydrolysis is faster under acidic conditions than in neutral or weak alkaline media, whereas the condensation is faster under neutral or alkaline conditions. The specific conditions of the hydrolysis and condensation steps profoundly affect the structure of the sol-gel silicates *(2)*. Indeed, these reactions are extremely dependent on the H_2O/TEOS ratio, the TEOS concentration, the proportion and type of organic solvents, the pH, the catalyst and its amount, the temperature, etc. *(3)*. This makes sol-gel syntheses attractive to researchers, since it is possible to manipulate readily the final structure of the materials.

In addition to the primary hydrolysis and polycondensation, leading to the formation of a series of siloxane linear and cyclic pre-polymers, polymer clusters or primary particles (sols), a sol-gel synthesis involves other important steps such as gelation (accompanied by a sharp increase in viscosity), aging, drying, stabilization (thermal or chemical) and densification (occurring between 1000 and 1700°C) *(2)*. Gelation may be defined as a collision process of growing sol particles leading to the formation of an infinite silica network - a gel. The further densification of this silica network occurs during aging, which involves the polymerization, coarsening and phase transformation of the gels. This process may take months at room temperature. During aging, the proportions of tertiary ($Si(O)_3OH$ or Q^3) and quaternary ($Si(O)_4$ or Q^4) structures increase dramatically. The finalization of condensation reactions and silica gel shrinkage occur when all solvents are removed during the drying process. The latter can be carried out at higher temperatures (50-150°C) to produce structurally compact microporous silica materials (xerogels), or under supercritical conditions to produce a silica material possessing a structure similar to that of the original sol (aerogels). The surface dehydroxylation of silica is accomplished at around 700°C (thermal stabilization) and its transformation into solid ceramics of different polymorphs occurs at >1000°C.

The SiO_4 tetrahedral unit is the fundamental building block of the majority of silicates (Fig.1a). SiO_4 tetrahedra can be presented in a simplified mode as rigid units (the Si-O bond length is 1.62 Å with the ideal tetrahedral angle of 109.5°), which are linked through their corners to form pairs, rings, chains, sheets, or frameworks. When the tetrahedra are linked by their corners, the Si-O-Si bond angle θ (Fig. 1b) can vary between around 91° (two-membered rings) and 180° (linear chain). Amorphous silica (as in the case of xerogels and aerogels) contains usually four-, five-, six-, seven- and eight- membered rings *(2)*. Due to the low thermal resistance of organic polymers, in most of the synthesized OIHs (bulk materials and films) the bound silica is in the form of xerogel or aerogel.

124

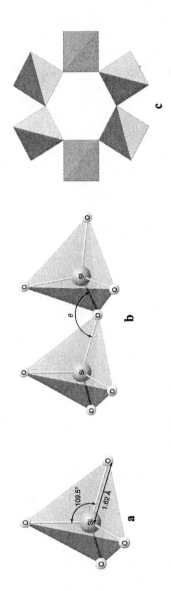

Figure 1. Polyhedral-filling representation of silica building blocks: SiO₄ tetrahedral unit (a), two SiO₄ tetrahedra linked by their corners (b) and six-membered cyclic structure (c).

However, an organic polymer matrix, and in particular a lignocellulosic material, can also be used as a template to produce solid ceramics with specific microstructure *(4)*. In addition to silicon alkoxides, titanium and zirconium alkoxides are used in the sol-gel synthesis of OIH materials *(1)*.

Cellulose-Based Organic/Inorganic Hybrids

Cellulose, the most abundant renewable biopolymer, possesses several unique properties required in different technical areas and biomedicine *(5)*. However, the serious drawback of cellulosic fibers is their hydrophilic nature, which results in a poor compatibility with hydrophobic polymer matrices in composite fabrications. The other disadvantages are the relatively low degradation temperature of cellulose-based materials and the dramatic loss of mechanical properties after moistening *(6)*. One of the promising ways to overcome these deficiencies is a surface modification with an inorganic component. Silica derivatives of cellulose are particularly attractive, since silica substantially improves the thermal stability of the parent polymer, its lipophilic behavior and its affinity towards specific substrates *(5, 7)*. The incorporation of silica into cellulose or cellulose derivatives has been accomplished with a wide range of silylation reagents, the most important being trimethylsilyl chloride *(8)*. However, these reactions usually require an organic solvent, such as pyridine, xylene or DMF, and high temperatures (80-160°C). Alternatively, silica incorporation by milder sol-gel processes has been reported for cellulose derivatives such as hydroxypropyl cellulose *(9, 10)* and cellulose acetate *(11, 12)*, with the main objective of increasing the mechanical and thermal resistance of the ensuing materials. Tanaka and Kozuka *(13)* described a sol-gel method for producing hybrid materials from cellulose acetate (CA), which has bioresorption properties, and silica, a bioactive and biocompatible inorganic material, with the aim of achieving mechanical properties similar to those of cortical bones. This work is one of several examples of applications of cellulose derivatives in biomedicine. Similarly, titanium was incorporated into ethyl cellulose, yielding a transparent material with interesting dielectric properties *(14)*. The cellulose acetate/Al_2O_3 hybrid was obtained by a sol-gel procedure using aluminium tetraisopropoxide *(15)*. After the Al_2O_3 modification with γ-aminopropyltrimethoxysilane, the mixed Al_2O_3/SiO_2 hybrid containing amine groups showed promising results on transition metals adsorption from ethanol solutions. Zeolite (Y and L) /cellulose hybrids were prepared from bleached kraft pulp and zeolite powders *(16)*. These OIHs can be used in the preparation of soft membranes with molecular sieving properties. An interesting approach to produce molecular imprinted cellulose/silica composites for the selective binding of organic substrates from solutions was recently reported *(17)*. A sol-gel method has also been applied for the deposition of hydrophobic polysiloxane coatings on wood *(18)* and for the modification of sulphite pulp with water-soluble silicon-containing compounds *(19)*, to increase the hydrophobicity of the final hybrid materials. The modification of cellulose fibers with pre-hydrolyzed trialkoxysilanes bearing variable functions (propylamine, allyl, alkyl, etc.) significantly increases their compatibility to hydrophobic polymer matrices in thermoset composites *(20)*. Modified fibers may be considered as OIHs as well, since they contain bound inorganic components in the form of a discontinuous xerogel film.

Based on this research, no information is available on OIHs for composite materials derived from unmodified cellulosic fibers (pulp) and silica. The pulp industry produces annually more than 200 million tons of bleached/unbleached pulp worldwide. Although the majority of pulp fibers are used for papermaking, cellulose/silica hybrid materials with attractive mechanical, thermal and sorption properties might be considered as potentially interesting alternative applications.

Sol-Gel Synthesis of Cellulose/Silica Hybrids from Pulp

Synthesis Procedure and Raw Material Requirements

Prior to their use in the sol-gel synthesis, the pulp was disintegrated (swollen) in water (1g of oven dried pulp/100 ml of water) and then washed with ethanol to remove the water excess. The ensuing filtered pulp contained only a small amount of residual solvent. This pre-activation step is necessary to improve the accessibility of cellulose. A pre-determined proportion of TEOS, distilled water, ethanol, and catalyst were added to the pre-activated pulp. The reaction proceeded over 24 hours at room temperature (~20°C) with constant stirring. The final hybrid material was filtered, washed with ethanol to remove the excess of TEOS and dried first at 40°C (24 h) and then at 105°C (24 h). The percentage of silica incorporation was determined by weighing the final hybrid material and/or by weighing the residue after calcination at 525°C (3 h).

The success of the cellulose/silica hybrids (CSHs) synthesis depends critically on the state of the fibers' surface *(21)*. Among the studied *Eucalyptus globulus* kraft pulps (bleached, unbleached and primary sludge fibers) the best results were obtained with the bleached pulp (BP) containing a minimum of residual lignin, extractives, and ash. In the case of the primary sludge fibers, the yield of CSH was minimal (< 5% weight increment) because of the strong surface contamination with precipitated lignin, extractives and minerals (Fig. 2). Pulp beating did not influence significantly the percentage of incorporated silica in the CSHs *(21)*.

Figure 2. SEM images of bleached kraft pulp (left) and fibers from the primary sludge (right).

Catalyst Selection

An acid-catalyzed sol-gel synthesis normally involves strong mineral acids, which may damage the acid-labile host polymer matrix during the aging and drying steps. Additionally, mineral acids (especially volatile and thermally unstable compounds) can induce equipment corrosion during production and limit the final material applications due to environmental concerns. In this context the use of strong non-volatile and thermally stable solid acids, such as heteropoly acids (HPAs), deserves attention. The efficiency of sol-gel catalysis with HPAs had not been assessed before this study.

In addition to different conventional acids (HCl, HNO_3, H_3PO_4 and H_2SO_4), several heteropoly acids bearing the general formula $H_x[XM_{12}O_{40}]^{x-}$ (X=Si, P; M= Mo, W) were employed in the preparation of CSHs based on BP. HPAs possess a very low negative charge on the surface bridging oxygens (low basicity) and are therefore strong Brønsted acids, even stronger than mineral acids *(22)*. HPAs are thermally stable up to 350-400°C, and unlike most strong mineral acids, have more affinity to silica than to the cellulosic material.

The molar concentration of the HPAs was about 10^3 times lower than that used with conventional acids and corresponded to 30-50 times lower catalyst weight load. Under the same reaction conditions, nitric acid was the most effective catalyst among the examined conventional acids, whereas among the HPAs, the highest catalytic activity was displayed by $H_3PW_{12}O_{40}$ (or simply PW_{12}), which was even higher than that of nitric acid (Fig. 3). The efficiency of the TEOS hydrolysis/condensation reactions and of the interaction of siloxane pre-polymers with cellulose in the presence of HPAs increased in the following order: $H_3PW_{12}O_{40}$ (PW_{12}) > $H_4SiW_{12}O_{40}$ (SiW_{12}) > $H_4SiMo_{12}O_{40}$ ($SiMo_{12}$) > $H_3PMo_{12}O_{40}$ (PMo_{12}). This is in fair agreement with the reported acidic strength in acetone (pK$_1$ values in parentheses), viz. PW_{12} (1.6) > SiW_{12} (2.0) ≈ PMo_{12} (2.0) > $SiMo_{12}$ (2.1) *(22)*.

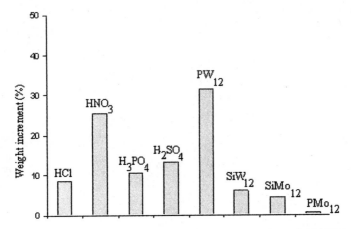

Figure 3. Pulp weight increment (%) after BP reaction with TEOS in the presence of mineral (ca. 0.3 mol/l) and heteropoly acids (ca. $3.0 \cdot 10^{-4}$ mol/l). Reaction condition: 24 h, 20°C, H_2O/ TEOS/EtOH molar proportion 3.2/1.0/8.7.

Effect of H_2O/TEOS Ratio and Catalyst Concentration

The water content influences the rate of TEOS hydrolysis, which in turn affects the series of hydrolysis/condensation reactions characteristic of a sol-gel process *(3)*. Theoretically, to convert completely silicon alkoxides to SiO_2, two moles of water are needed per mole of precursor (R = Alk):

$$n\mathrm{Si(OR)}_4 + 2n\mathrm{H}_2\mathrm{O} \rightarrow n\mathrm{SiO}_2 + 4n\mathrm{ROH} \qquad (3)$$

In practice, this molar H_2O/TEOS ratio (r) is not sufficient, probably because of the formation of non-cyclic intermediate species, and should preferably be higher than 2.5. Thus, an excess of water is required to prepare cellulose/silica hybrids. According to the practice of CSH syntheses with PW_{12} as catalyst, at r < 3, silica incorporation into the pulp fibers is too slow and at r >4.4 it is difficult to control the spontaneous silica gel formation in bulk. Thus, the acceptable r values varied between 3.0 and 4.4. It was observed that increasing r from 3.2 to 4.4, increased the silica incorporation into BP from 43 to 52% (w/w). These results were attributed to the increase in the molecular weight of the intermediate siloxane polymers, arising from the enhancement of the condensation reactions, promoted by the increase in the H_2O to alkoxysilane molar ration *(3)*.

The amount of catalyst in the reaction medium influences significantly both the hydrolysis (Eq.1) and the condensation reactions (Eq.2). Increasing the catalyst to alkoxysilane ratio normally favors the hydrolysis reactions, rather than the condensations. As a consequence, the rate of polymer cluster formation increases, but their intermediate size decreases. This explains the decrease in silica bound to the pulp when the PW_{12} charge was increased above $3.0 * 10^{-4}$ mol/l. It can be concluded that the size of the silica domains, related to each active growing centre on the pulp fiber surface, diminishes as the catalyst load is increased. A maximum incorporation of silica ($> 50\%$ wt.) was obtained with BP with a 4.4 H_2O/TEOS molar ratio, 8.3 EtOH/TEOS molar ratio and a PW_{12} concentration of $3.0 * 10^{-4}$ mol/l (optimized conditions).

Characterization of the Cellulose/Silica Hybrids

Image Analysis

Image analysis showed (Fig. 4) that silica was deposited on the fiber surface as isolated or aggregated mesoparticles (1-4 μm in diameter). Simultaneously, larger particles of 5-10 μm were detected, which were mostly localized at the intersection of neighboring fibers (Fig. 4), thus bounding them and reinforcing the composite structure. Large-scale particles were more abundant in hybrids obtained with higher H_2O/TEOS molar ratios.

Large areas of the fibers' surface, visually not covered by silica, nevertheless showed the presence of silicon, as revealed by EDS (Fig. 4). The AFM analysis revealed that the silica film comprised conjugated round-shape domains of 0.05-0.3 μm (Fig. 4). This film was discontinuous and interrupted by uncovered regions of the fiber's surface.

FTIR and NMR Analysis

The solid-state ^{13}C-NMR spectrum of BP/SiO$_2$ hybrids obtained under optimized conditions (Fig. 5a) did not show any signals at 16.9 and 58.5 ppm, assigned to methyl and methylene carbons, respectively, in ethoxy moieties, i.e. during the CSHs preparation, the hydrolysis of TEOS was complete. The search for possible cellulose linkages with siloxy moieties, using differential ^{13}C CP-MAS NMR spectra of the initial pulp and the hybrids, failed. However, this fact does not prove the absence of these linkages. Indeed, the extent of cellulose-silica covalent bonds was probably very low ($< 3\text{-}5$ mol-%) and, therefore, not detectable by solid-state ^{13}C-NMR. In the FTIR spectrum (not shown) of the BP/SiO$_2$ hybrids, the band at 1000-1150 cm^{-1}, corresponding to Si-O-C vibration, was not clearly detected, because it overlapped with a broad band at 1000 -1150 cm^{-1} assigned to the δ (O-H) mode of primary and secondary

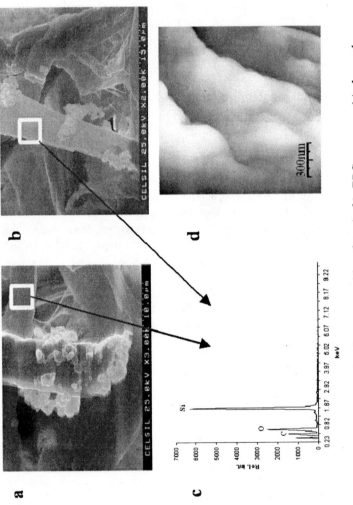

Figure 4. SEM images of pulp fibers covered by silica (a, b). EDS spectrum (c) shows the presence of silica film on pulp surface that was assessed by AFM (d) applied in a tapping mode.

alcohol groups of cellulose, and with a band at 1080 cm^{-1} assigned to the asymmetric stretching of the inter-tetrahedral oxygen atoms of silica *(23)*.

Additional structural information on the silica portion of the BP/SiO$_2$ hybrids was obtained by solid-state NMR. ^{29}Si MAS NMR spectra provided information about the proportions of Q^n species (Q^n designates a Si atom bonded to n other Si atoms via O-bridges) in silica materials allowing quantification of their crosslink density. A typical spectrum of a BP/SiO$_2$ hybrid (Fig. 5b) showed two major signals at -102 and -111 ppm (Q^3 and Q^4 structures, respectively), a small signal at -91 ppm (Q^2 structures) and the absence of a signal at -81 ppm (Q^1 structures). This indicated the presence of a silica network composed essentially of cyclic units (Q^3 structures) connected by oxygen bridges (Q^4 structures) *(2)*.

The crosslink density (η) may be determined from the NMR spectra. It is defined as the ratio of effective (f_{ef}) and potential (f_{pot}) functionalities of Si substituted by OSi moieties and is calculated as shown below (x_n is the mole fraction of Q^n structures calculated from corresponding peak areas):

$$\eta = \frac{f_{ef}}{f_{pot}} \quad \text{where} \quad f_{ef} = \sum_{n=1}^{4}(x_n \times n), \quad f_{pot} = 4 \times \sum_{n=1}^{4}(x_n)$$

For the BP/silica hybrids obtained under our optimized conditions, η was found to be close to 0.83.

Thermal Analysis

Thermogravimetric analyses of BP and BP/SiO2 hybrids (Fig. 6) revealed a first weight loss at 60°C, corresponding to water release. The thermal degradation of the organic material was observed at 305 °C for BP and at 345°C for the hybrid material. This increase in degradation temperature suggests strong organic-inorganic phase interactions (strong hydrogen bonding and, probably, covalent bonding) that greatly influence thermal resistance. The results of the DSC analyses of the hybrid materials corroborated TGA data.

Cellulose-Silica Interactions in Hybrids

According to the results obtained by spectroscopic and thermal analyses, it may be proposed that both the xerogel film and the silica particles are attached to the cellulose surface in pulp hybrids essentially by hydrogen bonding, although a small proportion of covalent bonding between cellulose and silica cannot be completely ruled out. X-ray scattering analysis (WAXRD) showed a significant

132

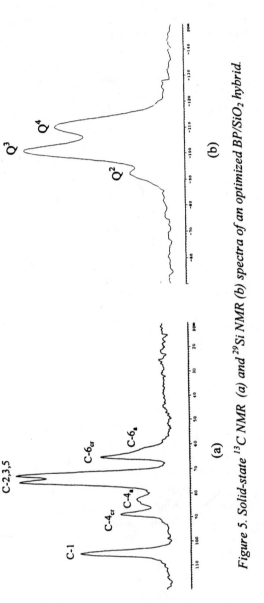

Figure 5. Solid-state ^{13}C NMR (a) and ^{29}Si NMR (b) spectra of an optimized BP/SiO$_2$ hybrid.

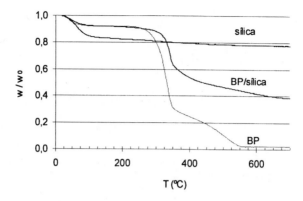

Figure 6. Thermogravimetric plots of SiO₂, BP and BP/SiO₂ hybrid.

decrease and broadening in the cellulose crystalline reflection peaks of the hybrid materials. A simplified scheme of cellulose-silica interaction in these hybrid materials is shown in Fig. 7.

Properties of Cellulose/Silica Hybrids

A series of composite materials based on CSHs from BP were prepared by molding (Fig. 8) and subsequent pressing at room temperature or at 140°C for 8 minutes, followed by a post-thermal treatment *(24)*. The density (d) of the final materials varied from 0.15 to 0.60 g/cm^3. Their thermal conductivity, mechanical properties, and dimensional stability were evaluated using 5-mm thick plates. The preparation of CSH formulations was done under optimized conditions as discussed above (material designated as BP/SiO$_2$). In a series of CSHs syntheses, 5% (vol.) of TEOS was substituted by triethoxyoctylsilane (TEOcS) in order to improve the hydrophobicity of the ensuing hybrid materials (designated as BP/SiO$_2$/C$_8$).

Some properties of these CSH materials and BP are compared in Table 1. The incorporation of silica into bleached kraft pulp changed radically its hydrophilicity (expressed as Water Retention Value or WRV) and its dimensional stability (expressed as relative weight (ΔW%) and volume (ΔV%) increment after soaking in water). This is especially notable in the case of the BP/SiO$_2$/C$_8$ hybrid, which was a highly hydrophobic material. The CSHs also displayed an enhanced bending strength (σ), which was similar to that of medium-density fiberboards, in the case of the BP/SiO$_2$ hybrids, and similar to that of high-density fiberboards in the case of the BP/SiO$_2$/C$_8$ hybrids.

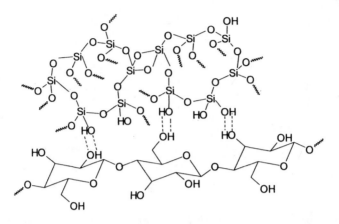

Figure 7. Schematic representation of cellulose-silica interactions in a bleached pulp/silica hybrid material. The fragment of cellulose chain sitting on the pulp surface is hydrogen bound to a silica fragment composed of two $(SiO)_6$ cyclic structures.

Figure 8. Pre-formed CSH formulation (right) prepared by molding from suspension (left).

Table 1. Some properties of bleached kraft pulp and CSH materials (d=0.45)

Hybrid	SiO_2 %	WRV (%)	σ (MPa)	λ (W/m.K)	Dimensional stability ΔW (%)	ΔV(%)
BP	0	118	2.0	0.08	235	59
BP/SiO_2	55	12	3.4	0.14	117	20
BP/SiO_2/C_8	54	< 5	5.4	0.13	67	5

The thermal conductivity (λ) of the hybrid materials, measured in the range 40-180°C, was slightly higher than that of BP, but comparable to values reported for commercial insulating polymer materials. It is also important to note that the hybrid materials showed heat release rates during ignition at ~ 800°C which were some three times lower than that of bleached kraft pulp and a thousand times lower than those of commercial insulating polymers, i.e. they were much less flammable materials. The acoustic tests on the low density CSHs are in progress.

Conclusions

Organic/inorganic hybrids based on *Eucalyptus globulus* bleached kraft pulps and silicon alkoxides were synthesized by a sol-gel method under acidic conditions. Heteropolyacids (HPAs) were used as catalysts for the first time and showed a higher catalytic efficiency than conventional mineral acids (HCl, HNO_3, H_3PO_4, H_2SO_4). Tungstophosphoric acid $H_3PW_{12}O_{40}$ was the best catalyst among the HPAs used in this study. The conditions of the sol-gel syntheses in terms of the efficiency of silica incorporation into the pulp were optimized. Between 40 and 60% of silica was incorporated into the cellulosic material, diminishing considerably its hydrophilicity and improving its thermal stability, as revealed by TGA and DSC analyses. The formation of the silica network bound to the cellulosic fibers was assessed by ^{29}Si MAS NMR and FTIR spectroscopy. However, no clear evidence was found of the existence of covalent bonding between cellulose and silica.

The preliminary results on the thermal conductivity suggested that these CSHs could be used as thermal insulation materials. The presence of a silica network also increased significantly the mechanical properties (bending strength) of the starting pulps. The hydrophobic nature and the thermal stability of these novel hybrids make them potentially suitable for the preparation of special packaging or insulating materials.

References

1. Yano, S.; Iwata, K.; Kurita, K. *Mat. Sci. Eng., C: Biomim. Supramol. Syst.*, **1998**, *6*, pp 75-90.
2. Brinker, C.J.; Scherer, G.W. *Sol-gel science: the physics and chemistry of sol-gels.* Academic Press, New York, **1990**.
3. McCormick, A. In: *Sol-gel processing and application*, Ed. Y.A. Attia, Plenum Press, New York, **1994**, pp 3-16.
4. Greil, P. *J. Eur. Cer. Soc.*, **2001**, *21*, pp 105-118.

136

5. Klemm, D.; Heublein, B.; Fink, H.-P.; Bohn, A. *Angew. Chemie. Int. Ed. Engl.,* **2005**, *44*, pp 3358-3393.
6. Mohanty, A. K.; Misraa, M.; Hinrichsen G. *Macromol. Mater. Eng.,* **2000**, *276/277*, pp 1–24.
7. Heinze, T.; Liebert, T. *Prog. Polym. Sci.,* **2001**, *26*, pp 1689-1762.
8. Klemm, D.; Philipp, B.; Heinze, T.; Heinze, U.; Wagenknecht, W. *Comprehensive cellulose chemistry: fundamentals and analytical methods.* Vol.1, Wiley-VCH, Weinheim, **1998**.
9. Yano, S. *Polymer,* **1994**, *35*, pp 5565-5570.
10. Yano, S.; Kodomari, M. *Nihon. Reoroji. Gakk.,* **1996**, *24*, pp 15-20.
11. Shojaie, S. S.; Rials, T. G.; Kelley, S. S. *J. Appl. Polym. Sci.,* **1995**, *58*, pp 1263-1274.
12. Zoppi, R. A.; Gonçalves, M. C. *J. Appl. Polym. Sci.,* **2002**, *84*, pp 2196-2205.
13. Tanaka, K.; Kozuka, H. *J. Mat. Sci.,* **2005**, *40*, 5199-5206.
14. Yoshinaga, I.; Katayama, S. *J. Sol-Gel Sci. Technol.,* **1996**, *6*, pp 151-154.
15. Lazarin, A.M.; Goshikem, Y. *J. Braz. Chem. Soc.* **2002**, *13*, pp 88-94.
16. Vu, D., Marques, M.; Larsen, G. *Micropor. Mesopor. Mat.,* **2002**, *55*, pp 93-101.
17. Gill, R.S., Marques, M.; Larsen, G. *Micropor. Mesopor. Mat.,* **2005**, *85*, pp 129-135.
18. Tshabalala, M. A.; Kingshott, P.; VanLandingham, M. R.; Plackett, D. *J. Appl. Polym. Sci.,* **2003**, *88*, pp 2828-2841.
19. Telysheva, G.; Dizhbite, T.; Arshanitsa, A.; Hrols, J.; Kjaviav, *J. Cell. Chem. Technol.,* **1999**, *33*, pp 423-436.
20. Abdelmouleh, M.; Boufi, S.; Belgacem, M. N.; Dufresne, A.; Gandini, A. *J. Appl. Polym. Sci.,* **2005**, *98*, pp 974-984.
21. Sequeira, S.; Portugal, I.; Evtuguin, D.V. *Book of Abstracts 9th Intern. Chem. Eng. Conference CHEMPOR,* Sept.21-23, Coimbra, **2005**, pp 83-84 (2005). Extended abstracts are available on CD-ROM.
22. Mizuno, N.; Misono, M. *J. Mol. Cat., A: Chem.,* **1994**, *86*, pp 319-342.
23. Fidalgo, A.; Nunes, T.G.; Ilharco, L.M. The structure of Hybrid Gels by DRIFT and NMR Spectroscopies. *J. Sol-Gel Sci. Technol.,* **2000**, *19*, pp 403-407.
24. Sequeira, S.; Evtuguin, D.V.; Portugal, I. *Proceedings of 8h EWLP,* Aug. 26-29, Riga, Latvia, **2004**, pp 145-148.

Chapter 9

Oxygen Barrier Films Based on Xylans Isolated from Biomass

Maria Gröndahl[1,2] and Paul Gatenholm[1]

[1]Biopolymer Technology, Department of Chemical and Biological Engineering, Chalmers University of Technology, SE–412 96 Göteborg, Sweden
[2]Current address: Xylophane AB, Stena Center 1B, SE–412 92 Göteborg, Sweden

Xylan films were prepared by casting from aqueous solution and their equilibrium moisture content, mechanical properties, morphology and oxygen permeability were measured. The films were transparent and homogeneous and the mechanical properties could be controlled by an addition of plasticizers such as xylitol and sorbitol. Tensile testing showed that barley husk arabinoxylan films had a higher stress at break and strain at break as compared to aspen glucuronoxylan films at corresponding plasticizer contents. It was also shown that water is a good plasticizer for xylans. The water content of the films depends on the chemical structure of the xylan, such as branching and its substituents. The glucuronoxylan films were semicrystalline, whereas the arabinoxylan films were mainly amorphous with small crystalline peaks detected by Wide Angle X-ray Scattering. Both the glucuronoxylan and arabinoxylan films had low oxygen permeability and can thus be used in packaging for oxygen-sensitive products.

Introduction

Today, the great majority of plastics are based on fossilic raw materials. Their disposal contributes to growing landfills and enhanced greenhouse effects when burned. Furthermore, the earth's fossil resources are limited. Interest in the use of renewable resources for the production of polymeric materials is growing as the problem of rapidly reducing oil resources gains more attention and oil prices increase. Recycling of synthetic polymers has been studied for many years and has become a vital part of the development of a sustainable society (*1*). However, sustainable development requires greater use of renewable materials. Using plant biopolymers to produce plastic materials has several advantages. In addition to their being a renewable resource, the products can be composted, recycled or incinerated after use with minimal environmental effect (*2*).

One typical application where the life time of the material is rather short is food packaging. The material should have good mechanical properties so that the food remains undamaged during storage, it should be resistant to water and it is important that it does not emit poisonous substances since it is in contact with food (*3*). In food packaging good barrier properties are necessary to prevent gases such as oxygen from degrading the product.

Polysaccharides such as cellulose, starch and hemicelluloses are produced by plants in vast quantities by the conversion of carbon dioxide and water using solar energy, which leads to a better carbon dioxide balance in our ecosystem. Cellulose, which is the reinforcing component of the cell wall, is used in many applications, such as paper, textile fibers, plastics, membranes, food additives and medicines. Starch is produced as an energy reserve in plants and has successfully been converted into plastic materials by conventional thermoplastic processing (*4*). Hemicelluloses are biosynthesized by the majority of plants and act as a matrix material that is present between the cellulose microfibrils and as a linkage between cellulose and lignin in the cell wall (*5*). In contrast to cellulose and starch, the commercial utilization of isolated hemicellulose has as yet been very limited. In the pulp and paper industry, retaining the hemicellulose in the pulp has been shown to improve both the mechanical properties of the paper and the yield (*6-7*). Hemicellulose has also been used as thickeners and emulsifiers in food (*5, 8*). There has been interest in the use of hemicellulose as a nutraceutical (*9*), in chiral separations (*10*) and as an HIV inhibitor (*11*). Hemicellulose can also be hydrolyzed to monosaccharides that are converted to chemicals such as furfural and xylitol or can be used as fermentation feedstock for making ethanol or lactic acid (*12-13*).

Hemicelluloses

Hemicelluloses were originally believed to be intermediates of cellulose biosynthesis but later proved to be a group of heterogeneous polysaccharides that are formed through their own biosynthetic routes (*14*). Hemicelluloses are synthesized in the Golgi apparatus. They are packed into vesicles and targeted to the plasma membrane where they become integrated with the newly synthesised cellulose microfibrils (*15-18*).

A general definition of hemicellulose is polysaccharides that can be extracted by water or aqueous alkali from plant tissue (*8, 19*). Hemicelluloses are heteropolysaccharides whose composition and amount vary between different plant species. In wood, the hemicellulose comprises between 20 and 35 % of the total material and the hemicellulose content in brans and hulls from annual plants, such as maize, can be as high as 40-50 wt. % (*8, 20*). The most common monosaccharides are D-xylose, L-arabinose, D-glucose, D-galactose, D-mannose, D-glucuronic acid, 4-*O*-methyl-D-glucuronic acid and D-galacturonic acid, see Figure 1. In comparison to cellulose, hemicellulose chains are rather short with an average degree of polymerization of only 200 (*14*).

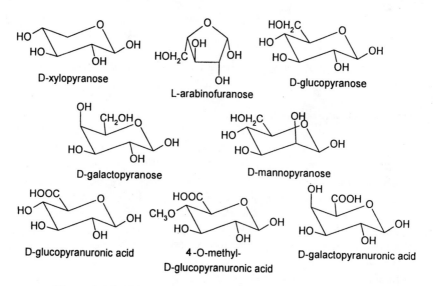

D-xylopyranose

L-arabinofuranose

D-glucopyranose

D-galactopyranose

D-mannopyranose

D-glucopyranuronic acid

4-O-methyl-
D-glucopyranuronic acid

D-galactopyranuronic acid

Figure 1. The most common building blocks in hemicelluloses.

The most abundant hemicelluloses, found mainly in hardwood and annual plants, comprise a 1,4-β-D-xylopyranosyl main chain with a varying number of side chains based on L-arabinofuranosyl, 4-*O*-methyl-D-glucuronopyranosyl, D-galactopyranosyl or D-glucurono-pyranosyl units. Xylans isolated from hardwood and annual plants differ from one another.

In hardwoods such as aspen, beech and birch, the hemicellulose consists chiefly of *O*-acetyl-(4-*O*-methylglucurono)xylan (*21*), often simply referred to as glucuronoxylan, see Figure 2. The backbone consists of β-(1→4)-linked D-xylopyranosyl residues substituted with one α-(1→2)-linked 4-*O*-methyl-D-glucuronic acid per approximately every tenth such residue (*22*). The xylopyranosyl residues are partially acetylated in the C2 and/or C3 positions (*21, 23*). The degree of acetylation in native aspen glucuronoxylan has been reported to be between 0.6 and 0.7 (*24-25*).

Xylan is generally considered to be amorphous in the native state but can crystallize after isolation (*26*).

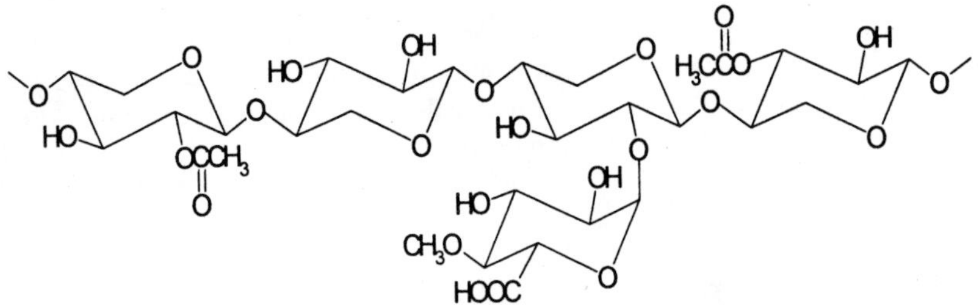

Figure 2. Schematic structure of hardwood xylan.

The hemicelluloses found in annual plants, such as maize, rice, oats, rye, barley and wheat, are generally more structurally diverse and complex. These plant hemicelluloses have a 1,4-β-D-xylopyranosyl main chain that can be heavily branched with xylopyranosyl, arabinofuranosyl and galactopyranosyl side chains and can also contain 4-*O*-methyl-D-glucuronopyranosyl or D-glucuronopyranosyl substituents. The two predominant monosaccharides in these annual plant hemicelluloses are generally xylose and arabinose, and they are thus termed arabinoxylans.

The degree of side-chain substitution determines the degree of solubility of the xylan; the higher the degree of substitution, the higher the water solubility (*27*). In addition to substitution with acetyl groups, 4-*O*-methylglucuronic acid, arabinose etc., as mentioned above, xylan has covalent bonds to lignin in so called lignin-carbohydrate complexes (LCCs) (*28*).

One of the possible reasons for the lack of utilization of hemicellulose as a material has been the shortage of access to high molecular weight hemicellulose on an industrial scale. For example, in traditional pulping processes designed for isolation of cellulose from wood, hemicelluloses are degraded to a low degree of polymerization (*14*). A pilot plant for the isolation of hemicellulose from plant tissue has been built, however (*19, 26, 29*). In order to utilize the enormous renewable resource that hemicellulose constitutes, the availability must be improved by developing efficient processes for isolation. Nevertheless, these processes will not be fully developed until we see potential applications.

The liberation of xylan is restricted by the presence of lignin, which acts as a capturing network, and some sort of delignifying treatment is necessary before xylan can be extracted (30). Hot water can be used to isolate xylans, but extraction with alkali is more efficient. However, this method involves hydrolysis of ester linkages and thus deacetylation of the xylan. The isolation of xylan on the laboratory scale has been extensively studied (31-54).

The aim of this work was to evaluate the potential use of xylans as oxygen barrier materials with possible applications in food packaging.

Experiments

Materials

Aspen glucuronoxylan was isolated by alkali extraction, resulting in deacetylation of the material. The separation procedure is described in detail elsewhere (19, 26). The isolated material contained 87 mol % xylose, 10 mol % 4-O-methyl-glucuronic acid, 2 mol % mannose and less than 1 mol % of other sugars. The weight average molecular weight was 15 000 g/mol.

A beech glucuronoxylan sample was provided by Lenzing AG. The sample contained 4-O-methylglucuronoxylan with 95.2 mol % xylose and 4.8 mol % 4-O-methylglucuronic acid and had a weight average molecular weight of 11 800 g/mol.

Arabinoxylan was isolated from barley husks by alkali extraction after chlorite delignification. The isolation procedure is described in detail elsewhere (55). The sample had a sugar composition of 74.4 mol % xylose, 16.5 mol % arabinose and 7.5 mol % glucose. The xylose to arabinose ratio (Xyl/Ara) was thus 4.5. The weight average molecular weight was 36 000 g/mol. Further treatment of the sample resulted in an arabinoxylan with a lower arabinose content (Xyl/Ara = 5.2).

Methods

Film Preparation

Films were prepared by mixing xylan with water and, in some cases, additives such as xylitol or sorbitol during magnetic stirring at 95 °C for 15 minutes. The total amount of dry substance in each film was kept constant at 1 g, and 35 ml of water was added. The solutions were poured onto polystyrene petri dishes with a diameter of 14 cm, and films were allowed to form upon drying in a temperature of 23 °C and a relative humidity (RH) of 50 %.

Equilibrium Moisture Content

The equilibrium moisture content of the samples at 50 % RH and 23 °C was measured. Three pieces of the films were conditioned, the samples were weighed and, after equilibration of the water content, dried at 130 °C overnight. The equilibrium moisture content was measured gravimetrically and calculated as the weight of water in the sample compared to the total weight.

Tensile Testing

The mechanical properties of the conditioned glucuronoxylan and arabinoxylan films were measured with a tensile testing machine (Lloyd L2000R) with a load cell with a capacity of 100 N. The samples were cut into dogbone-shaped strips with a width of 7 mm. The initial distance between the grips was 20 mm, and the separation rate of the grips was kept constant at 5 mm per minute. Fifteen replicates were tested. The stress-strain curve was recorded for each sample, and Young's modulus, stress at break, σ_b, and strain at break, ε_b, were calculated. The measurements were made at 50 % RH and 23 °C after conditioning for at least one week.

Wide Angle X-ray Scattering

The crystallinity of the glucuronoxylan and arabinoxylan films was studied with wide angle x-ray scattering, WAXS. The films were milled in liquid nitrogen and investigated with a Siemens D5000 goniometric diffractometer. CuK_α radiation with a wavelength of 1.54 Å was used, and 2θ was varied between 5 and 30° at a rate of 1° (2θ) per min. and a step size of 0.1° (2θ).

Oxygen permeability

The oxygen transmission of the glucuronoxylan and arabinoxylan films was measured with Mocon Oxtran 2/20 equipment in accordance with ASTM D3985-81. The area of the samples was 5 cm^2 and the analysis was made in 50 % RH after conditioning of the samples. Data was collected for 24 hours. The permeability was calculated from the transmission and the thickness of the films.

Results and Discussion

Film Formation

Film properties of aspen glucuronoxylan and of barley husk arabinoxylan were studied. Films prepared from glucuronoxylan were very brittle and fragmented upon drying. After the addition of xylitol or sorbitol, however, continuous films were formed, see Figure 3.

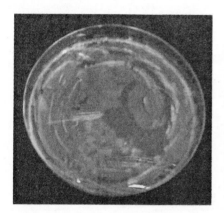

Figure 3. Glucuronoxylan film (left), film of plasticized glucuronoxylan (right).

Since arabinoxylan has a more substituted/branched structure than glucuronoxylan our hypothesis was that the substituents might give internal plasticization by creating more free volume and potentially by binding water. This would allow the formation of flexible films with smaller amounts of added external plasticizers. In agreement with this, the experiments showed that films prepared from barley husk arabinoxylan without added external plasticizers were homogeneous and transparent.

Equilibrium Moisture Content

The water content plays an important role in xylan films, since water acts as a plasticizer and softens the material (56-57). The equilibrium moisture contents of the aspen and beech glucuronoxylan films and of the barley husk arabinoxylan films are shown in Table I.

Table I. Equilibrium Moisture Content of Xylan Films at 50 % RH and
23 °C

Sample	Equilibrium Moisture Content (wt. %)	References
Aspen glucuronoxylan	18.0	
Beech glucuronoxylan	13.0	
Aspen glucuronoxylan with 20 wt. % xylitol	14.4	(57)
Aspen glucuronoxylan with 35 wt. % xylitol	12.5	(57)
Aspen glucuronoxylan with 50 wt. % xylitol	13.7	(57)
Aspen glucuronoxylan with 20 wt. % sorbitol	13.3	(57)
Aspen glucuronoxylan with 35 wt. % sorbitol	12.6	(57)
Aspen glucuronoxylan with 50 wt. % sorbitol	12.8	(57)
Arabinoxylan with Xyl/Ara=4.5	35.5	(55)
Arabinoxylan with Xyl/Ara=5.2	17.5	
Arabinoxylan with Xyl/Ara=5.2 and with 20 wt. % sorbitol	19.3	
Arabinoxylan with Xyl/Ara=5.2 and with 30 wt. % sorbitol	22.7	

The beech glucuronoxylan sample has a more linear structure than the aspen glucuronoxylan and the equilibrium moisture content at 50 % RH and 23 °C was 13 wt. % as compared to 18 wt. % for the former. The arabinose substitution also strongly affects the moisture content in the samples. Arabinoxylan with a xylose to arabinose ratio of 4.5 had a moisture content of 35.5 wt. % at 50 % RH and 23 °C. Determination of the moisture content in arabinoxylan with a xylose to arabinose ratio of 5.2 showed that this sample contained only 17.5 wt. % moisture at the same conditions, which clearly shows how the substituents control the moisture uptake in the xylan.

Mechanical Properties

The mechanical properties of the aspen glucuronoxylan films and the barley husk arabinoxylan films were evaluated using tensile testing. The results are summarized in Table II.

Table II. Mechanical Properties of Xylan Films Evaluated by Tensile Testing at 50 % RH and 23 °C

Sample	Strain at Break (%)	Stress at Break (MPa)	Young's Modulus (MPa)	References
Glucuronoxylan with 20 wt. % xylitol	2.1	39.4	2410	(57)
Glucuronoxylan with 35 wt. % xylitol	5.3	10.6	570	(57)
Glucuronoxylan with 50 wt. % xylitol	8.0	3.0	90	(57)
Glucuronoxylan with 20 wt. % sorbitol	2.0	35.4	2500	(57)
Glucuronoxylan with 35 wt. % sorbitol	5.8	13.5	690	(57)
Glucuronoxylan with 50 wt. % sorbitol	10.4	3.9	130	(57)
Arabinoxylan with Xyl/Ara=4.5	2.5	50.3	2930	(55)

Glucuronoxylan films with 20 wt. % xylitol or sorbitol are strong, with a stress at break above 40 MPa, but brittle, with a strain at break of around 2 %. The addition of more xylitol or sorbitol results in a reduction in strength but an increase in elongation at break. A greater amount of plasticizer increases the free volume by separating the glucuronoxylan chains from each other, making the material softer and more flexible. This effect is more pronounced for sorbitol than for xylitol. It can be concluded, however, that both xylitol and sorbitol act as plasticizers. The arabinoxylan films are strong, with a stress at break above 50 MPa, and have a strain at break of around 2.5 % and in contrast to the glucuronoxylan needed no plasticizer for film formation. Apart from the difference in molecular structure, the arabinoxylan had a higher molecular weight than the glucuronoxylan, 36 000 g/mol and 15 000 g/mol, respectively. The presence of proteins and other impurities can also affect the film formation.

Water acts as a plasticizer and softens xylan films. The total of the amount of added plasticizer, e.g. sorbitol, and the water content can be regarded as the total plasticizer amount in the films. In Figures 4-6, the amount of sorbitol +

water in arabinoxylan films is plotted against the mechanical properties of the films studied by tensile testing. Figure 4 shows the strain at break, Figure 5 shows the stress at break and Figure 6 shows Young's modulus.

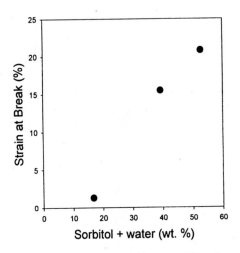

Figure 4. Strain at break of arabinoxylan films as a function of the total amount of added plasticizer (0 %, 20 % and 30 % sorbitol) and water.

Wide Angle X-ray Scattering (WAXS)

The morphology of the films was investigated using WAXS, see Figure 7. The diffractograms showed that the samples are semicrystalline. The arabinoxylan is primarily amorphous but has small crystalline peaks in the region of $2\theta = 17\text{-}21°$. This indicates that some of the arabinoxylan is present in crystalline form. It is probable that unsubstituted regions crystallize and that substituted regions are amorphous. The hindrance of crystallization by substituents has also been observed in the case of acetyl groups (*58*).

Oxygen Permeability

The oxygen transmission rates through both glucuronoxylan and arabinoxylan films were measured at 50 % RH and the oxygen permeability was calculated, as shown in Table III.

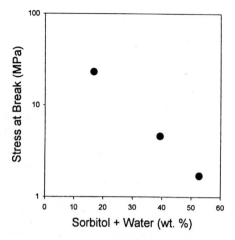

Figure 5. Stress at break of arabinoxylan films as a function of the total amount of added plasticizer (0 %, 20 % and 30 % sorbitol) and water.

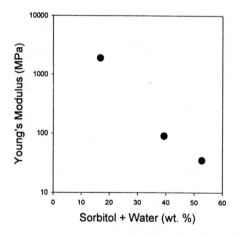

Figure 6. Young's modulus of arabinoxylan films as a function of the total amount of added plasticizer (0 %, 20 % and 30 % sorbitol) and water.

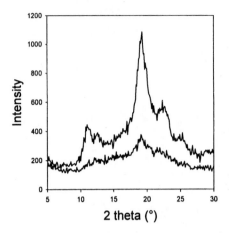

Figure 7. X-ray diffractograms of aspen glucuronoxylan (top) and arabinoxylan from barley husks (bottom).

The oxygen permeability of xylan films is very low and in the same range as the frequently used barrier plastic ethylene vinyl alcohol (EVOH). The permeability of the xylan films is also compared with that of the plasticized starch polymers (amylose and amylopectin), with poly vinyl alcohol (PVOH) and with low density polyethylene (LDPE), which is not an oxygen barrier material.

Conclusions

Xylans are among the most abundant biopolymers on earth, but the commercial utilization of isolated xylan is as yet very limited. There are several separation processes of xylans from biomass on the laboratory and pilot scales. The development of xylan-based products will promote up-scaling of the separation processes. Externally plasticized hardwood xylans and arabinoxylans from agricultural residues were found to form films with good mechanical properties that exhibited excellent oxygen barrier properties. Xylans are potential candidates for oxygen barriers in multilayer packaging.

Table III. Oxygen Permeability of Xylan Films, Amylose, Amylopectin, PVOH, EVOH, and LDPE

Material	Oxygen Permeability (cm3 μm)/(m2 d kPa)	Testing Conditions	References
Glucuronoxylan with 35 wt. % sorbitol	0.21	50 % RH	(57)
Arabinoxylan with Xyl/Ara=4.5	0.16	50 % RH	
Amylose with 40 wt. % glycerol	7	50 % RH	(59)
Amylopectin with 40 wt. % glycerol	14	50 % RH	(59)
PVOH	0.21	50 % RH	(57)
EVOH	0.1-12	70 % VOH, 0-95 % RH	(60)
LDPE	1870	50 % RH	(60)

References

1. Karlsson, S. *Adv.Polym. Sci.* **2004,** *169,* pp 201-229.
2. Gross, R. A.; Kalra, B. *Science* **2002,** *297,* pp 803-807.
3. Kaplan, D. L.; Mayer, J. M.; Ball, D.; McCassie, J.; Allen, A. L.; Stenhouse, P. In *Biodegradable Polymers and Packaging*; Thomas, E., Ed.; Technomic Publishing Company, Inc., 1993; pp 1-42.
4. Shogren, R. L.; Fanta, G. F.; Doane, W. M. *Starch/Stärke* **1993,** *45,* pp 276-280.
5. Popa, V. I. In *Polysaccharides in Medicinal Applications*; Dumitriu, S., Ed.; Dekker: New York, NY, 1996; pp 107-124.
6. Meller, A. *Holzforschung* **1965,** *19(4),* pp 118-124.
7. Molin, U.; Teder, A. *Nord. Pulp Pap. Res. J.* **2002,** *17(1),* pp 14-19.
8. Whistler, R. L. In *Industrial Gums*; Academic Press: New York, NY, 1993; pp 295-308.
9. Sugawara, M.; Suzuki, T.; Totsuka, A.; Takeuchi, M.; Ueki, K. *Starch/Stärke* **1994,** *46,* pp 335-337.
10. Okamoto, Y.; Kawashima, M.; Hatada, K. *J. Am. Chem. Soc.* **1984,** *106,* pp 5357-5359.
11. Magerstaedt, M.; Meichsner, C.; Schlingmann, M.; Schrinner, E.; Walch, A.; Wiesner, M.; Winkler, I.; Bader, H.; Paessens, A. DE 3,921,761, 1991

150

12. Heikkila, H.; Alen, R.; Kauko, S.; Lindroos, M.; Nurmi, J.; Sarmala, P.; Tylli, M. US 6,262,318, 2001.
13. Parajó, J. C.; Domínguez, H.; Domínguez, J. M. *Biores. Technol.* **1998**, *65(3)*, pp 191-201.
14. Sjöström, E. *Wood Chemistry: Fundamentals and Applications*, second edition; Academic Press, Inc.: San Diego, CA, 1993.
15. Carpita, N.; McCann, M. In *Biochemistry & Molecular Biology of Plants*; Buchanan, B.; Gruissem, W.; Jones, R., Eds.; American Society of Plant Physiologists: Rockville, MD, 2000; pp 52-107.
16. Gidley, M. J.; Chanliaud, E.; Whitney, S. In *Plant Biopolymer Science: Food and Non-food Applications*; Renard, D.; Della Valle, G.; Popineau, Y., Eds.; Royal Society of Chemistry: Cambridge, UK, 2002; pp 39-47.
17. Gregory, A. C. E.; O'Connell, A. P.; Bolwell, G. P. *Biotechnol. Gen. Eng. Rev.* **1998**, *15*, pp 439-455.
18. Whitney, S. E. C.; Brigham, J. E.; Darke, A. H.; Reid, J. S. G.; Gidley, M. J. *Plant J.* **1995**, *8(4)*, pp 491-504.
19. Gabrielii, I.; Gatenholm, P.; Glasser, W. G.; Jain, R. K.; Kenne, L. *Carbohydr. Polym.* **2000**, *43*, pp 367-374.
20. Chanliaud, E.; Saulnier, L.; Thibault, J.-F. *J. Cereal Sci.* **1995**, *21*, pp 195-203.
21. Timell, T. E. *Wood Sci. Technol.* **1967**, *1*, pp 45-70.
22. Jacobs, A.; Larsson, P. T.; Dahlman, O. *Biomacromolecules* **2001**, *2(3)*, pp 979-990.
23. Batista Chaves Corrêa, J.; Reicher, F.; Odebrecht, S.; Lourdes de Oliveira, M. *Anais da Academia Brasileira de Ciencias* **1979**, *51(3)*, pp 429-433.
24. Khan, A. W.; Lamb, K. A.; Overend, R. P. *Enzyme Microb. Technol.* **1990**, *12*, pp 127-131.
25. Teleman, A.; Lundqvist, J.; Tjerneld, F.; Stålbrand, H.; Dahlman, O. *Carbohydr. Res.* **2000**, *329*, pp 807-815.
26. Gustavsson, M.; Bengtsson, M.; Gatenholm, P.; Glasser, W. G.; Teleman, A.; Dahlman, O. In *Biorelated Polymers: Sustainable Polymer Science and Technology*; Chiellini, H. G. E.; Braunegg, G.; Buchert, J.; Gatenholm, P.; van der Zee, M., Eds.; Kluwer Academic/Plenum Publishers: New York, NY, 2001; pp 41-52.
27. Ishii, T.; Shimizu, K. In *Wood and Cellulosic Chemistry*, second edition; Hon, D. N.-S.; Shiraishi, N., Eds.; Marcel Dekker, Inc.: New York, NY, 2001; pp 175-212.
28. Björkman, A. *Svensk Papperstidning* **1956**, *59*, pp 477-485.
29. Glasser, W. G.; Jain, R. K.; Sjöstedt, M. A. US 5,430,142, 1995.
30. Ebringerová, A.; Heinze, T. *Macromol. Rapid. Commun.* **2000**, *21*, pp 542-556.
31. Annison, G.; Choct, M.; Cheetham, N. W. *Carbohydr. Polym.* **1992**, *19(3)*, pp 151-159.

32. Bataillon, M.; Mathaly, P.; Nunes Cardinali, A.-P.; Duchiron, F. *Ind. Crops Prod.* **1998**, *8*, pp 37-43.

33. Bengtsson, S.; Åman, P. *Carbohydr. Polym.* **1990**, *12(3)*, pp 267-277.

34. Buchanan, C. M.; Buchanan, N. L.; Debenham, J. S.; Shelton, M. C.; Wood, M. D.; Visneski, M. J.; Arumugam, B. K.; Sanders, J. K.; Lingerfelt, L. R.; Blair, L. US 6,352,845, 2002.

35. Buchanan, C. M.; Buchanan, N. L.; Debenham, J. S.; Gatenholm, P.; Jacobsson, M.; Shelton, M. C.; Watterson, T. L.; Wood, M. D. *Carbohydr. Polym.* **2003**, *52*, pp 345-357.

36. Doner, L. W.; Hicks, K. B. *Cereal Chem.* **1997**, *74(2)*, pp 176–181.

37. Ebringerová, A.; Hromádková, Z.; Alföldi, J.; Hříbalová, V. *Carbohydr. Polym.* **1998**, *37*, pp 231-239.

38. Fang, J. M.; Sun, R. C.; Salisbury, D.; Fowler, P.; Tomkinson, J. *Polym. Degr. Stab.* **1999**, *66*, pp 423–432.

39. Fang, J. M.; Sun, R. C.; Tomkinson, J. *Cellulose* **2000**, *7*, pp 87-107.

40. Faurot, A.-L.; Saulnier, L.; Berot, S.; Popineau, Y.; Petit, M.-D.; Rouau, X.; Thibault, J.-F. *Food Sci. Technol. (London)* **1995**, *28(4)*, pp 436-441.

41. Glasser, W. G.; Kaar, W. E.; Jain, R. K.; Sealey, J. E. *Cellulose* **2000**, *7*, pp 299-317.

42. Hromádková, Z.; Ebringerová, A. *Chem. Pap.* **1995**, *49(2)*, pp 97-101.

43. Hromádková, Z.; Kováčiková, J.; Ebringerová, A. *Ind. Crops Prod.* **1999**, *9*, pp 101-109.

44. Izydorczyk, M. S.; Biliaderis, C. G. *Carbohydr. Polym.* **1995**, *28*, pp 33-48.

45. Izydorczyk, M. S.; Macri, L. J.; MacGregor, A. W. *Carbohydr. Polym.* **1998**, *35*, pp 259–269.

46. Methacanon, P.; Chaikumpollert, O.; Thavorniti, P.; Suchiva, K. *Carbohydr. Polym.* **2003**, *54*, pp 335-342.

47. Nilsson, M.; Saulnier, L.; Andersson, R.; Åman, P. *Carbohydr. Polym.* **1996**, *30*, pp 229-237.

48. Oscarsson, M.; Andersson, R.; Salomonsson, A.-C.; Åman, P. *J. Cereal Sci.* **1996**, *24*, pp 161-170.

49. Puls, J.; Poutanen, K.; Körner, H. U.; Viikari, L. *Appl. Microb. Biotechnol.* **1985**, *22(6)*, pp 416–423.

50. Saake, B.; Kruse, T.; Puls, J. *Biores. Technol.* **2001**, *80*, pp 195-204.

51. Sun, R. C.; Lawther, J. M.; Banks, W. B. *Carbohydr. Polym.* **1996**, *29(4)*, pp 325–331.

52. Sun, R. C., Sun, X. F. *Carbohydr. Polym.* **2002**, *49(4)*, pp 415-423.

53. Viëtor, R. J.; Kormelink, F. J. M.; Angelino, S. A. G. F.; Voragen, A. G. J. *Carbohydr. Polym.* **1994**, *24(2)*, pp 113-118.

54. Zhuang, Q.; Vidal, P. F. *Cellulose Chem. Technol.* **1997**, *31* pp 37–49.

55. Höije, A.; Gröndahl, M.; Tømmeraas, K.; Gatenholm, P. *Carbohydr. Polym.* **2005**, *61(3)*, pp 266-275.

56. Salmén, L.; Olsson, A.-M. *J. Pulp Pap. Sci.* **1998,** *24(3),* pp 99-103.
57. Gröndahl, M.; Eriksson, L.; Gatenholm, P. *Biomacromolecules* **2004,** *5(4),* pp 1528-1535.
58. Gröndahl, M; Gatenholm, P. In *Polysaccharides: Structural Diversity and Functional Versatility,* second edition; Dumitriu, S., Ed.; Marcel Dekker, Inc., 2005; pp 509-514.
59. Rindlav-Westling, Å.; Stading, M.; Hermansson, A.-M.; Gatenholm, P. *Carbohydr. Polym.* **1998,** *36,* pp 217-224.
60. McHugh, T. H.; Krochta, J. M. In *Edible Coatings and Films to Improve Food Quality*; Krochta, J. M.; Baldwin, E. A.; Nisperos-Carriedo, M. O., Eds.; Technomic Publishing Company: Lancaster, 1994; pp 139-187.

Chapter 10

Hydrogels from Polysaccharides for Biomedical Applications

Margaretha Söderqvist Lindblad[1,2], John Sjöberg[1], Ann-Christine Albertsson[1], and Jonas Hartman[1]

[1]KTH Fibre and Polymer Technology, Royal Institute of Technology, Teknikringen 56–58, SE–100 44 Stockholm, Sweden
[2]Current address: Södra Cell AB, Research and Development, SE–430 24 Väröbacka, Sweden

Development of products made from renewable sources is considered to be a strategic research area by the international scientific community since it is generally accepted that the fossil fuels will be exhausted in the foreseeable future. Another related strategic area is the development of new bioactive and biocompatible polymers capable of exerting a temporary therapeutic function. Among other substances, polysaccharides have been proposed to be suitable materials as matrixes for the preparation of hydrogels, e.g., for use in key applications such as drug release systems and tissue engineering. Methods reported for preparation of hydrogels using renewable polysaccharides aimed for biomedical use will be reviewed.

The term hydrogel refers to a class of polymeric materials that swell in water but do not dissolve. Hydrogels can be classified as neutral or ionic. They include many materials of plant, animal and synthetic origin. Initial development of synthetic hydrogels designed for biomedical applications was made by Wichterle and Lim who described the potential use of crosslinked poly(2-hydroxyethyl methacrylate) in 1960 (1). They were looking for a material which could satisfy the following requirements:

- A structure permitting the desired water content

- Inertness to normal biological processes

- Permeability for metabolites

They established that such materials must have hydrophobic groups and a three-dimensional structure. From a large number of plastics they found crosslinked poly(2-hydroxyethyl methacrylate) to be most suitable.

Hydrogels are presently under investigation as matrixes for the controlled release of bioactive molecules, in particular pharmaceutical proteins and encapsulation of living cells. These applications often require that the gels degrade under physiological conditions meaning the original three-dimensional structure has to disintegrate into harmless products to ensure good biocompatibility. Hydrogels of natural polymers, especially polysaccharides, have been widely used because of their unique advantages. Polysaccharides are, in general, non-toxic, biocompatible, biodegradable and abundant (2).

Hydrogel preparation from polysaccharides requires chemical or physical crosslinking. Chemical crosslinking is a highly versatile method to create hydrogels with good mechanical stability. However, the crosslinking agents are often toxic compounds which have to be extracted from the gels before they can be applied. Other disadvantages with covalently crosslinked hydrogels are their low degree of swelling and absence of pH-controlled release in basic conditions (3). Further, crosslinking agents can undergo unwanted reactions with the bioactive substance present in the hydrogel matrix. Ionic crosslinking is a simple and mild method and the hydrogels can be used for controlled release in both acidic and basic media and as a thermogelling system. Hydrogels can also be formed by aggregation or complexation, e.g., between a polysaccharide and chitosan (4). This is also a mild method for hydrogel preparation and those hydrogels can be used for pH-controlled drug delivery but dissolution can occur.

Numerous polysaccharides have been used for hydrogel preparation employing different methods. The aim of this review is to give examples of methods reported for preparation of hydrogels using renewable polysaccharides targeted for biomedical use.

Dextran

Dextrans are polysaccharides often used for hydrogel synthesis. They may be defined as glucans containing a backbone of $(1\rightarrow6)$-linked α-D-glucopyranosyl residues *(5)*. $(1\rightarrow3)$-linked chain residues have been found and, in addition, branching to O-2 and O-4 may occur. The percentage of branching residues may vary widely. Dextrans are synthesized by bacteria.

Research on polymerizable dextran was pioneered by Edman, Ekman and Sjöholm *(6)*. They reacted dextran dissolved in water with glycidyl acrylate. A hydrogel was formed but, since the reaction was carried out in water, the degree of substitution was low and difficult to control.

An alternative method is to derivatize dextran by glycidyl methacrylate in dimethyl sulfoxide *(7)*. Almost quantitative incorporation of glycidyl methacrylate was found and the degree of substitution was controlled by varying the molar ratio of glycidyl methacrylate to dextran. Hydrogels were then prepared by radical polymerization in aqueous solution of derivatized dextran and obtained within 20 minutes. Analysis by NMR of the derivatized dextran showed that the reaction of glycidyl methacrylate with dextran was a transesterification resulting in a dextran derivative with the methacrylate group directly attached to the dextran chain *(8)*, see Figure 1.

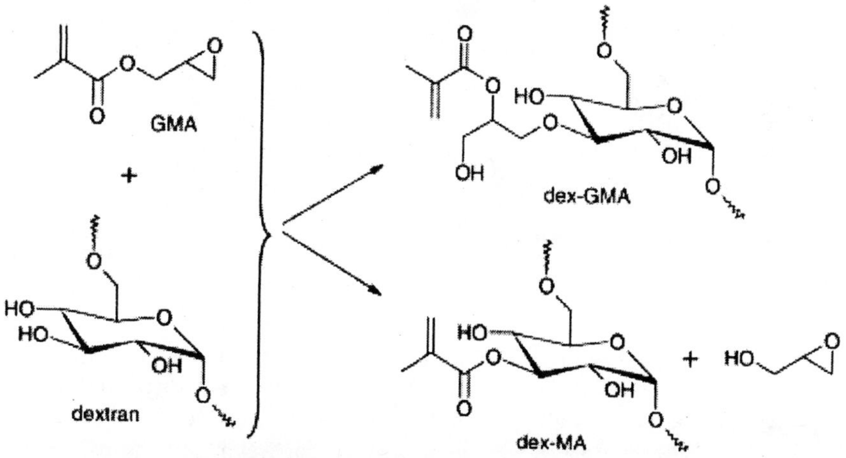

Figure 1. Reactions of dextran with glycidyl methacrylate (9).

The release of model proteins e.g. lysozyme from these hydrogels varying in water content and degree of substitution was found to be dependent on the size of the proteins and the equilibrium water content of the gel *(10)*. However, these gels are resistant toward hydrolysis and in order to accomplish degradation endo-dextranase was incorporated *(11)*. Model protein release was studied and it was found that the release profile was highly dependent on the concentration of dextranase present in the gel. At a relatively high concentration, the release of the model protein started immediately and was completed within 5 to 10 days, while at a low concentration, a delay of up to 150 days was observed. The degradation rate was strongly affected by both concentration of dextranase and the crosslink density. Full degradation only occurred when the degree of methacrylate substitution was less than 4 (methacrylate groups per 100 glucose units). Incorporation of a hydrolyzable group between the polymerizable group and the dextran backbone is another approach to increase degradability. L-lactide has been grafted onto 2-hydroxyethyl methacrylate followed by activation of terminal hydroxyl groups *(12)*. These compounds were coupled to dextran, as shown in Figure 2. The dextran derivatives can be polymerized in aqueous solution to form hydrogels, which hydrolyze under physiological conditions due to hydrolysis of lactate and/or carbonate ester groups in the crosslinks *(13)*. The degradation rate varied from 1 day to 3 months. These dextran-based hydrogels were implanted subcutaneously in rats and no toxic effects on the surrounding tissue were found *(14)*. A method for rheological characterization has also been developed *(15)*.

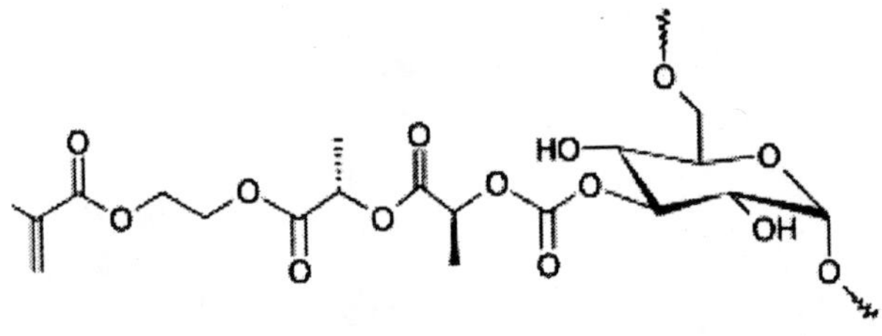

Figure 2. Structure of dextran modified with L-lactide grafted to 2-hydroxyethyl methacrylate (9).

Dextran has also been modified with maleic anhydride and the hydrogel was then made by irradiation with a long wave UV lamp *(16)* which produced a hydrogel with a high swelling ratio (maximum ca 1500%) in water. The magnitude of swelling depended on the pH of the water and the degree of substitution by maleic anhydride.

To obtain a hydrogel with both hydrophobic and hydrophilic components hydrophobic poly(D,L)lactic acid (PDLLA) and hydrophilic dextran were combined by UV photopolymerization *(17)* after unsaturated vinyl groups first were introduced onto the PDLLA and dextran polymer backbones. Swelling ratios between 50 and 850% were obtained depending on the dextran/PDLLA ratio and pH of the swelling solution. Release of bovine serum albumin from dextran-allyl isocyanate/PDLLA diacrylate hydrogels was suggested to be controlled by diffusion through swelling of the hydrophilic phase and degradation of the hydrophobic phase *(18)*. Thermal data indicated that no phase separation was observed in these hydrogels *(19)*. This concept was further developed by synthesizing a thermosensitive and partially biodegradable hydrogel based on dextran-allylisocyanate and poly(*N*-isopropylamide) *(20)*. The glass transition temperature showed an increase, from 138°C to 168°C, with an increase in the content of dextran. Hydrogels have been synthesized with dextran as the main chain and poly(*N*-isopropylacrylamide-*co-N,N*-dimethylacrylamide) as graft chain, also resulting in a thermosensitive hydrogel *(21)*. Poly(propylene glycol)-grafted dextran has been synthesized and inclusion complexes between the poly(propylene glycol) grafts and cyclodextrins have been formed *(22)*. This thermoreversible hydrogel system showed rapid gelation properties and is expected to be useful for biomedical applications, especially injectable drug delivery systems. Multifunctional hydrogels have been synthesized from isoproylacrylamide acting as a thermoresponsive component, poly(L-lactic acid) acting as a hydrolytically degradable and hydrophobic component, and from dextran as an enzymatically degradable and hydrophilic component *(23)*. Macroporous dextran-based hydrogels were obtained from an enzymatically catalyzed transesterification, between dextran and divinyladipate *(24)*. These hydrogels showed a unimodial distribution of interconnected pores and a higher elastic modulus for a given swelling ratio than chemically synthesized dextran-based hydrogels. This may offer advantages in biomedical applications ranging from tissue engineering to controlled drug delivery. Injectable and degradable dextran-based hydrogels have also been prepared by using adipic acid dihydrazide as crosslinker with acceptable mechanical properties *(25)*.

Scleroglucan

Scleroglucan, produced by fungi, consists of linearly linked (1→3)-β-D-glucose residues with side chains of (1→6)-β-D-glucose residues every three glucose units *(26)*. Scleroglucan has general industrial applications because of its water solubility, interesting rheological properties and resistance to hydrolysis. Scleroglucan has been proposed for many different applications in the pharmaceutical field. A polycarboxylated derivative of scleroglucan and 1,6-hexanedibromide has been crosslinked to a hydrogel. Water uptake by the hydrogel with a low degree of crosslinking was largely affected by ionic strength.

Pullulan

Pullulan is a linear polysaccharide consisting of three repeating glucose units joined by $(1\rightarrow6)$-α-D-linkages *(27)*. Pullulan is produced by the fungus *Aureobasidium pullulans*. Pullulan is widely accepted for use in medicine.

Partially methacrylated pullulan has been crosslinked by both a conventional and a "living" radical polymerization technique *(28)*. The reversible addition-fragmentation chain-transfer technique (RAFT) was used for the "living" radical polymerization. It was found that hydrogels prepared by the "living" radical polymerization always swelled to a greater extent than conventional gels having the same double bond conversion. Scanning electron microscopy showed that the porosity of the gels could be effectively regulated with the "living" polymerization.

Alginate

Sodium alginate is a naturally derived linear polysaccharide comprised of β-D-mannuronic acid and α-L-guluronic acid units arranged in blocks *(29)*. Sodium alginate is an attractive material to form hydrogels for biomedical applications since alginate is considered biocompatible and ionic hydrogels form easily in the presence of multivalent cations. These hydrogels generally possess limited mechanical properties. In order to overcome this problem alginate was covalently crosslinked with poly(ethylene glycol). The elastic modulus increased gradually with an increase in crosslinking density.

Macroporous scaffolds for migration of cells throughout the scaffold are another interesting application in the biomedical field. A method to form scaffolds with an interconnected pore structure from alginate has been developed by incorporating gas pockets, stabilizing the gas bubbles with surfactants and then removing the gas *(30)*. Those macroporous alginate beads support cell invasion both in vitro and in vivo.

The effect of alginate hydrogel architecture has been investigated by comparing rat bone marrow cell proliferation and differentiation on calcium crosslinked sodium alginate discs and 1 mm internal diameter tubes *(31)*. It was found that the thickness of the alginate gel is an important parameter in determining cell behavior. Poly(aldehyde guluronate) was prepared from α-L-guluronate, isolated from sodium alginate, and hydrogels were then formed with poly(acrylamide-*co*-hydrazide) as multifunctional crosslinker or adipic acid dihydrazide as a bifunctional crosslinker *(32)*. It was found that degradation and mechanical stiffness could be controlled with the multifunctional crosslinker.

Starch

Starch is composed of two polysaccharides with (1→4)-α-linkages, the essentially linear amylose and the extensively (1→6) branched amylopectin. Starch normally contains 20-30% amylose and 70-80% amylopectin *(33)*. A biodegradable hydrogel based on corn starch/cellulose acetate blends, produced by free-radical polymerization with methyl methacrylate monomer and/or acrylic acid monomer has been developed *(34)*. The materials were sensitive to pH showing a clear reversible transition in a relatively narrow interval which is in the range of physiological conditions. The diffusion mechanism was predominantly of Fickian-type. The results obtained in the mechanical tests were in the range of those reported for typical poly(methyl methacrylate) bone cements, showing that it is possible to develop partially degradable cements with an adequate mechanical behavior. The same group has also developed a starch-based hydrogel, with potential application as drug delivery systems, by free radical polymerization of acrylamide and acrylic acid, and in some formulations also bis-acrylamide, in the presence of a corn starch/ethylene-*co*-vinyl alcohol copolymer blend *(35)*.

Cellulose Ethers

The hydroxyalkyl cellulose ethers are obtained in the reactions of cellulose, e.g., dissolving pulp with alkene oxides or their corresponding chlorohydrins. Sodium carboxymethylcellulose is prepared from alkali cellulose with sodium chloroacetate as reagent *(36)*.

Hydroxypropylcellulose is one of the most important derivatives of cellulose due to its considerable hydrophilicity, complex phase behavior and ease of production. Aqueous hydroxypropylcellulose solutions exhibit lower critical solution behavior while hydrogels produced from hydroxypropylcellulose, (by crosslinking with divinyl sulfone and sodium hydroxide as the catalyst), are thermoresponsive, swell at low temperature and contract at high temperature *(37)*. Consequently, nonporous and microporous hydroxypropylcellulose hydrogels were prepared depending on the conditions for the crosslinking reaction.

Super absorbent cellulose derivatives, with the potential to adsorb large amounts of body water used for treatment of edemas, have been prepared from carboxymethylcellulose sodium salt and chemically crosslinked hydroxyethyl-cellulose *(38,40)*. The results obtained from the biocompatibility studies were consistent with the hypothesis that this gel may represent an alternative to diuretic therapies.

Glucomannan from Spruce

Galactoglucomannans have a backbone built of (1→4)-linked β-**D**-glucopyranose and β-**D**-mannopyranose units. The α-**D**-galactopyranose residue is linked as a single-unit side chain to the backbone by (1→6)-bonds *(39)*. The mannose and glucose units are partially substituted by acetyl groups, on the average one group per 3-4 hexose units.

Hemicellulose/poly(2-hydroxyethyl methacrylate)-based hydrogels were prepared by the radical polymerization of 2-hydroxyethyl methacrylate with hemicellulose purposely modified with well-defined amounts of methacrylic functions, as shown in Figures 3 and 4 *(40)*.

Oligomeric hydrosoluble hemicelluloses, consisting of 80% galacto-glucomannan and the remainder mainly xylan, produced from spruce wood chips in a steam explosion process, were used for the hydrogel preparation.

In a rheological characterization it was shown that those hemicellulose-based hydrogels had similar behavior as hydrogels prepared from pure poly(2-hydroxyethyl methacrylate) *(41)*.

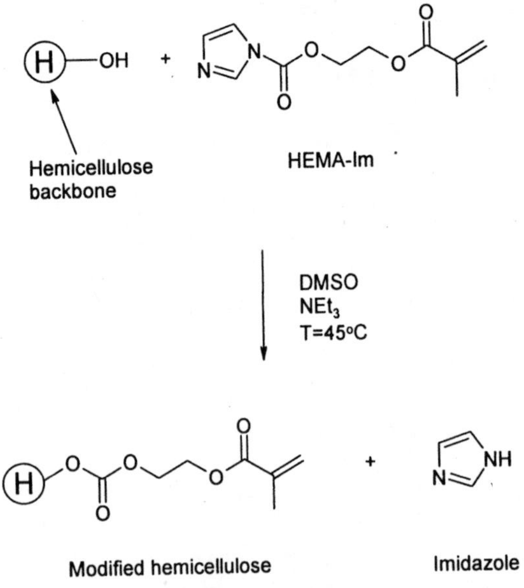

Figure 3. Synthetic pathway for modification of hemicellulose with HEMA-Im (2-[(1-Imidazolyl)formyloxy]ethyl methacrylate).

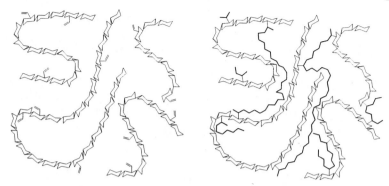

Figure 4. Schematic representation of methacrylolated hemicellulose (left) and a hemicellulose-based hydrogel (right). The chair conformation denotes an anhydroglucose unit in the hemicellulose. Polymerizable methacrylate groups are denoted with a double bond (left). Polymerized methacrylate groups and comonomers are denoted with bold lines (right). (Reproduced with permission from reference 40. Copyright 2005.)

Konjac Glucomannan

Konjac glucomannan is a high-molecular weight water-soluble nonionic random copolymer of (1→4)-β-linked D-mannose and D-glucose with a low degree of acetylation, which is the main constituent of the plant bulb of *Amorphophallus konjac*. Hydrogels, designed for colon-targeting drug delivery, have been prepared from konjac glucomannan copolymerized with acrylic acid and crosslinked by the aromatic agent bis(methacryloylamino)-azobenzene *(42)*. The swelling ratio was inversely proportional to the degree of acrylic acid grafting and the content of bis(methacryloylamino)-azobenzene. Konjac glucomannan has also been copolymerized with acrylic acid and crosslinked by *N,N*-methylene-bis-(acrylamide) *(43)*.

Guar Gum

Guar gum, from the seed of the guar plant *Cyamopsis tetragonolobus*, is a linear (1→4)-β-D-mannan backbone with (1→6)-α-D-galactose side chain on approximately every other mannose unit *(44)*. The molecular weight has been estimated to be about 220,000 *(45)*. Poly(vinyl alcohol)-guar gum interpenetrating network microspheres were prepared by linking with glutaraldehyde *(46)*. An *in vitro* release study of the antihypertesive drug Nifedipine, indicated that the release was dependent upon the extent of crosslinking, amount of loaded drug and method of drug loading.

Xylan

The major component in xylan is O-acetyl-4-O-methylglucurono-β-D-xylan *(47)*. Xylan, isolated from birchwood, was converted into hydrogels by dissolving xylan together with chitosan in acidic conditions *(48)*. Complexation between glucuronic acid functionalities of xylan and amino groups of chitosan was suggested to be responsible for network formation. A sponge-like microporous structure was formed when the hydrogel was freeze-dried.

Chitosan

Chitosan is a biocompatible polymer primarily derived from chitin, a natural polymer commercially extracted from crab shells through alkaline deacetylation. Chitosan is partly acetylated or nonacetylated chitin. Chitin is a linear 4-linked 2-acetamido-2-deoxy-β-D-glucopyranan of regular structure *(49)*. Being an analog of cellulose, chitin resembles cellulose in adopting highly ordered chain conformations.

Semi-interpenetrating polymer network hydrogels composed of β-chitosan and poly(ethylene glycol) diacrylate macromer were synthesized, for potential biomedical applications *(50)*. Those photocrosslinked hydrogels exhibited relatively high equilibrium water contents, ca 80%, which were mainly attributed to the free water content rather than to bound water.

A way to prepare polysaccharide-based pH-sensitive hydrogels with hydrophobic polyester side chains has been developed. The amino groups in chitosan were grafted with D,L-lactic acid without any catalyst *(51)*. The structure is shown in Figure 5. The crystallinity of chitosan gradually decreased after grafting because the amino groups of chitosan were randomly substituted which destroyed the regularity of packing between the chitosan chains. The polyester substituents provided the basis for hydrophobic interaction which contributed to the formation of hydrogels *(52)*. Reversible water uptake of the materials was observed when pH was switched between 2.2 (simulated gastric fluid) and 7.4 (simulated intestinal fluid). The lowest pH provided the highest swelling due to charge repulsion by protonation of the free amine groups on the polymer backbone.

Rheological properties were examined in hydrogels formed from a mixture of chitosan, β-glycerol phosphate and hydroxyethylcellulose *(53)*. The hydrogels were investigated as potential drug vehicles by studying the release of pilocarpine hydrochloride. An interpenetrating polymer network hydrogel composed of poly(vinyl alcohol) and chitosan exhibited electric-sensitive behavior *(54)*. When a swollen hydrogel was placed between a pair of electrodes, it exhibited bending behavior in response to the applied electric

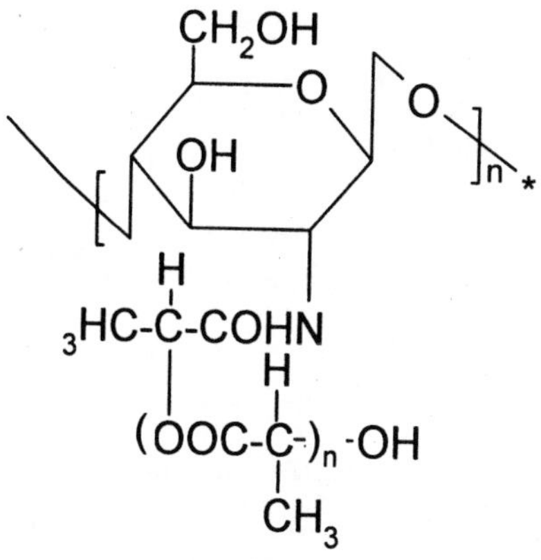

$$CH_2OH$$

Figure 5. Chitosan grafted with D,L-lactic acid.

field. A polyionic hydrogel has been obtained by complexation of chitosan and xanthan *(55)*. The swelling degree of the hydrogel was found to be influenced by the time of coacervation, the pH of the solution of chitosan used to form the hydrogel, the pH of the solution used for the swelling experiment, the molecular weight and the degree of acetylation of the chitosan. A homogeneously deacetylated chitosan with a deacetylation degree between 35% and 50% was required to prepare transparent chitosan/β-glycerophosphate pseudo-thermosetting hydrogels *(56)*. Poly(ethylene glycol) has been grafted to chitosan in such a formulation that its solution undergoes a thermally reversed phase transition from an injectable solution at a low temperature to gel at body temperature *(57)*. This hydrogel is potentially suitable for a wide range of biomedical applications. The hydrogel of chitosan and glycerophosphate was studied with the aim to examine its morphology *(58)*. Laser scanning confocal microscopy indicated that the gel is quite heterogeneous, which is important to know for the intended application.

Hyaluronan

Hyaluronan consists of repeating disaccharide units of D-glucuronic acid and (1→3)-β-*N*-acetyl-D-glucosamine *(59)*. It is distributed throughout the extracellular matrix, connective tissues and organs of all higher animals. Hyaluronan has been crosslinked by activating the carboxylate moieties by 2-chloro-1-methylpyridinium iodide and using 1,3-diaminopropane as a chemical bridge. Morphology of the gels was studied and the pore size was determined in order to find a relationship with the swelling properties. Hyaluronan microspheres have been prepared using adipic dihydrazide mediated crosslinking chemistry to avoid the use of organic solvents *(60)*. The results showed that release of encapsulated plasmid DNA can be sustained for months and is capable of transfection *in vitro* and *in vivo*. Rheological properties of thiol-functionalized hyaluronan crosslinked with poly(ethylene glycol) diacrylate were investigated *(61)*. Formation of a stable network depended strictly on the concentration of poly(ethylene glycol) diacrylate.

Conclusions

This review shows the versatility of the reactions of polysaccharides for potential biomedical uses. The solubility and degradability of polysaccharides can easily be modified by esterification, etherification and crosslinking reactions. However, there is little discussion in the literature about selecting the most suitable polysaccharide, from the structural point of view, for a specific reaction and/or application. The challenge for increased use of polysaccharides and other renewable materials in medicine is to a large extent within purification,. The potential for using polysaccharides from plants, instead of polysaccharides produced by bacteria, is remarkable, since this avenue offers more effective purification methods.

References

1. Wichterle, O.; Lim, D. *Nature* **1960**, *185*, p 117.
2. Chen, J.; Jo, S.; Park, K. *Carbohydr. Polym.* **1995**, *28*, p 69.
3. Berger, J.; Reist, M.; Mayer, J.M.; Felt, O.; Peppas, N.A.; Gurny, R. *Eur. J. Pharm. Biopharm.* **2004**, *57*, p 19.
4. Berger, J.; Reist, M.; Mayer, J.M.; Felt, O.; Gurny, R. *Eur. J. Pharm. Biopharm.* **2004**, *57*, p 35.
5. Kenne, L.; Lindberg, B. In *Molecular Biology, Volume 2*; G.O. Aspinall, ed.; Academic Press: New York, **1983**, p 346.
6. Edman, P.; Ekman, B.; Sjöholm I. *J. Pharm. Sci.* **1980**, *69*, p 838.
7. van Dijk-Wolthuis, W.N.E.; Franssen, O.; Talsma, H.; van Steenbergen, M.J.; Kettenes-van den Bosch, J.J.; Hennink, W.E. *Macromolecules,* **1995**, *28*, p 6317.
8. van Dijk-Wolthuis, W.N.E.; Kettenes-van den Bosch, J.J.; van der Kerk-van Hoof, A.; Hennink, W.E. *Macromolecules* **1997**, *30*, p 3411.
9. Hennink, W.E.; van Nostrum, C.F. *Adv. Drug Delivery Rev.* **2002**, *54*, p 13.
10. Hennink, W:E.; Talsma, H.; Borchert, J.C.H.; De Smedt, S.C.; Demeester, J. *J. Controlled Release* **1996**, *39*, p 47.

11. Franssen, O.; Vos, O.P.; Hennink, W.E. *J. Controlled Release* **1997**, *44*, p 237.
12. van Dijk-Wolthuis, W.N.E.; Tsang, S.K.Y.; Kettenes-van den Bosch, J.J.; Hennink, W.E. *Polymer* **1997**, *38*, p 6235.
13. van Dijk-Wolthuis, W.N.E.; Hoogeboom, J.A.M.; van Steenbergen, M.J.; Tsang, S.K.Y.; Hennink, W.E. *Macromolecules* **1997**, *30*, p 4639.
14. Cadée, J.A.; van Luyn, M.J.A.; Brouwer, L.A.; Plantinga, J.A.; van Wachem, P.B.; de Groot, C.J.; den Otter, W.; Hennink, W.E., *J. Biomed. Mater. Res.* **2000**, *50*, p 397.
15. Meyvis, T.K.L.; De Smedt, S.C.; Demeester, J.; Hennink, W.E. *J .Rheol.* **1999**, *43*, p 933.
16. Kim. S.-H.; Won, C.-Y.; Chu, C.-C. *J. Biomed. Mater. Res.* **1999**, *46*, p 160.
17. Zhang, Y.; Won, C.-Y.; Chu, C.-C. *J. Polym. Sci., Part A: Polym. Chem.* **1999**, *37*, p 4554.
18. Zhang, Y.; Chu, C.-C. *J. Biomed. Mater. Res.* **2001**, *54*, p 1.
19. Zhang, Y.; Chu, C.-C. *J. Mater. Sci.: Mater. Med.* **2002**, *13*, p 773.
20. Zhang, X.-Z.; Sun, G.-M.; Wu, D.-Q.; Chu, C.-C. *J. Mater. Sci.: Mater. Med.* **2004**, *15*, p 865.
21. Huh, K.M.; Hashi, J.; Ooya, T.; Yui, N. *Macromol. Chem. Phys.* **2000**, *201*, p 613.
22. Choi, H. S.; Kontani, K.; Huh, K.M.; Sasaki, S.; Ooya, T.; Lee, W.K.; Yui, N. *Macromol. Biosci.* **2002**, *2*, p 298.
23. Huang, X.; Nayak, B.R.; Lowe, T.L. *J. Polym. Sci., Part A: Polym. Chem.* **2004**, *42*, p 5054.
24. Ferreira, L.; Gil, M.H.; Cabrita, A.M.S.; Dordick, J.S. *Biomaterials* **2005**, *26*, p 4707.
25. Maia, J.; Ferreira, L.; Carvalho, R.; Ramos, M.A.; Gil, M.H. *Polymer* **2005**, *46*, p 9604.
26. Coviello, T.; Grassi, M; Rambone, G.; Alhaique, F. *Biomaterials* **2001**, *22*, p 1899.
27. Spiridon, I.; Popa V.I. In *Polysaccharides: Structural Diversity and Functional Versatility, 2nd ed.*; S. Dumitriu, ed.; Marcel Dekker: New York, **2005**, pp 479-480.
28. Crescenzi, V.; Dentini, M.; Bontempo, D.; Masci, G. *Macromol. Chem. Phys.* **2002**, *203*, p 1285.
29. Eiselt, P.; Lee, K.Y.; Mooney, D.J. *Macromolecules* **1999**, *32*, p 5561.
30. Eiselt, P.; Yeh, J.; Latvala, R.K.; Shea, L.D.; Mooney, D.J. *Biomaterials* **2000**, *21*, p 1921.
31. Barralet, J.E.; Wang, L.; Lawson, M.; Triffitt, J.T.; Cooper, P.R.; Shelton, R.M. *J. Mater. Sci.: Mater. Med.* **2005**, *16*, p 515.
32. Lee, K.Y.; Bouhadir, K.H.; Mooney, D.J. Biomaterials **2004**, *25*, p 2461.
33. Rindlav Westling, Å. *Crystallinity and Morphology of Starch Polymers in Films.* PhD thesis, Chalmers University of Technology, 2002.

166

34. Pereira, C.S.; Cunha, A.M.; Reis, R.L.; Vázquez, B.; San Román, J. *J. Mater. Sci.: Mater. Med.* **1998**, *9*, p 825.
35. Elvira, C.; Mano, J.F.; San Román, J.; Reis, R.L. *Biomaterials* **2002**, *23*, p 1955.
36. Sjöström. E. *Wood Chemistry – Fundamentals and Applications;* Academic Press: New York, 1981, pp 180-181.
37. Hirsch, S.G.; Spontak, R.J. *Polymer* **2002**, *43*, p 123.
38. Sannino, A.; Esposito, A.; De Rosa, A.; Cozzolino, A.; Ambrosio, L.; Nicolais, L. *J. Biomed. Mater. Res., Part A* **2003**, *67A*, p 1016.
39. Sjöström. E. *Wood Chemistry – Fundamentals and Applications;* Academic Press: New York, 1981, pp 60-61.
40. Söderqvist Lindblad, M.; Ranucci, E.; Albertsson, A.-C. *Macromol. Rapid Commun.* **2001**, *22*, p 962.
41. Söderqvist Lindblad, M.; Ranucci, E.; Albertsson, A.-C.; Laus, M.; Giani, E. *Biomacromolecules* **2005**, *6*, p 684.
42. Liu, Z.-L.; Hu, H.; Zhuo, R.-X. *J. Polym. Sci., Part A: Polym. Chem.* **2004**, *42*, p 4370.
43. Chen, L.-G.; Liu, Z.-L.; Zhuo, R.-X. *Polymer*, **2005**, *46*, p 6274.
44. Spiridon, I.; Popa V.I. In *Polysaccharides: Structural Diversity and Functional Versatility, 2nd ed.*; S. Dumitriu, ed.; Marcel Dekker: New York, **2005**, pp 479-481.
45. Sandford, P.A.; Baird, J. in *Molecular Biology, Volume 2*; G.O. Aspinall, Ed.; Academic Press: New York, **1983**, p. 462.
46. Soppimath, K.S.; Kulkarni, A.R.; Aminabhavi, T.M. *Biomater. Sci. Polymer Edn.* **2000**, *11*, p 27.
47. Sjöström. E. *Wood Chemistry – Fundamentals and Applications;* Academic Press: New York, 1981, p 62.
48. Gabrielii, I.; Gatenholm, P. *J. Appl. Polym. Sci.,* **1998**, *69*, p 1661.
49. Gorin, P. A. J.; Barreto-Bergter, E. *Molecular Biology, Volume 2;* G.O. Aspinall, Ed. ; Academic Press: New York, **1983**, p 386.
50. Lee, Y.M.; Kim, S.S.; Kim, S.H. *J. Mater. Sci.: Mater. Med.* **1997**, *8*, p 537.
51. Qu, X.; Wirsén, A.; Albertsson A.-C. *J. Appl. Polym. Sci.* **1999**, *74*, p 3193.
52. Qu, X.; Wirsén, A.; Albertsson A.-C. *J. Appl. Polym. Sci.* **1999**, *74*, p 3186.
53. Li, J.; Xu, Z. *J. Pharm. Sci.* **2002**, *91*, p 1669.
54. Kim, S.J.; Park, S.J.; Kim, I.Y.; Shin, M.-S.; Kim, S.I. *J. Appl. Polym. Sci.* **2002**, *86*, p 2285.
55. Magnin, D.; Lefebvre, J.; Chornet, E.; Dumitriu, S. *Carbohydr. Polym.* **2004**, *55*, p 437.
56. Berger, J.; Reist, M.; Chenite, A.; Felt-Baeyens, O.; Mayer, J.M.; Gurny, R. *Int. J. Pharm.* **2005**, *288*, p 17.
57. Bhattarai, N.; Matsen, F.A.; Zhang, M. *Macromol. Biosci.* **2005**, *5*, p 107.

58. Crompton, K.E.; Prankerd, R.J.; Paganin, D.M.; Scott, T.F.; Horne, M.K.; Finkelstein, D.I.; Gross, K.A.; Forsythe, J.S. *Biophysical Chemistry* **2005**, *117*, p 47.
59. Magnani, A.; Rappuoli, R.; Lamponi, S.; Barbucci, R. *Polym. Adv. Technol.* **2000**, *11*, p 488.
60. Yun. Y.H.; Goetz, D.J.; Yellen, P.; Chen, W. *Biomaterials* **2004**, *25*, p 147.
61. Ghosh, K.; Shu, X.Z.; Mou, R.; Lombardi, J.; Prestwich, G.D.; Rafailovich, M.H.; Clark, R.A.F. *Biomacromolecules* **2005**, *6*, p 2857.

Chapter 11

Room Temperature Ionic Liquids as Lubricants for Wood–Polyethylene Composites

Kaichang Li and Katie Zhou

Department of Wood Science and Engineering, Oregon State University, Corvallis, OR 97331

Room temperature ionic liquids (RTILs) may be used as green solvents and catalysts for various chemical reactions, as anti-electrostatic agents for wood, as electrolytes for batteries, and as extraction agents for reduction of hazardous pollutants because of their unique properties such as negligible vapor pressure, high thermal stability, and high electric conductivity. In this study, we demonstrated that RTILs were excellent lubricants for making wood-plastic composites (WPCs). A lubricant is essential for making acceptable WPCs and a compatibilizer is typically added to improve their mechanical properties. However, a lubricant and a compatibilizer normally have opposite effects on the strength and stiffness of WPCs. This study revealed that RTILs had better lubrication effect than Struktol, a commonly used commercial lubricant for making WPCs. When used with MAPE, some RTILs reduced the strength-enhancing effect of MAPE (maleic anhydride-grafted polyethylene), one of the most effective compatibilizers, less than Struktol. Some RTILs were also superior to Struktol in terms of increasing or retaining the stiffness of the resulting wood-PE composites. Among five RTILs studied, RTIL **F** (triisobutylmethylphosphonium tosylate) was the best RTIL in terms of providing the lubrication effect and maximizing retention of the strength-enhancing effects of MAPE.

Introduction

Room temperature ionic liquids (RTILs) are organic salts that are liquids at room temperature or slightly above *(1)*. As salts, RTILs are electrically conductive. Because of strong ionic bonding, RTILs also have no measurable vapor pressure even at very high temperature. Most RTILs have very high thermal stability, and some RTILs are stable even above 350 °C. Most RTILs are inert to various chemicals. Therefore, RTILs may be used as green solvents or reaction media for various chemical transformations, as solvents for dissolution of cellulose, as electrolytes for various applications such as batteries, as extraction agents for reduction of hazardous pollutants, and as anti-electrostatic agents for wood *(2-12)*. Many other new applications of RTILs are being discovered. In this study, we demonstrated that phosphonium-based RTILs were superior lubricants for the production of WPCs.

WPCs consist of wood in the form of wood flour or wood fibers, a thermoplastic, and some additives for improving the product properties or assisting the production process *(13-15)*. Commonly used thermoplastics include polypropylene, polyethylene, polyvinylchloride, and polystyrene *(13-15)*. Wood is lighter, less abrasive, and less expensive than commonly used inorganic plastic fillers such as glass fibers. Wood is hydrophilic and susceptible to biodegradation, whereas thermoplastics are hydrophobic and resistant to biodegradation. WPCs as a mixture of two materials with opposite properties not only have the good properties of wood and thermoplastics, but also have less undesirable properties of wood and thermoplastics. For example, wood is a reinforcing filler for thermoplastics, and WPCs are thus stronger than thermoplastics. Lignin and polyphenolics in wood are antioxidants and are able to quench free radicals, thus alleviating the aging and photodegradation accompanied by use of thermoplastics. Wood as a hydrophilic material can absorb water and then swell, which are undesirable properties for some applications. Wood as a natural material is also susceptible to biodegredation. In WPCs, wood is surrounded by thermoplastics which greatly reduces the water uptake, swelling and biodegradation of wood. Since 1995, WPCs have been one of the fastest growing sectors in wood composites industry *(16)*. At present, WPCs are widely used as decking materials, fencing materials, interior door panels, window moldings, automobile interior parts, and many other molded products *(16)*.

WPCs are typically produced through an extrusion process, i.e., wood, a thermoplastic, and additives are thoroughly mixed at high temperature in an extruder and then are extruded into products with different shapes *(13)*. A compatibilizer is typically added to improve the interfacial adhesion between hydrophilic wood and hydrophobic thermoplastic *(13,17)*. Without a compatibilizer, WPCs typically have low strength and low stiffness, and cannot meet property requirements of many applications *(13,17)*. At present, maleic anhydride-grafted polyethylene (MAPE) is one of the most effective compatibilizers for wood-PE (polyethylene) composites.

A lubricant is an essential process additive for the production of WPCs. Without a lubricant, wood flour cannot be uniformly distributed in thermoplastic matrix, and more importantly, the surface of extruded WPC products is unacceptably rough. Struktol is one of the most commonly used lubricants. It is believed that one of the key components in Struktol is zinc stearate, although the exact composition of Struktol is proprietary. A lubricant and a compatibilizer typically have opposite effect, i.e., a lubricant typically reduces the strength-enhancing effect of a compatibilizer *(18)*. In this study, we found that RTILs had an excellent lubrication effect, but reduced the strength-enhancing effect of MAPE much less than Struktol in making wood-PE composites.

Materials and Methods

Materials

The following RTILs were provided by Cytec Industries, Inc. (West Paterson, NJ): **A)** tributyltetradecylphosphonium dodecylbenzenesulfonate;

B) trihexyltetradecylphosphonium dodecylbenzenesulfonate;

C) trihexyltetradecylphosphonium methanesulfonate;

D) trihexyltetradecylphosphoniumbis-2,4,4-(trimethypentyl)phosphinate;

E) trihexyltetradecylphosphonium dicyanamide;

F) triisobutylmethylphosphonium tosylate.

Pine flour (40 mesh and 2.04% moisture content) and MAPE (maleic anhydride-grafted polyethylene, A-C OptiPak$^{(TM)}$ 200) were provided by American Wood Fibers (Schofield, WI) and Honeywell International, Inc. (Morristown, NJ), respectively. Pine flour was re-dried at 104 °C for 20 hours prior to use, and the resulting pine flour had the moisture content of 0.50%. HDPE (high density polyethylene) (melt flow index: 0.55 g/10 min) and Struktol were provided by BP Solvay Polyethylene North America (Houston, TX) and Louisiana Pacific Corporation (Nashville, TN), respectively.

Compositions of Wood-PE Composites

The weight ratios of different components in wood-PE composites are as follows: wood/PE/MAPE/Struktol/RTIL=40/P/M/S/R. Where P+M+S+R=60. For the control, i.e., wood-PE composites without any additive, P=60 and M=S=R=0. For wood-PE composites containing Struktol as a sole additive,

P+S=60 and M=R=0. For wood-PE composites containing both Struktol and MAPE, P+S+M=60 and R=0. For wood-PE composites containing a RTIL as a sole additive, P+R=60 and M=S=0. For wood-PE composites containing both MAPE and a RTIL, P+M+R=60 and S=0. In this study, the usage of MAPE, if used, was 2 wt% based on the total weight of the wood-PE composites, i.e., M=2 in the previous weight ratio formula. For Struktol and RTILs, the following usages (the weight percent based on the total weight of the wood-PE composites) were studied respectively: 0.5 wt%, 1.0 wt%, 2.0 wt% and 3.0 wt%, i.e., S=0.5, 1.0, 2.0, and 3.0 or R=0.5, 1.0, 2.0, and 3.0.

Preparation of Wood-PE blends and Measurement of the Lubrication Effect of RTILs

Re-dried wood flour, PE, and additives were mixed in a Brabender Plasticorder with a mixing bowl (60 mL) and two roller blades (C. W. Brabender Instruments, Inc., South Hackensack, NJ). The total weight of wood, PE and additives for each mixture was 44.00 g where the weight of wood flour was 17.60 g. Here is an example blending procedure for making wood-PE composites containing 2.0 wt% MAPE and 3.0 wt% RTIL A (the weight ratio wood/PE/MAPE/Struktol/RTIL = 40/55/2.0/0.0/3.0). Pine flour (17.60 g), HDPE (24.20 g), MAPE (0.88 g), and RTIL A (1.32 g) were mixed in a 400 mL beaker by mechanical stirring for 2 minutes. The bowl of the Brabender Plasticorder was preheated to 170 °C and the speed of the rotors in the bowl was set to 100 rpm. The mixture of pine flour, HDPE, MAPE, and RTIL A was added to the bowl and mixed for 15 minutes. The torque value was taken at the mixing time of 9.0 minutes. The blend was then removed from the Brabender bowl, cut into small pieces with chisel while the blend was still hot, and stored for compression molding.

Preparation of wood-PE composite board

The wood-PE blend was added to a stainless steel mold with dimensions of 101.6 x 101.6 x 2 mm, and the blend-filled mold was placed on the lower platen of an automatic benchtop Carver press (Carver, Inc., Wabash, IN) where the platens were preheated to 185 °C. The mold was maintained on the lower platen for 10 minutes before the press was closed slowly, allowing the wood-PE blend to flow into the mold shape. The press pressure was increased from 34.5 to 344.8 kPa over two minutes and was at 344.8 kPa for an additional 10 minutes. The mold was then removed from the hot press and cooled in a cold press at 344.8 kPa under ambient conditions. The resulting wood-PE board was cut into 10 small test specimens. Each specimen was 54.50-56.50 mm long, 13.00-14.30 mm wide, and 2.10-2.40 mm thick.

Evaluation of strength properties of the wood-PE composites

Strength properties of wood-PE composites were evaluated in accordance with ASTM D790-02. Ten specimens were tested for each wood-PE composite. Three-point bending test with support span of 43 mm and the crosshead speed of 1.0 mm/min was performed on a Sintech machine (MTS Systems Corporation, Enumclaw, WA). The load-deflection curve was recorded and used for calculation of the modulus of rupture (MOR) and modulus of elasticity (MOE). The maximum force at the first point on the load-deflection curve to show a slope of zero was used to calculate the MOR. The MOE was determined from the slope in the initial elastic region of the load-deflection curve.

Results and Discussion

It has been demonstrated that alkylimidazolium tetrafluoroborates can serve as promising lubricants for reducing friction between metals and between metals and ceramics *(19)*. However, these fluoride-containing imidazolium-based RTILs are not stable at high temperature under aerobic conditions and some hazardous or corrosive chemicals such hydrogen fluoride may be released from these RTILs *(20)*. Since WPCs are made at high temperature, halogen-containing RTILs and imidazolium-based RTILs might not be used for making WPCs. In this study, we chose halogen-free phosphonium-based RTILs, and these RTILs are not corrosive and are much more stable at high temperature than imidazolium-based RTILs *(21)*.

A torque value during the mixing of wood-PE blend is a very good indicator for the effectiveness of a lubricant. The lower the torque value the better the lubrication effect. A typical torque vs. mixing time curve is shown in Figure 1. As the PE melted, the wood-PE mixture became viscous and sticky, thus requiring higher torque for the blades to compound the mixture. This was why the torque value rapidly increased at the beginning of the mixing. As a lubricant took effect and a more uniform wood-PE blend is formed, the torque value then rapidly decreased and began to flatten out at the mixing time of around five minutes. Therefore, the torque value at the mixing time of nine min. was used to evaluate the lubrication effect of each RTIL.

The effect of the usage of a RTIL or Struktol on the torque value at the mixing time of nine min. is shown in Figure 2. At 0.5 wt% and 1.0 wt%, Struktol had basically the same torque values as the control, i.e., having little lubrication effect. At 2.0 wt%, Struktol had significantly lower torque values than the control. The torque value rapidly decreased when the usage of Struktol was increased from 2.0 wt% to 3.0 wt%. Three percent is a commonly used usage for Struktol in commercial production of WPCs. For RTIL **A**, the torque remained statistically the same when the usage was increased from 0.5 wt% to

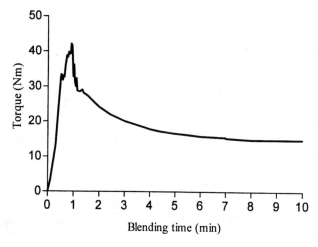

Figure 1. A typical Torque vs. mixing time curve in blending wood and PE

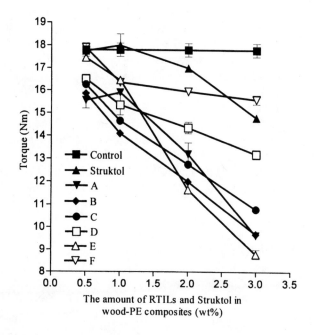

Figure 2. Effect of RTILs and Struktol on torque. (Control: wood-PE without any additive. Error bars show two standard errors of the means from two independent experiments)

1.0 wt%, and then decreased almost linearly when the usage was further increased from 1.0 wt% to 3.0 wt%. For RTILs **B, C, D, E,** and **F**, the torque value for each RTIL significantly decreased when the usage of each RTIL was increased from 0.5 wt% to 3 wt%. At 0.5 wt%, RTILs **A, B, C,** and **D** resulted in significantly lower torque values than Struktol, whereas RTILs **E** and **F** were as ineffective as Struktol in lowering the torque value (Figure 2). All RTILs at 1.0 wt% were more effective in lowering the torque value than Struktol at 1.0 wt % and even at 2.0 wt%. At 2.0 wt%, all RTILs resulted in lower torque values than Struktol. At 3.0 wt%, all RTILs except RTIL **F** all had lower torque values than Struktol (Figure 2). RTIL **D** at 1.0 wt% was as effective as Struktol at 3.0 wt%, and RTILs **B** and **C** at 1.0 wt% were even more effective than Struktol at 3.0 wt% in terms of lowering the torque values (Figure 2). All these results indicated that almost all RTILs were more effective lubricants than Struktol in the same usage. Some RTILs were three times more effective than Struktol.

The effect of RTILs and Struktol on the MOR of wood-PE composites is shown in Figure 3. At 0.5 wt%, Struktol had little effect on the MOR. However, increase in the usage from 0.5 wt% to 3.0 wt% resulted in linear reduction of the MOR (Figure 3). For RTIL **A**, the MOR remained statistically unchanged when the usage was increased from 0.5 wt% to 2.0 wt%, but significantly decreased when the usage was further increased from 2.0 wt% to 3.0 wt% (Figure 3). For RTILs **B** and **D**, the MOR decreased when the usage was increased from 0.5 wt% to 2.0 wt%, and then remained statistically unchanged when the usage was further increased from 2.0 wt% to 3.0 wt%. For RTIL **C**, the MOR significantly decreased when the usage was increased from 0.5 wt% to 1.0 wt%, and then flattened out when the usage was further increased from 1.0 wt% to 3.0 wt%. For RTIL **E**, the MOR rapidly decreased when the usage was increased from 0.5 wt% to 3.0 wt%. Interestingly, RTIL **F** at 0.5 wt% resulted in significantly higher MOR than the control. For RTIL **F** at 1.0-3.0 wt%, the MOR was statistically the same as that of the control (Figure 3). Except RTIL **F**, all RTILs resulted in much lower MOR than Struktol at the same usage.

The effect of RTILs and Struktol on the MOE of wood-PE composites is shown in Figure 4. Interestingly, RTIL **F** had higher MOE than the control at all usages studied. At 0.5 wt%, Struktol and RTIL **E** had higher MOE than the control; RTIL **D** had the same MOE as the control; RTILs **A, B,** and **C** had lower MOE than the control. At 1.0 wt% and 2.0 wt%, MOE values were statistically the same for Struktol and the control. At 3.0 wt%, Struktol had lower MOE than the control. With the exception of RTIL **F**, all RTILs at the usage of 1 wt% to 3.0 wt% had much lower MOE than the control and Struktol (Figure 4).

The effect of combinations of RTILs and MAPE and combinations of Struktol and MAPE on the torque value is shown in Figure 5. When compared with M2 (wood-PE composites containing 2.0 wt% MAPE only), increase in the usage of Struktol from 0.5 wt% to 1.0 wt% had little effect on the torque value

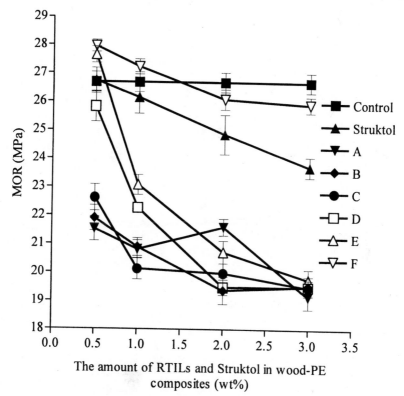

*Figure 3. Effect of RTILs and Struktol on the MOR of wood-PE composites.
(Control: wood-PE without any additive. Error bars show two standard errors
of the means from ten measurements.)*

(curve Struktol-M2, Figure 5). However, further increase in the dosage of
Struktol from 1.0 wt% to 3.0 wt% resulted in linear reduction of the torque
value, which means that Struktol had decent lubrication effect in making wood-
PE composites. At 0.5 wt%, RTIL **A** had a significantly lower torque value than
Struktol, whereas all other RTILs had the same torque values as Struktol (Figure
5). At 1.0 wt%, RTILs **A** and **C** had better lubrication effects, i.e., lower torque
values, than Struktol, whereas all other RTILs had the same lubrication effect as
Struktol. At 2.0 wt% and 3.0 wt%, all RTILs had better lubrication effect than
Struktol. At 2.0 wt%, the lubrication effect of RTILs had the following order: **B**
> **A** > **C** > **E** ≈ **F** > **D**. At 3.0 wt%, the lubrication effect of RTILs had the
following order: **E** ≈ **A** ≈ **C** > **D** ≈ **F** (Figure 5). Interestingly, the order of the
lubrication effect of RTILs changed with the usage of RTILs. Therefore, it is

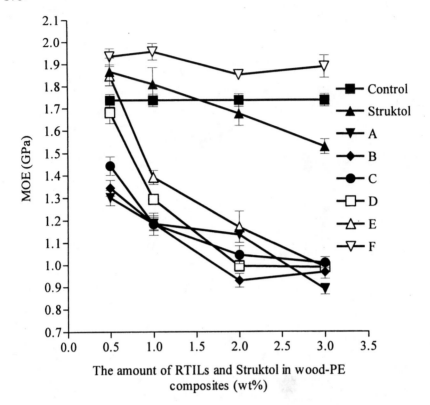

Figure 4. Effect of RTILs and Struktol on the MOE of wood-PE composites. (Control: wood-PE without any additive. Error bars show two standard errors of the means from ten measurements.)

difficult to speculate the relationship between molecular structures of RTILs and their lubrication effect. When compared with results in Figure 2, the lubrication effect of RTILs had different orders at the same usage, which means that MAPE affected the lubrication effect of each RTIL in a different way.

The effect of combinations of RTILs and MAPE and combinations of Struktol and MAPE on the MOR is shown in Figure 6. At 0.5 wt%, Struktol had a significantly lower MOR than the control, i.e., M2, whereas RTIL **F** had a significantly higher MOR than the control. Other RTILs (**A**, **C**, **D**, and **E**) had the same MOR as the control at 0.5 wt%. At 1.0 wt%, Struktol had a lower MOR than all RTILs and the control. At 1.0 wt%, RTILs **C**, **D**, and **E** eac h had a lower MOR than the control, whereas RTILs **A**, **B**, and **F** each had the same MOR as the control (Figure 6). At 2.0 wt% of Struktol and RTILs, the order of

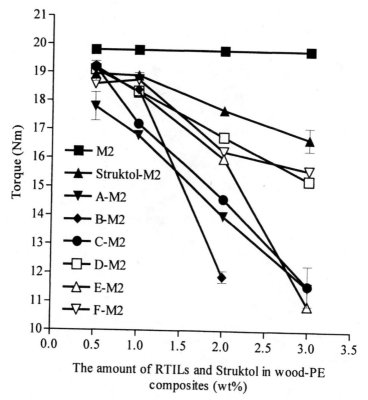

Figure 5. Effect of RTILs-M2 and Struktol-M2 on torque. (M2: 2 wt% MAPE. Error bars show two standard errors of the means from two independent experiments.)

the MOR for Struktol and RTILs is as follows: **F** > **A** ≈ **C** > **E** > Struktol ≈ **D** > **B**. In other words, four RTILs resulted in a higher MOR than Struktol. At 3.0 wt of Struktol and RTILs, the order of the MOR for Struktol and RTILs is as follows: **F** > **D** > **E** ≈ **A** > Struktol > **C**. The order of the MOR for Struktol and RTILs varied with the usage (Figure 6). The order of the MOR was not in accord with the order of the torque value at each usage (Figs. 5 and 6). Overall, RTIL **F** reduced the strength-enhancing effect of MAPE much less than other RTILs and Struktol at the usage of 2.0 wt% and 3.0 wt% (Figure 6).

The effect of combinations of RTILs and MAPE and combinations of Struktol and MAPE on the MOE is shown in Figure 7. At 0.5 wt%, Struktol and RTILs **A**, **C**, **D**, **E**, and **F** all increased the stiffness of the wood-PE composites. At 1.0 wt%, Struktol and RTILs **A**, **B**, **C**, and **E** each had a higher

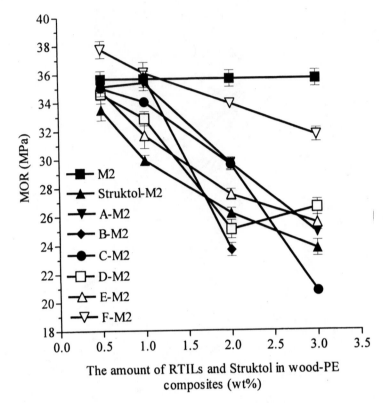

Figure 6. Effect of RTILs-M2 and Struktol-M2 on the MOR of wood-PE composites. (M2: wt% MAPE. Error bars show two standard errors of the means from ten measurements.)

MOE than the control, i.e., M2, whereas RTILs **D** and **F** had the same MOE as the control. At 2.0 wt%, RTILs **F** and **C** had a higher MOE than Struktol, whereas RTILs **B**, **D**, and **E** each had a much lower MOE than Struktol. At 3.0 wt%, Struktol had a much lower MOE than RTIL **F**, but had a much higher MOE than RTIL **C**. RTIL **F** was the only RTIL that resulted in a higher MOE than the control at all usages studied (Figure 7).

Conclusions

In the absence of MAPE, RTILs generally had better lubrication effect than Struktol at the same usage. Except RTIL **F**, RTILs also resulted in lower MOR and MOE than Struktol at the same usage. RTIL **F** had the lubrication effect

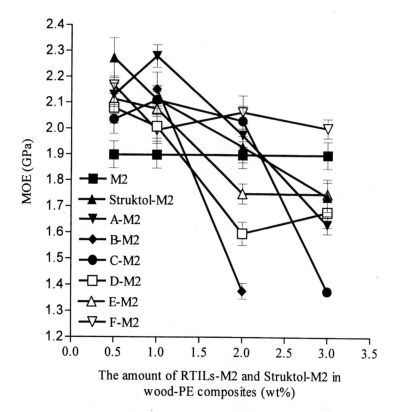

Figure 7. Effect of RTILs-M2 and Struktol-M2 on the MOE of wood-PE composites. (M2: wt% MAPE. Error bars show two standard errors of the means from ten measurements.)

better than or comparable to Struktol, but had higher MOR and MOE than Struktol at each usage studied. In the presence of 2.0 wt% MAPE, all RTILs at a practical usage level (2.0 wt% or 3.0 wt%) had better lubrication effect than Struktol. RTILs **A**, **C**, **E**, and **F** at 2.0 wt% usage and RTILs **A**, **D**, **E**, and **F** each reduced the strength-enhancing effect of MAPE less than Struktol. When used with 2.0 wt% MAPE, RTIL **F** was far superior to other RTILs and Struktol in terms of providing the lubrication effect and retaining the strength-enhancing effect of MAPE. At 0.5 wt% or 1.0 wt% usage level, a combination of a RTIL and MAPE or a combination of Struktol and MAPE all increased the stiffness of the resulting wood-PE composites. At 2.0 wt%, RTILs **B**, **D**, and **E**, when used with 2.0 wt% MAPE, significantly reduced the stiffness of the resulting wood-PE composites, whereas other RTILs either increased or did not change the stiffness of the resulting composites. At 3.0 wt%, Struktol and all RTILs except

RTIL **F** significantly decreased the stiffness of the resulting wood-PE composites. Combinations of RTIL **F** and 2.0 wt% MAPE were superior to combinations of Struktol and 2.0 wt% MAPE in terms of increasing the stiffness of the resulting wood-PE composites.

Acknowledgments

We appreciate American Wood Fibers (Schofield, WI), BP Solvay Polyethylene North America (Houston, TX), Honeywell International, Inc. (Morristown, NJ), and Cytec Industries, Inc. (West Paterson, NJ) for donating wood flour, high-density polyethylene, MAPE, and room temperature ionic liquids, respectively. This research was supported by a grant from the national research initiative competitive grants program of USDA (award number: 2003-35103-13864).

References

1. Welton, T. *Chem. Rev.* **1999**, *99*, pp 2071-2083.
2. Blancard, L. A.; Hancu, D.; Beckman, E. J.; Brennecke, J. F. *Nature (London)* **1999**, *399*, pp 28-29.
3. Blanchard, L. A.; Brennecke, J. F. *Ind. Eng. Chem. Res.* **2001**, *40*, pp 287-292.
4. Branco, L. C.; Afonso, C. A. M. *Tetrahedron* **2001**, *57*, pp 4405-4410.
5. Carmichael, A.J.; Earle, M.J.; Holbrey, J.D.; McCormac, P.B.; Seddon, K.R. *Org. Letters* **1999**, 1, pp 997-1000.
6. Pernak, J.; Czepukowicz, A.; Pozniak, R. *Ind. Eng. Chem. Res.* **2001**, 40, pp 2379-2383.
7. Li, X.; Geng, Y.; Simonsen, J.; Li, K. *Holzforschung* **2004**, 58, pp 280-285.
8. Shelton, R. *Chem. Commun.* **2001**, pp 2399-2407.
9. Huddleston, J.G.; Willauer, H.D.; Swatloski, R.P.; Visser, A.E.; Rogers, R.D. *Chem. Commun.* **1998**, pp 1765-1766.
10. DSwatloski, R.P.; Spear, S.K.; Holbrey, J.D.; Rogers, R.D. *J. Am. Chem. Soc.* **2002**, 124, pp 4974-4975.
11. *Ionic liquids as green solvents: progress and prospects,* Rogers, R.D.; Seddon, K.R., Eds.; ACS symposium series 856; American Chemical Society: Washington, DC, 2003.
12. Kim, K.S.; Choi, S.; Demberelnyamba, D.; Lee, H.; Oh, J.; Lee, B.B.; Mun, S.J. *Chem. Commun.* **2004**, pp 828-829.
13. Wolcott, M.P.; Englund, K. In *Proceeding of the 33rd Washington State University International Particleboard/Composites Materials Symposium* **1999**, pp 103-111.

14. Clemons, C. *For. Prods. J.* **2002**, 52, pp 10-18.
15. Sellers, Jr., T.; Miller, G D.J.; Katabian, M. *For. Prods. J.* **2000**, 50, pp 24-28.
16. Morton, J.; Quarmley, J.; Rossi, L. In *The seventh international conference on woodfiber-plastic composites*; Madison, Wisconsin, 2003, pp 3-6.
17. Gauthier, R.; Joly, C.; Coupas, A.C.; Gauthier, H.; Escoubes, M. *Polym. Compos.* **1998**, 19, pp 287-300.
18. Harper, D.; Wolcott, M. *Composites: Part A* **2004**, 35, pp 385-394.
19. Ye, C.; Liu, W.; Chen, Y.; Yu, L. *Chem. Commun. (Cambridge, U. K.)* **2001**, pp 2244-2245.
20. Swatloski, R.P.; Holbrey, J.D.; Rogers, R.D. *Green Chem.* **2003**, 5, pp 361 - 363.
21. Bradaric, C.J.; Downard, A.; Kennedy, C.; Robertson, A.J.; Zhou, Y. In *Ionic liquids as geen solvents: progress and prospects,* Rogers, R.D., Seddon, K.R., Eds.; ACS symposium series 856; American Chemical Society: Washington, DC, 2003, pp 41-56.

Chapter 12

Lignins as Emulsion Stabilizers

Orlando J. Rojas[1], Johnny Bullón[2], Fredy Ysambertt[3],
Ana Forgiarini[4], Jean-L. Salager[4], and Dimitris S. Argyropoulos[1]

[1]Department of Forest Biomaterials Science and Engineering,
North Carolina State University, Raleigh, NC 27695–8005
[2]Lab. LMSSI, Universidad de Los Andes, Mérida, Venezuela
[3]Lab. Petroquímica y Surfactantes, LUZ, Maracaibo, Venezuela
[4]Lab. FIRP, Universidad de Los Andes, Mérida 5101, Venezuela

This is a report on the use of (kraft and soda) lignins as polymeric amphiphiles for the stabilization of emulsions. The lignins' phase behavior in oil/water systems is presented and explained in terms of their molecular affinities. Emulsions with various oils (including crude oils) were formulated and their properties were rationalized in terms of lignin surface activity as a function of pH and salinity. It is concluded that lignins are an effective emulsion stabilizer; the resulting emulsions behave according to the balance of affinities between oil and water phases and lignin-based emulsions can be tailored to meet specific demands.

A by-product of chemical digestion of lignocellulosics, "black liquor," is highly concentrated in organic materials, with lignin being the main component of the total dry mass. Only about 1.5 % of the ca. 70 million tons of lignin yearly available in the pulp and paper industry is commercially used (about 1 million tons per year) (1-4). Furthermore, these 70 million tons represent only a very small fraction of the lignin produced by nature in biomass every year.

New opportunities which appear in industrial production appear in other contexts such as bioethanol production for the replacement of transportation fuel of fossil origin. In all cases, there is an intensive use of the carbohydrate fraction of the lignocellulosic resources while lignin is a non-negligible part of the raw material that cannot be transformed into e.g., sugars. Therefore, these bio-energy efforts and other forest-based industrial initiatives lead to the generation of even larger amounts of lignin side streams. For achieving economic viability in such current and future processes, lignin's highest possible co-product value becomes one of the key success factors.

Lignins as Polymeric Surfactant

Since the pioneering work of Landoll (5), studies on naturally-occurring polymeric surfactants have been centered on derivatives of polysaccharides. The various polymer types and the characteristic property profiles distinguishing polymeric surfactants have been discussed recently (6). Such polymers seem attractive in many fields because of their good performance in lowering surface tension. Despite the fact that lignins from black liquors don't have the chemical structure of conventional amphiphilic molecules, they exhibit surface activity. It is hypothesized that lignin adsorbs at the air/water and oil/water interfaces forming a condensed, viscoelastic surface or interfacial film. In the literature, lignin molecules have been categorized as polymeric surfactant when separated from spent black liquors (e.g., lignin sulphonates). Surface activity has been demonstrated in the case of lignins obtained directly from black liquors, with little or no modification (7,8), after chemical derivatization (9-16) and also in combination with other species (17-19). All in all, as is the case of alkyl polyglucosides and other biosurfactants, lignin polyelectrolytes derived from black liquors can be classified as a natural, renewable polymeric amphiphile.

Emulsion Formulation

Only very few studies have been reported on the use of lignin in emulsion technology (8,20-24). The preparation of emulsions allows many degrees of freedom with respect to the formulation and the emulsification protocol. These degrees of freedom could be classified as variables that influence the properties of the final emulsion. Furthermore, the actual effect of these variables is very complex (25).

Three different factors or variables are important in emulsion manufacture: (1) "*formulation variables*" that depend on the nature of the components; (2) "*composition variables*" that quantitatively describe the *composition* of the system in terms of, e.g., volume fractions; and (3) the ones that describe the way the emulsion is prepared, that is, the stirring and "*emulsification*" conditions.

The physicochemical variables that depend on the nature of the different ingredients are those that are able to alter the prevailing physicochemical condition at the liquid-liquid interfaces. There are at least four such variables, i.e., (1) the *alkane carbon number (ACN)* when the oil is an alkane or the equivalent ACN (EACN) when it is not (26-29); (2) the *ionic strength or salinity* of the aqueous phase, which in turn depends on the nature and content of the electrolyte; (3) the *surfactant* and, (4) the *temperature*. The implications of this multiplicity in the number of variables are overwhelming. A systematic study with a few values of each variable would require an astounding number of tests. Obviously, a systematic approach is needed to rationalize the involved phenomena.

Winsor's pioneering theoretical work (30) showed that the formulation concept could be described through a single parameter that gathered all effects. This parameter, R, is the ratio of the interaction energy of the surfactant with the oil phase, to the interaction energy of the surfactant with the aqueous phase, $R=(A_{co}-0.5A_{oo})/(A_{cw}-0.5A_{ww})$, where A_{co} is the interaction energy between the surfactant (c) and the oil phase (o) molecules; A_{cw}, the interaction energy between the surfactant and water; A_{oo}, the interaction energy between two oil molecules and so forth. The 0.5 coefficients are introduced because cross interactions $(A_{co}$ and $A_{cw})$ involve one surfactant molecule and both an oil molecule and a water molecule, while the self interactions $(A_{oo}$ and $A_{ww})$ involve two molecules of oil, and two of water.

Winsor's research showed that the phase behavior at low surfactant concentration, say from 0.1% to 5%, as in most applications, is directly linked with the R value (30,31). For $R<1$, a so-called Winsor I phase behavior is exhibited in which a surfactant-rich aqueous phase is in equilibrium with an essentially surfactant-free oil phase. In this case, the interaction of the surfactant for the aqueous phase exceeds the interaction of the surfactant for the oil phase. In Winsor II phase behavior, that is attained for $R>1$, the opposite occurs, i.e., the surfactant-rich phase is the oil phase, which is in equilibrium with an aqueous phase that contains essentially no surfactant, and the dominant interaction of the surfactant is for the oil phase.

When $R=1$, the surfactant interactions for the oil and water phases are equal and the system splits into three phases in equilibrium: a microemulsion that contains most of the surfactant and that co-solubilizes large amounts of oil and water, and two excess phases that contain essentially pure oil and pure water (30,31), which is a complex but not uncommon phase behavior situation.

Since R varies with one or more changes in interaction energy, those factors that affect the A's are likely to change R, and consequently, the phase behavior. For instance, an increase in surfactant lipophilic group length tends to increase the A_{co} and thus increases R. If the starting R value is smaller than unity, and the final value is $R>1$, or vice versa, then a complete phase behavior transition is exhibited. Similar transitions may be attained by changing any of the formulation variable that can affect any of the A's (salinity, pH, T, etc.).

A more quantitative description of the physical-chemical formulation is available through the so-called surfactant affinity difference (SAD) and the Hydrophilic-Lipophilic Deviation (HLD) (see Refs. 32-36) which translate Winsor's conceptual approach into numerical values. At present, there is an accumulated knowledge for emulsion formulation mainly for conventional, monomeric surfactants. In this collection of information it is evident that correlations for lignin-based surfactants are lacking. Therefore, the objective in this effort is to specifically address this lack of scientific data in anticipation that such information will eventually permit the development of new, lignin-based emulsion formulations that may effectively compete with their synthetic counterparts.

Lignins and Characterization Methods

In order to illustrate the abilities of lignins to act as an emulsion stabilizer, different lignin sources are cited in this chapter. Industrial black liquor was obtained from Smurfit-Mocarpel pulp mill (Yaracuy, Venezuela), where a low-sulfidity Kraft process is used in the digestion of *Pinus caribaea* (*KBL-P*). In this case, the black liquor samples were collected in the recovery boiler (pH 13.5, 31% solids content and 18% Klason lignin) and suspended solids were separated by filtration with a Whatman filter No. 1. Industrial black liquor was also obtained from the same pulp mill after soda digestion of sugar cane bagasse (*SBL-B*). Finally, there is reference to two commercially-available lignin samples, both from Westvaco Corp. (Charleston, SC, USA), namely, *Indulin C*, a sodium salt of Kraft pine lignin and *Indulin AT*, a purified form of pine Kraft lignin, free of hemicellulosic materials. The details about the chemical and physical characteristics of the industrial and laboratory black liquors can be found elsewhere (19,37,38).

The surface tension (γ) of aqueous solutions of lignins were measured by employing the Wilhelmy plate technique in a Dataphysics tensiometer, model DCAT11 (Germany). A thermostated bath with circulating water at 25 °C was used to ensure constant temperature conditions throughout the measurements. Phase Behavior and lignin partitioning between oil and aqueous phases was studied by using formulation scans in equilibrated systems (31,34,39). Lignin concentration in both organic and aqueous phases was obtained by UV spectroscopy (280 nm).

Oil-in-water emulsions with different internal phase content and different pHs of the aqueous phase were prepared by using standard emulsification protocols. The aqueous phase consisted of distilled water with adjusted salinity and pH to which lignin (as 1% solids or as otherwise specified) was added as an emulsifier. Different oleic phases were employed. Depending on the experiments ,kerosene (Aldrich) was used with a measured Equivalent Alkane Carbon Number of 9 or an extra heavy crude oil (Cerro Negro) from the Orinoco Belt (Venezuela) of 8.5 °API at 60 °C (Viscosity 500-2500 mPa.s at down hole pressure of 700-1300 psi and 50-70 °C) (40). Additional phases consisted of mixtures as specified in the respective sections. In most cases, a Water-to-Oil Ratio (WOR) of 30/70 was used or as specified otherwise. In the case of the extra heavy crude oil, the preparation of the emulsion involved heating of both phases to 60 °C, and the emulsification was carried out in a turbine blender (Taurus) at 12,000 rpm for 30 seconds. For all other oil phases emulsification was conducted at 25 °C. In all cases emulsion properties were measured at 25 °C.

The resulting emulsions were analyzed in terms of drop size distribution by laser light diffraction (Malvern) and stability. The stability was estimated both by drop size evolution with time and by a phase separation technique (41). The rheological behavior of the emulsions was studied with a Rheometric Scientific SR-5000 using parallel plates geometry (40 mm diameter, 1 mm gap), and the flow curves for less viscous emulsions were obtained in a parallel cylinder geometry using a Rheomat 30 (Contraves AG, Zurich).

Lignin Surface Activity

When a surface active substance is added to water (or oil), it spontaneously adsorbs at the surface, and decreases the surface tension γ. In the case of small surfactant molecules, a monolayer is formed with the polar parts of the surface-active molecules in contact with water, and the non-polar parts in contact with air (or oil). The surface activity of lignin solids separated after the evaporation of black liquors at low temperature is illustrated in Fig. 1. For comparison purposes, the surface tension curves for a commercial lignin (*Indulin C*) and a typical anionic surfactant (petroleum sulphonate) are also included. Minimum surface tensions of ca. 30 mN/m were observed for the lignin solutions. Despite the fact that the reduction in surface tension is not as large as for conventional surfactants, it is low enough to unequivocally suggest the adsorption and packing of lignin molecules at the interface. The reduction in surface tension is not enough for lignin to act as typical emulsifiers work, by facilitating the creation of surface area or droplets due to the lowering in interfacial energy. However, as will be demonstrated in later sections, lignin derivatives perform effectively as an emulsion stabilizer since it prevents droplet coalescence by means of electrostatic and steric repulsion as well as by the formation of a viscoelastic skin or interfacial film around the droplets of the dispersed phase (22).

A break in the surface tension tendency is observed at solids concentrations between 1 and 10%. In the case of monomeric surfactants this break indicates the critical micelle concentration (*cmc*). In the case of lignin molecules it is unlikely the association of the molecules in the form of micelles and therefore this concentration is rather described as a pseudo-*cmc*. These figures are similar to those reported by Rojas and Salager for *SBL-B* (7). It can be concluded that given the right conditions, the lignins considered here behave as polymeric surfactants in aqueous media.

Adsorption of conventional surfactants is generally very fast, but for larger molecules, such as lignins, the adsorption process is slower (see evidence in (7)) and adsorption is generally irreversible. Furthermore, it is hypothesized that lignin adsorbs in the form of aggregates, the surface layers become thick, and a solid-like "skin" is formed.

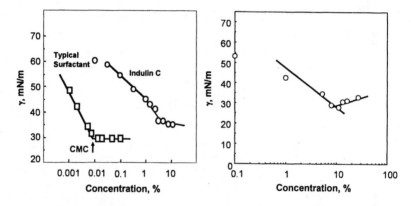

Figure 1. Left: Surface tension γ (25 °C) of solutions of commercial lignin (Indulin C) compared to a typical anionic surfactant (petroleum sulphonate). Right: Surface tension for solids separated from kraft black liquor, KBL-P.

In our earlier efforts it has been demonstrated that lignin can be separated from spent pulping liquors in high yields and purity by acid precipitation and ultrafiltration, UF (7,42). These methods of lignin separation offer the significant advantage of simplicity over other alternative routes (43-45). Furthermore, one can take advantage of the fact that separation by molecular size is coupled with the selective concentration of functional groups, O/C ratios and H/C ratios in the separated fractions (46). These results indicate that the concentration of phenolic groups in the lignin fractions obtained by UF of industrial kraft lignins decreased by increasing mass-weighted hydrodynamic radii of the diffusants. Interestingly, this leads to differences in the surface activity of the separated molecules, as can be observed in Fig. 2 (left). The lignin samples used in Fig. 2 were obtained by ultrafiltration (UF) using a Minitan[TM] (Millipore) unit with polysulfone semi permeable membranes in a plate-and-frame set up. The initial feed consisted of black liquor solution diluted to 15 % total solids content. This solution was fed to the membrane of MWCO 100 kDa. The retentate in each UF step was recycled to the feed stream. The permeate from the membrane of MWCO 100 kDa was filtered using the membrane of MWCO 30 kDa and the same was done successively for the membranes of smaller MWCOs. It is seen that the permeate solutions for the membranes with MWCO 100 and 30 kDa show high surface activity. However, this is not the case for the permeate of the membrane of 1 kDa MWCO. From these results it can be concluded that lignin fractions of higher molecular weight are more surface active than the lower molecular weight counterparts. The amphiphilic nature of these anionic polyelelectrolytes is explained by the hydrophobicity of the molecule balanced by a high number of anionically-charged groups. As the lignin size is reduced, the hydrophilic/hydrophobic balance is changed making the molecule less hydrophobic and less surface-active.

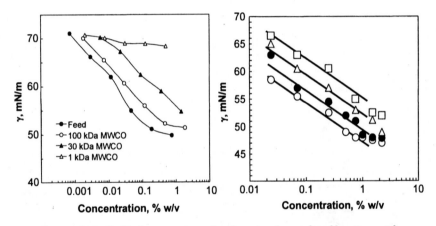

Figure 2. Left: Surface tension of permeates from ultrafiltration with polysulfone membranes of KBL-P with membranes of nominal MWCO 1, 30 and 100 kDa. Right: Surface tension of SBL-B lignin dissolved in aqueous media of pH 7 (squares), 8(triangles), 9 (filled circles) and 10 (open circles).

The way lignin adsorbs at the interface is still an issue under scrutiny. In the bulk solution a randomly branched model for lignin derivatives has been proposed with a hydrophobic backbone and hydrophilic side chains as well as flat, fractal structures (47-49). It is envisioned that the conformation at the interface involves molecular orientation according to the affinities of the involved structures. Evidently, more research is needed to confirm or elucidate these issues.

It is worth mentioning that the knowledge of surface tension alone is not enough to understand emulsion properties (see later sections), and the surface viscoelasticity plays an important role in a number of dynamic processes. There is research in laboratories underway to explore this aspect. Nevertheless, there is evidence provided here of the interesting properties obtained when lignin is used as emulsion stabilizer.

Finally, the surface activity of lignin solids when dissolved in aqueous media of different pH is addressed as follows. As can be seen in Fig. 2, the surface activity of lignin is increased when dissolved in high pH aqueous solution. This effect is explained by the ionization of functional groups, mainly carboxylic and phenolic hydroxyl groups. A right molecular size is predicted to give better surface activity if the molecule contains enough hydrophilic groups, depending on the pH of the aqueous system. A fully dissociated group such as sulphonate would be required to provide surface activity less dependant on the pH of the solution. Furthermore, chemical modification of lignin can be used for optimizing its surface activity.

Phase Behavior in Oil-Water Systems

The investigation of the phase behavior of lignin in oil/water systems is required in order to understand the affinity of the molecule towards these phases as a function of the pH, salinity and oil type (oil's Equivalent Alkane Carbon Number, EACN). Figure 3 shows a typical phase behavior for a kraft lignin in a kerosene/water system. In this case the lignin was first dissolved in the aqueous phase (1.5% lignin concentration) and then an equal amount of oil (kerosene) was placed in contact with the aqueous solution. After gentle mixing and equilibration for at least 24 h at constant temperature (25 °C), the partitioning of the surface active molecule is followed as a function of a formulation variable (pH of the aqueous phase in the case of Fig. 3).

As expected, it is observed that at high pHs, the system can be described as a Winsor I ($R<1$), that is, lignin remains in the aqueous phase. However at pH 7 and below, the lignin becomes less soluble and two phases are separated, namely, an oleic phase and an O/W (micro)emulsion. At pH 4.9, the lignin molecules are distributed in two (micro)emulsion phases, O/W and W/O, respectively. At even lower pH the affinity between lignin and the aqueous phase is minimum (better affinity towards the oil phase) and a gel is formed. Similar behaviors were observed with the other lignins studied.

An alternate way to increase the affinity of the lignin towards the oleic phase is to modify the molecule by, e.g., alkylation. This makes the molecule more hydrophobic and increases surface activity (12). Accordingly, acetylation was conducted following the protocol explained elsewhere (50). Different degrees of acetylation were studied and this study reports the cases of 97% and 57% acetylation degree.

Figure 3. Phase behavior for KBL-P/water/kerosene system at pH of the aqueous phase of 11.5, 10, 7.4, 5.2, 4.9, 4.1 and 2 (from left to right).

A pH scan as that presented in Fig. 3 was undertaken with the acetylated lignins. In these cases an aqueous pseudo-phase was used (water/DMSO, 60:40) to ensure better solubility. The result of this scan is depicted in Fig. 4. As can be seen, at high pH the lignin has better affinity towards the aqueous phase while at low pH the lignin has a favorable affinity towards the oil phase. In fact, lignin molecules are almost fully partitioned in the respective phases at extreme pHs. At pH between 3 and 5 there should be a condition in which the affinity or partitioning of the molecules towards both phases is identical.

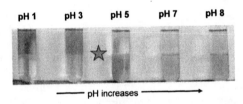

Figure 4. Phase behavior in a pH scan for acetylated (97%) Indulin AT/water-DMSO/ethyl acetate system. Sec-butanol was added to reduce gel formation. Note lignin is preferentially located in the darker (yellow) phase.

The salinity of the aqueous phase was also used as the formulation variable. Figure 5 (left) illustrates the case of *Indulin AT* of 97% and 57% degree of acetylation. It is evident that at low salinities lignin has higher affinity towards the aqueous phase whereas at high salt concentration the opposite occurs (higher affinity with the oil phase, i.e., salting out). This behavior is simply explained by changes in electrostatic interactions and solvency that occur as the ionic strength is varied. In the case of lignin of 97% acetylation degree the interactions are "balanced" at 0.02-0.04% salt concentration. Figure 5 (right) confirms the partitioning of lignin (as measured by its relative concentration via UV spectrophotometry) between the two phases.

A similar scan as that explained before but with lignin of lower degree of acetylation (57%) shows that lignin affinity with the aqueous and oil phases is balanced at higher salt concentration (between 0.08 and 0.1% salt) as compared to the high acetylation case (see Fig. 5, left).

The previous observations for pH and salinity scans validate the possibility of tuning the chemical affinity of this polyelectrolyte and also demonstrate that

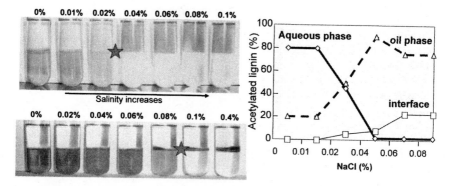

Figure 5. Left: Phase behavior in a salinity scan of acetylated Indulin AT/Water-DMSO/Ethyl acetate system. Lignin acetylation % was 97(top) and 57 (bottom). Right: Partitioning of acetylated Indulin AT (97% acetylation). Note that lignin concentrations are equal in both phases at 0.03% salinity.

lignin follows the predictions set forth in Winsor's theory. As the lignin becomes more hydrophilic (for lower degree of acetylation, e.g., 57%) more salt is needed to "push" the lignin molecules to the oil phase.

Lignins as Emulsion Stabilizer and Rheological Behavior

The next step after understanding the phase behavior of lignin in oil/water systems is to emulsify the bi-phase mixture and to measure the properties as they vary with changes in formulation and composition. As expected, after stirring the systems depicted in Fig. 4 and 5, O/W emulsions were obtained in the case of systems with Winsor's $R<1$ (lignin with more affinity towards the water phase, e.g., low salinity or high pH). On the other hand, W/O emulsions were obtained in the case of systems with Winsor's $R>1$ (lignin with higher affinity with the oil phase, e.g., high salinity or low pH).

In order to focus on more meaningful, practical systems, there is focus on systems involving either kerosene (as reference oil cut from refineries) or extra heavy crude oils which were emulsified with unmodified lignins. In these cases, as predicted by the phase behavior such as that shown in Fig. 3, the emulsions to be obtained are of the O/W type. All emulsions were characterized by small drop size, excellent rheological properties, and high stability.

Table 1 summarizes the drop size distribution for O/W (kerosene-in-water) emulsions prepared with *KBL-P* as emulsifier. The drop size distribution gives a statistical inventory of the dispersed phase fragmentation and it is therefore the best way to describe the emulsion. The technique used here is based on the fact that within the usual macroemulsion range (1-100 μm drop size) the diffraction angle increases as particle size decreases. The distribution frequency obtained from the light scattering analyzer is in volume and the distribution followed a log-normal statistics (as it is often the case). A summary of statistical drop sizes are reported in Table 1 as the median (symbolized as D[0.5]), which represents the size so that half the internal phase volume is fragmented in smaller drops and half in larger ones and also the mean distribution in volume, symbolized by D[4,3].

Table 1. Properties (mean diameters and apparent viscosity at a shear rate of 200 s^{-1} and 25 °C) of kerosene-in-water (O/W) emulsions (a) with different internal phase concentrations (different water-to-oil ratios, WOR) at pH 11 and, (b) with different pH of the aqueous phase (at WOR=40/60). The employed emulsifier was *KBL-P* at a concentration of 1.5 %.

WOR	90/10	80/20	70/30	60/40	50/50	40/60	30/70
D[0.5], μm	2.95	3.62	3.29	3.36	3.69	3.81	4.59
D[4,3], μm	3.17	5.31	3.6	3.77	4.15	4.43	5.17
Visc, mPa.s	4.12	5.11	5.89	8.71	16.7	97.18	760.3

pH	12.2	10	9.4	7.2	6.3	4.6	2.6
D[0.5], μm	4.03	3.59	3.54	3.84	4.89	6.29	12.78
D[4,3], μm	4.62	4.02	3.93	4.29	6.38	8.61	16.39
Visc, mPa.s	88.0	61.1	57.2	70.1	79.1	760.2	795.7

A slight displacement of the average particle size towards larger drop sizes is observed as the internal phase concentration is increased. This behavior can be simply explained by the fact that as the internal phase concentration is increased less lignin is available to stabilize the higher surface area produced and drop coalescence is more likely to occur. The ability of lignin in emulsion stabilization is explained by the electrostatic and steric barriers provided by the adsorbed lignin molecules which behave as a protecting viscoelastic "skin".

The apparent viscosities of the emulsions reported in Table 1 are in agreement with the expected behavior, i.e., emulsion viscosity increases as the internal phase content is increased (WOR is reduced in O/W emulsions). This is due to a tighter packing and increased inter-drop interactions. A relatively much higher apparent viscosity was observed for emulsions of the highest internal phase concentration (e.g., WOR of 30/70 in Table 1). This is related to the formation of macrostructures or gels (51-54) as well as the growth of large, stable associated complexes from smaller lignin moieties as the pH is reduced (55). It is worth noting that emulsions of low internal phase content (WOR > 50/50) show a Newtonian behavior whereas a departure from this behavior is observed at higher internal phase content.

The concept of emulsion stability is linked either to the persistence or the decay of the dispersed system under certain circumstances. The only absolute way to define stability is to count the number of drops in the system and to evaluate how it changes as time elapses. However, this is impractical because most of the measurement techniques require dilution and thus modification of the emulsion. The emulsion stability was measured using the time required to observe phase separation. In all cases it was observed that emulsion stability as measured by 1/3 phase separation was of the order of 47 days. This implies that the studied emulsions can be classified as highly stable.

Emulsions with WOR 40/60 were prepared to investigate the effect of pH of the internal aqueous phase on the emulsion drop size (and size distribution) and viscosity (see Table 1). The average drop size is significantly increased as the pH of the aqueous phase is reduced. This behavior agrees with the fact that as the system becomes more acidic, the surface activity of lignin molecules is reduced. On the other hand, the viscosity is reduced as the pH is lowered from pH 9. However, at the lowest pH, a steep increase in the apparent viscosity is noticed. This effect is most probably caused by the formation of structures triggered by more favorable inter-drop interactions as electrostatic repulsion between the molecules is reduced. These structures produce a more elastic behavior as confirmed by rheometric measurements (not reported here). Also, a departure from the Newtonian flow behavior is evident as pH is reduced. The emulsion stability was of the order of +40 days, which again, confirms the fact that the lignins from the black liquor are excellent stabilizers of liquid interfaces.

Figure 6 shows the flow curves for kerosene-in-water (O/W) emulsions of high internal phase content (WOR 30/70) and stabilized with *KBL-P* at a concentration of 1 %. The pH of the aqueous phase was varied to study its effect on the rheological behavior. In general, all emulsions show a shear thinning behavior, but at high pH, the emulsions are close to Newtonian. Furthermore, the apparent viscosity is reduced as the pH of the aqueous phase is increased. There were similar behaviors observed in the case of *Indulin C* and the other industrial lignins.

Emulsions with Extra Heavy Crude Oils

The synthetic surfactants used in the stabilization of heavy crude oils include nonylphenol ethoxylates. One of the aims of this work was to test the viability of lignins from black liquors as a substitute for such additives. Thus, the properties of emulsions prepared with a reference extra heavy crude oil, a Cerro Negro from the Orinoco Belt (Venezuela) of 8.5 API at 60 C. For this purpose, emulsions with high internal phase content (WOR = 30/70) and 1% lignin concentration (in aqueous phase) were prepared.

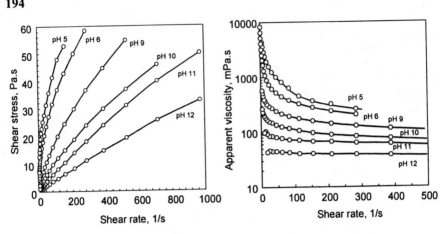

Figure 6. Left: Flow curves for kerosene -in-water emulsions (WOR 30/70) stabilized with KBL-P (1% w/v concentration). The pH of the aqueous phase was varied between pH 5 and 12. Right: Respective apparent viscosities.

Figure 7 (left) shows the oil drop size evolution with time for emulsions prepared with aqueous solutions of different pH. The high stability of the emulsions (in this case remarkably better stabilities than for emulsions with kerosene) is evident. It is also interesting to note the variation in drop size with pH. As the pH increases the drop size is reduced. Similar behaviors were obtained for other lignins tested (such as *Indulin C* and *KBL-P*, data not shown). Figure 7 (right) includes the flow curves for these emulsions. It is important to note that the apparent viscosity of the emulsions is up to three orders of magnitude lower than the neat heavy crude oil. Also, Fig. 7 suggests a rheological behavior for these emulsions (shear thinning) which is similar to that observed for other oil phases presented previously. Most importantly, it is seen that in contrast with earlier findings, a lower apparent viscosity is attained at lower pH. This effect makes sense if the drop size is considered: at high pH the surface activity of the lignin molecule is increased and the ability of a mixing device to produce new surface area is also increased. As a consequence, the average size of the formed drops is reduced, as shown in Fig. 7. On the other hand, an emulsion with same internal phase content but with smaller drop size will be subject to higher interdrop interactions and therefore a more viscous emulsion is expected. In fact this agrees with the observed trends showed in Figure 7.

As can be seen from the results presented so far, there is an intricate relationship between pH, salinity, WOR and the resultant properties of the emulsions (viscosity, drop size and stability). This underscores the need for a generalized description of emulsion behaviors stabilized by lignin molecules, for example, by using bidimensional maps as will be shown in the next section.

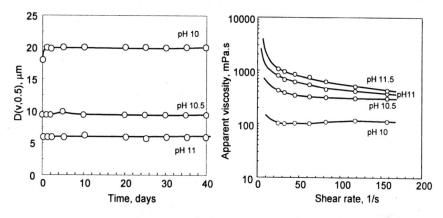

Figure 7. Left: Change with time of average drop size for emulsions produced with extra heavy crude oil and SBL-B lignin solutions at different pH and WOR of 30/70. Right: Apparent viscosities as a function of the shear rates.

Emulsion Bidimensional Maps with Lignins

Figure 8a,b summarizes the apparent viscosity in a bidimensional scan (formulation-composition). Here the salinity was used as the formulation variable in O/W (kerosene-in-water) emulsions prepared with *Indulin C* as emulsifier. As can be seen, the behavior of this reference lignin follows qualitatively the trend observed for conventional surfactants: very high viscosities at high internal phase ratios (close to the so-called catastrophic inversion for conventional surfactants) (56). It is also observed an increase in the drop size as the salinity is increased (a balanced system or Winsor's *R* close to unity). Figure 8b,c shows similar maps for the case of *KBL-P* lignins. The pH was used here as the formulation variable. A slight displacement of the average particle size towards higher drop sizes is observed as the internal phase concentration is increased (lower water fractions). What is more remarkable is that the trends observed in these bidimensional scans are in general agreement with the behavior of conventional surfactants (56). Such behaviors were explained earlier in the context of Fig. 6 and 7 and Table 1 and are also related to the affinity of the molecule with the water and oil phases (see Phase Behavior section). It is important to note that in the presented cases of bidimensional maps, the emulsion inversion lines (either transitional or catastrophic) are ill-defined (no such inversion has been observed) but this terminology is used in this case for comparison purposes and also with the expectation that other lignin derivatives may show such behaviors.

196

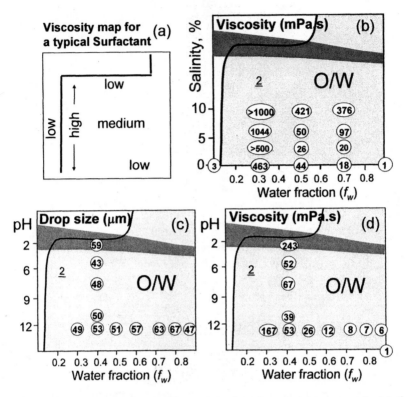

Figure 8. Apparent viscosity map for a typical surfactant (a) compared with that of kerosene-in-water (O/W) emulsions stabilized with Indulin C (1.5 % conc) (shear rate of 200 s⁻¹, 25 °C) (b). The emulsion average drop size (c) and apparent viscosities (at a shear rate of 200 s⁻¹, 25 °C)(d) for similar emulsions stabilized with KBL-P (1.5 % conc) is also shown.

Conclusions

In view of the presented results, it can be concluded that lignins are effective emulsion stabilizers even in the case of heavy crude oils. The resulting emulsions behave according to the affinity of lignin with the oil and water phases and lignin-based emulsions can be tailored to meet specific demands. All in all, there is convincing evidence of the potential for lignins separated from spent liquors for formulating oil-in-water emulsions with target properties.

Acknowledgments

We gratefully acknowledge the financial support provided by FONACIT-Venezuela (Grant No. S1-2000000816); by Universidad de Los Andes (CDCTH-ULA) and by NCSU Faculty Internationalization Grant. Professor Maria Briceño (FIRP) and the following undergraduate students who have been involved in our research program are also acknowledged: J. Jimenez, J. Sánchez, L. Aldana, A. Lorenzo, P. Contreras, A. Valdivieso, J. García, V. Padilla, M. Alayón, and L. Tolosa.

References

1. A new starting point for powerful lignin promotion: Eurolignin and linked activities, ILI: URL www.ili-lignin.com/Progress_ILI.doc.
2. Lora, J. H.; Glasser, W. G. *J. Polymer and Environ.t* **2002**, *10*, p 39.
3. Chum, H. L.; Parker, S. K.; Feinberg, D. A.; Wright, J. D.; Rice, P. A.; Sinclair, S. A.; W.G., G. *SERI/TR* **1985**, *231–488*, p 1.
4. Gargulak, J. D.; Lebo, S. E.; . *ACS Symp. Ser* **2000**, *742*, pp 304–320.
5. Landoll, L. M. *J. Polym. Sci. Polym. Chem.* **1982**, *20*, p 443.
6. Laschewsky, A. *Tenside Surfactants Detergents* **2003**, *40*, p 246.
7. Rojas, O. J.; Salager, L. *Tappi* **1994**, *77*, p 169.
8. Askvik, K. M.; Gundersen, S. A.; Sjoblom, J.; Merta, J.; Stenius, P. *Colloids Surfaces A* **1999**, *159*, p 89.
9. Keirstead, K. F.; Caverhill, L.; Dekee, D. *Canadian J. Chem. Eng.* **1982**, *60*, p 680.
10. Owusu, G.; Peters, E.; Dreisinger, D. B. *Canadian J. Chem. Eng* **1992**, *70*, p 173.
11. Joniak, D.; Polakova, M.; Kosikova, B.; Demienova, V. *Drevarsky Vyskum* **1999**, *44*, p 60.
12. Kosikova, B.; Duris, M.; Demianova, V. *European Polymer J.* **2000**, *36*, p 1209.
13. Duris, M.; Kosikova, B.; Demianova, V. *Drevarsky Vyskum* **1997**, *42*, p 25.
14. Yanhua, J.; Weihong, Q.; Zonghi, L.; Lubo, C. *Tenside Surfactants Detergents* **2003**, *40*, p 77.
15. Jiao, Y. H.; Qiao, W. H.; Li, Z. S.; Cheng, L. B. *Tenside Surfactants Detergents* **2003**, *40*, p 77.
16. Wang, R.; Chen, C. L.; Gratzl, J. S. *Holzforschung* **2004**, *58*, p 631.
17. Askvik, K. M.; Hetlesaether, S.; Sjoblom, J.; Stenius, P. *Colloids Surfaces A* **2001**, *182*, p 175.
18. Norgren, M.; Edlund, H. *Colloids Surfaces A* **2001**, *194*, p 239.
19. Tolosa, L. I.; Rodriguez-Malaver, A. J.; Gonzalez, A. M.; Rojas, O. J. *J. Colloid Interface Sci.* **2006**, *294*, p 182.

198

20. Gundersen, S. A.; Sjoblom, J. *Colloid and Polymer Sci.* **1999**, *277*, p 462.
21. Zaki, N. N.; Ahmed, N. S.; Nassar, A. M. *Petroleum Sci. and Technol. and* **2000**, *18*, p 1175.
22. Gundersen, S. A.; Ese, M. H.; Sjoblom, J. *Colloids Surfaces A* **2001**, *182*, p 199.
23. Gundersen, S. A.; Saether, O.; Sjoblom, J. *Colloids Surfaces A* **2001**, *186*, p 141.
24. Selyanina, S. B.; Makarevich, N. A.; Tel'tevskaya, S. E.; Afanas'eva, N. I.; Selivanova, N. V. *Russian J. Appl. Chem.* **2002**, *75*, 1873.
25. Shah, D. O. *Surface Phenomena in Enhanced Oil Recovery*; Plenum Press: N.Y., 1981.
26. Cash, R.; Cayias, J. L.; Fournier, G.; McAllister, D.; Shares, T.; Schechter, R. S.; Wade, W. H. *J. Colloid Interface Sci.* **1977**, *59*, p 39.
27. Cayias, J. L.; Schechter, R. S.; Wade, W. H. *J. Colloid Interface Sci.* **1977**, *59*, p 31.
28. Cayias, J. L.; Schechter, R. S.; Wade, W. H. *Soc. Petroleum Eng. J.* **1976**, *16*, p 351.
29. Cayias, J. L.; Schechter, R. S.; Wade, W. H. *Advances in Chemistry Series* **1975**, p 234.
30. Winsor, P. *Solvent Properties of Amphiphilic Compounds*; Butterworth: London, 1954.
31. Bourrel, M.; Schechter, R. S. *Microemulsions and Related Systems*; Marcel Dekker, 1988.
32. Salager, J. L.; Bourrel, M.; Schechter, R. S.; Wade, W. H. *Soc. Petroleum Eng. J.* **1979**, *19*, p 271.
33. Bourrel, M.; Koukounis, C.; Schechter, R.; Wade, W. *Journal Dispersion Sci. & Technol.* **1980**, *1*, p 13.
34. Bourrel, M.; Salager, J. L.; Schechter, R. S.; Wade, W. H. *J. Colloid Interface Sci.* **1980**, *75*, p 451.
35. Anton, R. E.; Garces, N.; Yajure, A. *Journal Dispersion Sci. & Technol.* **1997**, *18*, p 539.
36. Salager, J. L.; Antón, R.; Andérez, J. M.; Aubry, J.-M. Formulation des microémulsions par la method du HLD. In *Techniques de l'Ingénieur, Traité Génie des Procédés*, 2001; Vol. J 2157; pp 1 to 20.
37. Alayón, M.; Valdivieso, A.; Bullon, J.; Forgiarini, A.; Rojas, O. J. "Efecto de la fuerza iónica en estabilidad emulsiones con ligninas" Jornadas de Inv. Facultad Ingenieria UCV, 11/29-12/3, 2004, Caracas, Venezuela.
38. Alayón, M.; Padilla, V.; Valdivieso, A.; García, J.; Bullon, J.; Forgiarini, A.; Rojas, O. J. "Comportamiento de fases de ligninas de licores negros"; III Congreso Iberoamericano de Investigación en Celulosa y Papel CIADICYP 04, 11/10-12, 2004, Cordoba, Spain.

39. Salager, J. L. In *Handbook of Detergents. Part A*; Broze, G., Ed.; Marcel Dekker: N.Y., 1999; p 253.

40. Salager, J. L.; Briceño, M. I.; Bracho, C. L. Heavy Hydrocarbon Emulsions - Making use of the State of the Art in Formulation Engineering. In *Encyclopedic Handbook of Emulsion Technology*; Sjöblom, J., Ed.; Marcel Dekker: N.Y., 2001; Vol. 20; p 455.

41. Vinatieri, J. E. *Soc. Pet. Eng. J.* **1980**, *20*, p 402.

42. Padilla, V.; Rangel, M. G.; Bullon, J.; Rodríguez-Malaver, A.; Gonzalez, A. M.; Rojas, O. J. "Surface activity of lignin fractions obtained by membrane-separation technologies of black liquors", Congreso Iberoamericano de Inv. Celulosa y Papel CIADICYP 2002, 10/10-12, 2002, São Paulo, Brasil.

43. Liu, G. L.; Liu, Y. S.; Ni, J. R.; Shi, H. C.; Qian, Y. *Desalination* **2004**, *160*, p 131.

44. Li, J. X.; Sanderson, R. D.; Hallbauer, D. K.; Hallbauer-Zadorozhnaya, V. Y. *Desalination* **2002**, *146*, p 177.

45. Carlsson, D. J.; Dal-Cin, M. M.; Black, P.; Lick, C. N. *J. Membrane Sci.* **1998**, *142*, p 1.

46. Li, L.; Zhao, Z. Y.; Huang, W. L.; Peng, P.; Sheng, G. Y.; Fu, J. M. *Organic Geochemistry* **2004**, *35*, p 1025.

47. Goring, D. A. I.; Vuong, R.; Gancet, C.; Chanzy, H. *J. Appl. Polym. Sci.* **1979**, *24*, p 931.

48. Pla, F.; Robert, A. *Holzforschung* **1984**, *38*, p 213.

49. Vainio, U.; Maximova, N.; Hortling, B.; Laine, J.; Stenius, P.; Simola, L. K.; Gravitis, J.; Serimaa, R. *Langmuir* **2004**, *20*, p 9736.

50. Koda, K.; Gaspar, A. R.; Argyropoulos, D. S. *Holzforschung* **2005**, *59*, p 612.

51. Lindström, T. *Colloid & Polymer Sci.* **1979**, *257*, p 277

52. Lindström, T. *Colloid & Polymer Sci.* **1980**, *258*, 168

53. Lindström, T.; Westman, L. *Colloid & Polymer Sci.* **1980**, *260*, p 594

54. Lindström, T.; Westman, L. *Colloid & Polymer Sci.* **1980**, *258*, p 390.

55. Woerner, D. L.; McCarthy, J. L. *Macromolecules* **1988**, *21*, p 2160.

56. Salager, J. L.; Forgiarini, A.; Marquez, L.; Pena, A.; Pizzino, A.; Rodriguez, M. P.; Rondo- Gonzalez, M. *Adv. Colloid & Interface Sci.* **2004**, *108*, p 259.

Chapter 13

Modifying the Functionality of Starch Films with Natural Polymers

Lucian A. Lucia[1], Dimitris S. Argyropoulos[1], Weiping Ban[2], and Jianguo Song[3]

[1]Department of Forest Biomaterials Science and Engineering, North Carolina State University, Raleigh, NC 27695–8005
[2]Georgia Institute of Technology, Institute of Paper Science and Technology, 500 Tenth Street, N.W., Atlanta, GA 30332–0620
[3]Dalian Institute of Light Industry, Dalian, China 116034

The physical and chemical properties of composite starch-based films containing cellulosic fiber, chitosan, and gelatin were investigated. Films containing both cellulosic fibers and chitosan demonstrated tremendous enhancements in film strength and gas permeation. The water absorbency of composite films could be greatly reduced in film composites containing cellulosic fibers and gelatin, but the inclusion of chitosan into these films provided a higher hydrophilicity, increasing water absorbency. Film transparency was not noticeably affected in the composite films that were made. These films may have wide application in the food packaging, agricultural mulching, and the biomedical industries.

The Importance of Films in Society

Background to Agricultural Applications

Films may be defined as two-dimensional solid substrates of multivariate functionality whose physical and chemical characteristics are derived in part or whole from the chemical components used in their manufacture. In addition, the physical and chemical properties of these films can be judiciously modulated or tuned by affecting the composition of the material. For example, the use of composite plastic films having modulated degradability characteristics for agricultural applications has burgeoned into an extremely attractive value proposition to increase crop productivity without sacrificing traditional planting and harvesting techniques.[1,2] The importance of the role of plastic films in the daily needs of modern society cannot be overemphasized. They have been widely applied in food and merchandise packaging, agricultural mulching, gas and liquor separation, and medical technologies. In 2000, more than 40 million metric tons of petroleum-based plastic material was used worldwide in the packaging industry alone. Indeed, the versatility of these materials for current engineering and technological applications is almost limitless; they have enjoyed wide usage in the biomedical (dialysis, separations, gauzes, surgical films), textiles (wound dressings, sound insulation), food (wrappings, storage media), building (house insulation, window films), computer & electronics (electronic films, protective coatings), and dental (teeth bleaching, sealants) industries.

However, the bulk of these materials was for single-use applications and was thus discarded, contributing to growing landfills. Given real and pressing environmental concerns, the rising cost of petroleum, and the eventual depletion of our oil reservoirs, many research efforts have focused on developing renewables-based or composite biodegradable films. Such films would overcome the intrinsic deficiencies of petroleum-based films given their renewability, biodegradability, and biocompatibility.

As alluded to above, an economically and environmentally powerful and immediately ostensible application of renewables-based films is for the development of agricultural mulching films. The use of plastic mulches is common in agricultural and horticultural industries where plant growth can be twice as great on mulched soil compared to non-mulched soil. There are several factors which contribute to this improved growth. First, soil temperature in the planting bed is raised, promoting faster crop development and earlier harvest. Black plastic mulch can give a harvest earlier by some 7-14 days, while clear plastic may advance the harvest date by 21 days. Second, because black and white-on-black mulches will reduce light penetration to the soil, weeds cannot generally survive under these mulches. Therefore, cultivation is eliminated, except in the area between the mulched strips where weed growth can be controlled by cultivation or by the use of a chemical herbicide.[3] Third, soil water loss is reduced under a plastic mulch. As a result, more uniform soil moisture is maintained and irrigation frequency can be reduced. Fourth, excess water runs off the impervious mulch. Fertilizer beneath the mulch is not lost by leaching, so that fertilizers are optimally used and not wasted. Finally, the soil under plastic mulch remains loose, friable, and well aerated. The roots will have access to adequate oxygen, and microbial activity is enchanced, contributing to a thriving crop. The edible product from a mulched crop is cleaner and less subject to rot, since no soil is splashed onto the plants or fruit. To keep plants clean, they are grown in a raised bed that is firm and slightly convex, with the highest point down the center of the row, while the plastic is stretched tightly to encourage the run-off of water.

Currently available mulch films are nearly impervious to movement of many gasses including carbon dioxide. This has both advantages and disadvantages. Mulches increase the effectiveness of chemicals applied as soil fumigants. Because of the impervious nature of the plastic mulch, it acts as a barrier to gas escape and keeps gaseous fumigants in the soil. Water is shed from the row area by the raised tapered bed so that excess water runs off the field, thus reducing stress from excess soil water. Research has shown that high levels of carbon dioxide may build up under the plastic. Because the film does not allow the gas to penetrate, it can escape through the holes punched for the plants creating a "chimney effect" that results in CO_2 for the actively growing leaves and contributing to improved growth. However, plastic mulches can also result in inadequate gas exchange and poor root growth conditions.

Many available agricultural film mulches have undesirable end-of-life characteristics, such as low degradability, poor mechanical properties, and environmental incompatibility. For example, polyethylene mulching films must be collected and burned after they have served their purpose. Therefore, this is an added expense that detracts from the value of polyethylene as a useful mulching film since it increases cultivation costs and contributes to air pollution.

Development of Biobased Films

There have been many research efforts to address the requirements of agricultural mulching films in ways that have utilized the attractive features of plastics mentioned above, but also to minimize their environmental incompatibility. Lahalih had the ingenious idea of developing a novel plastic film that could be photo-oxidatively degraded and provide a slow micronutrient release during degradation.[3] However, the surrounding environmental issues of plastic manufacture and soil biocompatibility mitigate the benefits afforded in this work. To better address environmental issues, Saulnier decided to develop cellulosic films derived from agricultural by-products, wheat bran and maize, through esterification by lauroyl chloride. Yet, this work is complicated by chemical transformation issues such as removing the heteroxylans and lignins in the by-products by acid prehydrolysis and a final esterification step.[4] Indeed many efforts have explored mulch film biocomposites that are composed of natural materials such as cellulosics coupled with synthetics such as polyesters or polyvinyl alcohols.[5,6] Although unique in their approaches, they nevertheless suffer from the ultimate drawbacks of facility of manufacture, cost, and biocompatibility. Most films these days belong to the family of polyolefins and their copolymers such as poly(ethylene-co-vinyl acetate) as monolayers or multilayers with a certain amount of poly(vinyl chloride) for mulching or ensilage.[7] Obviously, as discussed earlier, the amount of plastics at the end of their useful life for crops or fodder poses serious disposal problems due to regulatory restrictions for land filling or incineration. In fact, recycling may be impossible for photodegraded materials and composites since they may aggregate in the soil to restrict nutrient uptake, reduce moisture levels, and reduce the carbon value of the soil.

Therefore, the development of environmentally compatible, recyclable, inexpensive, and functional mulch films is a high priority for US farming efforts and is the focus of this chapter. Agrobiomaterials were developed that not only provide the typical useful functions of mulching films, but that have superior nutrient release and end-of-life properties. These features may be accommodated using pulp and associated biomaterials which will be explored.

There are two realistic ways to develop biodegradable films: one is to incorporate biodegradable synthetic materials, such as polylactide, polyesters, and polycaprolactone into the films. However, given the high cost of these materials, their use is limited. Thus, the alternative way, use of abundant, natural materials, is becoming more and more attractive since unlike the latter polymers they are renewable.

Starch is one among a number of naturally occurring biomaterials which are abundant and have a low cost for commercial applications. It has been widely used in fermenting as well as other chemical manufacturing industries. In packaging and agricultural mulching industries, starch-based films have significant potential to replace synthetic films such a polyethylene (PE). In the last several decades, great efforts have been made to develop starch-based films that display improved film properties. However, two disadvantages of starch-based films that have limited their use are inherently low moduli of elasticity (strength) and high hydrophilicity (leads to enhanced biodegradability). To overcome these properties, two approaches have been typically adopted: the copolymerization of starch with synthetic polymers or the admixture of starch with polymers.[8-12] However, synthetic polymers such as PE are non-biodegradable even after the starch has degraded. Therefore, such a film would impact the environment unfavorably.

Recently, there has been a research effort that has reported that cellulosic fibers may be added to modified starches such as hydroxypropylated starch to enhance film mechanical properties.[13] Cellulose fibers were found to improve starch film performance by decreasing water uptake.[14] Blending starch with polyester polymers demonstrated good biodegradability for agricultural mulch applications,[8] while gelatin, another biopolymer, was reported to function as a strength reinforcement agent for improved mechanical performance.[15]

Hence, the objective of the present chapter is to provide a simple study on the functionality of several natural biopolymers on the physical and chemical properties of starch-based films. The effects of these polymers on film strength, water absorbency, transparency, and gas permeation are presented.

Experimental

Materials
Starch: Commercial corn starch was used as is.

Chitosan: Commercial chitosan was dissolved in 1% acetic acid to make a 2% m/v solution.

Gelatin: Analytical chemistry purity (purissimo) class gelatin was dissolved in water to make a 20% solution.

Fiber: Bleached mechanical pulp and kraft Southern pine pulp from laboratory manufacture were dispersed in a blender with sufficient water to make a pulp slurry of 10% solids content.

Film preparation
The materials were mixed in a quantitative manner with starch while the mixture of starch was gelatinized at 80°C for 25 minutes. Glycerol as a plasticizer was added at a 25% ratio based on total dry weight of composite materials. The total mixture of starch and other materials was 20 g and kept at a concentration of 4g/100ml. Composite material design was varied according to the described experiment plan. After cooling and observing the dispersion was homogeneous, 50 ml of the sample (2g dry material) was poured into a Petri dish and then dried at 60°C in an oven in order to cast the films.

Strength measurement
Tensile tests were performed with an ALWETRON TH1 horizontal tensile tester in accordance with TAPPI Method T-404 cm-94.

Water absorption
Film samples (80mm×15mm) were conditioned at 25°C in a desiccator containing sodium sulfate to ensure a relative humidity ratio of 95% for a designated time. The excess water was then dried with tissue paper applied to the film surface. Water uptake was calculated as follows:

$$\text{Water absorption \%} = W\text{-}W_0/W_0 \times 100$$

where: W_0 is dried film sample weight, and W is the weight of the film sample after exposure to 95% RH, respectively.

Opacity measurements
Film opacity was measured with a BNL-3 Opacimeter. The measurements were carried out on a black and white background.

Gas permeation
A 20 ml headspace vial was sealed with a crimp cap and septa, and its bottom was cut off and polished. The film sample was covered and sealed on the exposed bottom of the vial. A 1.0 ml slug of carbon dioxide was injected into the sealed vial and the vial was allowed to remain in a bulk air environment to allow a free gas exchange through the film. After 10 minutes, the content of carbon dioxide in the vial was determined by gas chromatography. A control sample without the film was also tested to establish a baseline for the changes in the content of carbon dioxide.

Results and Discussion

Film Strength Properties

As already cited, poor strength properties (elastic moduli, tensile, compression) of neat starch films are major drawbacks for their use in diverse commercial applications. Improvements in film mechanical properties are therefore mandatory; Figure 1 provides an illustration of how the addition of cellulosic fiber impacts film strength. An appreciable enhancement by increasing the cellulosic fiber content was immediately observed, in which as

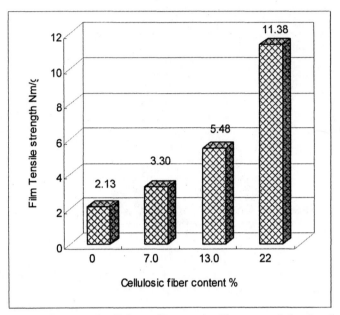

*Figure 1. The influence of cellulosic fiber on the film strength for the short fiber fraction in which bleached mechanical pulp was used as short fiber (refer to **Experimental** for details on film development).*

much as five-fold increases in film tensile strength were achieved by incorporating up to 22% fiber content to these films.

To investigate how different fiber types influence film strength, long fibers (up to 10 mm) from softwood kraft pulps were applied to starch films. Figure 2 shows film strength development from using Southern softwood long fibers. Although no obvious differences in the tensile properties of film were observed relative to the short fiber, applying long fibers in starch had a negative effect on film uniformity. Because of the entanglement induced from long fiber-fiber interactions, it was difficult to uniformly disperse such fibers in a starch dispersion to result in a homogeneous film.

Chitosan, another biopolymer that has a cellulose-like structure, was mixed in starch with the purpose of enhancing the film strength since its tertiary structure (having a glucose-based monomer unit) is similar to that of starch and cellulose and thus should undergo hydrogen bonding. The chitosan compound dissolved in weak acidic acid solution and was easily incorporated into the starch dispersion to form a homogeneous film. The influence of chitosan on the strength of these starch-based films is plotted in Figure 3.

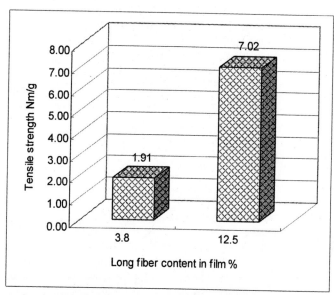

Figure 2. The influence of cellulosic fiber on the film strength using the long fiber fraction of kraft softwood fibers.

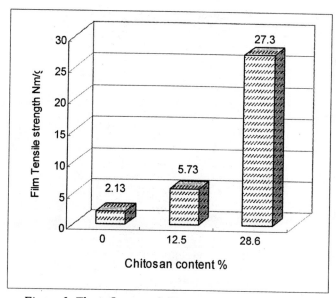

Figure 3. The influence of chitosan on the film strength.

In a similar vein to cellulosic fiber, chitosan significantly enhanced starch film strength. A tremendous tensile increase was achieved when the chitosan content comprised 12.5% of the starch film. Compared to the original starch film, an almost ten-fold increase in tensile strength was reached with 28% content of chitosan in the film. The results clearly indicate that both cellulosic fiber and chitosan can easily be incorporated into the films while significantly enhancing their strength. Figure 4 shows the impact of gelatin on starch film strength. Gelatin has been previously reported to improve starch film strength.[9] However, in these experiments, gelatin did not show any obvious enhancements in starch-based film strength.

Water Absorbency

Owing to the strong hydrophilicity of starch molecules, a neat starch film displays high water absorbency (54.7% is saturation limit). This factor impedes the application of pure starch films for most non-absorbency applications. However, incorporation of varying amounts of cellulosic fibers into the starch matrix was able to reduce the water absorbency of these films as shown in Figure 5. Although cellulose is typically a hydrophilic polymer, the

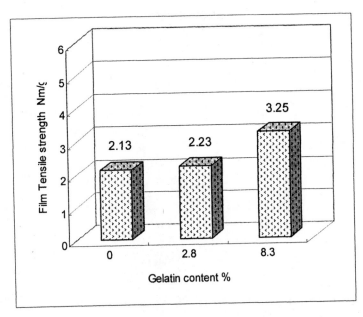

Figure 4. The influence of gelatin on the film strength.

high crystallinity of the cellulose polymer and tight microfibrils structure within the fiber tend to discourage water absorbency as compared to amorphous starch. Approximately 40% reduction in total water absorbency was achieved by a 6.7% fiber addition in the film, while the film water absorbency was further reduced to more than 50% when the cellulosic fiber content reached to 12.5%. However, continuing to increase fiber percentage within the film did not achieve any further improvement in water absorbency. When the fiber content of the films reached 22%, the water uptake increased to 36.1% compared to 26.7% of water uptake in the case of 12.5% fiber content. This result indicates that the combination of cellulosic fiber with starch improves water resistance to a certain degree since, as already known, cellulosic fibers are hydrophilic materials, although their hydrophilicity is weaker than that of starch.

Contrary to cellulosic fibers, chitosan had a reverse role on film water absorbency. Figure 6 illustrates the influence of chitosan on film water uptake. With increasing chitosan content to 28%, water uptake of film linearly increased. When the chitosan content reached one third of the total film, as much as a two-fold increase in water uptake was observed compared to a pure starch film. A higher water absorbency of chitosan is ascribed to its chemical features since besides the hydrophilic hydroxyl groups on the chitosan molecule, it possesses strong hydrophilic amine groups on the chitosan molecule, which encourage a

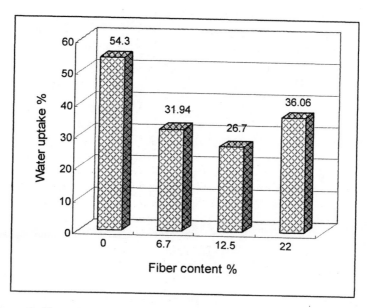

Figure 5. The influence of mechanical pulp fiber on film water absorbency after 24 hours had elapsed.

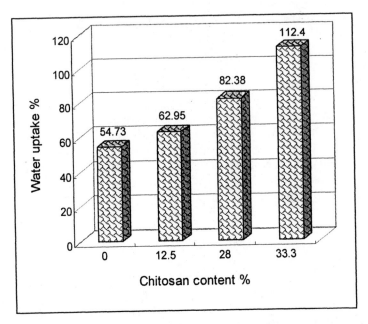

Figure 6. The influence of chitosan on film water absorbency after 24hrs. had elapsed.

higher water absorbency. Furthermore, unlike the cellulosic fibers, chitosan is freely soluble which allows these molecules more degrees of freedom with respect to water absorption.

Gelatin functioned as an efficient water resistant agent: its influence on film water uptake is plotted in Figure 7. A 2.8% total incorporation of gelatin to the film resulted in a 40% reduction in water uptake, while almost 60% lower water uptake was achieved by 8.3% gelatin levels in the film.

To further investigate the influence of various non-starch components on film water absorbency, the changes in water uptake over a discrete time window were plotted in Figure 8. Although different fiber content resulted in differences in water absorbency levels, the water uptake over an extended time at all fiber levels did not exhibit much change. The latter result certainly demonstrates how the cellulosic fiber component influences water uptake. Even though cellulosic fibers were not expected to impart very high water resistance to the film, it was able to obviously improve water resiliency of starch-based films despite changes in water conditions. Because starch films are very sensitive to changes in moisture that results in changes in their mechanical properties, controlling film moisture content is highly important. Hence, the moisture stabilization provided

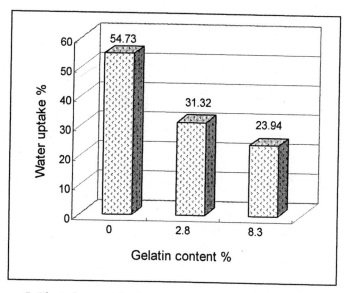

Figure 7. The influence of gelatin on film water absorbency after 24hrs. had elapsed.

by cellulosic fibers is significant for new applications that are moisture sensitive such as gas selective membranes, ion channels, etc.

A film containing chitosan also demonstrated a high sensitivity to water absorbency. Water uptake not only increased with an increase in the chitosan content, but kept increasing over a long time period. High water sensitivity would tend to impair film strength. To avoid adverse impacts on film strength, high chitosan content was avoided. Gelatin may be used to enhance water resistance of the films (Figure 7). However, at a low gelatin content (2.8%), water absorbency displayed an obvious change over time. The data indicated that a low gelatin level would not be enough to provide high moisture stabilization though it did impart lower water uptake. When a higher gelatin level was added to the starch-based films, 8.3% in the current experiments, the films were more resistant to water uptake.

Opacity of Films

Figure 9 illustrates the influence of kraft softwood cellulosic fibers on the transparency of starch-based films. The transparency of the film can be decreased with an increase in the fiber content of the film. As opposed to cellulosic fibers, as shown in Figure 10, chitosan did not impair film transparency.

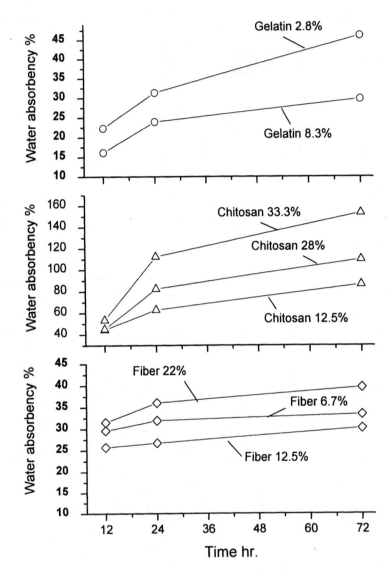

Figure 8. The influence of various non-starch components on the water uptake rate of various films.

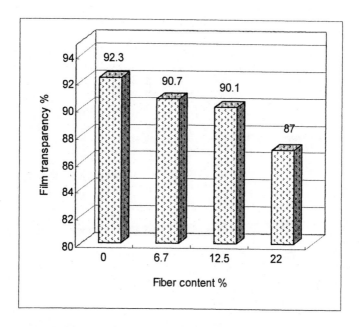

Figure 9. The influence of kraft softwood fibers on film transparency.

When light strikes a film surface, it is reflected, absorbed, or transmitted and as a result of the light-product interaction a color, gloss and transparency is exhibited by the film. The Kubelka-Munk theory[10] proposed a model for the explanation of multi-components materials optical properties. The light diffusion occurs at the interface of these different components and greatly impairs the light transmission. Scallan[11-13] investigated the influence of the fundamental properties of paper as well as fillers on paper optical property. Adding filler into cellulose fiber paper increases the interface area and leads to an increase in the light diffusion within the internal structure of paper, and as a result, paper opacity is increased. In the case of multi-component films, a chitosan solution mixed with a starch matrix formed a homogeneous matrix. After drying, this homogeneous matrix formed a continuous phase in the film state. On the other hand, cellulosic fibers are not water soluble and randomly disperse in the starch matrix, and as such, exist in film as a discontinuous phase after drying. Therefore, cellulosic fibers provide more interfacial area within a film structure. The light diffusion is increased and results in a reduction in film transparency.

Overall, starch-based films maintain relatively high transparency while additions of various biopolymers do not have significant effects on the transparency.

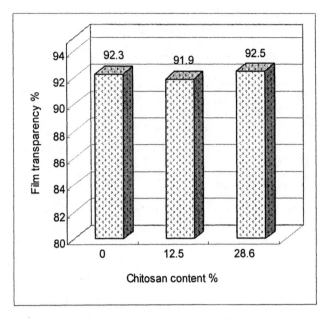

Figure 10. The influence of chitosan on film transparency.

Influence of Fiber and Chitosan on Film Gas Permeation

Many different films have been widely applied in separation processes including gas and liquor separation. In the packaging and agricultural industries, the gas permeability of films are also important for specific functions. In the current work, preliminary results for the gas permeation of starch-based films are provided. In Figure 11, the changes of carbon dioxide gas permeation through different cellulosic fiber contents of the composite films are illustrated. Interestingly, the permeation rate of carbon dioxide was influenced by varying the cellulosic fiber ratios in the starch film. Over a certain time period, the concentration of carbon dioxide in a film-sealed vial increased with the increase of fiber content. For a pure starch film, the carbon dioxide concentration in the vial reached almost the same level as that in air after ten minutes. However, as compared to a starch film, a 50% increase in the carbon dioxide concentration was achieved by using a film having 19% cellulosic fiber content. Similar features for chitosan composite films were also observed (Fig.12). When the chitosan content in these films reached 33% of the total film composition, almost a two-fold increase in the remaining carbon dioxide concentration could be

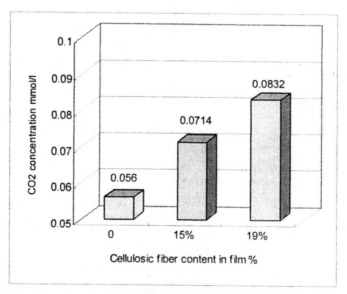

Figure 11. The influence of kraft softwood cellulosic fiber content on film gas permeability under the following conditions: 10 min, 1.0ml CO_2 in a 20 ml vial.

achieved in the vial as compared to a pure starch film. This phenomenon could not be explained by a simple variation in the film densities by changing the film components. In the fiber and chitosan films, films densities decreased with an increase in the concentration of the biopolymers as shown in Figure 13. Both cellulosic fiber and chitosan lower film density. Certainly, a lower film density could result in larger film pore size and faster gas permeation rate. Therefore, biopolymer incorporation in these films has a function that is currently not understood for gas permeation and will serve as the subject of future investigations.

Conclusions

The incorporation of biopolymers with starch can modify the physical and chemical properties of starch-based films. There are some important improvements for starch-based film which could be achieved while maintaining the natural advantages of starch films. The inclusion of cellulosic fibers into starch film greatly enhances film strength by as much as five-fold in the film

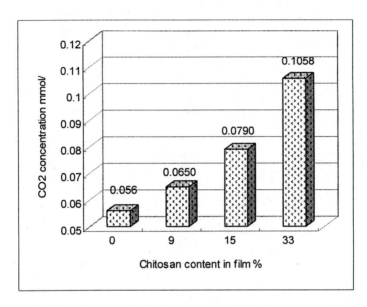

Figure 12. The influence of chitosan content on film gas permeability under the following conditions: 10 min, 1.0ml CO_2 in a 20 ml vial.

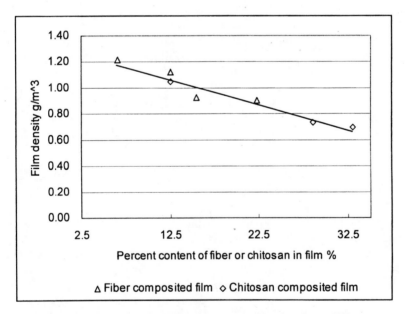

Figure 13. The film density as a function of kraft softwood fiber or chitosan content.

tensile strength by adding 22% of cellulosic fiber in film compared to neat starch films. Furthermore, cellulosic fibers also demonstrated an important influence on film water absorbency. A 12.5% of fiber content could decrease film water uptake more than 50% than that of the neat starch film. Another significant effect of fiber on water absorbency is its stabilization of film water sensitivity although there is a slight decrease in film transparency.

Similarly to the cellulosic fibers, chitosan contributed a tremendous improvement to film strength. A ten-fold increase in film strength was observed with 33% chitosan content. However, chitosan demonstrated a negative impact on film water absorbency. The increase in film water uptake was observed with an increase in the chitosan content, especially in the high chitosan content.

Gelatin imparted good water resistance to the starch-based films. A small amount of gelatin effectively reduced film water uptake and did not show any obvious impact on other film properties.

Both cellulosic fiber and chitosan demonstrated a profound influence on film gas permeation. The gas permeation rate decreased with an increase in both fiber and chitosan content although the mechanism needs more investigation.

References

1. Copinet, A.; Betrand, C.; Longieras, A.; Coma, V.; Couturier, Y. "Photodegradation and Biodegradation Study of a Starch and Poly(Lactic Acid) Coextruded Material." *J. Polym. Environ.* **2003**, *11*, pp 169-179.

2. Laverd, G. "Agricultural Films: Types and Applications." *J. Plast. Film Sheet.* **2002**, *18*, pp 269-277.

3. Lahalih, S.M.; Akashah, S.A.; Al-Hajjar, F.H. "Development of Degradable Slow Release Multinutritional Agricultural Mulch Film." *Ind. Eng. Chem. Res.* **1987**, *26*, pp 2366-2372.

4. Chauvelon, G.; Gergaud, N.; Saulnier, L.; Lourdin, D.; Buléon, A.; Thibault, J.-F.; Krausz, R. "Esterification of Cellulose-Enriched Agricultural By-Products and Characterization of Mechanical Properties of Cellulosic Films." *Carb. Polym.* **2000**, *42*, pp 385-392.

5. Halley, P.; Rutgers, R.; Coombs, S.; Kettels, J.; Gralton, J.; Christie, G.; Jenkins, M.; Beh, H.; Griffin, K.; Jayasekara, R.; Lonergan, G. "Developing Biodegradable Mulch Films from Starch-Based Polymers." Starch/Stärke **2001**, *53*, pp 362-367.

6. De Prisco, N.; Immirzi, B.; Malinconico, M.; Mormile, P. ; Petti, L. ; Gatta, G. "Preparation, Physico-Chemical Characterization, and Optical Analysis of Polyvinyl Alcohol-Based Films Suitable for Protected Cultivation." *J. Appl. Polym. Sci.* **2002**, *86*, pp 622-632.

7. Pacini, L. Proceedings of the 13th International Congress of C.I.P.A., Verona, Italy, 8-11 March, 1994.

8. Otey, F.H.; Mark, A.M.; Mehltretter, C.L.; Russell, C.R. "Starch-Based Film for Degradable Agricultural Mulch." *Ind. Eng. Chem.* **1974**, *13*, pp 90-92.

9. Shamekh, S.; Myllärinen, P.; Poutanen, K.; Forssell, P. "Film Formation Properties of Potato Starch Hydrolysates." *Starch/Stärke* **2002**, *54*, pp 20–24.

10. Chauvelon, G.; Gergaud, N.; Saulnier, L. "Esterification of Cellulose-Enriched Agricultural By-Products and Characterization of Mechanical Properties of Cellulosic Films." *Carbohyd. Polym.* **2000**, *42*, pp 385–392.

11. Nerın, C.; Tornes, A.R.; Domeno, C.; Cacho, J. "Absorption of Pesticides on Plastic Films Used as Agricultural Soil Covers." *J. Agric. Food Chem.* **1996**, *44*, pp 4009-4014.

12. Chiellinia, E.; Cinellia, P.; Cortia, A.; Kenawyb, E.-R. "Composite Films Based on Waste Gelatin: Thermal–Mechanical Properties and Biodegradation Testing." *Polym. Degrad. Stabil.* **2001**, *73*, pp 549–555.

13. Kim, D.-H.; Na, S.-K.; Park, J.-S. "Preparation and Characterization of Modified Starch-Based Plastic Reinforced with Short Pulp Fiber. 1. Structural Properties." *Appl. Polym. Sci.* **2003**, *88*, pp 2100-2107.

14. Dufresne, A.; Vignon, M.R. "Improvement of Starch Film Performances Using Cellulose Microfibrils." *Macromolecules* **1998**, *31*, pp 2693-2696.

15. Halley, P.; Rutgers, R.; Coombs, S. "Developing Biodegradable Mulch Films from Starch-Based Polymers." *Starch/Stärke* **2001**, *53*, pp 362–367.

16. Kong, X.; Zheng, Z. "Soluble and Edible Corn Starch Package Film and Its Preparation, Chinese patent, CN, 2002-115925 20020604.

17. Kubelka, P. and Munk, F. "Ein beitrag zur optik der farbanstriche." *Z. Tech. Phys. (Leipzig)* **1931**, *12*, pp 593-601.

18. Scallan, A. M.; Borch, J. Fundamental parameters affecting the opacity and brightness of uncoated paper. Fundam. Prop. Pap. Relat. Its Uses, Trans. Symp. (1976), Meeting Date 1973, 1, pp 152-171.

19. Middleton, S. R.; Desmeules, J.; Scallan, A. M.. The Kubelka-Munk coefficients of fillers. *J. Pulp Pap. Sci.* **1994**, *20*, pp 231-235.

20. Middleton, S. R.; Scallan, A. M. The optical properties of bleached kraft pulps. *Nord. Pulp Pap. Res. J.* **1992**, *7*, pp 22-24, p 29.

Chapter 14

Modeling Water Transport in Saturated Wood

Aaron J. Jacobson and Sujit Banerjee

Institute of Paper Science and Technology and School of Chemical and Biomolecular Engineering, Georgia Institute of Technology, 500 Tenth Street N.W., Atlanta, GA 30332–0620

The diffusion rate of tritiated water into water-saturated pine and aspen particles was studied in order to establish a benchmark for the transport of solutes into wood. The diffusion follows a Fickian mechanism. The tortuosity for the diffusion of water into wood is quite low at about 1.6 and increases with decreasing particle size. The tortuosity for aspen is higher because the shorter fiber structure in aspen gives rise to a more extensive network of pores. Diffusion into free and bound water occurs at the same rate. Also, diffusion into and out of the particles is nearly identical, demonstrating the reversibility of the process. No hysteresis was evident, in contrast to behavior for water adsorption on unsaturated wood. The implications of these findings to pulping and wood drying are discussed.

The main components of wood are hemicelluloses, cellulose, and lignin. Each constituent can be processed further into other products through chemical or biological processes. In native wood these components are integrated into structures that require energy to separate. In kraft pulping, the action of chemicals at high temperature and pressure separates most of the lignin from the cellulose, but removes most of the hemicelluloses as well. Even if the lignin were removed selectively, there would still be issues with the cellulose because of its crystallinity and high degree of polymerization [1], which would restrict the ingress of chemical or biological reagents.

Chemicals can transport into wood through penetration and diffusion. Penetration is the liquid transfer into the air-filled cavities of the chips and is driven by capillary forces and pressure. Once penetration is complete, the only way for chemicals to enter wood is through diffusion, which is governed by Fick's law, equation 1,

$$\frac{dm}{dt} = -D\frac{dc}{dx}dy\,dz \qquad (1)$$

where dm/dt is the transfer rate of dissolved material, D represents the diffusion constant, dc/dx is the concentration gradient in the x-direction, and dydz stands for the available cross section.

For both diffusion and penetration, the dimensions of the wood are key components in describing the mass transport. For the production of chemicals from wood biomass the shape and size of the wood should not have an effect on the quality of the product. In fact, very small wood pieces such as sawdust could be preferred because of the shorter diffusion path lengths involved and the higher degree of cellulose exposure.

In this paper the results are reported on the diffusion of water into saturated wood in order to establish a reference point against which the uptake of other compounds could be benchmarked.

Diffusion in Porous Media

Diffusion in porous media such as saturated soils is very similar to diffusion in saturated wood particles. There is no flow of water, so any mass transfer through the soil must occur through diffusion. Solutes can only move through the liquid-filled areas of the solid [2]. Thus, the effective diffusion coefficient will relate to the actual path taken by the solute[3].

The shape and size of the solid is very important. In soils research, the material may be manipulated to give a particular shape. Rao et al. [4] successfully modeled soil aggregates as spheres, and used chloride ions and tritiated water to characterize the diffusion process. The equations governing diffusion into solids are complex and cannot be solved analytically for some geometries. In diffusion studies the particles are usually modeled as cylinders or spheres since the differential equations governing their shapes have been solved. In some instances creative modifications are needed. Novakowski and van der Kamp[5] used a radial diffusion method for analyzing soil samples, where their soil samples were removed as large cylindrical samples, and the center was cored out to allow for a diffusion reservoir.

One method that applies to both spherical and non-spherical particles is to reduce them to an equivalent sphere. This can be accomplished in several ways, Rao et al.[6] modeled cubic aggregates with spheres that had an equivalent volume. Thus an effective spherical radius could be determined and used in the equations. Because diffusion equations are dependent on the size of the solid, it is helpful to have a narrow size distribution. If the particles have a wider range of sizes, use of a weighted average is effective.

The use of tritiated water (HTO) allows diffusion to be measured in saturated media, and has been applied to the study of sediments, soils, and membranes [7, 8]. Tritiated water is a conservative tracer; *i.e.* it moves with water and can be used to track water flow. It has been extensively used for this purpose in hydrogeology and other fields[9-11].

Diffusion of Tritiated Water into Wood

Materials and Methods

Freshly cut Southern pine (*Pinus taeda*) particles were obtained from the Georgia-Pacific particleboard plant in Vienna, GA. Aspen (*Populus tremuloides*) chips from Alberta, Canada were obtained from Millar Western and then processed into small particles in a Wiley mill. They were fractionated with multiple Tyler sieves mounted to an automated shaker and stored airtight in a cold room. The screen openings ranged from 0.6 to 4.0 mm.

The particles were stirred in deionized water for at least 24 hours, at which point no floating material was observed. They were then washed at least ten times to remove any particulate matter or extractives that may have diffused out of the wood, filtered, and stored in airtight vessels. Experiments were completed shortly thereafter to minimize moisture loss from the wood. A 40 ml aliquot of tritiated water (from Amersham Biosciences) was added to about 20 g. of the wet wood particles in a Pyrex bottle. Samples (100 µl) were removed at timed intervals, added to vials containing 10 ml of Scintiverse E scintillation cocktail (from Fisher Scientific), and the radioactivity determined with a liquid scintillation counter. Measurements were continued until the tritium concentration in water leveled off, indicating that equilibrium had been reached. This point corresponded to within one percent of the equilibrium value expected from the amount of water added and the water content of the wood. In some experiments, the wood was removed after equilibrium was established, resuspended in water, and the diffusion of the tritiated water out of the particles measured by periodically sampling the bulk water.

The geometry of particles of each size fraction was determined. Wet particles (not used in the experiment) were placed on a white background and painstakingly aligned into rows and columns. Images were then captured with a digital camera and analyzed with ImageJ software version 1.32 from the National Institute of Health. The software converted the images to 8-bit thresholded images. The particle analysis yielded perimeter, area, longest and shortest dimensions, and the aspect ratio. The thickness dimension could not be imaged and was measured manually from at least 150 particles from each size fraction. This dimension was at least six times smaller than the longest dimension. An effective radius was used in the model. A similar approach was taken in analogous experiments using a mixture of particles of different shapes and sizes[6].

Data Analysis

The aspect ratios of the particles suggested that they could be modeled as cylinders. Fick's second law describes mass transfer in cylindrical coordinates through Eq 2, where C_a represents the concentration of species a, D_e is the

$$\frac{\partial C_a}{\partial t} = D_e \left[\frac{\partial^2 C_a}{\partial r^2} + \frac{1}{r} \frac{\partial C_a}{\partial r} \right] \tag{2}$$

effective diffusion coefficient, r is the distance from the center axis of the cylinder, and t represents time. The boundary conditions are

$$C(a,t) = C_b(t); t > 0$$

$$C(r,t) = 0; 0 \leq r \leq a; t = 0$$

$$\frac{\partial C}{\partial r} = 0; r = 0, \text{ for all } t$$

M_t and M_∞ are defined as the mass of solute inside the cylinder at time t and at infinity, respectively. M_t/M_∞ is a reaction coordinate, which is initially zero and reaches one at completion. Using this convention and solving the differential equation, two solutions emerge, one for short periods where $M_t/M_\infty < 0.5$ and one for longer times [12]. Our data points were all taken after $M_t/M_\infty > 0.5$ and only the solution for long time periods was used. This is given by Eq 3 where α_n's are the positive roots of the Bessel function of order zero. The variable r is the radius of the particle. Equation 3 was programmed into Excel,

$$\frac{M_t}{M_\infty} = 1 - \sum_{n=1}^{\infty} \frac{4}{r^2 \alpha_n^2} \exp\left(-D_e \alpha_n^2 t\right) \qquad (3)$$

and the summation taken to the fifth term, beyond which the solution did not change with additional terms. A least squares approach was used to obtain the best fit of the model to the experimental data. Excel solver was used to minimize the residuals and an effective diffusion coefficient (D_e) was determined. This parameter is related to the porosity of the material (ε), the self-diffusion coefficient (D^o), and the tortuosity (τ) through Eq 4. The tortuosity is a dimensionless number that describes how the diffusion of a substance is impeded

$$D_e = \frac{\varepsilon}{\tau} D^o \qquad (4)$$

in a material. It is sometimes referred to as a retardation factor [2]. Porosity is also a dimensionless number that represents the void fraction of a material where diffusing molecules can migrate.

Results and Discussion

The approach to equilibrium could not be measured for pine particles of <1 mm as equilibrium was reached before the first sample could be taken. Results from measurements made for 2.0-2.4 mm particles are shown in Figure 1, from which an effective diffusion coefficient of water of 1.0×10^{-5} cm^2 s^{-1} is obtained. Use of 2.5×10^{-5} cm^2 s^{-1} as the self-diffusion coefficient of water at 25°C [13, 14], and a porosity of 0.69 for pine [15] leads to a tortuosity of 1.6. In other words the diffusion of water into these saturated particles will only be about 40% slower than the diffusion of water in water itself. Figure 2 illustrates results for pine particles in the 1.7-2.0 mm size range at two temperatures. If an Arrhenius relationship is assumed, then an activation energy of 21.7 kJ mol^{-1} results. This

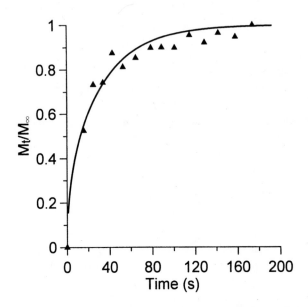

Figure 1. Approach to equilibrium for pine particles in the size range of 2.0 – 2.4 mm at 25° C. (Reproduced with permission from Holzforshung: *2006, 60, 509–563. Copyright 2006 Walter de Gruyter.)*

is in good agreement with published values of 20 kJ/mol [16] and 20.5 kJ mol[-1] [13] for the self diffusion of water. As expected, the tortuosity remains unchanged with temperature. Diffusion parameters for pine and aspen particles are summarized in Table I; tortuosity values are illustrated in Figure 3.

The tortuosity is higher for hardwood than for softwood particles, probably because the hardwood fibers are much shorter. In addition to fiber length differences, hardwoods also have a more diverse composition of structures that will facilitate mass transport. *Pinus taeda* is primarily composed of wood fibers, at approximately 91% by volume [15]. The remaining material is mostly rays with a small percentage occupied by resin canals. *Populus tremuloides* is approximately 55% wood fibers, 34% vessel elements, and 11% rays [15]. The vessel elements in hardwood should help increase diffusion, but these structures may not be fully intact in wood that has been processed into small particles. It is very significant that the tortuosity decreases with increasing particles size for both hardwood and softwood. In other words, as the wood particles increases, the relative resistance to diffusion decreases.

The Figure 1 data also allows our assumption of Fickian diffusion to be tested. Fickian diffusion requires a plot of $\ln(1 - M_t/M_\infty)$ to be linear with time for $M_t/M_\infty \geq 0.5$. For aspen particles in the 2.0 – 2.4 mm size range the r^2 value

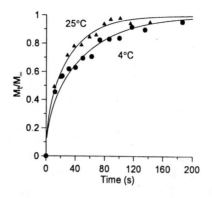

Figure 2. Approach to equilibrium for pine particles in the size range of 1.7-2.0 mm at two temperatures. (Reproduced with permission from Holzforshung: *2006, 60, 509–563. Copyright 2006 Walter de Gruyter.)*

for such a plot was 0.98. This degree of linearity was typical for all the experiments. For aspen, the diffusion into and out of the particles is nearly identical, demonstrating that diffusion of water into saturated wood particles is completely reversible. There was no hysteresis for the saturated wood as is common for water adsorption into unsaturated wood. Therefore, all the regions in fully saturated wood are equally accessible to the diffusing water molecules. Water that is bound to the wood structure readily exchanges with the loosely bound bulk water.

The observation that tortuosity increases with decreasing particle size has been reported in other areas. Cadmium uptake into chitosan particles showed a similar relationship [17]. The authors suggest that although they modeled the chitosan particles as spheres, the larger particles were actually flat discs, which would increase the uptake efficiency and would be reflected in the tortuosity factor. The sulfation reaction on different limestone sorbents also showed that tortuosity increased as the particle size decreased[18].[1818] This occurred only for limestone particles with a unimodal pore size distribution. Particles with a wider distribution showed no change in tortuosity with changing particle size. The authors speculate that the increase in tortuosity for smaller particles may be due to variations in porosity.

It is possible that the movement of tritium into the system could occur through proton transfer rather through the movement of the tagged water molecule. If this were the case, then the diffusion coefficient would have been much larger than that typical for water. The self-diffusion coefficient of HDO, HTO, and $H_2^{18}O$ at 25°C are all nearly identical [13]. If proton transfer were significant, there would be appreciable differences among these coefficients.

Figure 3 shows that the hardwood tortuosity increases to a greater extent than that of softwood with decreasing particle size. The reasons for this difference are not understood, but tortuosity would increase if pore closure occurred while the particles were being cut. Hardwood fibers are shorter in length and smaller in diameter than their softwood counterparts [19], and the tortuosity of hardwood could, therefore, be more sensitive to pore closure.

There was no apparent difference between diffusion into bound water and into free water, *i.e.* the diffusion coefficients remained constant throughout the process. This differs from the isothermal adsorption of the water vapor into dry wood [20], where Fick's law is not obeyed, since the diffusion coefficient changes as wood picks up more moisture. Also, in similar diffusion experiments in soil, all the water was accessible to diffusion [2].

The lower tortuosity of pine as compared to aspen may have applications in pulping and wood drying. Uniform chips are required for consistent pulping; small chips tend to overcook. However, the tortuosity of aspen chips increases with decreasing chip size, which will tend to offset the size effect. Such a benefit will not apply to pine, which will overcook. Our preliminary studies confirm this behavior. Our results also apply to wood drying. Liquid flow is significant when sapwood boards are dried [21]. The flow exiting the wood during early drying should follow the same pathways through which water diffuses into wood. Although wood drying is affected by other variables, the more tortuous path in aspen will make drying more difficult in the initial and constant rate regimes, as is the case for most hardwoods compared with softwoods [22].

Conclusions

In conclusion, the diffusion of water into small water-saturated wood particles can be modeled by using the sorption kinetics equation for cylinders. The diffusion coefficients are temperature-dependent and remain unchanged for diffusion into and out of the particles. As expected, the tortuosity of water is temperature-independent. The tortuosity for hardwood particles is higher than that for softwood, most likely because of the smaller fibers in hardwood. An increase in tortuosity with decreasing particle size was observed for both hardwood and softwood, with the effect being more pronounced for hardwood. Work on the tortuosity of components in pulping liquor is currently underway.

References

1. Thomas, M.; Malcolm, G.; Malcolm, E. W., *Alkaline Pulping*. ed.; Tappi Press: 1989; 'Vol.' 5.

Table I. Diffusion parameters of water into particles.

Size fraction (mm)	Direction	Temperature (°C)	$D_{eff} \times 10^6$ $cm^2 s^{-1}$, (σ)	Tortuosity
Pine				
2.0 - 2.4	In	25	11.1 (0.5)	1.6
1.7 - 2.0	In	25	6.8 (0.8)	2.5
1.7 - 2.0	In	4	3.5 (0.5)	2.2
1.4 - 1.7	In	25	6.0 (0.9)	2.9
Aspen				
2.0 - 2.4	In	25	7.2 (2.8)	3.0
2.0 - 2.4	Out	25	8.3 (1.4)	2.4
1.7 - 2.0	In	25	4.9 (1.2)	4.1
1.7 - 2.0	Out	25	4.9 (1.2)	4.1
1.4 - 1.7	In	25	3.1 (1.2)	6.8
1.4 - 1.7	Out	25	2.9 (0.3)	6.6

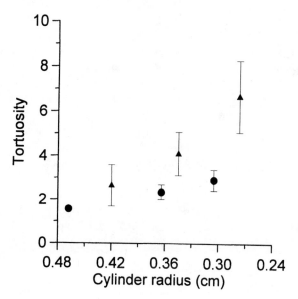

Figure 3. Tortuosity of particles. The circles and triangles represent pine and aspen, respectively. (Reproduced with permission from Holzforshung: **2006**, 60, *509–563. Copyright 2006 Walter de Gruyter.)*

2. Van Der Kamp, G.; Stempvoort, D. R.; Wassenaar, L. I., *Water Resources Res.* **1996**, *32*, (6), pp 1815-1822.

3. Moldrup, P.; Olesen, T.; Komatsu, T.; Schjonning, P.; Rolston, D. E., *Soil Sci. Soc. Am. J* **2001**, *65*, pp 613-623.

4. Rao, P. S. C.; Jessup, R. E.; Rolston, D. E.; Davidson, J. M.; Kilcrease, D. P., *Soil Sci. Soc. Am. J* **1980**, *44*, pp 684-688.

5. Novakowski, K. S.; Van Der Kamp, G., *Water Resources Res.* **1996**, *32*, (6), pp 1823-1830.

6. Rao, P. S. C.; Jessup, R. E.; Addiscott, T. M.,*Soil Sci.* **1982**, *133*, (8), pp 342-349.

7. Kohne, J. M.; Gerke, H. H.; Kohne, S., *Soil Sci. Soc. Am. J* **2002**, *66*, (September-October), pp 1430-1438.

8. Nitta, K.; Natsuisaka, M.; Tanioka, A., *Desalination* **1999**, *123*, pp 9-14.

9. Hutchison, J. M.; Seaman, J. C.; Aburime, S. A.; Radcliffe, D. E.,*Vadose zone J.l* **2003**, *2*, pp 702-714.

10. Kendall, C.; McDonnell, J. J., *Isotope Tracers in Catchment Hydrology.* ed.; Elselvier: Amsterdam, 1998; p 839.

11. Livingston, H. D.; Povinec, P. P., *Health Phys.* **2002**, *82*, pp 656-668.

12. Crank, J., *The Mathematics of Diffusion.* 2nd ed.; Clarendon Press: Oxford, 1975; p 414.

13. Pruppacher, H. R.,*The J. of Chem. Phys.* **1972**, *56*, (1), pp 101-107.

14. Mahoney, M. W.; Jorgensen, W. L., *J. of Chem. Phys.* **2001**, *114*, (1), pp 363-366.

15. Isenberg, I. H., *Pulpwoods of the United States and Canada - Conifers.* Third ed.; The Institute of Paper Chemistry: Appleton, 1980; 'Vol.' 1, p 219.

16. Price, W. S.; Ide, H.; Arata, Y., *J. of Phys. Chem. A* **1999**, *103*, pp 448-450.

17. Evans, J. R.; Davids, W. G.; MacRae, J. D.; Amirbahman, A., *Water Res.* **2002**, *36*, pp 3219-3226.

18. Adanez, J.; Gayan, P.; Garcia-Labiano, F., *Ind. Eng. Chem. Res.* **1996**, *35*, (7), pp 2190-2197.

19. Isenberg, I. H., *Pulpwoods of the United States and Canada - Hardwoods.* Third ed.; The Institute of Paper Chemistry: Appleton, 1981; 'Vol.' 2, p 168.

20. Stamm, A. J., *Forest Prod. J.* **1959**, *9*, pp 27-32.

21. Pang, S.; Haslett, A. N., *Drying Tech.* **1995**, *13*, (8&9), pp 1635-1674.

22. Martin, M.; Canteri, L., *Drying Tech.* **1997**, *15*, (5), pp 1293-1325.

Chapter 15

A Brief History of Lignin-Containing Polymeric Materials Culminating in X-ray Powder Diffraction Analyses of Kraft Lignin-Based Thermoplastic Polymer Blends

Yi-ru Chen and Simo Sarkanen

Department of Bio-based Products, University of Minnesota, St. Paul, MN 55108

The past 30-year history of lignin-containing polymeric materials has been fashioned by changing perceptions of macromolecular lignin structure. The first formulations originated in a view of lignins as "three-dimensionally branched network" polymers. Whether incorporated covalently or noncovalently into polymeric materials, lignin preparations were, with few exceptions, limited to maximum contents of 30–45% (w/w). However, once the significance of noncovalent interactions between the constituent molecular components had been recognized, it was possible to generate promising polymeric materials composed exclusively of ethylated and methylated kraft lignin, and other equally simple lignin derivatives. Plasticization of these materials can be achieved in a predictable way by blending with miscible low-T_g polymers. X-ray powder diffraction analyses have helped to identify characteristic variations in the separation distances between aromatic rings that accompany such plasticization effects.

Introduction

The year 1997 witnessed a paradigm shift in the field of lignin-based polymeric materials. A full-length account (*1*) was published about thermoplastics with 85% (w/w) kraft lignin contents that displayed promising mechanical behavior (tensile strengths reaching 26 MPa and Young's moduli as high as 1.5 GPa). Moreover, a preliminary characterization (*2*) of ethylated and methylated 100% kraft lignin-based polymeric materials with encouraging tensile strengths (~15 MPa) and Young's moduli (~1.3 GPa) appeared in the same year, and significant improvements in mechanical properties were shortly to be achieved through more complete alkylation of the kraft lignin preparations (*3*). These new lignin-based plastics had been preceded by over 20 years of research that produced lignin-containing polymeric materials with maximum lignin contents typically confined to levels between 30 and 45%.

Some 22 years before, it had been clearly recognized that the successful development of useful lignin-based plastics and related kinds of materials must be founded on a "sound scientifically based understanding of lignin as a macromolecule" (*4*). This concern was prompted by an expectation that has proven to be fundamentally correct: "It is quite possible that, in the attempted uses of lignin to meet polymer or materials needs, one should not just try to 'replace' a synthetic component, but to take new innovative approaches where the uniqueness of lignin as a macromolecule should be exploited" (*4*).

Early Assumptions about Crosslinking in Lignins

During the latter half of the 1950's and throughout the 1960's, a picture was developed of lignin macromolecules as three-dimensionally crosslinked species (*5*). The idea was largely based upon a compactness of these molecular entities that was apparent from their viscosimetric, sedimentation and diffusion behavior (*5*). The crosslinking of protolignins (namely, lignin biopolymers *in situ*) was investigated indirectly through applications of Flory-Stockmayer theory to the results of delignifying wood. Realistic models sought to distinguish between the inherent reactivities of different lignin substructures; the first led to an estimate of 18 for the average number of monomer residues in the crosslinked primary chains that were assumed to make up spruce protolignin structure (*6*). An extension of this model, which incorporated a most probable distribution of primary lignin chains joined through tetrafunctional branchpoints, led to a very similar estimate of crosslink density in native spruce lignin (*7*). Extended Flory-Stockmayer analyses have also been employed to examine the delignification of western hemlock and black cottonwood (*8, 9*).

In regard to experimental verification, the Flory-Stockmayer treatment of delignification has focused primarily on the variations in degree of polymerization of the dissolved lignin components (7). However, analytical estimates of crosslink density (related pre-eminently to alkyl aryl ether content in the scheme employed) have been reported to decrease, as the average molecular weight was increasing, for lignin components progressively dissolving from spruce wood into acidic aqueous dioxane (10). The two trends are mutually inconsistent in the context of Flory-Stockmayer theory, which basically considers how an infinite polymer network (insoluble "gel") can be formed either by crosslinking preformed polymer chains or by polymerizing monomers with functionality greater than two. As the (pregel) reaction proceeds, the average molecular weight of the soluble components ("sol") and their crosslink density increase until the gel first materializes. Beyond this point (i.e., postgel), the average molecular weight and crosslink density of the sol fraction decrease as the larger molecules are preferentially incorporated into the gel.

These parameters would be expected to exhibit the opposite behavior during delignification if the process is viewed as being the reverse of gelation. Thus, a decrease in the apparent crosslink density of the dissolved lignin components during acidolytic spruce wood delignification is "pregel-like", but a concomitant increase in their average molecular weight is "postgel-like" (10). This conceptual contradiction was dealt with by suggesting that linkages to polysaccharides could account for the gel-like behavior of lignin domains, which would not themselves have to be crosslinked (10). Indeed, the invariance of polydispersity as the hydrodynamic volume of the molecular components varies in kraft lignin preparations argues against the existence of crosslinking and long-chain branching in these lignin derivatives, and by extension, in the native lignin biopolymers themselves (vide infra: The Physicochemical Behavior of Simple Lignin Derivatives).

Crosslinked Lignin-Containing Polymeric Materials

Almost 30 years ago, a serious effort was mounted to develop a range of worthwhile lignin-based engineering plastics. In this connection, it had been declared that "lignin is a high molecular weight three-dimensionally branched network polymer" (11), where the concept of crosslinking seems to have been embraced through the term "network". Such ideas had spawned the following vision about how lignin-containing polymeric materials might usefully be produced. "It is quite obvious that [the] three-dimensionally branched structure of lignin requires the flexibilizing complementation of a linear chain polymer, yet the extent to which lignin's structural rigidity needs to be softened will also determine the percentage of lignin that can be utilized in the manufacture of a comercial polymeric material" (12). Indeed, this perspective was consistent with strategies for creating combinations of high modulus and strength "by crosslinking or reinforcing a soft segment matrix with hard segment domains" (*13*). Conveniently in accord with this approach was the belief that the "polyfunctional character of lignin limits its use to thermosetting polymers" (*14*).

Thus the stage was set for incorporating lignin derivatives into a variety of crosslinked polymeric materials. The formation of homogeneous lignin-containing polymer networks required that phase separation during curing be held back until crosslinking had occurred to such an extent that the demixing of individual molecules could no longer occur (*15*). This could be achieved either by improving the solubility of the lignin components through suitable derivatization (thus reducing the enthalpy of mixing) or by using lower molecular weight lignin components (and thus increasing the entropy of mixing). The former of the two alternatives was illustrated by an early series of five polyurethanes synthesized from hexamethylene diisocyanate and hydroxypropyl kraft lignin (*16*). These materials contained 26–44% kraft lignin (corresponding to reactant isocyanate/hydroxyl group ratios between 4.7 and 0.9), the presence of which engendered a perceptible increase in Young's modulus. In contrast, the tensile strength remained fairly steady around 77 MPa at kraft lignin contents from 26 to 38% but then fell abruptly ~5-fold as the amount of kraft lignin was raised to 44% (Figure 1). At the same time, there was a more uniform reduction in ultimate strain from ~15% to 1% over the same range of lignin content. These results were unexpected because the apparent crosslink density decreased as the polyurethane isocyanate/hydroxyl group ratio was lowered (*16*). It is reasonable to suppose that the nonbonded orbital interactions between the aromatic rings in the lignin derivative counteract the impact that a smaller number of crosslinks will have on these polymeric materials (*17*).

The introduction of soft-segment components into these hydroxypropyl-kraft-lignin-containing polyurethanes was expected to engender a considerable improvement in mechanical behavior (*13*). Thus, the inclusion of just 18% polyethylene glycol (number-average molecular weight 400) brought about a ~4-fold decrease in Young's modulus, ~3-fold lowering in tensile strength and a ~6-fold increase in ultimate strain. Concomitantly, however, the kraft lignin content was reduced 1.3-fold to 28% (*13*). Such effects exemplified a general observation that had few exceptions: crosslinked polymeric materials with high lignin contents were almost always extremely brittle. Nevertheless there was one case during the past 30 years where it proved possible to incorporate a lignin derivative covalently into a crosslinked polymeric material and produce something that was actually lignin-based (*18*). A more-or-less linear positive correlation between Young's modulus and lignin content was encountered with cured lignin–polyether epoxy resins; the tensile strength increased with lignin content up to ~40%, where it leveled off and then fell appreciably (Figure 1). All of the epoxide copolymers were fairly rigid, but one of them upheld quite acceptable tensile properties (1.0 GPa Young's modulus, 42 MPa tensile strength, 6% ultimate strain) in the face of a 57% organosolv lignin content (*18*).

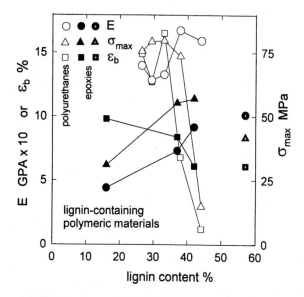

Figure 1. Effects of lignin content on the mechanical properties of hydroxypropyl pine kraft-lignin-containing polyurethanes (open symbols) and alkoxylated-lignin-containing epoxy resins crosslinked with m-phenylene-diamine (closed symbols). Epoxies with 16%, 37% and 42% (steam exploded poplar) lignin contents were fractions of increasing molecular weight produced by consecutive propoxylation and ethoxylation followed by endcapping with epichlorohydrin; that with a 57% (unspecified organosolv) lignin content was prepared by ethoxylation alone before epichlorohydrin endcapping. E denotes Young's modulus, σmax tensile strength and εb the ultimate strain (16, 18).

The effect of lignin-derivative molecular weight on the entropies of reactant mixtures used to create homogeneous crosslinked lignin-containing polymeric materials (15) can be seen in a series of kraft lignin–polyether triol–polymeric MDI polyurethanes (19). Four kraft lignin fractions embracing an 8-fold variation in apparent number-average molecular weight had been respectively solvent extracted for these purposes from a parent preparation. The tensile strengths of the polyurethanes produced from the first three fractions (exhibiting successively increasing molecular weights) rose with kraft lignin content up to 25–30%, where they leveled off or began to fall (Figure 2). In contrast, the tensile strength of the polyurethane made with the highest molecular weight fraction already reached its maximum value at a kraft lignin content of only ~10%, after which it steadily declined (Figure 2). This anomalous behavior was ascribed to the presence of increasing quantities of unincorporated lignin derivative in the cured polyurethane (19). Hence, as far as the kraft lignin is concerned, molecular weights beyond a particular threshold characteristic of

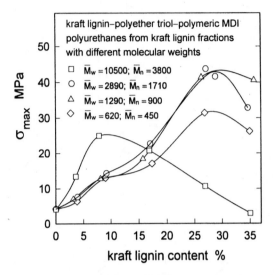

Figure 2. Tensile strengths (σmax) of cured kraft lignin–polyether triol–polymeric MDI polyurethanes with reactant isocyanate/hydroxyl group ratio of 0.9. Effects of softwood kraft lignin content in the form of four fractions with different molecular weights solvent-extracted from parent preparation (19).

these formulations may lead to phase separation in the polyurethanes that are formed (*15*). Furthermore, with kraft lignin contents greater than 30%, whatever their component molecular weights, the polyurethanes produced were glassy and brittle, regardless of the isocyanate/hydroxyl group ratio in the reactant mixture (*19*). Phenol-formaldehyde resins (*20*) and polyacrylic materials (*21*) have, to some extent, tolerated the covalent incorporation of lignin derivatives into them also, but lignin contents have again been restricted to levels of 40% or lower.

Multiphase Lignin-Containing Polymer Blends

As an alternative approach to developing useful lignin-containing materials, compatible polymer blends first took the form of two-phase mixtures (*22*) of hydroxypropyl kraft lignin and poly(methyl methacrylate). At levels corresponding to lignin contents below 20%, raising the proportion of hydroxypropyl kraft lignin led to an increase in Young's modulus but decreases in tensile strength and ultimate strain. The situation did not fundamentally improve in two-phase interpenetrating polymer network composites produced from a lignin-containing polyurethane and poly(methyl methacrylate). When the proportion of lignin in the polyurethane was fixed at 30%, both Young's modulus and the tensile strength of the composite decreased dramatically as the lignin content was raised (*23*). However, when the poly(methyl methacrylate) content was held constant at 50%, Young's modulus and the tensile strength both increased conspicuously with lignin content, although their values did not reach those characteristic of poly(methyl methacrylate) itself (*24*).

Thus, the tensile behavior of these interpenetrating polymer network composites depended pre-eminently on the nonlignin-containing constituent (*23, 24*). Indeed, it is very seldom that the lignin derivative in a multiphase blend has been capable of improving the mechanical properties of the composite overall. Thus, organosolv lignin carboxylate esters (ranging from acetate to laurate), at 50% incorporation levels, have been found to cause a greater than 5-fold reduction in the tensile strength of cellulose acetate butyrate (Figure 3), and the ultimate strain was even more markedly affected (*25*). Such findings inevitably led to the question of whether a simple lignin derivative could ever act as the principal active component in any kind of potentially useful (multiphase or homogeneous) polymer blend.

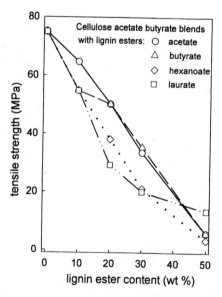

Figure 3. Tensile strengths of multiphase blends of cellulose acetate butyrate with organosolv lignin esters (25).

The Physicochemical Behavior of Simple Lignin Derivatives

Before 1997 no convincing rationale had been proposed in attempting to explain why it had not been possible to formulate polymeric materials containing

236

lignin preparations as the predominant constituents. Chain stiffness arising from the presence of the aromatic rings in lignin-containing polymeric materials (26) had been invoked as a possible cause of the difficulty, but it was never clear why the ensuing brittleness exceeds that of polystyrenes. Apart from their inherent chemical configurations, the most conspicuous property of simple lignin derivatives (other than the polyanionic sulfonates) is the tendency of the individual molecular components to associate with one another extensively (27–34). A typical result is depicted in Figure 4, where the apparent molecular weight distributions of a series of acetylated and methylated kraft lignin preparations in DMF embody just a handful of distinguishable supramacro-molecular complexes composed of many ($\sim 10^3$–10^4, according to their molecular weights) individual components (33). The huge entities undergo interconversion as a result of incubation, *before* derivatization, at different concentrations in aqueous alkaline solutions: under these conditions, there appears to be a remarkable degree of selectivity in the associative processes, the rates of which are thought to be controlled by changes in the conformations of individual molecular components (28–30).

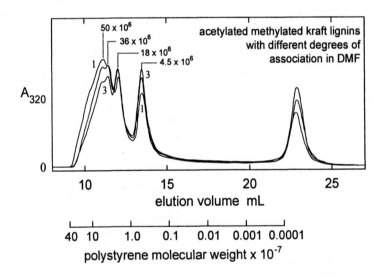

Figure 4. Apparent molecular weight distributions in DMF of acetylated and methylated kraft lignin preparations differing only in degree of association (33). Elution profiles from 10^7 Å pore-size poly(styrene-divinylbenzene) column monitored at 320 nm. Samples were fractionated through Sephadex LH20 in aqueous 35% dioxane after association at 195 gL⁻¹ for (1) 6740 h, (2) 3910 h and (3) 1630 h in aqueous 1.0 M ionic strength 0.40 M NaOH.

Their relatively small number testifies to the fact that the enormous supramacromolecular kraft lignin complexes are assembled in a well-defined (*i.e.*, non-random) way. The largest entities visible in transmission electron micrographs of acetylated and methylated kraft lignin preparations are about 0.25 μm across in size (*31*). Dimensions like these are comparable to those of 20 x 10^6 molecular weight polystyrene; they suggest that the intermolecular forces in simple lignin derivatives could organize the individual components into cohesive domains without the need for further covalent bond formation (*31*). Moreover, such effects should not be substantially influenced by acylation and/or alkylation of the lignin derivatives since intermolecular association here is primarily governed by nonbonded orbital interactions between aromatic rings (*17*). This is the reason why the molecular species in lignin preparations are hydrodynamically compact (*5*).

The discrete kraft lignin components themselves, on the other hand, exhibit neither long-chain branching nor crosslinking: the polydispersity of narrow fractions derived size-exclusion chromatographically from (dissociated) kraft lignin preparations in aqueous 0.10 M NaOH does not vary with the hydrodynamic volume of the solute species (*35, 36*). This conclusion about the configurations of individual kraft lignin components can also be applied to the native biopolymer if it is true that the characteristic inter-unit linkages are not positioned in any particular order along the macromolecular chain (*37*).

Simple Alkylated Kraft Lignin-Based Plastics

Initially, formulations for promising polymeric materials with the highest attainable lignin contents were realized in (partially optimized) homogeneous blends of 85% (w/w) underivatized kraft lignin with 12.6% poly(vinyl acetate), 1.6% diethyleneglycol dibenzoate and 0.8% indene (*1*). Interesting as these thermoplastics were, they remained partially soluble in aqueous alkaline solutions. Thus, their primary importance lay in what they implied about the feasibility of fabricating polymeric materials with very high lignin contents. Nevertheless, through a revealing aspect of the way in which their mechanical properties were affected by composition, these kraft lignin-based plastics established for the first time that, even in solid materials, there is a fundamental functional difference between individual lignin components which are incorporated into supramacromolecular associated complexes and those which are not (*1, 33*).

It was then discovered that, without any further chemical modification, ethylated and methylated kraft lignin derivatives can form polymeric materials which are remarkably similar to polystyrene in tensile behavior (*3*). In contrast to kraft lignins themselves, these alkylated kraft lignin-based materials are readily plasticized by blending with miscible low-T_g polymers (*38, 39*) such as aliphatic main-chain polyesters with methylene/carboxylate group ratios of 2.0–4.0 (*38*). Analogous windows of miscibility had been previously observed in blends of unbranched aliphatic polyesters with poly(bisphenol A carbonate), poly(bisphenol A glycerolate), poly(tetramethylbisphenol A carbonate), poly(vinyl chloride) and a novolak-type phenolic resin (*40–44*).

Figure 5 depicts the plasticization of an ethylated methylated softwood kraft lignin preparation and a (similarly alkylated) higher molecular weight kraft lignin fraction derived from it (*38*) as they are blended with progressively increasing proportions of poly(butylene adipate). The plasticization threshold in both cases corresponds to a poly(butylene adipate) content slightly below 30% (w/w). Beyond this point, the ultimate stresses borne by blends with the same composition based on the alkylated parent kraft lignin preparation and corresponding higher molecular weight fraction, respectively, are almost identical in magnitude (*38*). The similarity is noteworthy because the parent kraft lignin preparation contains many more lower molecular weight components

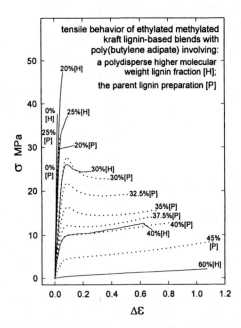

Figure 5. Plasticization of ethylated methylated kraft lignin-based materials by poly(butylene adipate) reflected in tensile behavior of their homogeneous blends (38); impact of lower molecular weight kraft lignin components on plastic deformation. Polydisperse higher molecular weight fraction: $\overline{M}_w = 23.0 \times 10^3$, $\overline{M}_n = 8.8 \times 10^3$; *parent preparation:* $\overline{M}_w = 18.0 \times 10^3$, $\overline{M}_n = 5.5 \times 10^3$.

than the higher molecular weight fraction. The proximity of the tensile strengths arises from the fact that most of the individual kraft lignin components in these plastics are embedded in huge well-defined supramacromolecular complexes (*31, 33*). Yet there is a marked disparity in how much plastic deformation can be sustained before fracture. Past the threshold of plasticization, the ultimate strains tolerated by the alkylated parent kraft lignin-based materials are ~30% greater (Figure 5) than those characteristic of the plastics into which the higher molecular weight fraction has been blended (*38*). Accordingly, lower molecular weight ethylated methylated kraft lignin components (of which there are more in the parent preparation) synergistically enhance the effect of the aliphatic polyester acting in its capacity as a plasticizer in these homogeneous blends (*39*).

Naturally, the strengths of the corresponding intermolecular interactions will affect how much miscible low-T_g polymer is needed to reach the plasticization threshold for an alkylated kraft lignin-based blend. When the individual components in the associated kraft lignin complexes interact more strongly with the polymeric plasticizer, the huge supramacromolecular entities tend to be (counterproductively) dismantled to a greater extent; consequently, more low-T_g polymer is required in the blend composition prevailing at the threshold of plasticization. For the blends of alkylated kraft lignin preparations with aliphatic main-chain polyesters, the strongest intermolecular interactions occur near the middle of the miscibility window circumscribed by the range of 2.0–4.0 for the methylene/carboxylate group ratio. These are the conditions, therefore, where the polymeric plasticizer is least efficacious (*38, 39*).

X-Ray Powder Diffraction Analyses

In their wide-angle X-ray powder diffraction patterns, simple lignin derivatives typically give rise to diffuse halos which, as exemplified in Figure 6 (*45*), assume the forms of broad peaks in diagrams of diffracted intensity *versus* 2θ (where θ is the angle of diffraction). When the quantity of alkylated kraft lignin is raised in a homogeneous aliphatic polyester blend, a point is reached where the characteristic reflections from the crystalline domains of the low-T_g polymer vanish; the blend composition at this threshold depends upon the relative strengths of the intermolecular interactions between the polymeric components involved (*38*). For instance, in blends of a methylated higher molecular weight kraft lignin fraction with poly(trimethylene succinate) and poly(butylene adipate), respectively, the peaks from the crystalline domains disappear at alkylated kraft lignin contents of < 30% and 45% (*38*).

As the proportion of alkylated kraft lignin in such blends is increased further, the maximum in the intensity of the amorphous halo in the X-ray diffraction diagram is progressively shifted toward smaller scattering angles,

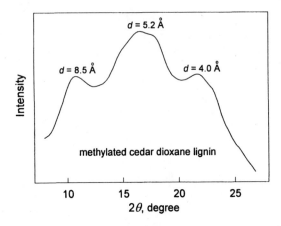

Figure 6. Wide angle X-ray diffraction diagram for methylated cedar dioxane lignin, with d-*spacings given at the broad peak maxima (45).*

i.e., larger *d*-spacings (*39*). The polyester affects the overall diffuse scattered intensity in a manner that cannot be easily predicted, and therefore, the underlying reason for the trend may be more clearly discernible from any series of alkylated kraft lignin preparations that can exhibit an analogous result in the absence of other blend components. Such a systematic effect is observed (Figure 7) with paucidisperse methylated kraft lignin fractions possessing successively increasing molecular weights (*39*). It is fitting that the trend in question is manifested in this context since alkylated low molecular weight kraft lignin components enhance the action of miscible low-T_g aliphatic polyesters as plasticizers (Figure 5). In other words, the X-ray powder diffraction patterns seem to bear a structure–function relationship to particular concomitants of plasticization.

As shown in Figure 7, two Lorentzian component peaks (centered at $2\theta = 16.9°$, $d = 5.25$Å and $2\theta = 22.1 ± 0.6°$, $d = 4.0 ± 0.1$Å) are necessary to account for the diffuse scattered intensity in the 5–35° (Cu K$_\alpha$) range of 2θ in the X-ray diffraction patterns of the methylated paucidisperse kraft lignin fractions (*39*). These two peak maxima occur at the same scattering angles as previously observed with a methylated cedar dioxane lignin (*45*), although an inner halo centered at $d = 8.5$Å was also present in the earlier diffraction diagram (Figure 6). It is usually difficult to interpret the diffuse scattered intensity in the X-ray diffraction pattern of an amorphous polymeric material. However, as far as partially crystalline polymeric materials are concerned, when there is a correspondence between the amorphous peaks and some of the reflections from the crystalline domains, reasonable assignments can be readily proposed (*46*).

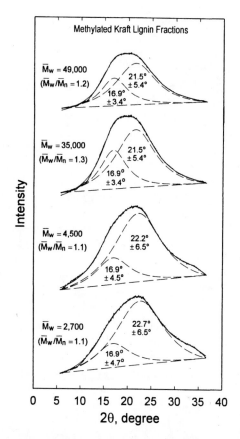

Figure 7. Fits of two principal Lorentzian functions to diffuse scattered intensity in the X-ray diffraction patterns of methylated paucidisperse kraft lignin fractions. (Reproduced from reference 39. Copyright 2005 American Chemical Society.)

Of course, lignins and lignin derivatives are not themselves crystallizable, and so any insights about the origins of distinguishable features in the X-ray diffraction diagrams of the methylated paucidisperse kraft lignin fractions (Figure 7) can only be gained from the crystal structures of appropriate lignin model compounds (*39*). The aromatic moieties are expected to make dominant contributions to the X-ray diffraction patterns of lignin preparations.

The relative placement of these rings varies from parallel offset arrangements to edge-on orientations (*47*). The former encompass a broad range of (intra- and intermolecular) separation distances centered around 4.0Å (*48–50*), while the latter have been observed to prefer intermolecular separations of

~5.3Å (*51*). The close match between such values and those for the respective Lorentzian peak maxima in the amorphous halos emanating from the methylated kraft lignin fractions (Figure 7) is quite remarkable. Thus, these component features may be provisionally assigned, in part, to relationships between the aromatic rings that are parallel and edge-on with respect to one another (*39*).

The peak maximum at $2\theta = 16.9°$, which reflects the ~5.25Å spacings between edge-on pairs of aromatic rings in the methylated paucidisperse kraft lignin fractions, does not vary with the component molecular weight, but the corresponding peak width is over 25% smaller for the higher molecular weight preparations (Figure 7). The underlying cause is not difficult to uncover: most of the higher molecular weight kraft lignin components participate in the formation of supramacromolecular associated complexes, where variations in intermolecular registration are much more restricted than for the individual molecular species. For the same reason the $2\theta \approx 22°$ peak maximum reflecting the ~4.0Å spacings between parallel pairs of aromatic rings appears at slightly lower scattering angles, while its width is over 15% narrower with the higher molecular weight methylated paucidisperse kraft lignin fractions (Figure 7).

The $2\theta = 21.5 \pm 5.0°$ Lorentzian peak that encompasses the scattering from parallel aromatic rings in a methylated (polydisperse) higher molecular weight kraft lignin fraction used to make thermoplastics (Figure 8) possesses a maximum at the same 4.1Å *d*-spacing as found with the paucidisperse high molecular weight fractions (Figure 7). On the other hand, the Lorentzian component peak embodying reflections from edge-on aromatic rings exhibits a maximum at a slightly smaller 5.1Å *d*-spacing than observed with the methylated paucidisperse high molecular weight kraft lignin fractions; the peak width is, moreover, about 30% narrower (Figure 8). This presumably results from the fact that the associated complexes are more completely assembled in the polydisperse than in the paucidisperse higher molecular weight kraft lignin fractions (*39*).

Blending with 30–40% poly(trimethylene succinate) has been found (*39*) to engender ~10% and 60% broadening, respectively, of the two Lorentzian peaks with maxima at 4.1Å and 5.1Å in the amorphous halo emanating from the methylated polydisperse higher molecular weight kraft lignin fraction (Figure 8). The consequences of introducing the polymeric plasticizer are not completely systematic, and thus it is not possible to predict in a simple way how the reflections to which the polyester contributes are integrated into the overall X-ray diffraction pattern of the lignin-based blend. However, it is very clear that the broadening in the distribution of separation distances between edge-on aromatic rings is much greater than that between parallel rings in the methylated kraft lignin components (Figure 8). The effects of adding poly(trimethylene succinate) and reducing lignin component molecular weight (Figure 7) on the corresponding X-ray diffraction diagrams are qualitatively similar; this is consistent with the enhancement in plasticization of alkylated kraft lignin-based polymeric materials by miscible aliphatic polyesters (Figure 5) that occurs in the presence of lower molecular weight lignin components (*39*).

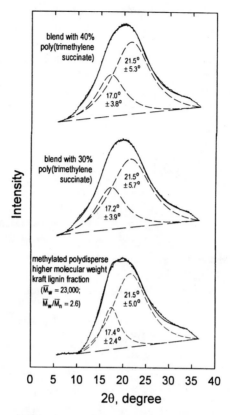

Figure 8. Fits of two principal Lorentzian functions to diffuse scattered intensity in the X-ray diffraction patterns of a methylated polydisperse higher molecular weight kraft lignin fraction and its blends with poly(trimethylene succinate). (Reproduced from reference 39. Copyright 2005 American Chemical Society.)

Conclusions

As one contemplates recent developments in the field of lignin-containing polymeric materials, one cannot help being reminded again of Falkehag's prescient admonition of over 30 years ago: "….in the attempted uses of lignin to meet polymer or materials needs, one should not just try to 'replace' a synthetic component, but to take new innovative approaches where the uniqueness of lignin as a macromolecule should be exploited" (4). Curiously, this statement is probably as true today as it was then: it is largely through further fundamental studies that the full range of useful lignin-based plastics will be finally realized.

Nevertheless, the mechanical properties of alkylated kraft lignin-based thermoplastics as a whole seem to fulfill the requirements for a number of worthwhile applications. The fact that they have not yet reached the threshold of commercialization is a consequence of the hurdles that need to be overcome in producing any new commodity, when existing operations must contend with disruptive changes in raw materials' supply and processing conditions. Thus, the first successful creation of a niche market for lignin-based plastics will play a crucial role in attracting the level of investment that this field deserves.

Acknowledgments

Support from the U.S. Department of Agriculture (Grant 98-35103-6730), the U.S. Environmental Protection Agency through the National Center for Clean Industrial and Treatment Technologies, the Vincent Johnson Lignin Research Fund, the Minnesota State Legislature, and the Minnesota Agricultural Experiment Station (Project No. 43-68, maintained by Hatch Funds) for the research performed at the University of Minnesota is gratefully acknowledged.

References

1. Li, Y.; Mlynár, J.; Sarkanen, S. *J. Polym. Sci., Part B: Polym. Phys.* **1997**, *35*, pp 1899–1910.
2. Li, Y.; Sarkanen, S. Ninth International Symposium on Wood and Pulping Chemistry Proceedings; CPPA: Montreal, PQ, Canada, 1997; 63, pp 1–6.
3. Li, Y.; Sarkanen, S. In *Lignin—Historical, Biological, and Materials Perspectives*; Glasser, W. G., Northey, R. A., Schultz, T. P., Eds.; ACS Symposium Series No. 742; American Chemical Society: Washington, DC, 2000; pp 351–366.
4. Falkehag, S. I. *Appl. Polym. Symp.* **1975**, *28*, pp 247–257.
5. Goring, D. A. I. In *Lignins—Occurrence, Formation, Structure and Reactions*; Sarkanen, K. V., Ludwig, C. H., Eds.; Wiley-InterScience: New York, 1971; pp 695–768, and references therein.
6. Bolker, H. I.; Brenner, H. S. *Science* **1970**, *170*, pp 173–176.
7. Yan, J. F.; Pla, F.; Kondo, R.; Dolk, M.; McCarthy, J. L. *Macromolecules* **1984**, *17*, pp 2137–2142.
8. Dolk, M.; Pla, F.; Yan, J. F.; McCarthy, J. L. *Macromolecules* **1986**, *19*, pp 1464–1470.
9. Pla, F.; Dolk, M.; Yan, J. F.; McCarthy, J. L. *Macromolecules* **1986**, *19*, pp 1471–1477.
10. Smith, D. C.; Glasser, W. G.; Glasser, H. R.; Ward, T. C. *Cellulose Chem. Technol.* **1988**, *22*, pp 171–190.

11. Hsu, O. H.-H.; Glasser, W. G. *Wood Sci.* **1976**, *9*, pp 97–103.
12. Hsu, O. H.-H.; Glasser, W. G. *Appl. Polym. Symp.* **1975**, *28*, pp 297–307.
13. Saraf, V. P.; Glasser, W. G.; Wilkes, G. L.; McGrath, J. E. *J. Appl. Polym. Sci.* **1985**, *30*, pp 2207–2224.
14. Rials, T. G.; Glasser, W. G. *Holzforschung* **1986**, *40*, pp 353–360.
15. Glasser, W. G. In *Adhesives from Renewable Resources*; Hemingway, R. W., Conner, A. H., Branham, S. J., Eds.; ACS Symposium Series No. 385; American Chemical Society: Washington, DC, 1989; pp 43–54.
16. Saraf, V. P.; Glasser, W. G. *J. Appl. Polym. Sci.* **1984**, *29*, pp 1831–1841.
17. Sarkanen, S.; Chen, Y.-r. 59th Appita Proceedings incorporating 13th Internat. Symp. Wood Fibre Pulp. Chem.; Appita: Carlton, Victoria, Australia, 2005; Vol. 2, pp 407–414.
18. Hofmann, K.; Glasser, W. G. *J. Wood Chem. Technol.* **1993**, *13*, pp 73–95.
19. Yoshida, H.; Mörck, R.; Kringstad, K. P.; Hatakeyama, H. *J. Appl. Polym. Sci.* **1990**, *40*, pp 1819–1832.
20. Muller, P. C.; Kelley, S. S.; Glasser, W. G. *J. Adhesion* **1984**, *17*, pp 185–206.
21. Glasser, W. G.; Wang, H.-X. In *Lignin—Properties and Materials*; Glasser, W. G., Sarkanen, S., Eds.; ACS Symposium Series No. 397; American Chemical Society: Washington, DC, 1989; pp 515–522.
22. Ciemniecki, S. L.; Glasser, W. G. *Polymer* **1988**, *29*, pp 1021–1029.
23. Kelley, S. S.; Ward, T. C.; Glasser, W. G. *J. Appl. Polym. Sci.* **1990**, *41*, pp 2813–2828.
24. Kelley, S. S.; Glasser, W. G.; Ward, T. C. *Polymer* **1989**, *30*, pp 2265–2268.
25. Ghosh, I.; Jain, R. K.; Glasser, W. G. *J. Appl. Polym. Sci.* **1999**, *74*, pp 448–457.
26. Yoshida, H.; Mörck, R.; Kringstad, K. P.; Hatakeyama, H. *J. Appl. Polym. Sci.* **1987**, *34*, pp 1187–1198.
27. Connors, W. J.; Sarkanen, S.; McCarthy, J. L. *Holzforschung* **1980**, *34*, pp 80–85.
28. Sarkanen, S.; Teller, D. C.; Stevens, C. R.; McCarthy, J. L. *Macromolecules* **1984**, *17*, pp 2588–2597.
29. Dutta, S.; Garver, T. M., Jr.; Sarkanen, S. In *Lignin—Properties and Materials*; Glasser, W. G., Sarkanen, S., Eds.; ACS Symposium Series No. 397; American Chemical Society: Washington, DC, 1989; pp 155–176.
30. Garver, T. M., Jr.; Iwen, M. L.; Sarkanen, S. Fifth International Symposium on Wood and Pulping Chemistry; TAPPI Proceedings; TAPPI Press: Atlanta, GA, 1989; Vol. I, pp 113–119.
31. Dutta, S.; Sarkanen, S. In *Materials Interactions Relevant to the Pulp, Paper, and Wood Industries*; Caulfield, D. F., Passaretti, J. D., Sobczynski, S. F., Eds.; MRS Symposium Proceedings; Materials Research Society: Pittsburgh, PA, 1990; Vol. 197, pp 31–39.

32. Sarkanen, S.; Teller, D. C.; Hall, J.; McCarthy, J. L. *Macromolecules* **1981**, *14*, pp 426–434.

33. Li, Y.; Sarkanen, S. In *Biodegradable Polymers and Plastics*; Chiellini, E., Solaro, R., Eds.; Kluwer Academic/Plenum Publishers: New York, 2003; pp 121–139.

34. Dutta, S.; Garver, T. M., Jr.; Iwen, M. L.; Sarkanen, S. Sixth International Symposium on Wood and Pulping Chemistry Proceedings; Appita: Parkville, Victoria, Australia, 1991; Vol. 1, pp 457–462.

35. Sarkanen, S.; Teller, D. C.; Abramowski, E.; McCarthy, J. L. *Macromolecules* **1982**, *15*, pp 1098–1104.

36. Mlynár, J.; Sarkanen, S. In *Strategies in Size Exclusion Chromatography*; Potschka, M., Dubin, P. L., Eds.; ACS Symposium Series No. 635; American Chemical Society: Washington, DC, 1996; pp 379-400.

37. Brunow, G.; Lundquist, K.; Gellerstedt, G. In *Analytical Methods in Wood Chemistry, Pulping and Papermaking*; Sjöström, E., Alén, R., Eds.; Springer-Verlag: Heidelberg, Germany, 1999; pp 77–124.

38. Li, Y.; Sarkanen, S. *Macromolecules* **2002**, *35*, pp 9707–9715.

39. Li, Y.; Sarkanen, S. *Macromolecules* **2005**, *38*, pp 2296–2306.

40. Cruz, C. A.; Paul, D. R.; Barlow, J. W. *J. Appl. Polym. Sci.* **1979**, *23*, pp 589–600.

41. Harris, J. E.; Goh, S. H.; Paul, D. R.; Barlow, J. W. *J. Appl. Polym. Sci.* **1982**, *27*, pp 839–855.

42. Fernandes, A.; Barlow, J. W.; Paul, D. R. *Polymer* **1986**, *27*, pp 1799–1806.

43. Woo, E. M.; Barlow, J. W.; Paul, D. R. *Polymer* **1985**, *26*, pp 763–773.

44. Ma, C.-C. M.; Wu, H.-D.; Chu, P. P.; Tseng, H.-T. *Macromolecules* **1997**, *30*, pp 5443–5449.

45. Hatakeyama, T.; Hatakeyama, H. *Polymer* **1982**, *23*, pp 475–477.

46. Murthy, N. S.; Correale, S. T.; Minor, H. *Macromolecules* **1991**, *24*, pp 1185–1189.

47. Hunter, C. A.; Sanders, J. K. M. *J. Am. Chem. Soc.* **1990**, *112*, pp 5525–5534.

48. Lundquist, K.; Stomberg, R. *Holzforschung* **1988**, *42*, pp 375–384.

49. Roblin, J.-P.; Duran, H.; Duran, E.; Gorrichon, L.; Donnadieu, B. *Chem.–Eur. J.* **2000**, *6*, pp 1229–1235.

50. Stomberg, R.; Lundquist, K. *Acta Chem. Scand. B* **1987**, *41*, pp 304–309.

51. Stomberg, R.; Lundquist, K. *Acta Chem. Scand. A* **1986**, *40*, pp 705–710.

Chapter 16

Functional Polyester and Copolymers by a One-Pot Acetylation–Oxidation of Cellulose

Mo Hunsen

Department of Chemistry, Kenyon College, Gambier, OH 43022

In an effort to convert renewable materials to value added products research was conducted on new green oxidation reactions. Chromium, a favorite oxidizing agent among organic chemists, is not suitable for any large scale or industrial oxidation due to its carcinogenicity. New environmentally friendly oxidation methods that retain the power of chromium oxidations are in great need. It has been shown that chromium catalyzed oxidations can be successfully carried out to prepare aldehydes and ketones from alcohols, and carboxylic acids from alcohols and aldehydes. This study presents the chromium-catalyzed oxidation of carbohydrates. This method involves using the recyclable periodic acid as the terminal oxidant and the solvent of choice is acetic acid. Using Chromium as a catalyst reduces the amount of waste generated by more than 100 fold. Starting from cellulose, the most abundant organic compound on earth, preparation of a novel functionalized polyester and copolymer in a one-pot acetylation/oxidation reaction.1 This method also works for furanosides and simple glycosides. The ketoesters and polyketoesters of glycosides are key intermediates in the effort to synthesize iminocyclitols, which are potent inhibitors of glycosidases and are shown to have anti-cancer and anti-viral activities.

Introduction

Carbohydrates are among the most abundant naturally occurring organic substances on earth. They are found in large storage polymers such as starch, in cell wall components such as cellulose and hemi-cellulose, in insect exoskeletons as chitin, as bacterial cell surface antigens, on viruses as the immunodominant species that govern infection, as the predominant substances on blood cell surfaces that determine blood type and in a large variety of other roles and functions. Nucleic acids contain a large proportion of carbohydrates in the form of ribose and deoxy-ribose. With the dwindling of the petroleum reserve, carbohydrates are expected to play a major role as the next generation renewable materials for the preparation of core intermediate building block chemicals. It will be from these building blocks that future pharmaceuticals, chemicals, and materials will be synthesized or manufactured. The aim is to tap the abundance, renewability, biodegradability and rich structural diversity of these biopolymers for high-end applications with the goal of taming carbohydrates for a bio-based economy.

Cellulose, a polysaccharide consisting of β-linked glucose units, is the basic structural component of plant cell walls. It comprises about 33 percent of all vegetable matter (90 percent of cotton and 50 percent of wood) and is the most abundant of all naturally occurring organic compounds. Cellulose acetate[2] is one of the most widely used derivatives of cellulose. It is a versatile material that can be used in a wide range of applications. For example, it is used as impact resistant plastics and as soft fabrics. It is soluble in most conventional solvents and it is hypoallergenic (safe in contact with skin and food). It easily bonds with plasticizers and it can be composted or incinerated. Camille and Henry Dreyfus developed the first commercial process to manufacture cellulose acetate in 1910 (Scheme 1). Since then the market for cellulose acetate has increased to a large extent.[2]

However, the core structure of cellulose and most of its derivatives (e.g. cellulose acetate) is the same. This limits the range of its structural and mechanical properties and its applications. In the current push towards a bio-based economy, there is a great need to convert cellulose (and other carbohydrates) to value added products. There is active pursuit of the development of methods for the preparation of functionalized monomers, macromers, and polymers - suitable for the chemicals, materials, and pharmaceutical industry - from cellulose and other carbohydrates using chemical and enzymatic catalysis. One approach involves developing new green oxidation reactions and green chemical synthesis - towards developing methods for converting biopolymers into value-added products. The specific goal of the work described in this chapter was to convert cellulose to a functionalized polyester and copolyester in a one-pot acetylation/green oxidation reaction. The copolyesters would be prepared by a controlled partial oxidation reaction. Conversion of cellulose to functionalized polyester and copolymers will expand the current applications of cellulose and its derivatives and will open new markets. Subsequent chemical and/or enzymatic transformations will enable the preparation of novel drug candidates, pharmaceutical intermediates, and templates for drug discovery (Scheme 2).

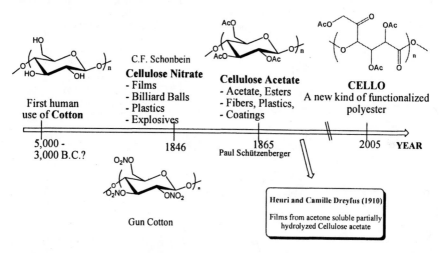

Scheme 1. Historical background on the use of cellulose and its inorganic or organic ester derivatives.

One class of such pharmaceutical intermediates is the iminocyclitols. Iminocyclitols, which are potent inhibitors of glycosidases, are also shown to have anti-cancer and anti-viral activities.[3-10] Glycosidase catalyzes the hydrolysis of the glycosidic linkages in carbohydrates and glycoconjugates. They are involved in digestion, biosynthesis of glycoproteins, and lysosomal catabolism of glycoconjugates. Finding a cure for cancer, diabetes, and AIDS is a matter of national and international interest. In 2003, the number of diagnosed American adults with diabetes was more than 14 million (6.6%)[11] and it is estimated that 90-95% have type II diabetes. On the other hand while cancer is the second leading cause of death in the United States,[12] today, 40.3 million people are also estimated to be living with HIV/AIDS.[13] The ketoesters (Schemes 3,4) are key intermediates in the synthesis of glycosidase inhibitors and hence of interest to the pharmaceutical industry. The preparation of ketoesters and polyketoesters using only a catalytic amount (0.02 equiv) of chromium hence reducing the amount of chromium waste is reported in this work. In addition, the recyclable periodic acid is used as the cooxidant.

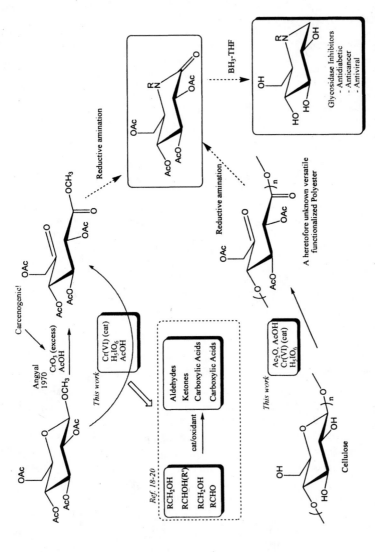

Scheme 2. *Preparation of ketoesters and polyketoesters via a green oxidation reaction and their proposed use in the synthesis of glycosidase inhibitors (Ac_2O = Acetic anhydride, AcOH = Acetic acid).*

Results and Discussion

Oxidation, one the most fundamental reactions in synthetic organic chemistry, has been the subject of extensive studies.[14-17] In light of the importance of oxidation reaction in the transformation of renewable materials to value added products, there is still a need for new environmentally friendly versatile oxidation methods. Chromium, a favorite powerful oxidant among organic chemists, causes major problems to the environment due to its carcinogenicity. To address this problem, we have recently[18,19] shown that Cr (VI) can be used catalytically to prepare aldehydes and ketones from primary and secondary alcohols, respectively, using one equivalent of periodic acid as the terminal oxidant. We have also shown[19,20] that carboxylic acids can be prepared effortlessly from aldehydes and directly from primary alcohols using Cr (VI) as a catalyst and one and two equivalents periodic acid, respectively. This is a more environmentally friendly oxidation as periodic acid is a recyclable oxidant and less waste is generated since only a catalytic amount of Cr (VI) is used. Iodate can be reoxidized to periodate electrochemically.[21] The hypothesis was that the Cr (VI) - periodic acid combination may form chromatoperiodate[22] that is capable of oxidations beyond what is possible by the individual reagents.

One of the most direct routes to iminocyclitols involves oxidation of glycosides to ketoesters followed by reduction of their oximes[8] or reductive amination with a primary amine or ammonia (Scheme 2). It has been shown that[23,24] oxidation of glycosides to ketoesters could be achieved by using excess CrO_3 in acetic acid. However as mentioned above, the carcinogenicity of chromium reagents in combination with the need to use much more than the stoichiometric amount is a major problem for any large scale or industrial application. To further study the scope and limitations of our Cr (VI)-catalyzed oxidations, and in an effort to find ways of preparing the above mentioned ketoesters in an environmentally compatible way we have investigated the oxidation of carbohydrates using Cr (VI) as a catalyst and periodic acid as the terminal oxidant and the results are reported here.

In a model study,[25] we investigated the oxidation of alkyl glycosides protected by acetyl groups using a catalytic amount of Cr (VI) and 2.2 equivalent of periodic acid in acetic acid and in the presence of acetic anhydride. For example, a mixture methyl 2,3,5-tri-O-acetyl-α- and β-D-ribofuranoside was successfully oxidized to its ketoester (Scheme 3). Methyl 2,3,4,6-tetra-O-acetyl-β-D-glucopyranoside was also oxidized to its ketoester in excellent yields. However, our attempt to oxidize methyl 2,3,4,6-tetra-O-acetyl-α-D-glucopyranoside was unsuccessful. This indicates that, similar to the chromic acid oxidation,[23,24] the CrO_3 catalyzed periodic acid oxidation was selective for β-linked glycosides.

The selectivity of our oxidation protocol for β-linked glycosides inspired us to attempt the Cr (VI) catalyzed oxidation on cellulose. If successful this would lead to functionalized polyester (Scheme 4). Furthermore our solvent of choice for the oxidation reaction, i.e. acetic acid, turns out to be the same solvent used for acetylation of cellulose to prepare cellulose acetates. This prompted us to carry out a one-pot acetylation-oxidation of cellulose. Standard acetylation of cellulose using the 'acetic acid process' where cellulose is acetylated with acetic anhydride in acetic acid, in the presence of a catalytic amount of sulfuric acid, was performed. A catalytic amount of chromic acid and 2.2 equivalent of periodic acid per glucose units were then added to the reaction mixture in ice-bath and stirred for two hours and then stirring was continued at room temperature overnight. Aqueous precipitation of the product delivered a heretofore unknown polyketoester (CELLO, Scheme 4) in quantitative yields. By limiting the amount of periodic acid used to less than 2 equivalents, we have also successfully prepared a new kind of copolymer CTA-CELLO with an ester and acetal linkages (Scheme 5). Cellulose triacetate was obtained by taking an aliquot of the reaction mixture right before the addition of the oxidants followed by aqueous precipitation, filtration, and drying overnight in vacuum oven at 40°C. It is worth mentioning that the above discussed oxidation of cellulose also works when halochromates (chloro- and fluorochromates) are the catalysts. It also works with chromic acid as the terminal oxidant (i.e. without periodic acid) but the reaction is sluggish.

The products were characterized using ^{1}H and ^{13}C nuclear magnetic resonance (NMR) (Figures 1-2) and infrared (IR) spectroscopy (spectra not shown). The infrared spectrum of the polyketoester CELLO showed a strong peak at 1756 cm^{-1} compared to that of cellulose triacetate at 1752 cm^{-1} corresponding to the C=O band of esters and ketones. The other strong bands at 1376, 1210, and 1064 cm^{-1} for CELLO are also shifted as compared to 1369, 1232, and 1049 cm^{-1} for cellulose triacetate. Carbonyl overtone bands were also observed at 3480 cm^{-1} and 3501 cm^{-1} for CELLO and cellulose triacetate, respectively. The shifts in IR spectrum signals of the copolymer CTA-CELLO with respect to cellulose triacetate were smaller since the majority (58%) of the copolymer CTA-CELLO was cellulose triacetate itself. The major peaks of the copolymer CTA-CELLO were at 1748, 1373, 1230, and 1046 cm^{-1}.

On the other hand, the ^{1}H and ^{13}C NMR signals of the products were more informative. The ^{1}H NMR spectrum of cellulose triacetate aliquot collected before further oxidation to CELLO matched those shown in literature[2] and showed 7 hydrogens on the glucose unit relative to 9 on the acetyl groups indicating complete acetylation of cellulose. After the oxidation and isolation of CELLO, its ^{1}H NMR spectrum (Figure 1) showed 5 hydrogens on the glucose unit compared to 9 on the acetyl groups confirming that the substrate has been completely oxidized to a polyketoester. The ^{13}C NMR spectrum was also very conclusive. The peak at 100.5 ppm corresponding to the anomeric carbon in

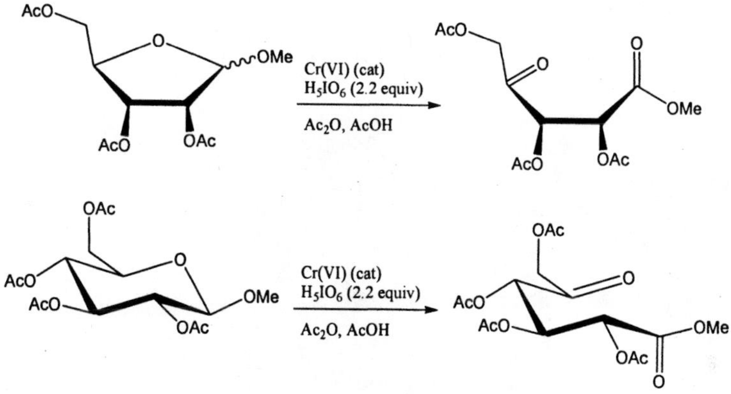

Scheme 3. Preparation of ketoesters via a green oxidation reaction of alkyl furanosides and alkyl pyranosides.

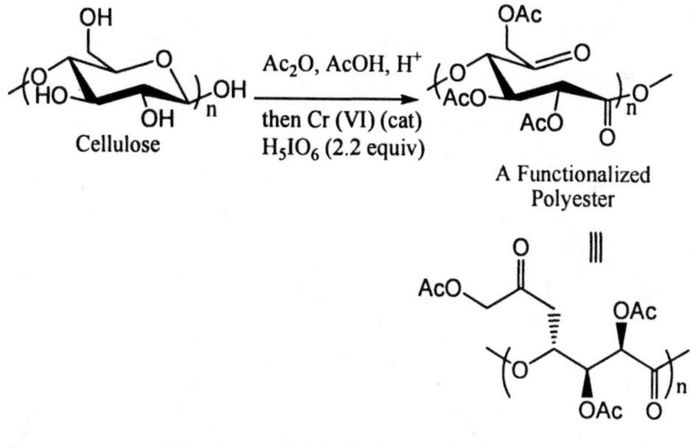

Scheme 4. One-pot acetylation and Cr (VI) catalyzed periodic acid oxidation of cellulose to a polyketoester.

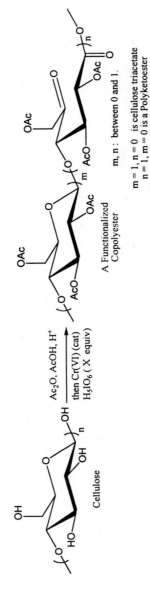

Scheme 5. One-pot acetylation and Cr (VI) catalyzed periodic acid oxidation of cellulose to a copolymer, CTA-CELLO, of cellulose triacetate (CTA) and ketoester (CELLO) with a tunable ratio determined by the amount of periodic acid used.

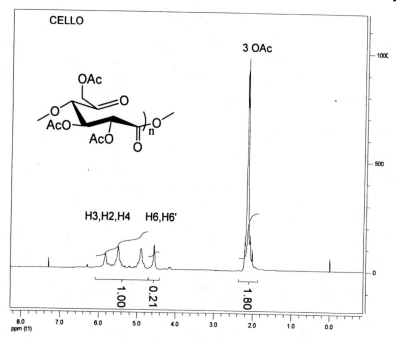

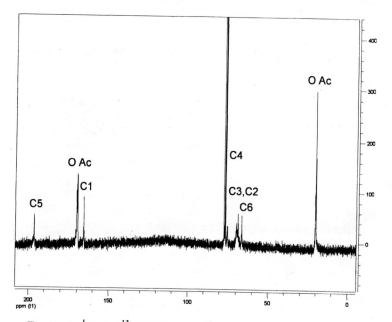

Figure 1. ^1H and ^{13}C NMR spectra of a polyketoester (CELLO).

cellulose triacetate is shifted to 166.3 ppm indicating that it has become an ester functional group in CELLO. The C5 peak of cellulose triacetate also shifted from 72.0 to 196.6 ppm corresponding to a ketone functionality, further confirming the complete oxidation. This is similar to what is observed to that of the glucose analog.[23-25] The C6 peak of CELLO, being next to a C=O has also shown a significant shift compared to that of cellulose triacetate.

As expected, the NMR spectra of the copolymer CTA-CELLO (Figure 2) showed the characteristic signals of both cellulose triacetate (CTA) and CELLO. In the ^{13}C NMR spectrum, for example, while some of the anomeric carbon still showed a peak at 100.5 ppm corresponding to the CTA component of the copolymer, some of it showed a peak at 166.3 ppm similar to CELLO. Similarly, a peak at 196.6 ppm characteristic of the ketone functionality at C5 was also observed. The ^{1}H NMR spectrum of CTA-CELLO showed an overlap of that of cellulose triacetate and CELLO.

The integration ratio of the ^{1}H NMR signals of the hydrogens on the glucose unit to those on the acetyl groups is 7:9 in cellulose triacetate and 5:9 in CELLO. Hence, from the corresponding integration ratio in CTA-CELLO, the rough estimate of the percent of glucose units that are opened to a polyketoester was calculated. Thus, the percentage of CELLO turns out to be $\{[7 - 9/R]/2\}$ X 100%, where R is the ratio of the OAc integration to that of the H's on the glucose unit. R is 1.80 for CELLO, 1.28 for cellulose triacetate, and between 1.28 and 1.80 for CTA-CELLO. In Figure 2, the ratio R = 1.43, indicating about 35% CELLO and 65% CTA as components of the copolymer.

Conclusions

A one-pot acetylation/oxidation of cellulose to a functionalized polyester is achieved via acetylation using acetic anhydride in acetic acid catalyzed by sulfuric acid, followed by oxidation with a catalytic amount of Cr (VI) and 2.2 equivalent of the recyclable oxidant periodic acid. A copolymer was also prepared using the same protocol but by limiting the amount of periodic acid used to deliver a cellulose triacetate/polyketoester copolymer. The method also works well for the oxidation of simple glycosides to ketoesters. The ketoesters and polyketoesters are key intermediates in the synthesis of iminocyclitols, potent antidiabetic, anticancer, and antiviral agents.

This one-pot preparation of keto functionalized polyester (CELLO) by direct acetylation/oxidation of cellulose will enable a two step preparation of lactones and lactams starting from cellulose. Further work is being done on transformations of this (poly) ketoesters to value added monomers and polymers that could serve as powerful templates for building a library of drug intermediates, important commodities for the pharmaceutical industry. The functionalized polyesters and copolymers are also expected to find applications

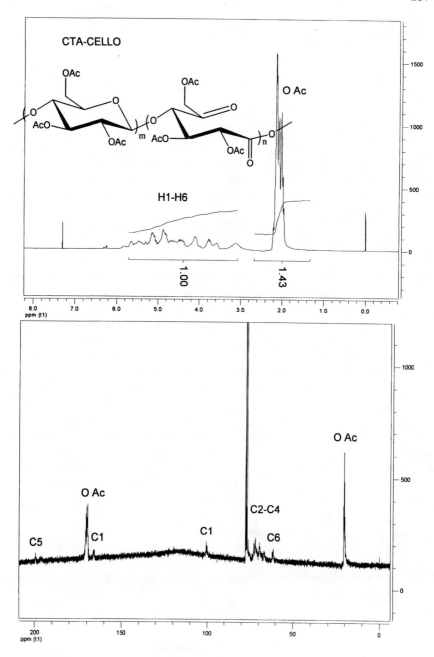

Figure 2. ^1H and ^{13}C NMR spectra of the copolymer CTA-CELLO.

in the pharmaceutical industry, for example, in drug delivery, similar to the polyesters and polyhydroxyalkanoates (PHAs).

Experimental

General. All solvents and reagents were obtained commercially at the highest purity available and used without further purification. ^{1}H and ^{13}C NMR spectra were recorded in CDCl$_3$ on a Bruker NMR spectrometer at 300 and 75 MHz, respectively. The chemical shifts in parts per million (ppm) for ^{1}H NMR spectra were referenced relative to tetramethylsilane (TMS, 0.00 ppm) as the internal reference. Infrared spectra were recorded on a Perkin Elmer Paragon 500 FTIR spectrometer.

One-pot acetylation/oxidation of cellulose. A mixture of 125 mL acetic acid, 62.5 mL acetic anhydride, and 8 drops of concentrated sulfuric acid were added into 16.25 g of Cotton. The reaction mixture was stirred at 105 $^\circ$C for 3 h and left standing overnight. An aliquot of the mixture was then taken out and poured into cold water (to precipitate out the product), filtered, washed with water, dried overnight in vacuum oven and its NMR spectrum was recorded in CDCl$_3$ and IR spectrum recorded as a CHCl$_3$ cast to indicate that the cellulose was completely acetylated to deliver cellulose triacetate as shown in the results section. The reaction mixture was then diluted with 75 mL AcOH and cooled in ice-bath, followed by addition of 52.3 g of periodic acid and 0.5 g chromic acid. The reaction mixture was stirred in ice-bath for 2 h and at r.t. overnight. The HIO$_3$ precipitate was recovered by filtration under vacuum and the solvents evaporated in a rotary evaporator until a small amount of the solvents remained. The residue was then poured into 300 mL cold water and the white precipitate that formed was filtered under vacuum, washed with aq. NaHSO$_3$ and water, successively, and dried in vacuum oven overnight to deliver CELLO (26.6 g). The NMR spectra were recorded in CDCl$_3$ and are shown and discussed in the results and discussion section. The IR spectrum were recorded as CHCl$_3$ cast and are discussed in the results and discussion section.

One-pot acetylation/oxidation of cellulose into copolyesters. The copolymer (CTA-CELLO) was prepared and isolated in a similar manner to that of CELLO except that 1 equiv of periodic acid instead of 2.2 equivalents was used. The NMR spectra were recorded in CDCl$_3$ and its IR spectrum as CHCl$_3$ cast and are shown and discussed in the results and discussion section.

Acknowledgements

Great appreciation is offered to Kenyon College for the generous start up fund.

References

1. Some of the results of this work were previously presented at an ACS meeting: See Hunsen, M. *Abstracts of Papers of the American Chemical Society* **2005**, 229, U282-U282.
2. Rustemeyer, R. Ed. *Cellulose Acetates: Properties and Applications*, Wiley-VCH, Weinheim, Germany, 2004.
3. Watson, A. A.; Fleet, G. W. J.; Asano, N.; Molyneux, R. J.; Nash, R. J.; *Phytochemistry* **2001**, 56, p 265.
4. Asano, N.; Nash, R. J.; Molyneux, R. J.; Fleet, G. W. J. *Tetrahedron Asymm.* **2000**, 11, p 1645.
5. Kajimoto, T.; Liu, K. K. C.; Pederson, R. L.; Zhong, Z.; Ichikawa, Y.; Porco, J. A. Jr.; Wong, C. H. *J. Am. Chem. Soc.* **1991**, 113, pp 6187-96.
6. Takebayashi, M,; Hiranuma, S.; Kanie, Y.; Kajimoto, T.; Kanie, O.; Wong, C. -H. *J. Org. Chem.* **1991**, 64, pp 5280-5291.
7. Heck, M. -P.; Vincent, S. P.; Murray, B. W.; Bellamy, F.; Wong, C. -H.; Mioskowski, C. *J. Am. Chem. Soc.* **2004**, 126, pp 1971-1979.
8. Pistia, G.; Hollingsworth, R. I. *Carbohydr. Res.* **2000**, 328, pp 467–472.
9. Moris-Varas, F.; Qian, X. -H.; Wong, C. -H. *J. Am. Chem. Soc.* **1996**, 118, pp 7647-7652.
10. Mitchell, M. L.; Lee, L. V.; Wong, C. -H. *Tet. Lett.* **2002**, 43, p 5691.
11. http://www.cdc.gov/nchs/fastats//diabetes.htm 12/12/2005.
12. http://www.cdc.gov/nchs/fastats/lcod.htm. 12/12/2005.
13. http://www.unaids.org/en/default.asp. 12/12/2005.
14. Sheldon, R. A.; Kochi, J. K. *Metal-Catalyzed Oxidations of Organic Compounds*. In Academic Press: New York, 1981.
15. Hudlicky, M., *Oxidations in Organic Chemistry*. American Chemical Society: Washington, DC, 1990.
16. Bowden, K.; Heilbron, I. M.; Jones, E. R. H.; Weedon, B. C. L. *J. Chem. Soc.* **1946**, p 39.
17. Bowers, A.; Halsall, T. G.; Jones, E. R. H.; Lemin, A. J. *J. Chem. Soc.* **1953**, p 2548.
18. Hunsen, M. *Tetrahedron Letters* **2005**, 46, pp 1651-1653.
19. Hunsen, M. *J. Fluor. Chem.* **2005**, 126, pp 356-1360.
20. Hunsen, M. *Synthesis-Stuttgart* **2005**, pp 2487-2490.
21. Khan, F. N.; Jayakumar, R.; Pillai, C. N. *J. Mol. Catal. A: Chemical* **2003**, 195, pp 139-145.
22. Okumura, A.; Kitani, M.; Murata, M. *Bull. Chem. Soc. Jpn.* **1994**, 67, p 1522.
23. Angyal, S. J.; James, K. *Chem. Comm.* **1969**, pp 517-518.
24. Angyal, S. J.; James, K. *Aust. J. Chem* **1970**, 23, pp 1209-1221.
25. Hunsen, M. unpublished results.

Chemicals from Forest Biomass

Chapter 17

Production of Chemicals from Cellulose and Biomass-Derived Compounds: Advances in the Oxidative Functionalization of Levulinic Acid

R. Saladino[1], T. Pagliaccia[1], D. S. Argyropoulos[2], and C. Crestini[3]

[1]Department of Agrobiology and Agrochemistry, University of Tuscia, 01100 Viterbo, Italy
[2]Department of Forest Biomaterials Science and Engineering, North Carolina State University, Raleigh, NC 27695–8005
[3]Department of Science and Chemical Technologies, University of Tor Vergata, 00133 Rome, Italy

This chapter presents new data regarding the oxidation of levulinic acid (4-oxopentanoic acid), which can be considered as one of the most significant cellulose-derived compounds. The various synthetic applications of these procedures for the preparation of fine-chemicals and commodities are also discussed, with particular attention on the reactions that use homogeneous and heterogeneous rhenium catalysts. Such systems are able to activate hydrogen peroxide under mild experimental conditions causing selective oxidation chemistries.

Introduction

Some of the most abundant chemicals on the planet that are intrinsic to fine chemical and commodities production *without concomitant toxic waste generation* are carbohydrates and lignin. These materials comprise a vast amount of biomass (in the range of hundreds of billions of tons) in which only 3% is used by humans. As opposed to many synthetic chemicals, these materials are readily renewable, inexpensive, and environmentally compatible. Despite a significant amount of recent research for augmenting the industrial use of readily available carbohydrates as raw materials (1-6) (e.g., car moldings, cosmetics, food), the systematic utilization of this vast resource is still in its infancy. Moreover, the fundamentally different chemistries of carbohydrates and lignins versus hydrocarbons must be addressed to efficiently use them as organic raw materials as opposed to petroleum-based materials. During the last years, a number of studies on the pyrolysis of lignocellulosic materials showed that a variety of interesting structures can be isolated from the pyrolysis oils (7). These oils have been shown to contain 5-hydroxymethyl furfural, 2-furaldehyde (furfural), 2-methylfurfural and levoglucosenone (1,6-anhydro-3,4-dideoxy-β-D-glycero-hex-3-enopyranos-2-ulose) (8). Continuing efforts involving acid hydrolysis have allowed the development of new technologies mainly from companies that claim the transformation of relatively low-grade waste cellulosic by-products from paper mills and other sources to levulinic acid **1** (4-oxopentanoic acid) (Figure 1) (9).

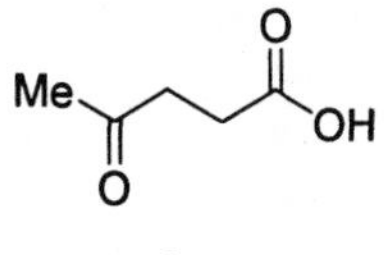

1

Figure 1. Structure of levulinic acid (4-oxopentanoic acid).

This compound can now be obtained in a very cost-effective manner (as low as $0.04-0.10/lb) in large scale (1000 (dry) ton/day). Feedstocks that have allowed the production of levulinic acid include paper mill sludge, non-recyclable waste paper, wood wastes, agricultural residues and municipal solid waste (MSW). This development is very significant since a major US city typically produces more than 700,000 tons/year of MSW, and the US generates more than 200 million tons of MSW annually. More than half of this waste consists of paper, cardboard, wood and herbaceous material; cellulosic materials that are difficult to recycle and naturally resistant to chemical breakdown.

Levulinic acid is usually produced from mixtures of hexoses generated *in situ* by acid hydrolysis of cellulose and hemicellulose. As a general reaction pathway, hexoses are initially dehydrated to 5-hydroxymethyl furfural which is in turn transformed to levulinic acid (10). The enolisation process of D-glucose, D-mannose or D-fructose in acid media affords the corresponding enediol. This intermediate is further dehydrated to a complex mixture of enol and dienediol derivatives that are readily converted to 5-hydroxymethyl furfural. Finally, the transformation of 5-hydroxymethyl furfural to levulinic acid is a result of a multi-step domino process involving addition of water to C(2)-C(3) double bond, ring-opening of the furan ring, dehydration, rearrangement and elimination of formic acid. Humin derivatives are the main side-products of these transformations (11). Recently, Biofine (BioMetics Inc, US) claimed an innovative technological process to obtain the production of levulinic acid from cellulose wastes with high conversion of starting materials and yield of desired product. The first industrial plant based on this process has been constructed in Italy by Le Calorie S.r.l. (12). In this latter case, paper mill wastes added with a lower amount of fruit and vegetable wastes are used as renewable substrates for the acid hydrolysis. As an example, the carbohydrate-containing material is supplied to a first reactor and treated with 1-5 % by weight mineral acid at 210-230°C for 15-25 seconds. The hydroxymethylfurfural formed in this first step is further supplied to a second reactor in which it is hydrolized at 195-215°C for 15-30 minutes to produce levulinic acid in 70% yield based on the hexose content of the starting material (13). Under favorable economics and developing markets for levulinic acid, the US Department of Energy projects energy savings of more than 75.6 trillion BTUs per year and waste reduction of nearly 26 million tons per year by 2020. Levulinic acid can be characterized as a versatile chemical since it can function as an intermediary to several other compounds. The derivatives of levulinic acid include methyl tetrahydrofuran (MTHF), a low volatility gasoline extender. In addition, various industrial chemicals can be manufactured from levulinic acid including tetrahydrofuran, 1,4 butanediol, diphenolic acid, succinic acid, gamma butyrolactone, angelicalactone, and *N*-methylpyrrolidone (14). Despite several studies reported on the chemical modification of levulinic acid, little attention has been devoted to the application of oxidative procedures.

Oxidation Chemistry of Levulinic Acid. The State of Art.

The distribution of the products of oxidation of levulinic acid **1** strongly depends on the nature of the oxidant and the actual conditions used for the oxidation (15). For example, the stoichiometric oxidation of levulinic acid esters (**2**) with peroxy compounds, under Baeyer-Villiger rearrangement conditions, yields succinates (**3**) as the main products (16). In the absence of the

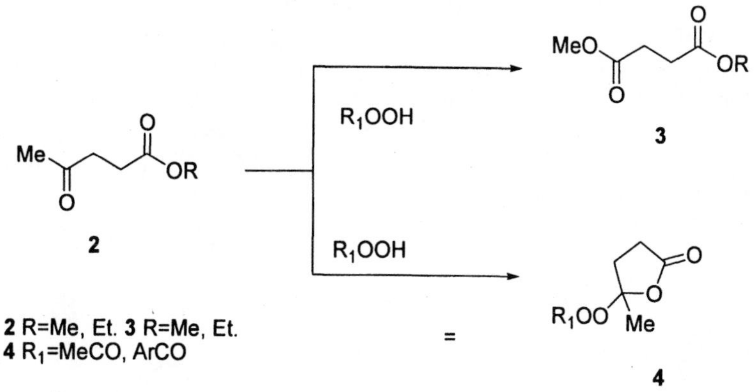

Scheme 1. Oxidation of levulinic acid esters with peroxy compounds.

carboxy protection, peroxyvalero-γ-lactones **4** of low stability were recovered (Scheme 1) (17).

The oxidation of **1** with selenium dioxide (SeO$_2$), under Riley reaction conditions, has been reported in the patent literature to give 2-oxo-glutaric acid **5** (Scheme 2) (18). However, the pronounced toxicity of SeO$_2$, together with the large excess of oxidant required for the complete conversion of **1** necessitates the development of alternative methods.

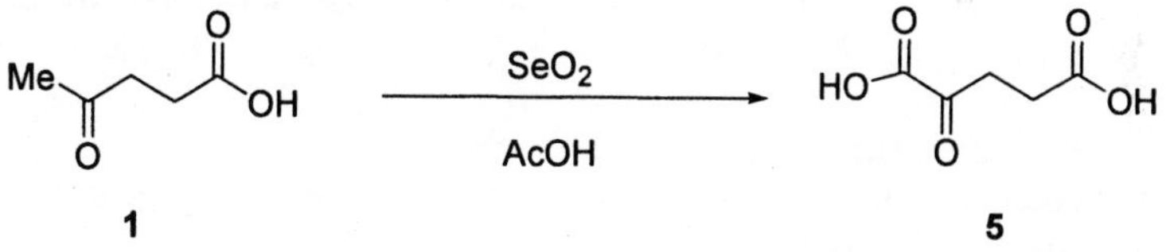

Scheme 2. Oxidation of levulinic acid with selenium dioxide.

The cyclization of phenylhydrazones of levulinic acid, levulinanilide and levulinic acid oxime with lead tetraacetate Pb(OAc)$_4$ has been reported. The phenylhydrazone of levulinic acid is cyclized to γ-phenylazo- γ-valerolactone, while the oxidation of levulinanilide phenylhydrazone leads, by analogous *O*-to-*C* ring closure, to γ-phenylazo- α-phenylimino- γ-valerolactone. Similarly, the oxime of levulinic acid is converted into γ-nitroso- γ-valerolactone, which is isolated as the dimer, trans-bis[γ -nitroso- γ -valerolactone] (19). The kinetics and mechanism of oxidation of levulinic acid with different stoichiometric reagents have been reported mainly with a focus on its degradation pathways (20).

To date, there are few literature accounts that describe the catalytic oxidation of levulinic acid. For example, the high-temperature oxidation of **1** with oxygen on vanadium pentaoxide (V_2O_5) allows the formation of succinic acid **6** in 81% yield. The reaction, which probably proceeds by a Bayer-Villiger rearrangement, requires hard conditions and is performed by passing vapors of **1** admixed with oxygen over a bed of V_2O_5 using a carrier gas at high temperatures (Scheme 3) (21).

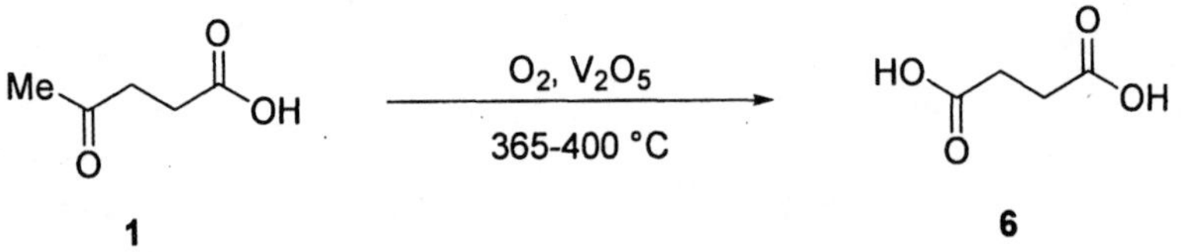

Scheme 3. Oxidation of levulinic acid with vanadium pentoxide.

Moreover, the oxidation of levulinic acid by vanadium (V) in aqueous solution with visible light is described to afford acetic acid with high conversion of substrate and yield of product (22).

Recently, the oxidation of **1** to a mixture of **6** and formic acid **7** with *N*-haloamides (where X = Br) in perchloric acid ($HClO_4$) and mercuric acetate [$Hg(OAc)_2$] (as a scavenger and ruthenium (III) chloride) has been reported (Scheme 4) (23). While these procedures offer the usual advantages associated with catalytic oxidations, the high temperatures necessary for the V_2O_5 transformation (Scheme 3) and the high toxicity of mercury salts (Scheme 4) are issues of immediate environmental concern.

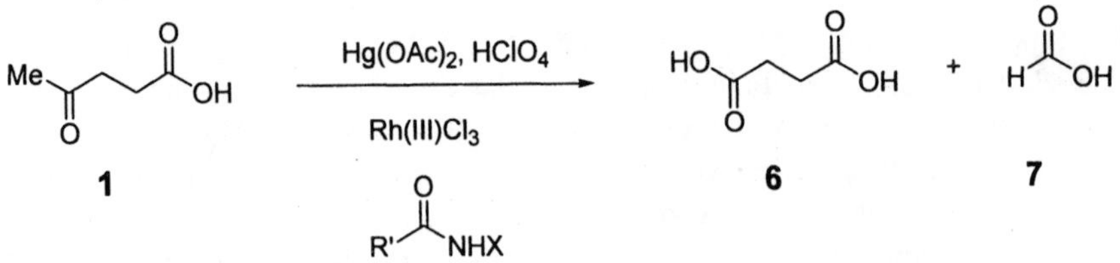

Scheme 4. Oxidation of levulinic acid with rhuthenium catalysts.

Homogeneous and Heterogeneous Methyltrioxo Rhenium (MTO) Compounds as Useful Catalysts for the Oxidation of Cellulose and Biomass-Derived Compounds

Methyltrioxo rhenium (MTO, CH_3ReO_3) **8** (Figure 2) (24) has become the subject of intense research activity during the past decade. This is due to its demonstrated efficiency and selectivity toward a variety of oxidative and other synthetic transformations (25). At room temperature, MTO is a colorless solid, crystallizing to colorless needles. It is readily sublimed and soluble in water and most common organic solvents. The merits of MTO in oxidation reactions have been well-documented for the oxidation of olefins (26), alkynes (27), aromatic derivatives (28), sulphur compounds (29), phosphines (30), organonitrogen compounds (31), Baeyer-Villiger rearrangements (32), and oxygen insertion into C-H bonds (33) with hydrogen peroxide (H_2O_2) or urea hydrogen peroxide adduct (UHP) as primary oxidants (34).

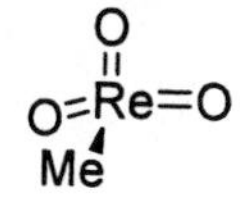

$$O=\overset{\overset{\displaystyle O}{\|}}{\underset{\underset{\displaystyle Me}{\diagup}}{Re}}=O$$

8

Figure 2. Structure of methyltrioxo rhenium 8 (MTO).

The active catalytic forms are a monoperoxo metal [$MeRe(O)_2(O_2)$] (**A**) and a bisperoxo metal [$MeRe(O)(O_2)_2$] (**B**) complexes, and/or their adducts with solvent molecules (35). A schematic representation of the mechanism of MTO oxidation with H_2O_2 is shown in Figure 3. Specifically, the oxygen atom transfer from complexes (**A**) and/or (**B**) to substrate is described by a concerted process requiring a butterfly-like transition state. It is worth noting that MTO systems are particularly appealing for the development of environmentally benign oxidative transformations since rhenium, in contrast to many other transition metals, does not catalyse the decomposition of hydrogen peroxide.

MTO is produced up to 50% yield by oxidation of tetramethylrhenium (VI) oxide $(CH_3)_4ReO$ or the decomposition of trimethyldioxorhenium (VII) $(CH_3)_3ReO_2$. Improvements of this synthesis have been proposed starting from commercially available rhenium heptaoxide Re_2O_7 and non-reducing transfer reagents such as tetraalkyltin $SnMe_4$ or dialkylzinc $ZnMe_2$ derivatives (36). The preparation of MTO in high yield (70-85%) from rhenium-containing compounds by reaction with silylating agent and organylating reagents has also been reported (37). These days, MTO can be produced directly from rhenium powder in kilogram quantities. Some efforts have been devoted to design novel

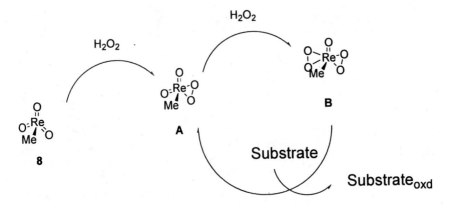

Figure 3. Schematic drawing of the mechanism of oxidation of MTO with H_2O_2 as primary oxidant.

heterogeneous MTO compounds. In fact, the heterogenation of MTO on polymeric supports is an important tool because it allows an easier recovery of the catalyst, decreases the toxicity of reaction wastes and sometimes improves the reactivity (38). Herrmann and co-workers first reported in patent literature the preparation of heterogeneous MTO compounds, of general formula [(polymer)$_f$/(MTO)$_g$] (the f/g quotient expresses the ratio by weight of the two components; polymer = not reticulated poly(4-vinylpyridine), poly(2-vinylpyridine), poly(vinylpyrrolidone), poly(acrylamide), and nylon 6), in which MTO was supposed to be bonded to the support by co-ordination with only one hetero atom, analogous to the homogeneous complexes (39). The structures of these (polymer)$_f$ /(MTO)$_g$ systems were not completely characterized and useful structure-activity relationships are still lacking. With the aim to develop clean oxidation processes, the preparation of novel heterogeneous rhenium compounds by heterogenation of MTO on poly(4-vinylpyridine) 2% and 25% cross-linked (with divinylbenzene) (PVP-2%/MTO **I** and PVP-25%/MTO **II**, respectively), applying an extension of the "mediator" concept (40) is described.

A further heterogeneous MTO catalysts, polystyrene/MTO **III** was prepared by a microencapsulation technique (41) with polystyrene as solid support (42). The structures of poly(4-vinylpyridine)/MTO and polystyrene/MTO catalysts **I**-**III** are schematically represented in Figure 4. Scanning electron microscopy (SEM) photographs showed that in the poly(4-vinylpyridine) family, a low value of the reticulation grade (2%) affords particles that are of different shapes and very irregular surfaces. In the presence of an increased value of the reticulation grade (25%), the transformation of irregular, spongy particles to regular

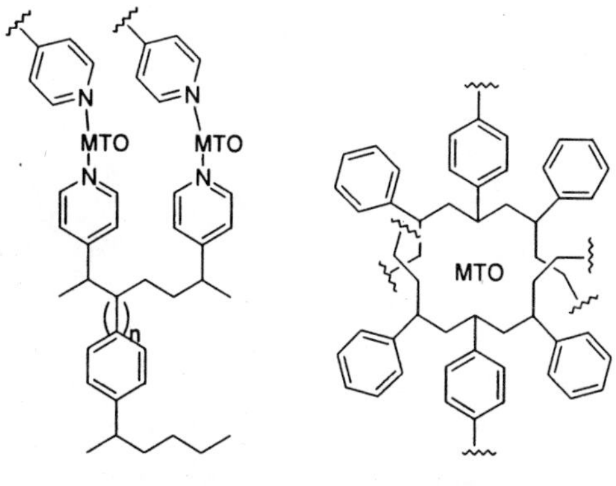

Figure 4. Heterogeneous MTO catalysts I-III.

spherical particles is observed. Polystyrene forms microcapsules that are characterized by an average value of the diameter that is smaller than that determined for the PVP-25%/MTO system. These systems showed high catalytic activity and selectivity in the oxidation of aromatic derivatives (43),pyrrolidines (44), alkenes and terpenes (45), including the oxygen atom insertion into the C-H sigma bond of hydrocarbons, both in molecular solvents (46) and ionic liquids (47), with environmental friendly H_2O_2 as primary oxidant.

Oxidation of Levulinic Acid with MTO and Heterogeneous MTO Catalysts. A New Entry to Fine-Chemicals and Commodities

Initially, the oxidation of levulinic acid **1** was investigated using MTO under homogeneous conditions. As a general procedure compound **1** (1.0 mmol) was added to a solution of MTO (5% in weight) in ethanol (2.0 mL).

Hydrogen peroxide (5.0 mmol, 30% aqueous solution) was added to the reaction mixture repeatedly for 4 days at 75 °C. In the absence of the catalyst, less than 2% conversion of substrate took place under otherwise identical conditions. The results of the oxidation are reported in Scheme 5 and Table 1.

The oxidation of **1** with MTO proceeded with 90% conversion of substrate to afford the β-lactone **9** (2-oxetanone) as the main reaction product in 45% yield, beside ethyl-3-hydroxy-4-oxopentanoate **10** and ethyl-2-hydroxy-4-oxopentanoate **11** as side-products (Table 1, entry 1). Traces of ethyl levulinate **12** (<4%) were also detected in the reaction mixture. Ester derivatives like

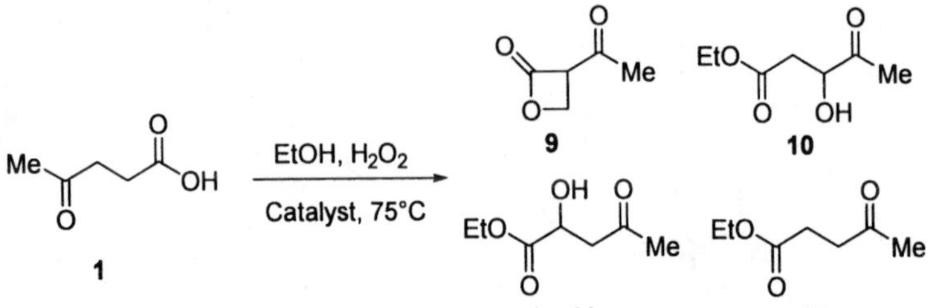

Scheme 5. Oxidation of levulinic acid 1 with MTO and heterogeneous MTO catalyst with H_2O_2 as primary oxidant.

compounds **10** and **11** are largely used for the manufacture of block thermoplastic polyether-polyesters (48), of cosmetics (49) and for the preparation of biologically active amino acids (50) and olefins (51). The presence of the β-lactone **9** is worthy of note. The β-lactone moiety is present in many biologically active natural products and can be used as a synthon for the preparation of high value chemicals (52). β-Lactones are usually synthesized by lactonization (*via* oxygen-alkyl bond formation or *via* oxygen-acyl bond formation) on activated intermediates or by [2+2] cycloaddition. This is most likely the first example dealing with the preparation of a functionalized β-lactone derivative by a domino oxidative C-H oxygen atom insertion/lactonization process from the corresponding acid derivative. The synthesis of the β-lactone **9** suggests a multifunctional catalytic behavior for MTO under these experimental conditions. Probably, compound **9** was obtained by initial oxygen atom insertion at secondary C(3)-H bond on newly formed ethyl levulinate **12** to give **10**, followed by intramolecular lactonization, due to the known Lewis and Brönsted acidity of MTO (Scheme 6).

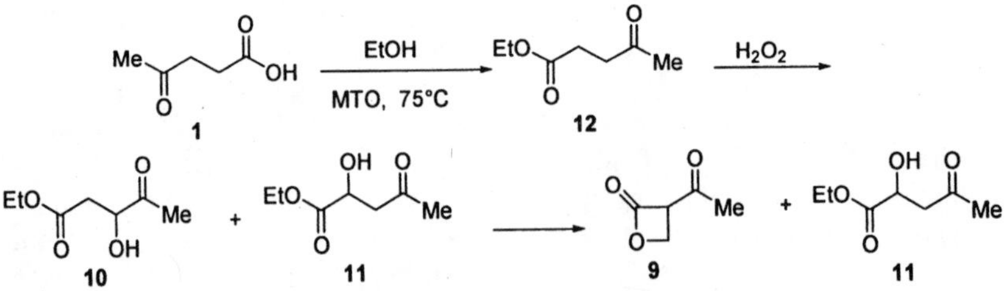

Scheme 6. Hypothesized mechanism of oxidation of levulinic acid 1 with MTO.

In agreement with this hypothesis, appreciable amounts of **9** were obtained by treating compound **10** with MTO in the absence of H₂O₂. The remarkable selectivity of *O*-atom insertion at two possible secondary C-H bonds, despite the high reactivity of MTO, is confirmed by the higher yield of compounds **9** and **10**, in comparison with **11** (necessarily produced as a consequence of the oxidation of the secondary C(2)-H bond).

Table 1. Oxidation of Levulinic Acid 1 with MTO and Heterogeneous MTO Catalyst I-III Using H₂O₂ as Primary Oxidant

Entry	Catalyst	Conversion %	Product(s)	Yield(s) %
1	MTO	90	9(10)[11]	45(15)[27]
2	PVP-2/MTO (I)	96	12(10)[11]	55(25)[15]
3	PVP-25/MTO (II)	>98	10(11)	50(47)
4	PS/MTO (III)	86	12(9)	79(21)
5	PS/MTO-L (IV)	>98	12(9)	38(50)

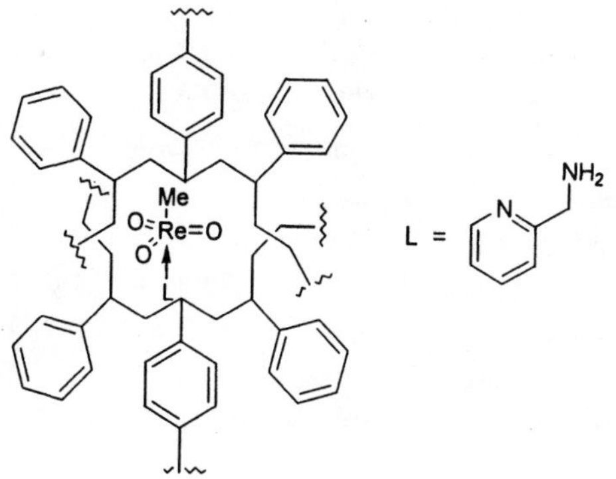

PS/MTO-L (**IV**)

Figure 5. Sketch of heterogeneous MTO catalysts IV.

All polymer-supported MTO systems **I-III** showed high efficiency in the oxidation of **1**, giving a different distribution of products depending on the catalyst used in the transformation. Briefly, levulinic acid **1** (1.0 mmol) dissolved in ethanol (2.0 mL) was added portionately with 40 mg of the appropriate heterogeneous catalyst (loading factor 1.0; the loading factor is referred as mmol of active species for gram of resin) and H_2O_2 (5.0 mmol) at 75°C. The oxidation of **1** with catalyst PVP-2/MTO **I** afforded ethyl levulinate **12** as the main reaction product in addition to minor amounts of hydroxyl levulinates **10** and **11** (Scheme 5, Table, entry 2). On the other hand, treatment of **1** with PVP-25/MTO **II** afforded levulinates **10** and **11** in high yield (Scheme 5, Table, entry 3). Thus, in accordance with data previously reported, the selectivity of the oxidation was found to be correlated to the reticulation grade of the poly(4-vinyl)pyridine support. Moreover, the acidity of MTO was probably tuned by the presence of the pyridinyl moieties, as suggested by the absence of **9** in the reaction mixture.

Notably, a different reaction pathway was observed with microencapsulated MTO catalyst **III**. In this latter case, ethyl levulinate **12** was recovered as the main reaction product in the presence of a little amount of **9** (Table, entry 4). A better result was obtained using a catalyst synthesized by microencapsulation on polystyrene of the complex between MTO and 2-aminomethyl pyridine (PS/MTO-L **IV**, Figure 5) (45a). MTO reacts with monodentate and bidentate ligands to give trigonal bipyramidal and distorted octahedral adducts, respectively (53). The activity of these complexes in several oxy-functionalizations was found to be dependent on different experimental parameters, such as the redox stability of the ligand, and on the reaction temperature (54). The oxidation of **1** with catalyst PS/MTO-L **IV** under the previously reported experimental conditions afforded **9** as the main reaction product (50% yield). Furthermore, the conversion of the substrate was quantitative, with an appreciable amount of **12** (Scheme 5, Table, entry 5). Irrespective of the catalyst used for the oxidation, heterogeneous catalysts **I-IV** showed a reactivity similar to that of MTO, suggesting the absence of a kinetic barrier to the approach of substrate to rhenium polymeric compounds.

Finally, the oxidation of some derivatives of ethyl levulinate **12** is presented.

It is well known from the literature that the MTO/H_2O_2 system reacts efficiently with silyl enol ethers to give the corresponding epoxides. These compounds, after successive ring-opening of the oxiranyl ring, can afford α-hydroxy ketones in high yield (55). To evaluate the generality of this procedure, ethyl levulinate **12** was treated with tert-butyldimethylsilyl triflate (TBDMS) in triethylamine and CH_2Cl_2 to give a mixture of corresponding silyl enol ether **13** and silyl ketene acetal **14** in comparable amounts (Scheme 7) (56).

The mixture was directly oxidized with MTO and H_2O_2 in CH_3CN/AcOH/pyridine mixture at 25°C. After treatment of the crude with KF in MeOH, levulinates **10** and **11** were obtained as the only recovered products in 40% and 10% yield, respectively (Scheme 7).

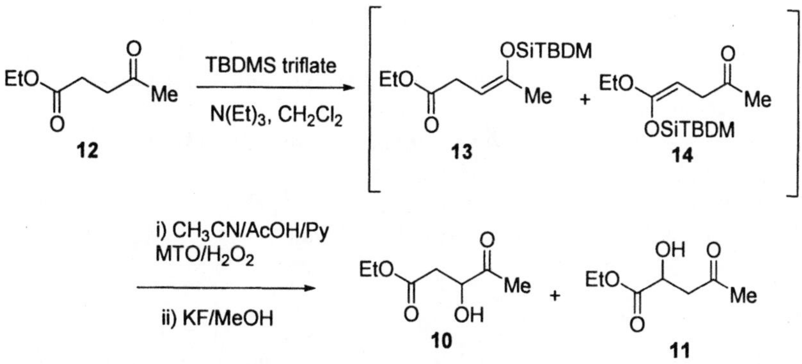

Scheme 7. *Oxidation of silyl derivatives of ethyl levulinate 12 with MTO.*

Conclusions

Methyltrioxo rhenium (MTO) and heterogeneous MTO compounds can be used as catalysts for the efficient oxidative functionalization of levulinic acid 1 and related compounds with environmental friendly H_2O_2 as the primary oxidant. The corresponding α- and β-hydroxyacids ethyl ester derivatives and a functionalized β-lactone were obtained from acceptable to high yield depending on the experimental conditions. The reaction proceeded by a concerted hydrogen peroxide oxygen atom insertion on secondary C-H bonds. Irrespective to the catalyst used, the C-H bond near the carbonyl group was oxidized faster than that adjacent to carboxylic moiety. As a general reaction pattern, during the oxidation with MTO and microencapsulated PS/MTO III, the formation of the β-hydroxyacid ethyl ester derivative was followed by an intramolecular lactonization probably due to the known Lewis and Brönsted acidity of MTO. A high value of the reticulation grade of the support played a relevant role in the oxidations performed, with poly(4-vinylpyridine) based catalysts, PVP-25/MTO II, being more reactive and selective than PVP-2/MTO I. PVP-25/MTO II was the best catalyst for the preparation of α- and β-hydroxyacids ethyl ester derivatives, while the highest yields of β-lactone were obtained with MTO and PS/MTO-L IV. It is worthy of note that the α- and β-hydroxyacids ethyl esters were synthesized in higher yield during the direct oxidation of 1 with respect to the oxidation of corresponding silyl enol ether and silyl ketene acetal derivatives. To the best of our knowledge this is the first example reported in the literature dealing with the catalytic oxidation of levulinic acid with H_2O_2 under mild experimental conditions. Due to its high reactivity and selectivity, this procedure can be a new and advantageous entry to the production of fine-chemicals and commodities from cellulose and biomass-derived compounds.

References

1. F.W. Lichtenthaler (Ed.), *Carbohydrates as Organic Raw Materials*, VCH Publ., Weinheim/ New York, 1991.
2. G. Descotes (Ed.), *Carbohydrates as Organic Raw Materials II*, VCH Publ., Weinheim/New York, 1993.
3. Van Bekkum, H.; Roper, H.; Voragen, A.G.J. (Eds.), *Carbohydrates as Organic Raw Materials III*, VCH Publ., Weinheim/New York, 1996.
4. Eierdanz, H. (Ed.), *Perspektiven nachwachsender Rohstofe in der Chemie*, VCH Publ., Weinheim New York, 1996.
5. Lichtenthaler, F.W.; Mondel, S. Perspectives in the use of low molecular weight carbohydrates as organic raw material. *Pure Appl. Chem.*, **1997**, *69*, pp 1853 - 1866.
6. Hugill, A. *Introductory Dedicational Metaphor, in Sugar and all That. A History of Tate and Lyle*, Gentry Books, London, 1978.
7. For some representative examples of pyrolysis of lignocellulosic materials see: (a) Shafizadeh, F.; Chin, P.P.S. Pyrolytic production and decomposition of 1,6-anhydro 3,4-dideoxy-β-Dglycero-hex-3-enopyranos-2-ulose. *Carbohydr. Res.* **1976**, *46*, pp 149-154. (b) Shafizadeh, F.; Ward, D.D.; Pang, D. Michael addition reaction of levoglucosenone, *Carbohydr. Res.* **1982**, *102*, pp 217-230.
8. For some representative examples of production of low molecular weight chemicals from cellulose and cellulose wastes see: (a) Seri, K.; Sakaki, T.; Shibata, M.; Inoue, Y.; Ishida, H. Lanthanum(III)-catalyzed degradation of cellulose at 250 °C. *Bioresource Technology* **2002**, *81*, pp 257-260, and references cited therein. (b) Sheldon, R.A.; Kochi, J.K. Metal Catalyzed Oxidation of Organic Compounds, Academic Press, New York, NY 1981. (c) Shafizadeh, F.; Chin, P. S.Preparation of 1,6-anhydro-3,4-dideoxy-D-glycero-hex-3-enopyranos-2-ulose (levoglucosenone) and some derivatives thereof. *Carbohydr. Res.* **1977**, 58, p 79.
9. For some example of patent claim the production of levulinic acid from cellulose and biomass-derived compounds see: (a) Taraban'ko, V. E.; Shambazov, V.K.; Kuznetsova, S.A.; Kuznetsov, B.N. Processing of wood from small-leaf trees in production of vanillin and syringyl aldehyde and levulinic acid. Russ. (2003). Application: RU 2002-124176 20020911. (b) Taraban'ko, V.E.; Kozlov, I.A.; Pervyshina, E.P.; Chernyak, M.Yu.; Kuznetsov, B.N. Chemical method of processing conifer wood by catalytic oxidation and acid hydrolysis yielding vanillin, levulinic acid and hydroxymethylfurfural. Russ. (2000). Application: RU 99-124884 19991125. (c) Taraban'ko, V.E.; Pervyshina, E.P.; Kuznetsov, B.N. Manufacture of vanillin and syringaldehyde by catalytic oxidation of small-leaved wood species with oxygen in alkaline medium and extraction of levulinic acid from the residue. Russ. (1998). Application: RU 97-101245

19970124. (d) Asahi Chemical Industry Co.,Ltd., Japan, Levulinic acid and esters. Jpn. Kokai Tokkyo Koho (1982), 7 pp., Application: JP 80-79672 19800613.

10. For some recent examples of patent claim the production of important industrial compounds from levulinic acid see: (a) Ochneva, V. A.; Trostyanetskaya, V. L.; Alenkin, A.V.; Silaeva, N.A.; Shapovalova, T.A.; Tudorovskii, E.L. Method for preparing 4,4'-azobis-(4-cyanopentanoic acid) as a radical polymerization initiator, Russ. (2004). Application: RU 2002-131371 20021122. (b) Shibuya, N.; Sato, A. Preparation of 4,4'-azobis-4-cyanovaleric acid. Jpn. Kokai Tokkyo Koho (1998), 4 pp. Application: JP 97-32994 19970131. (c) Narutomi, T.; Okamura, K.; Minami, S. Continuous manufacture of 5-aminolevulinic acid with methane-forming bacteria, (1994) 5 pp. Application: JP 92-33222619921119. (d) Natsuume, T.; Ueda, T. Stabilized ferrous compound composition. U.S. (1987), 8 pp. Application: US 85-807245 19851210.

11. For some examples of studies on the formation of levulinic acid see: (a) Horvat, J.; Klaic, B.; Metelko, B.; Sunjic, V. Tetrahedron Lett. 1985, 26, 2111. (b) Horvat, J.; Klaic, B.; Metelko, B.; Sunjic V. Croat. Chem. Acta1986, 59, p 429.

12. Fugalli S. La Chimica e L'Industria 2004, 86, pp 6-59.

13. (a) Fitzpatrick, S.W. Production of levulinic acid from carbohydrates-containing materials, Biofine Incorporated US patent 5,608,105, 1997. (b) Fitzpatrick, S. W. Lignocellulose degradation to furfural and levulinic acid, Biofine Incorporated US patent 4,897,497, 1990.

14. For some recent examples of patent claim the production of important industrial compounds from levulinic acid see: (a) Ochneva, V.A.; Trostyanetskaya, V.L.; Alenkin, A. V.; Silaeva, N.A.; Shapovalova, T.A.; Tudorovskii, E. L. Method for preparing 4,4'-azobis-(4-cyanopentanoic acid) as a radical polymerization initiator, Russ. (2004). Application: RU 2002-131371 20021122. (b) Shibuya, N.; Sato, A. Preparation of 4,4'-azobis-4-cyanovaleric acid. Jpn. Kokai Tokkyo Koho (1998), 4 pp. Application: JP 97-32994 19970131. (c) Narutomi, T.; Okamura, K.; Minami, S. Continuous manufacture of 5-aminolevulinic acid with methane-forming bacteria, (1994) 5 pp. Application: JP 92-332226 19921119. (d) Natsuume, T.; Ueda, T. Stabilized ferrous compound composition. U.S. (1987), 8 pp. Application: US 85-807245 19851210.

15. Timokhin, B.V.; Baransky, V.A.; Eliseeva, G.D. Levulinic acid in organic synthesis. *Russian Chemical Reviews* 1999, 68, pp 73-84.

16. Cubbon, R. C. P.; Hewlett, C. Organic peroxides containing functional groups. III Reaction of hydrogen peroxide and tert-butyl hydroperoxide with 4-oxovaleric acid. *J. Chem. Soc. C: Organic* 1968, p 2986.

17. (a) Cannon, G.; Garst, J.E. Preparation and Baeyer-Villiger reaction of certain 2-carbalkoxycyclopropyl methyl ketones. *Pharm. Sci.* **1975**, *64*, p 1059. (b) Hamprecht, G.; Ruchardt, C. Tert-alkylperoxy-γ-lactones and 3-(tert-alkylperoxy)phthalides. *Tetrahedron Lett.* **1968**, *28*, p 3195.

18. Jpn. Patent 4513; Ref. *Zh. Khim.* 16N 34P 1964.

19. Gubelt, G.; Warkentin, J. Oxidation with lead tetraacetate. IV. Cyclization of phenylhydrazones of levulinic acid, levulinanilide, 5-ketohexanoic acid, 4-keto-1-pentanol, and of levulinic acid oxime. *Canadian Journal of Chemistry* **1969**, 47(21), pp 3983-7.

20. For some representative examples see: (a) Cai, N.-C.; Wang, Y.-P.; Wang, E-F.; Peng, Z.-H. Degradation of levulinic acid in the new photodegradation reaction system. *Wuhan Daxue Xuebao, Ziran Kexueban* **1999**, 45(6), pp 814-816. (b) Venkateswarlu, K. Ch.; Rao, M.A. Kinetics and mechanism of oxidation of 4-oxopentanoic acid by cerium(IV) in acid medium. *Acta Ciencia Indica, Chemistry* **1996**, 22(4), pp 135-139. (c) Arora, S.D.; Prakash, A.; Mehrotra, R.N. Kinetics and mechanism of oxidations by metal ions. Part 16. Oxidation of 4-oxopentanoic acid by aquomanganese(III) ions. *Transition Metal Chemistry* **1993**, 18(4), pp 401-5. (d) Mehrotra, R.N. Kinetics and mechanisms of oxidations by metal ions. V. Oxidation of 4-oxopentanoic acid by the aquavanadium(V) ion. *Bulletin of the Chemical Society of Japan* **1985**, 58(8), pp 2389-94. (e) Joshi, U.R.; Limaye, P.A. Syntheses of biochemicals and industrial chemicals from levulinic acid .III. Application of Kolbe electroorganic process to derivatives of levulinic acid. *Journal of the University of Poona, Science and Technology* **1981**,54, pp 67-77. (f) Singh, B.; Singh, B.B.; Singh, S. Kinetics and mechanism of alkaline hexacyanoferrate(III) oxidation of levulinic acid.*Journal of the Indian Chemical Society* **1980**, 57(6), pp 662-3.

21. Smith, S.; Dunlop, A.P. Preparation of succinic acid US Patent 2,676,186; Ref. Zh. Khim. 16N 34P 1955.

22. Panwar, K.S.; Gaur, Jifendra N. Photo-induced oxidation of some organic carbonyl compounds by vanadium(V). *Talanta* **1967**, 14(1), pp 127-9.

23. (a) Singh, B.; Sahai, S. Mechanism of ruthenium(III) catalysis in oxidation of levulinic acid by N-bromoacetamide in perchloric acid. *J. Indian Chemical Society* **1991**, *68*, pp 208-209. (b) Pandey, S.K.; Yadav, S.P.S.; Prasad, M.; Prasad, J. Mechanism of Ru(III) catalysis in oxidation of levulinic acid by acidic solution of N-bromobenzamide. *Asian J. Chemistry* **1999**, *11*, pp 203-206.

24. Beattie, I.R.; Jones, P.J. Methyltrioxorhenium. An air-stable compound containing a carbon-rhenium bond. *Inorg. Chem.* **1979**, *18*, p 2318.

25. Romão C.C.; Kühn, F.E.; Herrmann W.A. Rhenium(VII) oxo and imido complexes: synthesis, structures, and applications *Chem. Rev.* **1997**, 97, pp 3197-3246.Herrmann W.A.; Fischer, R.W.; Marz, D.W.

Methyltrioxorhenium as catalyst for olefin oxidation *Angew. Chem., Int. Ed. Engl.* **1991**, *30*, p 1638.

26. Herrmann W. A.; Fischer, R. W.; Marz, D. W. Methyltrioxorhenium as catalyst for olefin oxidation *Angew. Chem., Int. Ed. Engl.* **1991**, *30*, 1638.

27. Zhu, Z.; Espenson, J.H. Oxidation of alkynes by hydrogen peroxide catalysed by methylrhenium trioxide *J. Org. Chem.* **1995**, *60*, p 7728.

28. (a) Saladino, R.; Fiani, C.; Belfiore, M.C.; Gualandi, G.; Penna, S.; Mosesso, P. Methyltrioxorhenium catalysed synthesis of highly oxidised aryltetralin lignans with anti-topoisomerase II and apoptogenic activities *Bioorg. Med. Chem.* **2005**, *13*, pp 5949-5960. (b) Crestini, C.; Pro, P.; Neri, V.; Saladino, R. Methyltrioxorhenium: A new catalyst for the activation of hydrogen peroxide to the oxidation of lignin and lignin model compounds" *Bioorg. Med. Chem.* **2005**, *13*, pp 2569-2578 (c) Saladino, R.; Neri, V.; Mincione, E.; Marini, S.; Coletta, M.; Fiorucci, C.; Filippone, P. A new and efficient synthesis of ortho- and para-benzoquinones of cardanol derivatives by the catalytic system MeReO$_3$-H$_2$O$_2$ *J. Chem. Soc., Perkin Trans 1* **2000**, pp 581-586.

29. Brown, K.N.; Espenson, J.H. Stepwise oxidation of thiophene and its derivatives by hydrogen peroxide catalysed by methyltrioxorhenium(VII) *Inorg. Chem.* **1996**, *35*, p 7211.

30. Kunkely, H.; Vogler, A. Photoreactivity of the triphenylphosphine adduct of methyl(trioxo)rhenium(VII) induced by ligand-to-metal charge transfer exitation *J. Organom. Chem.* **2000**, *606*, pp 207-209.

31. (a) Zauche, T.H.; Espenson, J.H. Kinetics and mechanism of the oxidation of secondary hydroxylamines to nitrones with hydrogen peroxide, catalysed by methylrhenium trioxide *Inorg. Chem.* **1997**, *36*, pp 5257-5261. (b) Goti, A.; Nannelli, L. Synthesis of nitrones by methyltrioxorhenium catalysed direct oxidation of secondary amines *Tetrahedron Lett.* **1996**, *37*, pp 6025-6028.

32. (a) Herrmann W.A.; Fischer, R.W.; Correira, J.D.G. Multiple bonds between main-group elements and transition metals. Part 133. Methyltrioxorhenium as a catalyst of the Baeyer-Villiger oxidation *J. Mol Catalysis.* **1994**, *94*, pp 213-223. (b) Bernini, R.; Mincione, E.; Cortese, M.; Aliotta, G.; Saladino, R. A new and efficient Baeyer-Villiger rearrangement of flavanone derivatives by the methyltrioxorhenium/H$_2$O$_2$ catalytic system *Tetrahedron Letters* **2001**, 42/32, pp 5401-5404.

33. Murray, R. W.; Iyanar, K.; Chen, J.; Wearing, J. T. Methyltrioxorhenium-catalyzed C-H insertion reactions of hydrogen peroxide *Tetrahedron Lett.* **1995**, *36*, 6415. Schuchardt, U.; Mandelli, D.; Shul'pin, G.B. *Tetrahedron Lett.* Methyltrioxorhenium catalyzed oxidation of saturated and aromatic hydrocarbons by H$_2$O$_2$ in air **1996**, *37*, pp 6487-6490.

34. J. Finlay, A. McKervey, H.Q.N. Gunaratne Oxidations Catalysed by Rhenium(V) Oxo Species 1. Conversion of Furans to Enediones using

Methyltrioxorhenium and Urea Hydrogen Peroxide. Tetrahedron Lett. **1998**, 39, pp 5651-5654.

35. Herrmann, W.A.; Fischer, R.W.; Scherer, W.; Rauch, M.U. Methyltrioxorhenium(VII) as catalyst for epoxidations: structure of the active species and mechanism of catalysis *Angew. Chem., Int. Ed. Engl.* **1993**, *32*, pp 1157-1160.

36. Herrmann, W.A.; Waguer, W.; Volkhardt, U. Organic derivatives of rhenium oxides and their preparation and use for the metathesis of olefins. US Patent 5,342,985 August 30, 1994.

37. Fischer, R. W., Herrmann, W.A.; Kratzer, R. Direct synthesis of organorhenium oxides from compounds containing rhenium US Patent 6,180,807 B1, 2001.

38. Thomas, J.M. and Thomas W.J. Principles and Practice of heterogeneous catalysis; VCH: New York, 1997.

39. Herrmann, W.A.; Fritz-Meyer-Weg, D.M.; Wagner, M.; Kuchler, J.G.; Weichselbaumer, G.; Fischer, R. Use of organorhenium compound for the oxidation of multiple C-C bonds, oxidation processes based thereon and novel organorhenium compounds U.S. Patent 5,155,247, Oct. 13, 1992.

40. Copèret, C.; Adolfsson, H.; Sharpless, K.B. A simple and efficient method for epoxidation of terminal alkenes *J. Chem. Soc. Chem. Commun.* **1997**, p 1915.

41. Donbrow, M. in Microcapsules and Nanoparticles in Medicine and Pharmacy; CRC Press: Boca Raton, FL, **1992**.

42. Saladino, R.; Neri, V.; Pelliccia, A.R.; Caminiti, R.; Sadun, C. Preparation and structural characterization of polymer-supported methylrhenium trioxide systems as efficient and selective catalysts for the epoxidation of olefins *J. Org. Chem.* **2002**, 67, pp 1323-1332.

43. (a) Saladino, R.; Neri, V.; Pelliccia, A.R.; Mincione, E. Selective epoxidation of monoterpenes with H_2O_2 and polymer supported methylrhenium trioxide systems *Tetrahedron* **2003**, *59*, pp 7403-7408. Saladino, R.; Mincione, E.; Attanasi, O. A.; Filippone, P. Microencapsulated methylrhenium trioxide MTO/H_2O_2 systems for the oxidation of cardanol derivatives *Pure Appl. Chem.* **2003**, 2, pp 261-268.

44. Saladino, R.; Neri, V.; Cardona, F.; Goti, A. Oxidation of N,N-disubstituted hydroxylamines to nitrones with hydrogen peroxide catalyzaed by polymer-supported methylrhenium trioxide systems *Adv. Synth. Catal.*, **2004**, *346*, pp 639-647.

45. a) Saladino, R.; Andreoni, A.; Neri, V.; Crestini, C. *Tetrahedron* **2005**, *61*, pp 1069-1075. b) Saladino, R.; Neri, V.; Mincione, E.; Filippone, P. Selective oxidation of phenol and anisole derivatives to quinones with hydrogen peroxide and polymer-supported methylrhenium trioxide systems *Tetrahedron* **2002**, *58*, pp 8493-8500.

46. Bianchini, G.; Crucianelli, M.; De Angelis, F.; Neri, V.; Saladino, R. A Novel Catalyzed C-H Insertion Reactions of Hydrogen Peroxide by Poly(4-vinylpyridine)/Methyltrioxorhenium *Tetrahedron Lett.* **2004**, *45*, pp 2351-2353.

47. Bianchini, G.; Crucianelli, M.; De Angelis, F.; Neri, V.; Saladino, R. Highly efficient C-H insertion reactions of hydrogen peroxide catalyzed by homogeneous and heterogeneous methyltrioxorhenium systems in ionic liquids *Tetrahedron Lett.* **2005**, *46*, pp 2427-2432.

48. PL Patent 87-269352 19871210, **1992**.

49. Slavtcheff, S.C.; Barrow, R.S.; Kanga, D.V.; Cheney, C.M.; Znaider, A. Cosmetic composition for treatment of pimples and redness CA Patent 94-2113232, 19940111, and US Patent 5482710, 19930730, **1995**.

50. Jin, Y.; Kim, D.H. Synlett **1998**, 11, pp 1189-1190.

51. Tanzawa, T.; Schartz, J. Catalytic conversion of hydroxy carboxylic acids to olefins by tungsten(VI) complexes: a new acyl group transfer catalyst. Organometallics **1990**, 9, pp 3026-3027.

52. For examples of reviews on β-lactone derivatives see: (a) Lowe, C.; Vederas, J.C. *Org. Prep. Proe. Int.* **1995**, 27, pp 305-346. (b) Pommier, A.; Pons, J.M. Recent Advances in β-lactone Chemistry *Synthesis* **1995**, 7, pp 729-744.

53. Ferriera, P.; Xue, W.M.; Bencze, E.; Herdtweck, E.; Kühn, F.E. *J. Inorg. Chem.* **2001**, *40*, 5834-5841, and references cited therein.

54. See for example: Takas, J.; Kiprof, P.; Riede, J.; Herrmann, W.A. *Organometallics* **1990**, *9*, pp 782-787.

55. Stankovic, S.; Espenson, J.H. *J. Org. Chem.* **2000**, 65, pp 5528-5530.

56. Mander, L.N. ; Sethi, S.P. *Tetrahedron Lett.* **1984**, 25, pp 5953-5956.

Chapter 18

Furan Chemistry at the Service of Functional Macromolecular Materials: The Reversible Diels–Alder Reaction

Alessandro Gandini[1] and Mohamed Naceur Belgacem[2]

[1]CICECO and Chemistry Department, University of Aveiro, 3810–193 Aveiro, Portugal
[2]Laboratoire de Génie des Procédés Papetiers, UMR 5518, École Française de Papeterie et des Industries Graphiques (INPG), BP65, 38402 Saint Martin d'Hères, France

This chapter describes the formation of various thermally reversible polymer architectures based on the Diels-Alder reaction occurring between furan and maleimide moieties. These functional materials are then discussed with respect to different areas of application.

There are two first-generation furan derivatives, furfural (**F**) and hydroxymethylfuraldehyde (**HMF**), readily obtained from a variety of renewable resources including sugars, hemicelluloses and other polysaccharides. Whenever pentoses are the saccharide units (as, *e.g.*, in sugarcane bagasse, rice hulls, corn cobs, etc.), well-known acid-catalyzed dehydration processes transform them into **F**, which has been an industrial commodity for about a century (*1-3*). When the units are hexoses (as, *e.g.*, in fructose, inulin, etc.), the same reaction mechanism yields **HMF** (*1-3*). Both compounds have been exploited as precursors to a wide spectrum of monomers and the ensuing polymers thoroughly studied (*1-3*). This comprehensive set of investigations represents an instructive example of the possibility of building an original family of macromolecular materials which is not (exclusively) based on fossil fuels chemistry and which can provide polymers with a vast array of properties and potential applications. The best example of an industrial process yielding a furan monomer is the catalytic reduction of **F** into furfuryl alcohol, which has been used to manufacture resins with a variety of applications (*2,3*).

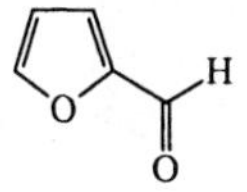

F 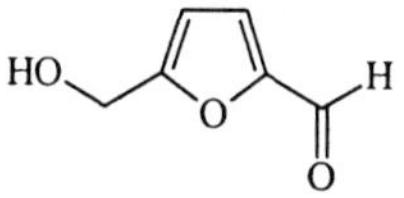   **HMF**

One of the most recent contributions to this alternative strategy arises from the exploitation of the highly dienic character of the furan heterocycle (compared with that of its thiophene and pyrrole homologues), which makes it particularly suited to generate adducts reversibly with many dienophiles through the Diels-Alder (DA) reaction and its retro-counterpart (RDA). The application of this approach to polymer synthesis, using maleimide moieties as dienophiles, is the subject of the following short critical review.

The Diels-Alder Reaction

This classical [4+2] cycloaddition reaction between a diene and a dienophile has received a great deal of attention since its discovery by the 1950 Nobel Prize winners Otto Diels and Kurt Alder. This reaction represents both a very useful synthetic tool and a challenging mechanistic and theoretical topic. There have been numerous reviews covering different aspects of the reaction, and a recent book provides a balanced view of the state of the art (*4*).

The features which are most relevant to the purpose of the present chapter are summarized in the following reaction scheme involving a simple furan derivative and a maleimide:

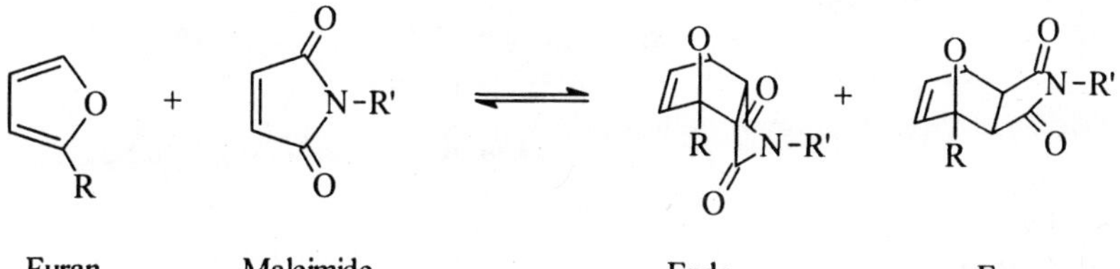

Furan   Maleimide   Endo   Exo

The DA cycloaddition always gives two stereoisomeric adducts (*5*), whose proportion depends on a number of specific intermolecular effects. However, the relative abundance of *endo* and *exo* isomers is irrelevant here, since what counts is the actual global yield of adducts. The nature of the substituents on both reagents plays an important role on the yield, in terms of both steric hindrance (e.g., number and bulkiness of the substituents attached to the furan ring) and electronic effects (e.g. a carbonyl group directly linked to the furan ring greatly reduces its reactivity). Finally, the reaction temperature is a key factor in determining the position of the equilibrium, which is progressively shifted to the left (Retro Diles Alder; RDA) as the temperature is raised. Typically, the forward reaction dominates up to ca. 60°C, whereas the reverse reaction is overwhelming above ca. 100°C (4). This thermal reversibility represents the basic working hypothesis for many of the systems which are discussed below, since this reversibility offers avenues for the creation of temperature-sensitive materials with promising applications.

Linear Step-Growth Systems

Bifunctional monomers bearing complementary DA functions have been utilized to synthesize linear oligomers and polymers. In some instances, the adducts in these structures were dehydrated to aromatic moieties, thus rendering them inaccessible to the RDA reaction, i.e. the materials became thermally stable. The basic mechanism of this thermal aromatization, typically carried out at temperatures above 200°C, is schematically depicted as follows, for the reactive moiety of the DA adduct:

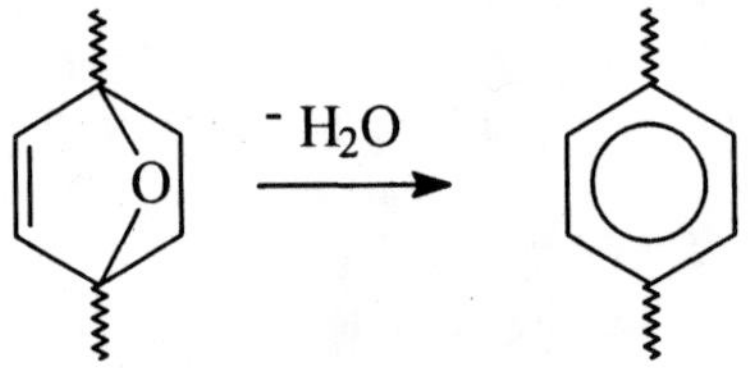

In other studies, the reversibility of the macromolecular architecture was examined in variable degrees of thoroughness. As with other polycondensation reactions, two classical synthetic approaches were adopted in the present context, as discussed below.

A-A + B-B Systems

It is most likely that the first investigation of the linear polymerization of a difuran compound with various bismaleimides was published some twenty years ago (6) and the same group complemented the original study a few years later with other difuran monomers (7). The ensuing polymers, whose reported molecular weights were surprisingly high, were not fully characterized nor submitted to RDA tests. They were instead aromatized in search of novel heat-resistant materials. The polycondensation of aromatic difurans and bismaleimides at 265°C in air (8) produced materials which could hardly be expected to display regular structures, since aromatization, as well as other oxidative degradation reactions, inevitably accompanied these DA polycondensations. Kuramoto *et al.* (9) were the first to carry out both the polymerization of a difuran monomer with a bismaleimide and the thermal depolymerization of the ensuing DA polyadduct, as shown in Scheme 1.

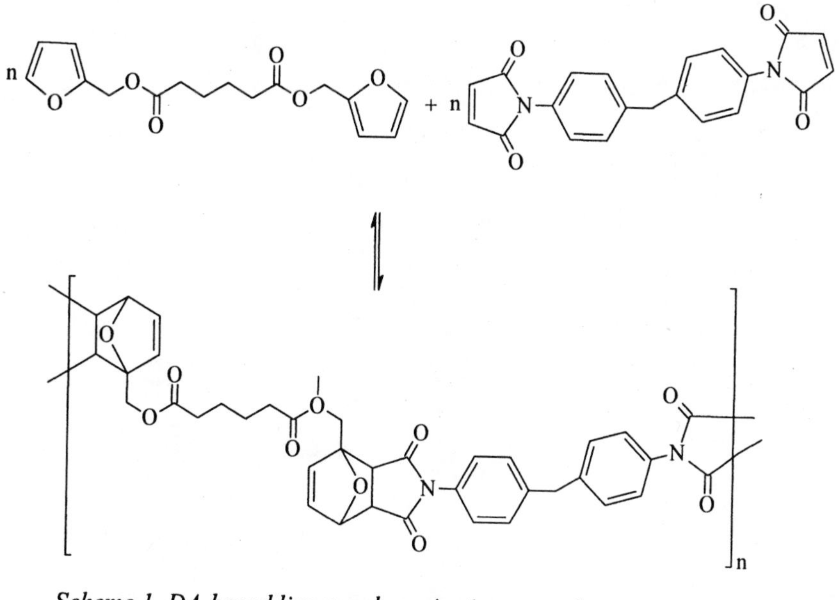

Scheme 1. DA-based linear polymerization according to reference 9.

The DA polycondensation at 60°C reached its equilibrium (~60% adduct formation) within a few hours and the isolated polymer, dissolved in DMF, underwent the RDA depolymerization at 90°C, giving an equilibrium ~80% adduct regeneration within less than an hour.

A more thorough approach to the reversibility of DA reactions involved in linear polymerizations (10) included ^{1}H NMR kinetic studies of both the forward and RDA reactions as well as viscosity measurements. Several monofunctional model compounds, including furans, maleimide and trifluorocrotonate derivatives were first inspected in order to assess the best structural combination of difunctional homologues. The aim of this research was to prepare materials displaying thermoreversible phase transitions.

In two separate investigations difurans and bismaleimides were used with the aim of synthesizing DA polyadducts. The polymerizations were preceded by a systematic inspection of the reactivity of monofunctional model compounds and of the structure, thermal stability and RDA behavior of their adducts. In the initial effort (*11*), relatively stiff monomer structures were employed (Scheme 2). These resulted in polymer precipitations of relatively low degrees of polymerization. However, these linear macromolecules showed the anticipated structure without any detectable anomaly, as verified by NMR spectroscopy.

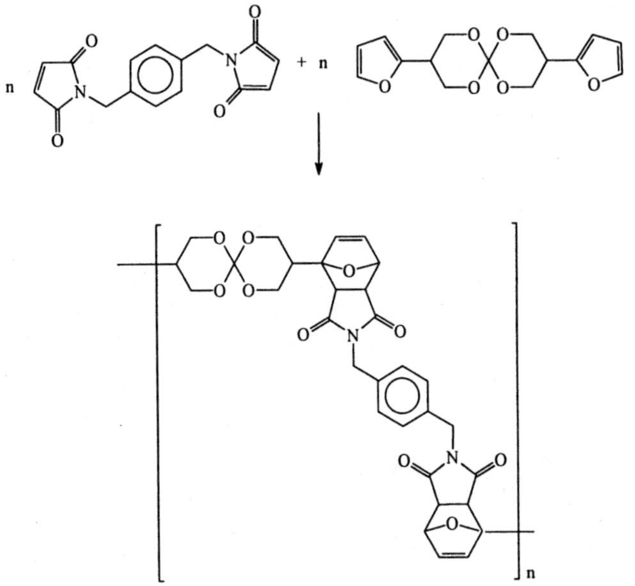

Scheme 2. DA-based linear polymerization according to reference 11.

In subsequent efforts (*12*), both monomers contained very flexible siloxane and methylene moieties (Scheme 3). Their DA polycondensation was conducted in bulk at room temperature and within a day, the medium had acquired a rubbery consistency. This polymer was entirely soluble in THF and methylene chloride and insoluble in ether and hexane. Again, its spectroscopic characterization displayed the anticipated regularity of the linear structure based on the DA-based chain growth.

The preparation of an optically active DA polyadduct of a difuran ester and an aromatic bismaleimide, based on asymmetric polymerizations induced by chiral Lewis acids, has also been reported (*13*).

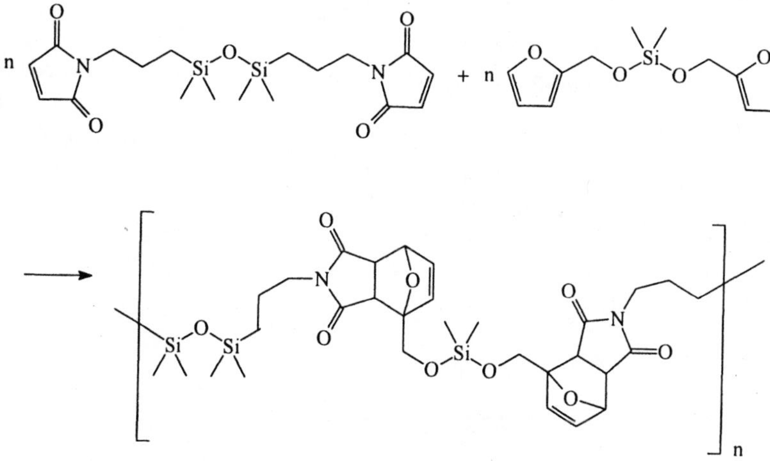

Scheme 3. DA-based linear polymerization according to reference 12.

A recent addition to these syntheses reported the DA polymerization of a difuran monomer bearing two urethane groups within its bridging unit with various aromatic bismaleimides (*14*). Unfortunately, the structures of the ensuing oligomers were not characterized.

A-B Systems

The possibility of synthesizing a monomer bearing both a furan and a maleimide end-function was first explored by Mikroyannidis (*8,15*) who investigated a series of such structures comprising different bridging moieties. The synthetic pathway adopted involved the conventional passage through the maleamic acid, which was then cyclized. The spectroscopic characterization of both the intermediates and the actual monomers did not provide convincing evidence in favor of the expected structures, and no NMR spectra of the final A-B compounds could be obtained due to "their limited solubility even in polar aprotic solvents". This observation casts serious doubts as to the actual nature of these products. The purpose of these studies was to prepare thermally stable polymers and this was achieved by the polymerization of either the carbamic acid or the supposed A-B monomer at high temperature, followed by the aromatization of the ensuing resins. None of these products were soluble in any solvent, despite the fact that they should all have had a linear macromolecular structure, given the difunctional character of their precursors. It seems clear that important side reactions could have induced substantial crosslinking, both during the syntheses of the monomers, and at all stages of the polymerization and aromatization processes.

The same issue has been investigated with the simplest A-B monomer of this series, viz. 2-furfurylmaleimide (**FM**), which was prepared by the reaction of 2-furfuryl amine and maleic anhydride (*16*) and the subsequent dehydration of the ensuing maleamidic acid. A complete characterization confirmed that the crystalline and very hygroscopic product obtained was indeed **FM**. Its solution polymerization in a sealed tube gave soluble oligomers with the expected polyadduct structure. However, this study was not continued due to the difficulties associated with handling **FM**.

FM

Gaina and Gaina (*17*) also tackled this topic with two original furan-maleimide A-B monomers and polymerized them by DA polycondensation, but focused their attention on the aromatization of the ensuing polymers.

Networks and Dendrimers

The use of complementary monomers bearing on average more than two functionalities in polycondensation reactions leads to crosslinked materials. This approach has been successfully applied to the DA reaction between furan and maleimide monomers with the purpose of preparing mendable materials, thanks to the thermal reversibility of the adduct formation (*18*). Indeed, heating these networks induces a shift in the equilibrium in favor of the regeneration of unreacted moieties through the RDA reaction, and allows a fracture in the material to be reabsorbed and subsequent cooling to restore the original intact material. In principle, barring unwelcome side reactions, this mending process can be repeated indefinitely. An example of a monomer combination which gave a crosslinked mendable material is shown in Scheme 4 (*19*).

An extension of this concept has recently been applied to thermally removable foams (*20*) and adhesives (*21*), prepared with epoxy resins incorporating furan/maleimide DA adducts. In the former study (20) the foam could be readily removed by 1-butanol at 90°C through the RDA decrosslinking reaction, while in the latter study (21), an elastomeric adhesive placed between

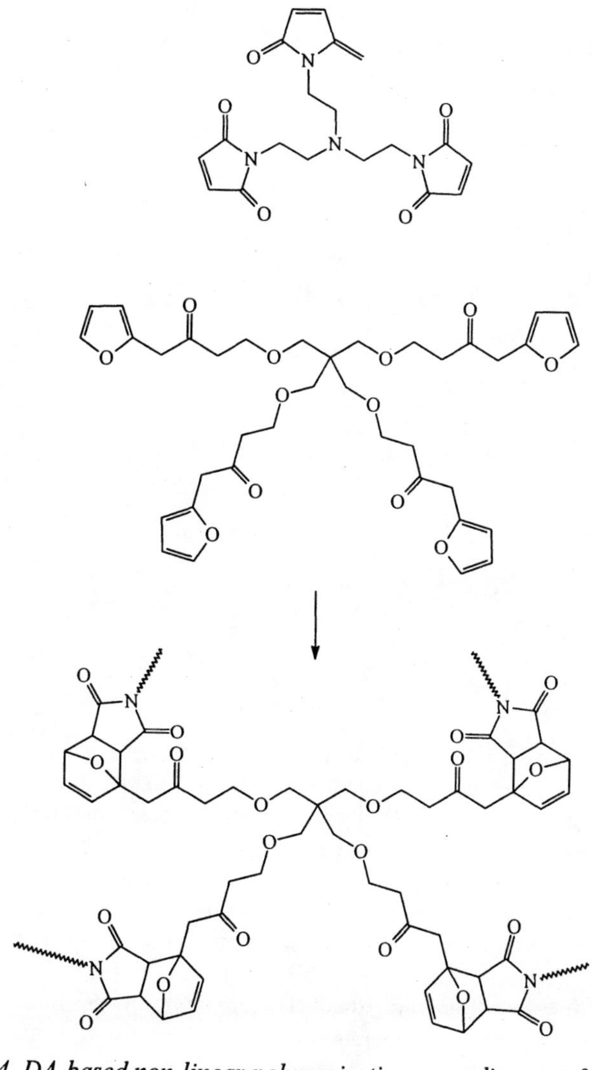

Scheme 4. DA-based non-linear polymerization according to reference 19.

two metal plates could be readily removed at temperatures above 90°C since the RDA reaction transformed it into a liquid material.

Another interesting application of the reversible DA reaction describes the synthesis of thermally responsive dendrons built through the formation of adducts (*22*). The third generation reversible architecture obtained with this approach had the following dendritic structure:

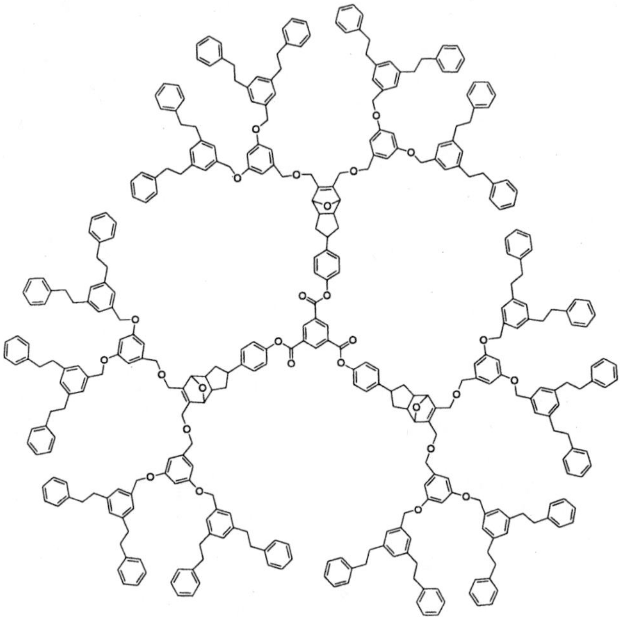

Although not strictly related to polymers, it is worth mentioning that the same research group has also examined the application of the furan-maleimide DA reaction with the aim to synthesize thermally cleavable surfactants. Such molecules lose their specific properties when heated in an aqueous medium, because the RDA reaction splits the hydrophilic from the hydrophobic moiety (*23*). Another interesting application of the DA/RDA principle was recently described in the context of the reversible fluorescence behavior of triaromatic maleimide electron-donating chromophores (non-emitting). Such molecules became fluorescent following their DA reaction with furan (*24*). The temperature-driven RDA regenerated the initial non-fluorescent structure by releasing the volatile furan.

Reversible Crosslinking of Linear Polymers

The first mention of the possibility of preparing thermally reversible crosslinked materials through the DA/RDA reaction appeared in the late seventies (25). However, it was only in the early nineties that investigations took on a systematic approach in this area. The report that poly(N-acetylethyleneimine)s bearing either maleimide or furancarbonyl side groups could be made to crosslink by interchain DA couplings, also claimed that the ensuing network was decrosslinked through the RDA reaction by heating at 80°C (26). Although sound in principle, these findings are questionable because it is not clear why the network, decrosslinked at high temperature, was not reformed, when the solution of the two regenerated linear polymers, still bearing the complementary reactive functions, was cooled to room temperature. The same reservation applies to two subsequent studies on novel materials based on the same system (27,28).

The possibility of reverting to the linear precursors, after crosslinking two polystyrene molecules modified with furan and maleimide functions, respectively, was verified by Canary and Stevens (29). These authors also discovered that the furfuryl moieties were not entirely stable at the processing temperature of 150°C. This drawback was confirmed in a study of the reversible crosslinking of acrylic copolymers bearing pendant furfuryl groups with bismaleimides (30) and attributed it to free radical reactions involving the methylene moiety linked to the furan heterocycle (1,2). Thus, a high concentration of appended furan rings, the presence of atmospheric oxygen and an excessive temperature are to be avoided in order to ensure the complete reversibility of these systems.

This was first achieved in a thorough study of the crosslinking of polystyrenes substituted in varying amounts with pendant furan moieties with an aromatic bismaleimide and model reactions with the corresponding monomaleimide (31). The DA reactions proceeded smoothly in refluxing methylene chloride and gave the expected DA adducts in high yields with the monomaleimide and 100% networks when the bismaleimide was used. The original strategy attempted to permanently decrosslink these networks by heating them in the presence of an excess of 2-methylfuran. Upon cooling, the mixture of regenerated furan-substituted polystyrene and bismaleimide would form. The latter would then be trapped by the 2-methyfuran present, thus preventing it from reacting again with the polystyrene furan rings. This idea was first applied to a model system. In all instances, after optimizing the experimental conditions, the original linear polymers were recovered in 95-98% yield. No evidence of side reactions associated with the furfuryl moieties were detected in these processes. Scheme 5 summarizes the adopted approach , which was then extended to other systems (32). Scheme 6 illustrates the reversible DA/RDA crosslinking of furan acrylic copolymers (**AcFu**), all possessing low Tg, with flexible bismaleimides, using the strategy that was applied to the furan-substituted polystyrenes (31). Scheme 7 shows the inverse process applied to the DA crosslinking of variably maleimide-substituted poly(dimethylsiloxane)s (**SiOM**) with a difuran compound and the subsequent high-temperature RDA decrosslinking in the presence of an excess of N-ethylmaleimide. Both systems were optimized to reach a high degree of reversibility without detectable side reactions.

The study of the dynamic mechanical properties of the furan acrylic networks provided a clear insight into the onset of the decrosslinking processes. This was revealed by the drop in the elastic modulus associated with the formation of a viscous liquid (*33*). Figure 1 shows this feature, which places the predominance of the RDA reaction at 90-100°C.

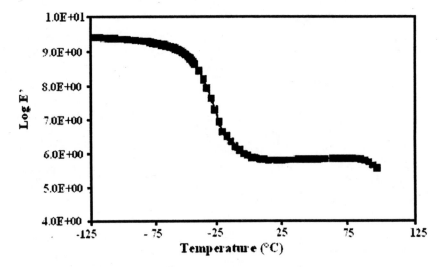

Figure 1. Variation of the elastic modulus as a function of temperature for a DA network based on a furan acrylic copolymer crosslinked with an oligoether bismaleimide (Reproduced from reference 33).

An organic-inorganic hybrid material based on the copolymer shown in Scheme 4 was recently reported (*34*). It was coupled through the DA reaction with a maleimide bearing a triethoxysilane moiety and the ensuing adduct subsequently crosslinked by the sol-gel process.

Huglin's group also tackled the DA/RDA cycle applied to the reversible crosslinking of a styrene-furfuryl methacrylate copolymer treated with a bismaleimide (*35*). The kinetics of the RDA reaction were studied by submitting the networks to soxhlet extraction with different solvents, all capable of dissolving the released non-crosslinked polymer, but at different temperatures (*35b*). The detection of the regenerated maleimide moieties in the extracted portion of the polymer was based on their UV absorbance at 320 nm, where the corresponding DA adduct did not absorb. Linear first-order plots were obtained at temperatures comprised between 77 and 110°C, while at 64°C no extracted material was detected after 13 hours. The Arrhenius plot for the RDA reaction gave good linearity for the temperature range of 77-110°C.

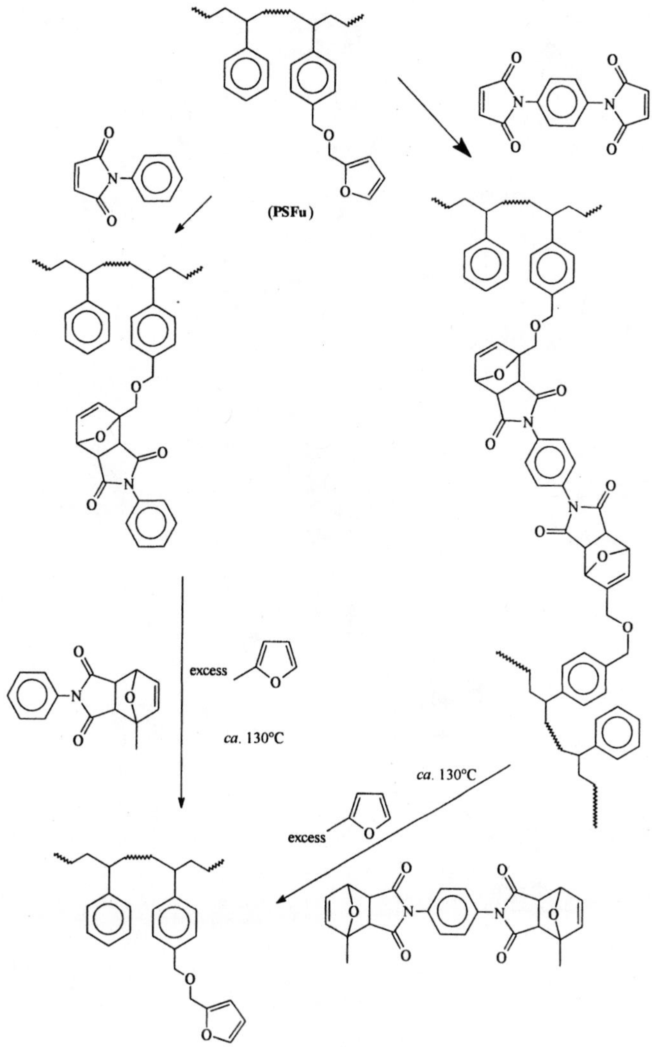

Scheme 5. DA/RDA-based reversible cross-linking mechanism, according to reference 31.

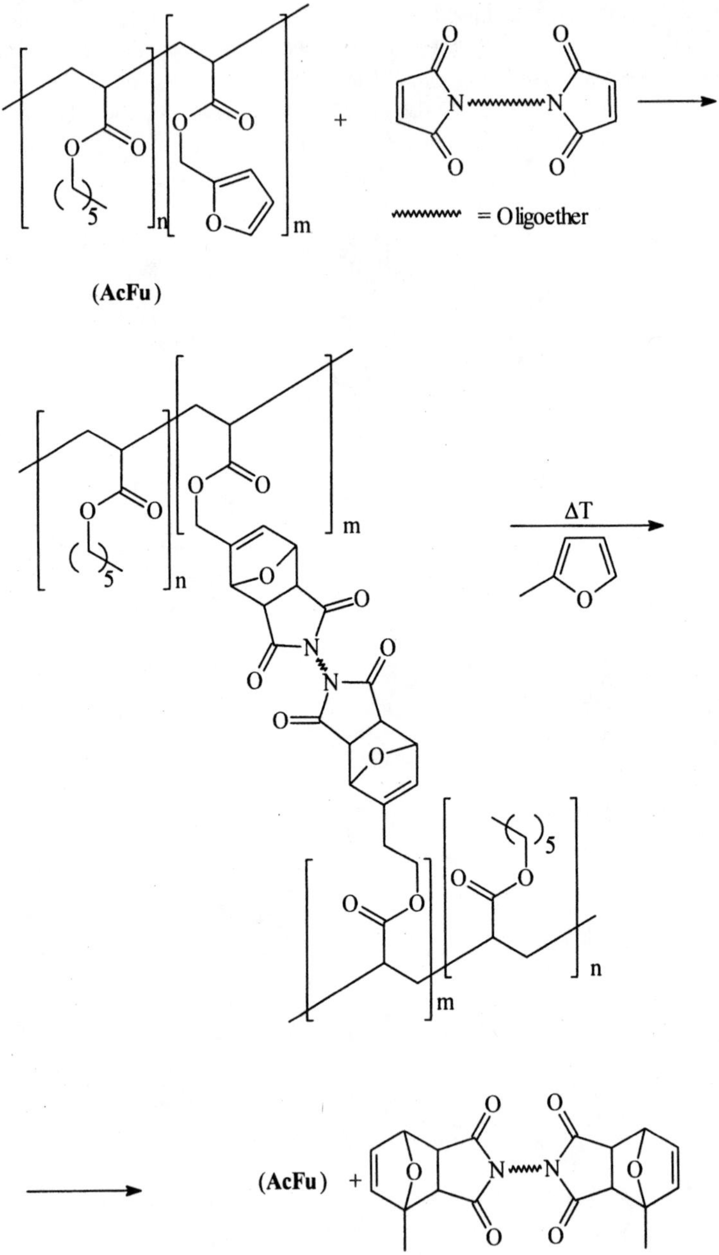

Scheme 6. *DA/RDA-based reversible cross-linking mechanism, according to reference 32.*

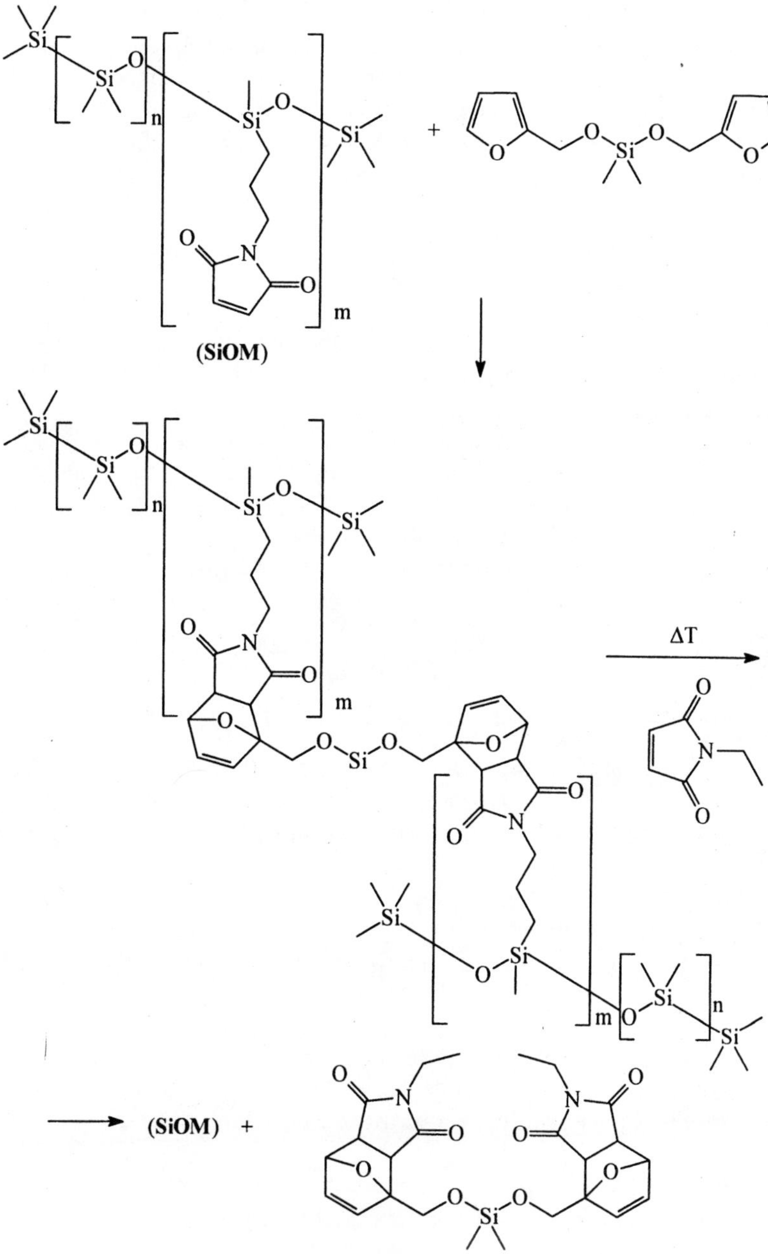

Scheme 7. DA/RDA-based reversible cross-linking mechanism, according to reference 32.

The kinetics of the DA crosslinking reaction (*35d*) and the characterization of the networks in terms of swelling, mechanical and thermal properties was also examined (*35c*). Finally the thermal reversibility associated with the DA-adduct-based polymers was applied to develop plastics with a "memory shape" effect (*36*) and to prepare poly(ethylene oxide)s bearing terminal furan/maleimide adducts (*37*).

Conclusions

One prominent feature in the behavior of the furan heterocycle is its ability to readily participate in the Diels-Alder reaction. This has created new avenues for the synthesis of novel materials. The DA reaction may lead to an irreversible transformation of the adducts into aromatic moieties, and the utilization of the ensuing materials, as thermally stable polymers. Alternatively, one may exploit the thermally reversible character of the adducts to prepare mendable, recyclable materials. These include the recuperation of otherwise intractable crosslinked polymers, as in the case of used tires.

References

1. Gandini, A. *Adv. Poly. Sci.* **1977**, *25*, p 47.
2. Gandini, A.; Belgacem, M.N. *Progr. Polym. Sci.* **1997**, *22*, p 1203.
3. C. Moreau ; Gandini, A.; Belgacem, M.N. *Topics Catal.* **2004**, *27*, p 11.
4. Fringuelli, F.; Taticchi, A. *The Diels-Alder Reaction*; John Wiley & Sons, Chichester, UK, 2002.
5. Rulisek, L.; Sebek, P.; Havlas, Z.; Hrabal, R.; Capek, P., Svatos, A. *J. Org. Chem.* **2005**, *70*, 6295.
6. Tesoro, G.C.; Sastri, V.R. *Ind. Eng. Chem. Prod. Res. Dev.* **1986**, *25*, p 444.
7. He, X.; Sastri, V.R.; Tesoro, G.C. *Makromol. Chem. Rapid Comm.* **1989**, *9*, p 191.
8. Diakoumakos, C.D.; Mikroyannidis, J.A. *J. Polym. Sci. Polym. Chem.* **1992**, *30*, 2559 and *Eur. Polym. J.* **1994**, *30*, p 465.
9. Kuramoto, N.; Hayashi, K.; Nagai, K. *J. Polym. Sci. Polym. Chem.* **1994**, *32*, p 2501.
10. Brand, T.; Klapper, M. *Des. Monom. Polym.* **1999**, *2*, p 287.
11. Goussé, C.; Gandini, A. *Polym. Intern.* **1999**, *48*, p 723.
12. Pérez-Berumen, C.; Gandini, A. *unpublished results.* C. Pérez-Berumem *Doctorate Thesis*, Grenoble National Polytechnic Institute, France, 2003.
13. Kamahori, K; Tada, S.; Ito, K., Itsuno, S. *Macromolecules*, **1999**, *32*, p 542.

14. Gaina, V.; Gaina, C.; Chiriac, C.; Rusu, M. *Macromol. Reports* **1995**, *A32*, p 121. Gaina, V.; Gaina, C.; Sava, A; Stoleriu, A.; Rusu, M. *J.M.S.-Pure Appl. Chem.* **1997**, *A34*, p 2435. Gaina, V.; Gaina, C. *Polym.-Plast. Technol. Eng.* **2002**, *41*, p 523.

15. Mikroyannidis, J.A. *J. Polym. Sci. Polym. Chem.* **1992**, *30*, 125 and 2017.

16. Goussé, C.; Gandini, A. *Polym. Bull.* **1998**, *40*, p 389.

17. Gaina, V.; Gaina, C. *Rev. Roum. Chim.* **2003**, *48*, p 881.

18. Chen, X.; Dam, M.A.; Ono, K.; Mal, A.; Shen, H.; Nutt, S.R.; Sheran, K.; Wudl, F. *Science* **2002**, *295*, p 1698.

19. Chen, X.; Wudl, F.; Mal, A.K.; Shen, H.; Nutt, S.R. *Macromolecules* **2003**, *36*, p 1802.

20. McElhanon, J.R.; Russick, E.M.; Wheeler, D.R.; Loy, D.A.; Aubert, J.H. *J. Appl. Polym. Sci.* **2002**, *85*, p 1496.

21. Aubert, J.H. *J. Adhes.* **2003**, *79*, p 609.

22. McElhanon, J.R.; Wheeler, D.R. *Org. Lett.* **2001**, *3*, p 2681.

23. McElhanon, J.R.; Zifer, T.; Kline, S.R.; Wheeler, D.R.; Loy, D.A.; Jamison, G.M.; Long T.M.; Rahimian, K.; Simmons, B.A. *Langmuir* **2005**, *21*, p 3259.

24. Zhang, X.; Li, Z-C.; Du, F.S.; Li, F-M. *J. Am. Chem. Soc.* **2004**, *126*, p 12200.

25. Stevens, M.P.; Jenkins, A.D. *J. Polym. Sci. Polym. Chem. Ed.* **1979**, *17*, p 3675.

26. Chujo, Y. ; Sada, K.; Saegusa, T. *Macromolecules* **1990**, *23*, p 2636.

27. Zewert, T.E.; Harrington, M.G. *Electrophoresis* **1999**, *20*, p 1339.

28. Imai, Y.; Itoh, H.; Naka, K.; Chujo, Y. *Macromolecules* **2000**, *33*, p 4343.

29. Canary, S.A.; Stevens, M.P. *J. Polym. Sci. Polym. Chem.* **1992**, *30*, p 1755.

30. Laita, H.; Boufi, S.; Gandini, A. *Eur. Polym. J.* **1997**, *33*, p 1203.

31. Goussé, C.; Gandini, A.; Hodge, P. *Macromolecules*, **1998**, *31*, p 314.

32. Gheneim, R.; Pérez-Berumen, C.; Gandini, A. *Macromolecules* **2002**, *35*, p 7246.

33. Gheneim, R.; Gandini, A. *unpublished results.* Gheneim, R. *Doctorate Thesis*, Grenoble National Polytechnic Institute, France, 2003.

34. Adachi, K.; Achimutu, A.K.; Chujo, Y. *Macromolecules* **2004**, *37*, p 9793.

35. (a) Goiti, E.; Huglin, M.B.; Rego, J.M. *Polymer* **2001**, *42*, p 10187; (b) *Macomol. Rapid Commun.* **2003**, *24*, p 692; (c) *Eur. Poly. J.* **2004**, *40*, p 219; (d) Goiti, E.; Heatley, F.; Huglin, M.B.; Rego, J.M. *Eur. Poly. J.* **2004**, *40*, p 1451.

36. Inoue, K.; Yamashiro, M.; Iji, M. *Kobunshi Ronbunshu* **2005**, *62*, p 261.

37. Sedaghat-Herati, R.; Chacon, A.; Hansen, M.E.; Yalaoui, S. *Macromol. Chem. Phys.* **2005**, *206*, p 1981.

Chapter 19

Catalysis of the Electrooxidation of Biomass-Derived Alcohol Fuels

Corey R. Anthony, Daniel Serra, and Lisa McElwee-White

Department of Chemistry and Center for Catalysis, University of Florida, Gainesville, FL 32611–7200

Oxidation of biomass-derived fuels such as methanol can serve as an energy source in applications such as the direct methanol fuel cell (DMFC). Electrocatalysis of methanol oxidation by heterobimetallic complexes provides a possible alternative to catalysis on the surface of bulk Pt/Ru anodes in DMFCs. Electrochemical oxidation of methanol has been demonstrated to be catalyzed by a series of RuPt, RuPd and RuAu complexes.

Introduction

Currently 85% of humanity's energy needs are supplied by fossil fuels. The use of fossil fuels has made possible the high standard of living in an industrialized nation. However, these benefits come at a cost, especially to the United States, which annually imports 50% of its fossil fuel needs. The adverse effects of an energy economy based on fossil fuels have been reviewed.[1-4]

These shortcomings have combined to make renewable energy sources appealing. One such energy source is biomass. With a few exceptions the cost of energy from biomass is currently on par if not cheaper than the energy derived from fossil fuels.[5] When properly used, biomass is an environmentally friendly energy source that can be converted into high energy biofuels such as hydrogen, methanol and ethanol. Due to its ease of transportation and high fuel efficiency, the electrochemical oxidation of methanol has received a significant amount of attention. Research in this area has been focused on the development of new catalytic systems and can be roughly grouped into two sets: heterogeneous (fuel cell) studies and homogeneous studies. The challenge is to link information obtained in these two sets of studies to improve catalysis in fuel cells that directly utilize methanol (DMFC) and other complex fuels.

Direct Methanol Fuel Cells and Heterogeneous Electrooxidation of Methanol

Fuel cells have been suggested as the power generation system of the immediate future,[6,7] poised to replace not only internal combustion engines but also advanced alkali batteries.[8] Fuel cells typically operate at 40-60% efficiency, making them much more efficient than conventional fossil fuel engines that operate at less than 20% efficiency. Due to their simplicity and high energy efficiency, direct methanol fuel cells (DMFCs) are especially suited for use in portable electronic devices. In DMFCs, aqueous methanol is electrochemically oxidized at the anode (Eq. 1) to CO_2 while oxygen is reduced at the cathode (Eq. 2) to form water. When combined, the two half reactions result in the overall cell reaction (Eq. 3).

Anodic Reaction:	$CH_3OH + H_2O$	$\longrightarrow$ $CO_2 + 6H^+ + 6e^-$	(1)
Cathodic Reaction:	$1/2\ O_2 + 6H^+ + 6e^-$	$\longrightarrow$ $3\ H_2O$	(2)
Overall Reaction:	$3/2\ O_2 + CH_3OH$	$\longrightarrow$ $CO_2 + 2\ H_2O$	(3)

The mechanism for this oxidation was reviewed by Parsons[9] in 1988 and as a result two key reactions were identified:

i)   Electrosorption of methanol to the electrode surface
ii)   Formation of CO_2 from adsorbed carbonaceous intermediates

Very few electrode materials are capable of performing both reactions, and of these only platinum and platinum-based electrodes display any significant activity and stability in an acidic medium.

When the electrooxidation of methanol is performed with platinum as the anode, the reaction is thought to occur at the surface and involve several adsorbed intermediates. After adsorption of methanol, the reaction mechanism is thought to proceed through a series of dehydrogenation products (Eq. 4-7). Adsorbed CO is a key intermediate of methanol oxidation reaction, and its presence has been observed with the aid of various in situ spectroscopic techniques. CO and other carbonaceous intermediates can then be oxidized to carbon dioxide reacting with adsorbed water (Eq. 8-10).

Presently DMFCs are faced with a few challenging problems, foremost of which is that the overall reaction is very slow. The electrochemical oxidation of methanol, although thermodynamically favored, is kinetically sluggish, due to the formation of stable intermediates[10,11] (Eq 5-7). Adsorbed CO in particular can quickly poison the Pt surface resulting in a high overpotential (approximately 0.5 V $vs.$ NHE) for a reasonable current density.

$$CH_3OH + Pt_{(s)} \longrightarrow Pt-CH_2OH + H^+ + e^- \qquad (4)$$

$$Pt-CH_2OH + Pt_{(s)} \longrightarrow Pt_2-CHOH + H^+ + e^- \qquad (5)$$

$$Pt_2-CHOH + Pt_{(s)} \longrightarrow Pt_3-COH + H^+ + e^- \qquad (6)$$

$$Pt_3-COH \longrightarrow Pt-CO + 2 Pt_{(s)} + H^+ + e^- \qquad (7)$$

$$H_2O + Pt_{(s)} \longrightarrow Pt-OH + H^+ + e^- \qquad (8)$$

$$Pt-CO + Pt-OH \longrightarrow Pt-COOH \qquad (9a)$$

$$Pt-CO + H_2O \longrightarrow Pt-COOH + H^+ + e^- \qquad (9b)$$

$$Pt-COOH \longrightarrow Pt_{(s)} + CO_2H^+ + e^- \qquad (10)$$

Although there has been moderate success with Pt anodes,[12-14] the high overpotentials required for these fuel cells have made Pt electrodes unacceptable for use in DMFCs. This conclusion has led to an intensive search for other materials that can improve the performance of the Pt anode during the methanol oxidation process. Several methods have been investigated for promoting the formation of CO_2. One method involves alloying Pt with a metal that readily adsorbs and dehydrogenates water within the potential range for methanol oxidation. Studies of Pt alloyed with such metals shows that Ru has by far the largest catalytic effect.[9]

Considerable effort has been expended to clarify the role of Ru in Pt/Ru binary electrodes. Of the mechanisms proposed, the bifunctional theory has gained general acceptance.[15] According to this mechanism the Pt sites in Pt/Ru alloys are responsible for the chemisorption and dehydrogenation of methanol, while the Ru serves as a source of "activated" oxygen, aiding the formation of CO_2 (Eq. 11-12).

$$H_2O + Ru_{(s)} \longrightarrow Ru-OH + H^+ + e^- \qquad (11)$$

$$Pt-CO + Ru-OH \longrightarrow CO_2 + H^+ + e^- \qquad (12)$$

Homogeneous Electrooxidation of Alcohols

In addition to heterogeneous studies, the homogeneous electrochemical oxidation of alcohols with ruthenium complexes has been extensively examined. This reaction can be made catalytic by applying a fixed potential capable of regenerating the active metal species. As an example, [1,3-bis(4-methyl-2-pyridylimino)isoindoline]RuCl$_3$ serves as a catalyst for both electrooxidation and autoxidation of several alcohols, including methanol.[16] The polypyridyl complexes [Ru(trpy)(dppi)(OH)]$^+$ and [Ru(4,4'-Me$_2$bpy)$_2$(PPh$_3$) (H$_2$O)](ClO$_4$)$_2$ catalyze the electrooxidation of benzyl alcohol to benzaldehyde. Cyclic voltammetry of these solutions exhibits an increase in anodic current when an alcohol is introduced; this increase in current is characteristic of the electrooxidation of alcohols by the catalytic species Ru^{IV}=O.[17]

Extensive mechanistic studies have been performed on the electrooxidation of aqueous alcoholic solutions with ruthenium complexes. Various mechanisms have been proposed, each containing a Ru-oxo complex as the catalytically active species. The Ru-oxo bond can either be generated in situ by reacting with water or be present in the pre-catalyst. The mechanisms which have been proposed include hydride transfer,[18,19] hydrogen atom abstraction,[20] and oxygen atom transfer.[21]

Oxo-bridged Ru dimers can also serve as catalysts for the electrooxidation of alcohols. Meyer reported rapid oxidation of a variety of alcohols, aldehydes, and carboxylates by [(bpy)$_2$(O)Ru^{IV}ORu^V(O)(bpy)$_2$].[22] Related binuclear Ru complexes such as [Ru$_2$(napy)$_2$(H$_2$O)$_4$Cl(OH)][ClO$_4$] (napy = 1,8-naphthyridine) have also been shown to be catalysts for the oxidation of primary and secondary alcohols, although the oxidation chemistry was complicated by the instability of the complexes.[23] The binuclear Ru complex (L$_{OMe}$)(HO)Ru^{IV}(μ-O)$_2$Ru^{IV}(OH) (L$_{OMe}$) where L$_{OMe}$ = [CpCo{P(O)(OMe)$_2$}$_3$]$^-$ also serves to oxidize alcohols, with a further electrooxidation step of formaldehyde to formate also accessible.[24]

Cooperative Effect in Bimetallic Catalysts

The introduction of binuclear complexes has contributed to the development of important applications due to the fact that the two metals can cooperate with each other and show different reactivities from the monometallic compounds. This effect can result in a significant modification of the individual metal properties where the catalytic activity of one metal is mediated by the presence of the other one. It is possible to observe effects of the chemical, electrochemical[25] or photochemical[26] modifications of one metal center on the properties of another. The intramolecular electronic interaction largely depends on the structure of the complex including the ancillary ligands and bridging ligands but also on the nature of the metals and their oxidation state. Information on the metal-metal interactions has been obtained from spectroscopy, crystallography and theoretical investigations of such model complexes.[27]

The cooperative bimetallic effect has been also demonstrated in many examples of heterobimetallic complexes with unsaturated hydrocarbon bridges,[28] complexes based on bridging ligands containing bipyridine, phenanthroline, and terpyridine chelating units,[26] as well as complexes containing μ-P-(CH_2)-Y (Y = P,S),[27,29,30] μ-thiolate,[31] μ-oxo[32] or μ-halide[33] bridged type ligands. Recently, Severin and co-workers have used the heterobimetallic complex (η^4-$C_4Ph_4CO)Rh(\mu$-Cl$)_3Ru(PPh_3)_2$(acetone) as catalyst in the Oppenauer-type oxidation of primary and secondary alcohols. The beneficial effect of two metal centers could be demonstrated since the homobimetallic Ru or Rh analogues are not active under the same conditions.[34]

One way to demonstrate metal-metal interaction in heterobimetallic complexes is to study their cyclic voltammetry (CV) processes, since the results can be compared with mononuclear model systems. Introduction of a second metal usually shows shifts in redox potentials which are strong evidence of metal-metal communication. There has been previous work which has examined several Mo/Pt and Ru/Pt heterobimetallic complexes in electrochemical processes.[35] The shifts in the oxidation potentials reflect the ability of the metal centers to communicate with each other through the bridging ligands. For example, the cationic $[Mo(CO)_3(\mu$-dppm$)_2Pt(H)]PF_6$ complex shows a 400 mV positive shift for both Mo(II/III) and Pt(II/IV) waves compared to the neutral $Mo(CO)_3(\mu$-dppm$)_2Pt(H)Cl$ complex. Since the only change in the compounds occurs at the Pt center, this result demonstrates the electronic interaction between the metal centers through the bridging ligand.

Product Detection and Analysis

The electrooxidation of methanol has been investigated since the 1960's and several reviews on methanol oxidation have been published.[11,36] As previously described by Parson and Iwasita for oxidation on an anode surface, the six-electron oxidation process involves a complicated multistep mechanism (Eq. 13-15), where methanol is converted to CO_2 via formation of formaldehyde and CO. However, when the process is performed in solution in presence of homogeneous catalysts, formic acid is formed as the 4e[-] oxidation product instead of carbon monoxide (Eq. 16, 17).

When methanol is used in excess during the homogeneous electrooxidation reaction, both formaldehyde and formic acid undergo condensation reactions with methanol to form dimethoxymethane and methyl formate as the 2e[-] and 4e[-] oxidation products, respectively (Eq. 18, 19). It is important to point out that the equilibria for these reactions are shifted to the right in the presence of an excess of methanol. As highlighted by Rand[37] and Sermon,[38] these reactions are also favored by the presence of acid catalysts.

$$CH_3OH \longrightarrow HCHO + 2H^+ + 2e^- \quad (13)$$

$$HCHO \longrightarrow CO + 2H^+ + 2e^- \quad (14)$$

$$CO + H_2O \longrightarrow CO_2 + 2H^+ + 2e^- \quad (15)$$

$$HCHO + H_2O \longrightarrow HCOOH + 2H^+ + 2e^- \quad (16)$$

$$HCOOH \longrightarrow CO_2 + 2H^+ + 2e^- \quad (17)$$

$$HCHO + 2 CH_3OH \rightleftharpoons H_2C(OCH_3)_2 + H_2O \quad (18)$$

$$HCOOH + CH_3OH \rightleftharpoons HCOOCH_3 + H_2O \quad (19)$$

It is possible to follow the oxidation process by analysis of the products. For example, a direct FTIR investigation of the methanol oxidation in a prototype direct methanol fuel cell (DMFC) demonstrated that dimethoxymethane, methyl formate and CO_2 were the products of the electrooxidation process with pure methanol in the anode feed.[39] However, the product distribution is dependent on different factors including the activity of the catalyst, the methanol/water ratio as well as the temperature of operation.

Heterobimetallic dppm-Bridged Catalysts for the Electrooxidation of Methanol

The development of heterobimetallic catalysts for electrooxidation of methanol was initially motivated by literature results on introducing a second metal into bulk metal anodes,[11,40-42] and also by the possibility that each metal center could exhibit cooperative activity or a unique mechanistic function. The choice of bis(diphenylphosphino)methane (dppm) as a bridging ligand was directed by the fact that metal-phosphorus bonds are often very strong and two metals can be locked together in close proximity by a bidentate phosphine.

The first generation of heterobimetallic complexes $Cp(PPh_3)Ru(\mu\text{-}Cl)(\mu\text{-}dppm)PtCl_2$ (1),[35] $Cp(PPh_3)Ru(\mu\text{-}Cl)(\mu\text{-}dppm)PdCl_2$ (2),[43] and $Cp(PPh_3)RuCl(\mu\text{-}dppm)AuCl$(3)[43] was prepared by the reaction of $CpRu(PPh_3)(\eta^1\text{-}dppm)Cl$ (4) with $Pt(COD)Cl_2$, $Pd(COD)Cl_2$ and $Au(PPh_3)Cl$, respectively. All of them possess a dppm linkage between Ru and the second metal center with a three-legged piano stool geometry at Ru (Figure 1). Complexes 1 and 2 possess a bridging chloride that links Ru centers and the quasi square planar Pt or Pd, in a

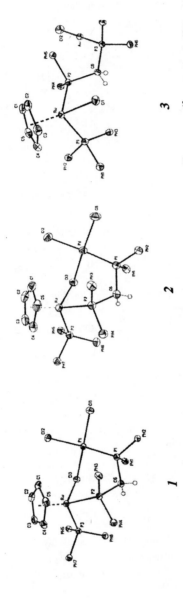

Figure 1. Thermal ellipsoid drawings of the molecular structures of complexes 1,[35] 2 and 3.[43] Thermal ellipsoids are plotted at 50% probability. Phenyl rings and most hydrogen atoms are omitted for clarity.

distorted six-membered ring. In contrast, complex **3** is linked only via dppm in a pendant fashion with a linear configuration at the Au center. Cyclic voltammetry of complexes **1-3** in the presence of methanol led to significant enhancement of oxidative current, consistent with a catalytic process.[35,43,44] Bulk electrolysis of methanol in the presence of the heterobimetallic complexes resulted in much higher current efficiencies than those obtained from the mononuclear models complexes CpRu(PPh$_3$)$_2$Cl (**5**), and CpRu(η^2-dppm)Cl (**6**) suggesting that the second metal center enhances the catalytic activity.

Table I. Formal Potentials of Complexes 1-3.

Complex	Couple	$E_{1/2}$ (V)^a	Couple	$E_{1/2}$ (V)^a
Ru/Pt (**1**)[43]	Ru(II/III)	1.13^b	Pt(II/IV)	1.78^c
Ru/Pd (**2**)[35]	Ru(II/III)	1.29	Pd(II/IV)	1.45^c
Ru/Au (**3**)[43]	Ru(II/III)	0.89	Au(I/III)	1.42^c
Ru^d	Ru(II/III)	0.61		
Ru$'^e$	Ru(II/III)	0.87^b		
Pt(η^2-dppm)Cl$_2$ [45]			Pt(II/IV)	2.21^c
Pd(η^2-dppm)Cl$_2$			Pd(II/IV)	2.16^c
Au(PPh$_3$)Cl$_2$ [46]			Au(I/III)	1.68^b

^{a} All potentials obtained in DCE/TBAT (tetrabutylammonium triflate) and reported *vs.* NHE. ^{b} Performed in CH$_2$Cl$_2$/TBAH. ^{c} Irreversible wave, E_{pa} reported. ^{d} Ru = CpRuCl(η^2-dppm). ^{e} Ru' = CpRu(PPh$_3$)$_2$Cl.

As mentioned above, cyclic voltammetry can be used to establish the metal-metal interactions in heterobimetallic complexes (Table I). Cyclic voltammetry of **1** and **2** demonstrates shifts in the formal potentials of Ru(II/III), Pt(II/IV) and Pd(II/IV) redox couples relative to the monomeric Ru, Pt, and Pd models, indicative of significant electron donation through the chloride bridge from the Ru to the electron deficient Pt or Pd centers. In contrast, the redox potentials of the Ru(II/III) and Au(I/III) couples in **3** are very similar to mononuclear complexes suggesting minimal interaction via the dppm bridge.

The study of metal-metal interactions and catalytic activities in these complexes was extended by preparing a series of similar complexes exhibiting systematic perturbations in the ancillary ligands as well as the metal centers.[47] Compounds **7-11** are similar in structure to compounds **1** and **2**, exhibiting a bridging halide between the two metal centers. The halide bridge varies from Cl to I, while the ligand on the second metal changes from Cl to I to an electron donating methyl group. Complexes **12-14** are related to the non-bridged halide complex **3**. The absence of a halide bridge leads to minimal interaction between Ru and Au centers. In these complexes the halide on the Ru centers varies from Cl to Br to I, and the bridging bidentate phosphine was changed from dppm to the longer carbon bridged bis(diphenylphosphino)butane (dppb).

	M	L	X^1	X^2	X^3
7	Pt	PPh$_3$	Cl	Cl	Me
8	Pt	PPh$_3$	I	Cl	Cl
9	Pt	PPh$_3$	I	I	I
10	Pd	PPh$_3$	Cl	Cl	Me
11	Pd	PPh$_3$	I	Cl	Cl

	L	X^1	X^2	Q
12	PPh$_3$	Br	Cl	CH$_2$
13	PPh$_3$	I	I	CH$_2$
14	PPh$_3$	Cl	Cl	(CH$_2$)$_4$

Electrochemical Oxidation of Methanol Using dppm-Bridged Catalysts

Cyclic voltammograms of the heterobimetallic complexes **7-14** generally exhibit three distinctive redox waves (Table II).[47] The first and third waves are assigned to the Ru(II/III) and Ru(III/IV) couples with formal potentials in the ranges 0.75-1.30 V and 1.76-1.98 V respectively, while the middle one is assigned to the redox potential of the second metal having a formal potential in the ranges 1.40-1.54 V. As expected from results on **1-3**, the oxidation potentials of each metal are dependant on the amount of electron donation from Ru to the second metal center through the bridging ligands. Significant electron donation from Ru to Pd or Pt through the halide bridge can be seen in all of the Ru/Pt and Ru/Pd bimetallics (Table II; **7-11**) when compared to the mononuclear models (Table I). Comparison of the I-bridged complexes **8** and **11** with their Cl-bridged analogues **1** and **2** shows negligible perturbations in the Ru(II/III) or Ru(III/IV) potential as the bridging halide is changed. However, more significant results can be observed for the Pt(II/IV) or Pd(II/IV) couples. First, these waves are irreversible for the Cl-bridged complexes **1** and **2** while compounds **8** and **11** exhibit reversible Pt(II/IV) and Pd(II/IV) waves, implying greater stability for the oxidized I-bridged complexes. The I-bridged Ru/Pt complex **8** exhibits a potential shift, with its Pt(II) center 150 mV easier to oxidize than the Cl-bridged analogue **1**, while I-bridged Ru/Pd complex **11** exhibits a 60 mV positive shift compared to the Cl-bridged **2**, making the Pd center more difficult to oxidize.

Interestingly, changing the ligand at the Pt or Pd center has a larger effect on the Ru(II/III) redox potentials. Substitution of a chloride in **1** or **2** by an electron donating methyl group affords the Pt or Pd methyl substituted complexes **7** and **10**. In both cases the Ru(II/III) waves became reversible and also shifted to lower potentials (by 50 mV for **7** and by 190 mV for **10**). A similar effect is observed for the I-bridged Ru/Pt complexes **8** and **9**, which exhibit a 190 mV negative shift at the Ru(II/III) couple when the substituents at the Pt centers change from Cl to I.

The dppm-bridged Ru/Au complexes **12** and **13** exhibit reversible waves for the Ru(II/III) couple and irreversible waves for the Au(I/III) couples. The redox potentials for this type of compound are very similar to the mononuclear model analogues (Table I), indicating limiting electronic interaction between the two metal centers through the dppm bridge. Changing the halide from Cl (**3**) to Br (**12**) and I (**13**) on the Ru center did not really affect the formal potential of the Ru(II/III) couple (a constant potential of 0.89 V is observed) while the anodic potential for the Au(I/III) couple is slightly shifted to positive values as the halide goes from Cl to Br to I.

The dppb-bridged Ru/Au compound **14** exhibits a reversible wave at 0.75 V and an irreversible wave at 1.76 V *vs.* NHE, attributed to the Ru(II/III) and Ru(III/IV) couples, respectively. No wave was observed for the Au(I/III) couple under the conditions of the experiment. In comparison, the Au(I/III) wave can be observed at 1.42 V for the dppm analogue **3**. A possible explanation can be overlap of the Au(I/III) wave with the Ru(III/IV) wave, since the mononuclear complex Au(PPh$_3$)Cl has been reported to oxidize at 1.68V *vs.* NHE in dichloromethane.[46] Another explanation involves the lack of suitable ligands in the electrolyte solution to coordinate with the Au(III) species formed upon oxidation. Cyclic voltammetry of **14** in presence of Bu$_4$NCl as a chloride source allowed the observation of a broad shoulder before the Ru(III/IV) wave.

Table II. Formal Potentials for Complexes 7-14.

Complex	Couple	$E_{1/2}$/V	Couple	$E_{1/2}$/V	Couple	E_{pa}/V
7	Ru(II/III)	1.08	Pt(II/IV)	1.40	Ru(III/IV)	1.95
8	Ru(II/III)	1.25	Pt(II/IV)	1.47	Ru(III/IV)	1.90
9	Ru(II/III)	1.17[b]	Pt(II/IV)	1.44	Ru(III/IV)	1.89
10	Ru(II/III)	1.10	Pd(II/IV)	1.43	Ru(III/IV)	1.95
11	Ru(II/III)	1.29[b]	Pd(II/IV)	1.50[b]	Ru(III/IV)	1.98
12	Ru(II/III)	0.89	Au(I/III)	1.48[b]	Ru(III/IV)	1.85
13	Ru(II/III)	0.89	Au(I/III)	1.54[b]	Ru(III/IV)	1.80
14	Ru(II/III)	0.75	Au(I/III)	-	Ru(III/IV)	1.76

[a] All potentials obtained in DCE/TBAT (tetrabutylammonium triflate) and reported *vs.* NHE. [b] Irreversible wave, E_{pa} reported.

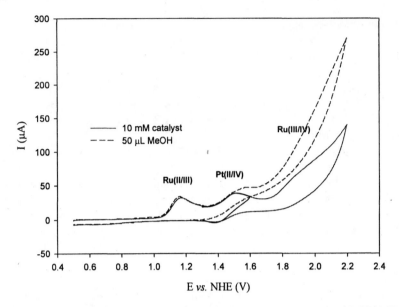

Figure 2. Cyclic voltammograms of 9 under nitrogen in 3.5 mL of DCE/0.7M TBAT; glassy carbon working electrode; Ag/Ag⁺ reference electrode; 50mV/s.

The cyclic voltammograms of the heterobimetallic complexes 7-14 after addition of methanol generally show a significant current increase at the Pt(II/IV), Pd(II/IV), Au(I/III) and Ru(III/IV) redox waves, indicating catalytic activity for methanol oxidation. As an example, Figure 2 shows the cyclic voltammogram of the Ru/Pt heterobimetallic complex 9 and the effect of the addition of 50 µL methanol to the system.

As for complexes 1-3, bulk electrolyses with complexes 7-14 were also performed for product identification and quantification.[33] A potential of 1.7 V *vs.* NHE for the bulk electrolysis was chosen in earlier studies, coincident with the redox potential of the second metal. As previously mentioned, the oxidation products observed during the bulk electrolysis are dimethoxymethane (DMM) and methyl formate (MF) as the 2e⁻ and 4e⁻ oxidation products, respectively. Methanol undergoes initial oxidation to formaldehyde and formic acid during the process, however, neither was observed in the reaction mixtures. Acid catalyzed condensation of these products with excess methanol yields DMM and MF (Eq. 16, 18, 19). The evolution of the product distributions as the reaction progresses is shown in Table III. At low conversion, all of the heterobimetallic catalysts afforded a much higher proportion of DMM. As the reaction progresses, the same tendency toward production of the more highly oxidized product (MF) can be observed for all of the heterobimetallic catalysts. This is presumably due to the water formed in situ during the condensation of formaldehyde and formic acid with excess methanol.

Table III. Product Distributions and Current Efficiencies for Methanol Oxidation by 7-14.

Charge/ C	Product ratios (DMM/MF)[a,b]							
	7	8	9	10	11	12	13	14
25	6.08	2.27	2.23	4.25	1.85	3.86	3.78	3.88
50	4.50	1.68	1.66	3.62	1.56	3.04	3.25	3.56
75	3.53	1.24	1.40	3.19	1.21	2.60	2.74	2.94
100	3.02	0.98	1.26	2.94	0.91	2.35	2.10	2.32
Current Efficiency (%)[c]	32	43	39	23	42	20	16	12

[a] Electrolyses were performed at 1.7V vs. NHE. A catalyst concentration of 10 mM was used. Methanol concentration was 0.35 M. [b] Determination by GC with respect to n-heptane as internal standard. Each ratio is reported as an average of 2-5 experiments. [c]Average current efficiencies after 75-100 C of charge passed.

The differences in the behaviors of complexes 7-14 are apparent in the product ratios. Oxidation with the I-bridged complexes (8, 9, 11) led to more of the four-electron oxidation product (MF), compared to the Cl-bridged complexes 7 and 10. The Ru/Au compounds 12-14 exhibited very similar behaviors for their product distributions as the reaction progressed, with all of them being very similar to the mononuclear analogue $CpRu(PPh_3)_2Cl$.

The current efficiencies for the oxidation processes are also summarized in Table III. These values are the ratio of the charge necessary to produce the observed yield of DMM and MF to the total charge passed during the bulk electrolysis. The I-bridged Ru/Pt 8, 9 and Ru/Pd 11 gave moderately higher current efficiencies (43, 39 and 42%) as compared to the Cl-bridged 1 and 2 (18.6 and 24.6), probably due to the higher stability of I-bridged catalysts during the process. Complexes with methyl substitution at the Pt and Pd centers (7 and 10) gave lower current efficiencies as compared to the I-bridged complexes, while the Ru/Au complexes 12-14 produced moderately low current efficiencies, (12 and 20%), similar to the values of the mononuclear Ru compound $CpRu(PPh_3)_2Cl$. The behavior for these complexes is consistent with their structure, in which only minimal interaction between the two metal centers is mediated by the bridging diphosphine ligand.

Conclusions

Single molecule catalysts could provide an alternative to surface catalysis on anodes composed of bulk Pt/Ru alloys. The use of discrete molecular catalysts could reduce the amounts of these expensive precious metals needed for DMFC applications while facilitating the direct utilization of methanol and other complex biomass-derived fuels in fuel cells. Heterobimetallic complexes of RuPt, RuPd and RuAu have been demonstrated to catalyze the electrochemical oxidation of methanol to DMM and MF. This is a promising start, but more active catalysts that oxidize methanol to CO_2 while immobilized on the anode surface will be required for adaptation of homogeneous catalysts for use in DMFCs. Toward this end, modified electrodes containing 1-14 and related complexes are currently being prepared and studied.

References

1. Chynoweth, D. P.; Owens, J. M.; Legrand, R. *Renewable Energy* **2000**, *22*, pp 1-8.
2. Orr, J. C.; Fabry, V. J.; Aumont, O.; Bopp, L.; Doney, S. C.; Feely, R. A.; Gnanadesikan, A.; Gruber, N.; Ishida, A.; Joos, F.; Key, R. M.; Lindsay, K.; Maier-Reimer, E.; Matear, R.; Monfray, P.; Mouchet, A.; Najjar, R. G.; Plattner, G.-K.; Rodgers, K. B.; Sabine, C. L.; Sarmiento, J. L.; Schlitzer, R.; Slater, R. D.; Totterdell, I. J.; Weirig, M.-F.; Yamanaka, Y.; Yool, A. *Nature* **2005**, *437*, pp 681-686.
3. Arakawa, H.; Aresta, M.; Armor, J. N.; Barteau, M. A.; Beckman, E. J.; Bell, A. T.; Bercaw, J. E.; Creutz, C.; Dinjus, E.; Dixon, D. A.; Domen, K.; DuBois, D. L.; Eckert, J.; Fujita, E.; Gibson, D. H.; Goddard, W. A.; Goodman, D. W.; Keller, J.; Kubas, G. J.; Kung, H. H.; Lyons, J. E.; Manzer, L. E.; Marks, T. J.; Morokuma, K.; Nicholas, K. M.; Periana, R.; Que, L.; Rostrup-Nielson, J.; Sachtler, W. M. H.; Schmidt, L. D.; Sen, A.; Somorjai, G. A.; Stair, P. C.; Stults, B. R.; Tumas, W. *Chem. Rev.* **2001**, *101*, pp 953-996.
4. Wuebbles, D. J.; Jain, A. K. *Fuel Processing Tech.* **2001**, *71*, pp 99-119.
5. Towler, G. P.; Oroskar, A. R.; Smith, S. E. *Environ. Prog.* **2004**, *23*, pp 334-341.
6. Yoshida, P. G. *Fuel cell vehicles race to a new automotive future.*; Office of Technology Policy, U.S. Dept. of Commerce, 2003.
7. Stone, C.; Morrison, A. E. *Solid State Ionics* **2002**, *152-153*, pp 1-13.
8. Kim, D.; Cho, E. A.; Hong, S.-A.; Oh, I.-H.; Ha, H. Y. *J. Power Sources* **2004**, *130*, pp 172-177.
9. Parsons, R.; VanderNoot, T. *J. Electroanal. Chem.* **1988**, *257*, pp 9-45.
10. Wasmus, S.; Vielstich, W. *J. Appl. Electrochem.* **1993**, *23*, pp 120-124.
11. Iwasita, T. *Electrochim. Acta* **2002**, *47*, pp 3663-3674.
12. Enyo, M.; Machida, K.-i.; Fukuoka, A.; Ichikawa, M. In *Electrochemistry in Transition*; Murphy, O. J., Ed.; Plenum Press: New York, 1992, pp 359-369.

13. Ross, P. N. *Electrochim. Acta* **1991**, *36*, pp 2053-2062.
14. Leger, J. M.; Lamy, C. *Ber. Bunsen-Ges.* **1990**, *94*, pp 1021-1025.
15. Watanabe, M.; Motoo, S. *J. Electroanal. Chem.* **1975**, *60*, pp 276-283.
16. Gagné, R. R.; Marks, D. N. *Inorg. Chem.* **1984**, *23*, pp 65-74.
17. Gerli, A.; Reedijk, J.; Lakin, M. T.; Spek, A. L. *Inorg. Chem.* **1995**, *34*, pp 1836-1843.
18. Roecker, L.; Meyer, T. J. *J. Am. Chem. Soc.* **1986**, *108*, pp 4066-4073.
19. Roecker, L.; Meyer, T. J. *J. Am. Chem. Soc.* **1987**, *109*, pp 746-754.
20 Muller, J. G.; Acquaye, J. H.; Takeuchi, K. J. *Inorg. Chem.* **1992**, *31*, pp 4552-4557.
21 Lebeau, E. L.; Meyer, T. J. *Inorg. Chem.* **1999**, *38*, pp 2174-2181.
22 Raven, S. J.; Meyer, T. J. *Inorg. Chem.* **1988**, *27*, pp 4478-4483.
23 Boelrijk, A. E. M.; Neenan, T. X.; Reedijk, J. *J. Chem. Soc.-Dalton Trans.* **1997**, pp 4561-4570.
24. Kelson, E. P.; Henling, L. M.; Schaefer, W. P.; Labinger, J. A.; Bercaw, J. E. *Inorg. Chem.* **1993**, *32*, pp 2863-2873.
25. Leschke, M.; Lang, H.; Holze, R. *J. Solid State Electrochem.* **2003**, *7*, pp 518-524.
26. Balzani, V.; Juris, A.; Venturi, M.; Campagna, S.; Serroni, S. *Chem. Rev.* **1996**, *96*, pp 759-833.
27. Xia, B. H.; Zhang, H. X.; Che, C. M.; Leung, K. H.; Phillips, D. L.; Zhu, N. Y.; Zhou, Z. Y. *J. Am. Chem. Soc.* **2003**, *125*, pp 10362-10374.
28. Ceccon, A.; Santi, S.; Orian, L.; Bisello, A. *Coord. Chem. Rev.* **2004**, *248*, pp 683-724.
29. Balch, A. L.; Catalano, V. J. *Inorg. Chem.* **1991**, *30*, pp 1302-1308.
30. Barranco, E. M.; Crespo, O.; Gimeno, M. C.; Laguna, A.; Jones, P. G.; Ahrens, B. *Inorg. Chem.* **2000**, *39*, pp 680-687.
31. Shapley, P. A.; Reinerth, W. A. *Organometallics* **1996**, *15*, pp 5090-5096.
32. Shapley, P. A.; Zhang, N. J.; Allen, J. L.; Pool, D. H.; Liang, H. C. *J. Am. Chem. Soc.* **2000**, *122*, pp 1079-1091.
33. Yang, Y.; McElwee-White, L. *Dalton Trans.* **2004**, pp 2352-2356.
34. Gauthier, S.; Scopelliti, R.; Severin, K. *Organometallics* **2004**, *23*, pp 3769-3771.
35. Orth, S. D.; Terry, M. R.; Abboud, K. A.; Dodson, B.; McElwee-White, L. *Inorg. Chem.* **1996**, *35*, pp 916-922.
36. Hamnett, A. *Catal. Today* **1997**, *38*, pp 445-457.
37. Rand, D. A. J.; Woods, R. *J. Electroanal. Chem.* **1972**, *35*, pp 209-218.
38. Sermon, P. A.; Bond, G. C. *Cat. Rev.* **1973**, *8*, pp 211-239.
39. Lin, W. F.; Wang, J. T.; Savinell, R. F. *J. Electrochem. Soc.* **1997**, *144*, pp 1917-1922.
40. Wasmus, S.; Kuver, A. *J. Electroanal. Chem.* **1999**, *461*, pp 14-31.
41. Jacoby, M. In *Chemical and Engineering News*, **1999**, pp 31-37.
42. Hogarth, M. P.; Hards, G. A. *Platinum Met. Rev.* **1996**, *40*, pp 150-159.

43. Matare, G.; Tess, M. E.; Abboud, K. A.; Yang, Y.; McElwee-White, L. *Organometallics* **2002**, *21* pp, 711-716.

44. Tess, M. E.; Hill, P. L.; Torraca, K. E.; Kerr, M. E.; Abboud, K. A.; McElwee-White, L. *Inorg. Chem.* **2000**, *39*, pp 3942-3944.

45. Brown, M. P.; Puddephatt, R. J.; Rashidi, M.; Seddon, K. R. *J. Chem. Soc.-Dalton Trans.* **1977**, pp 951-955.

46. Attar, S.; Nelson, J. H.; Bearden, W. H.; Alcock, N. W.; Solujic, L.; Milosavljevic, E. B. *Polyhedron* **1991**, *10*, pp 1939-1949.

47. Yang, Y.; Abboud, K. A.; McElwee-White, L. *Dalton Trans.* **2003**, pp 4288-4296.

Chapter 20

Oxidative Chemistry of Lignin in Supercritical Carbon Dioxide and Expanded Liquids

Dimitris S. Argyropoulos, Carl D. Saquing, Armindo R. Gaspar, Nestor U. Soriano, Lucian A. Lucia, and Orlando J. Rojas

Department of Forest Biomaterials Science and Engineering, North Carolina State University, Raleigh, NC 27695–8005

This chapter describes the use of supercritical carbon dioxide ($scCO_2$) and $scCO_2$-expanded liquids, as reaction media for the oxidative fragmentation of lignin using hydrogen peroxide as the oxidant. Kraft Indulin Lignin (KIL) is oxidized with peroxide in neat $scCO_2$ but more extensive oxidation occurs in $scCO_2$-expanded liquids, selectively inducing the formation of large amounts of carboxylic acids via the elimination of phenolic moieties. Interestingly, the weight average molecular weight of the lignin oxidized in neat $scCO_2$ decreased while it increased in $scCO_2$-expanded liquid, most likely via the preferential elimination of the lower molecular weight fragments. Notably, all these significant oxidative transformations occur in total absence of alkali, which is essential when such reactions occur in aqueous media.

Lignin, as the second most abundant natural polymeric material, presently finds little industrial and commercial utility. Most of the 50 million metric tons of lignin being produced in the U.S. alone, for example, are mainly used as fuel [1]. However, its potential as a source of high-value low molecular weight chemicals is attractive due to its chemical composition. It has a highly intricate aromatic polymer structure composed of partly oxygenated phenylpropane units. Several interunit carbon-carbon and carbon-oxygen bonds are present in the lignin structure and the relative abundance of these interunit linkages vary for different types of wood. The controlled fragmentation of these interunits in the complex polymer network can result in monomeric and/or oligomeric phenolic compounds. Studies have focused on the oxidative degradation of lignin aimed at producing useful low molecular weight aromatic compounds. Phenolic compounds such as vanillin have been produced from lignin using the alkaline-nitrobenzene method [2]. There were other processes which carried out the oxidation of lignin using nitroaromatics [3], air in alkaline media[4], ozone [5], electrochemical means [6], enzymes [7] or peroxide mediated by various metal ions [8] with the end in view of fragmenting lignin into smaller molecules.

There are additional promising biomimetic degradation systems which use dioxygen as the oxidant and various metal complexes as catalysts. Drago et al. [9] have shown that dioxygen was able to oxidize isoeugenol to vanillin in good yields, using [bis(salicylidene-γ-iminopropyl)methylamine]cobalt(II), [Co(SMDPT)], as catalyst; Bozell et al. [10,11] has reported the oxidation of para-substituted phenolic compounds to benzoquinones using [N,N'-bis (salicylidene)ethane-1,2-diaminato]cobalt(II), [Co(salen)], as the catalyst, while Haikarainen et al. [12] has reported the aqueous oxidation of lignin model compounds using a similar water soluble cobalt compound as a catalyst to obtain vanillin and other biphenolic products. All of these efforts involve oxidation chemistry on lignin or its models in aqueous media and sometimes under strongly alkaline conditions.

For a process that aims to produce useful chemicals using lignin as the feedstock, product separation technology is pivotal in determining the commercial feasibility and environmental superiority over traditional approaches. One attractive approach is the use of supercritical fluids (SCFs) as process solvents for biomass conversion. Since most SCFs exist as gas at room temperature and pressure, the solvent can readily be separated and recycled from the reaction mixture. This is accomplished via control of several key parameters such a temperature, pressure, flow rates and processing time [13]. Moreover, with higher diffusivity and lower viscosity of SCFs compared to a liquid, mass transfer rates are much higher than conventional solvents [14]. There is minimal literature that exists with regard to the use of SCFs to process biomass. Earlier publications dealt with mainly the extraction of low molecular compounds from plant materials [15,16], wood [17,18,19], or from sulfite pulping liquors [20] with carbon dioxide, nitrous oxide, nitrogen, and lower alkanes. Other works using SCFs for processing lignocellulosic materials involved the liquefaction or gasification of biomass. The liquefaction of wood using various organic solvents such as dioxane, toluene, tetrahydrofuran, methanol, acetone, ethyl acetate and other alcohols at supercritical conditions has been reported [21,22,23,24,25]. On the other hand, the gasification of lignocellulosics has been successfully achieved using near- and supercritical water [26,27,28]. The major gas products are H_2, CH_4, and CO. In general, two approaches in biomass gasification using supercritical water are employed: low temperature catalytic gasification where the reaction temperature ranges from 350 to 600 °C and high temperature gasification where the temperature ranges from 500 to 750 °C without catalyst or with nonmetallic catalyst [29]. Using supercritical water, lignin was also shown to be converted to low molecular weight phenolic compounds [30].

The two most prominent supercritical fluids are carbon dioxide and water because they are non-toxic, non-flammable and cheap. CO_2 has a low critical point (31°C, 73 atm), hence requiring moderate operating conditions, which is not the case for water (374 °C, 218 atm). However, water at the critical conditions is a powerful solvent while CO_2 is feeble compared to conventional organic solvents. In an extensive study by Li and Kiran [31] of the interactions and reactivity of lignocellulosics in various SCFs, they have concluded that the solubility of biomass in supercritical ethylene, nitrous oxide, CO_2 and n-butane is not substantial and the interactions are non-reactive. Meanwhile, the success of biomass gasification in supercritical water is believed to predominantly depend on the dissolution of the biomass in the supercritical solvent [32,33]. For successful applications of SCFs to process biomass, the understanding of the effect of the interactions between SCF and lignocellulosic materials is valuable.

The present study is focused on the SCF processing of lignin, in particular, the oxidation of kraft indulin lignin using hydrogen peroxide in the presence of $scCO_2$ or $scCO_2$-expanded ethylene glycol. A supercritical fluid-expanded liquid is a mixed solvent where a supercritical fluid is condensed in an organic solvent, usually resulting in the expansion of the organic solvent. It is used to exploit synergistically the solvent properties of a supercritical fluid and an organic liquid. Novel work reported by Subramanian's group has shown that various phenols can be oxidized with dioxygen when it is catalyzed by the cobalt Schiff base, Co(salen) in CO_2-expanded liquids[34]. Cyclohexene was also shown to be oxidized under similar conditions when catalyzed by tetraphenylporphyrinato iron (III) chloride, in CO_2-expanded CH_3CN. These reactions proceeded with impressive catalytic turn over frequencies, conversions, and selectivities. Subramanian's group has demonstrated that there are several advantages of conducting oxidations in homogeneous CO_2-expanded reaction mixtures including higher oxygen miscibility, compared to neat organic solvents. In addition, CO_2-expanded reaction mixtures allow for transition metal catalysts to be solubilized without modification, increasing catalyst solubility by about one to two orders of magnitude.

The oxidative changes in the starting material were probed with the aim of understanding the effect of pressure and H_2O_2 concentration on oxidation in neat CO_2 and $scCO_2$-expanded media. This study is part of the efforts of using $scCO_2$ in combination with an organic solvent instead of supercritical water to convert biomass to high-value chemicals.

Aqueous hydrogen peroxide is known to decompose to produce oxygen and water via various free-radical intermediates in the following manner [35, 36],

$$H_2O_2 + OH^- \leftrightarrow HO_2^- + H_2O$$

$$H_2O_2 + HO_2^- \rightarrow HO^\cdot + O_2^- + H_2O \qquad (1)$$

$$HO^\cdot + O_2^- \rightarrow OH^- + O_2$$

Hence, hydrogen peroxide can interact with lignin in two ways. Firstly, the perhydroxyl anions nucleophilically attack and eliminate lignin chromophores. Secondly, the radical species cause the oxidative degradation of phenolic structures via coupling and/or rearrangement reactions converting them to carboxylic acids [37]. It is hypothesized that conducting the lignin oxidation in high pressure by using $scCO_2$ can help stabilize the radical species by slowing the decomposition of H_2O_2 to O_2 and H_2O and consequently, induce oxidative changes in the lignin even without the use of acids or bases. The use of ethylene glycol, which dissolves lignin, hopes to show the effect of mass transfer during $scCO_2$ oxidation.

Experimental

Materials

The following chemicals were used without further purification: H_2O_2 (30%), ethylene glycol (99%), tetrahydrofuran (99.8%), ethanol (99%), acetic anhydride (99%), N,N-dimethylformamide (99%) and pyridine (99%) were obtained from Fisher Scientific Co. while carbon dioxide (Research Grade 4.8, 99.998%) and nitrogen (Prepurified Grade 4.8, 99.998%) were purchased from Airgas. Water was de-ionized and distilled. The Kraft Indulin Lignin (KIL) was isolated and purified using a method proposed by Argyropoulos et al. [38].

Lignin Oxidations in $scCO_2$ and CO_2-Expanded Liquid

Lignin oxidations were conducted in a Parr Reactor Assembly (Series 4590 with 4842 Controller) batchwise. In a typical experiment, a predetermined amount (typically ~200 mg) of KIL (for $scCO_2$ oxidations) or KIL in solution (for CO_2-expanded ethylene glycol (EG)) and H_2O_2 were placed in the reactor. The reactor was evacuated and charged with CO_2 from a syringe pump (ISCO, 260D). The reactor was then heated to the desired temperature with concomitant adjustment of pressure to the desired pressure setting. The reaction was kept at the desired conditions for an hour after which the system was depressurized to atmospheric conditions. The reactions were conducted at a temperature of 80 °C, a pressure range of 50-200 atm and an H_2O_2/KIL mole ratio of 2 to 20.

For $scCO_2$ oxidations, the reaction products were dried overnight in a vacuum oven at 40 °C. Meanwhile for CO_2-expanded EG oxidations, the products were dumped to an ample amount of acidic water (pH=2.0), inducing precipitation of the unreacted lignin and dissolution of the liquid solvent in water. The resulting suspension was centrifuged, where the supernatant was collected and the solid residue freeze-dried and oven-dried overnight in succession. The collected dried samples were weighed and subsequently subjected to gel permeation chromatography (GPC) and ^{31}P NMR spectroscopy analyses. In addition, larger scale reactions (~ 1g of KIL) were performed, after which the reaction mixture was subjected to a separation scheme to quantify the mass loss in the initial lignin.

Reactions were also conducted in pure water and pure EG under identical conditions to those performed in $scCO_2$ and CO_2-expanded EG (atmospheric pressure). The solubility behavior of the lignin and H_2O_2 in neat CO_2 and CO_2-expanded EG under conditions similar to those used in the oxidation reactions were observed directly using a windowed reactor cell. The cell built in-house was made of SS 316 equipped with two sapphire windows (Sapphire Engineering, dimensions).

Quantitative ^{31}P NMR Spectroscopy

Quantitative ^{31}P NMR spectra of KIL before and after the oxidation reaction were obtained using a Bruker 300 MHz spectrometer equipped with a Quad probe dedicated to ^{31}P, ^{13}C, ^{19}F and ^1H acquisition. Accurately weighed samples (~40 mg) were dissolved in 0.3 ml of N,N-dimethylformamide overnight, then 0.575 ml of deuterated pyridine/deuterated chloroform (1.6/1, V/V) was added to the solution. N-Hydroxynaphthalimide was used as the internal standard, the relaxation agent was Cr(III)acetylacetonate and the phosphitylation reagent was 2-chloro-4,4,5,5-tetramethyl-1,3,2-dioxaphospholane. N-Hydroxynaphthalimide (~24.0 mg) and Cr(III)acetylacetonate (~6.0 mg) were dissolved in a solution of pyridine/chloroform (1mL, 1.6/1). Next, 0.1 mL of each solution and 0.1 mL of phosphitylation reagent were added at room temperature to the lignin samples. The ^{31}P NMR spectra were obtained by using the inverse gated decoupling. An internal deuterium lock was acquired from the deuterium atom found in deuterated chloroform in the solvent. Due to the use of Cr(III)acetylacetonate as a relaxation agent, a time delay of only 5 seconds was adequate.

Gel Permeation Chromatography

Samples were acetylated using the method described by Koda et. al.[39] with minor modifications for GPC analyses. All GPC analyses were carried out on a GPC Waters system consisting of a U6K injector, a 510 pump, Ultrastyragel HR 5E and HR 1 Waters columns (connected in series), a UV 484 detector and a refractive index 410 detector, at ambient temperature using THF as eluent at a flow rate of 0.44 ml/min. The elution volume versus molecular weight calibration curve was constructed using a series of monodisperse polystyrene standards with molecular weights ranging from 800-186 x 10^4 g/mol. The collected data was processed using the Millenium^{32} GPC software (Waters).

Results and Discussion

Solubility Behavior of Lignin in scCO$_2$

Prior to conducting reaction experiments in the Parr reactor assembly (50 mL capacity), simulated reaction experiments were performed in a windowed reactor (5 mL capacity). The objective of this preliminary study was to determine by direct observation the qualitative solubility behavior of KIL in CO$_2$ and of the KIL dissolved in a liquid solvent at all pressure, temperature and H$_2$O$_2$ concentration ranges used in the reaction experiments. For the former, the results showed that at 80°C and pressure from 50 to 200 atm, 20 mg of KIL did not dissolve in 0.5-3.0 g of CO$_2$. Li and Kiran [31] previously measured quantitatively the solubility of different lignocellulosic materials in various SCFs and reported that kraft lignin (Indulin AT) has negligible solubility in CO$_2$ at 37°C and 276 atm. There were no other temperature and pressure conditions were reported by these authors.

In the case of the KIL in EG, the solubility of KIL in this solvent at room temperature and atmospheric pressure was determined to be 35 mg/ml using the method by Ni and Hu [40], however, a concentration of 13.5 mg/ml was found to maintain the solubility of KIL in EG (no precipitation observed) in the presence of scCO$_2$ over the entire range of pressure tested at 80°C. There is also an expansion of the solution after scCO$_2$ pressurization but this was not quantified. The impact of the contrasts in the solubility behavior became apparent when the reaction products were spectroscopically analyzed and are the subject of the following discussion.

Oxidations in scCO$_2$ and scCO$_2$-Expanded Ethylene Glycol

Table 1 provides a summary of the reaction conditions and parameters performed in this study including the percentage of recovered solids after precipitation and thorough drying procedures. The precipitation and drying procedure was determined to be accurate to ± 4.0 %. In general, the recovery of solids from the oxidation of KIL performed in neat scCO$_2$ was almost complete.

Table 1. Summary of oxidation conditions and the corresponding percentage recovery of solid residue.

Sample	Temperature (°C)	Pressure (atm)	H$_2$O$_2$ to KIL Mole Ratio	Recovery (%)
Oxidation of KIL in CO$_2$				
1	80°C	85	2:1	94.0
2	80°C	85	20:1	94.0
3	80°C	126	2:1	93.0
4	80°C	167	2:1	94.0
5	80°C	200	2:1	93.0
Oxidation of KIL in CO$_2$-Expanded Ethylene Glycol				
6	80°C	85	20:1	62.0
7	80°C	126	2:1	73.0
8	80°C	126	10:1	66.0
9	80°C	126	20:1	56.0
10	80°C	167	20:1	67.0
11	80°C	200	20:1	72.0
Oxidation of KIL in Water				
12	80°C	1	2:1	~
13	80°C	1	20:1	85.0
Oxidation of KIL in Ethylene Glycol				
14	80°C	1	2:1	~
15	80°C	1	20:1	81.0

The losses which are presumed to be low molecular weight compounds derived from the oxidative fragmentation account for about 6-7 %, and are independent of pressure. In the oxidation done in scCO$_2$-expanded liquid, the recovery was below 73 % and goes as low as 56 %. An experiment was performed to determine the reliability of the recovery method. 200 mg of KIL (not treated) was dissolved in the same amount of EG used in the SC reactions, and was recovered using the method previously described. The measured recovery was 6 % less than the initial amount.

Figure 1 depicts the change in the percentage recovery as a function of H_2O_2 concentration and pressure. It appears that the percentage recovery consistently decreased with an increase in H_2O_2 concentration (which is not unexpected), since this indicates that an increase in the amount of H_2O_2 leads to increased oxidation. Meanwhile, the recovery decreases as pressure increases to a minimum. Further pressure increases leads to increase in recovery although it is still much lower than the initial amount.

^{31}P NMR Spectroscopy

The functional groups distributions of the initial and the recovered oxidized KIL were probed using quantitative ^{31}P NMR spectroscopy. This form of spectroscopy monitors the changes in the aliphatic, phenolic (both condensed and non-condensed) and carboxylic OH functionalities in the lignin. The data at atmospheric pressure and zero H_2O_2 concentration in the plots refer to KIL before oxidation. Figure 2 provides the plots for the three OH functionalities as a function of H_2O_2 concentration for the oxidized KIL in neat CO_2. The data obtained show that there is slight decrease in both the total aliphatics and total phenolics and substantial increase in the total carboxylics with increase in H_2O_2 concentration. These data indicate oxidation of side chain aldehyde and alcohol moieties leading to decreased aliphatic OH functionalities and increased total carboxylics. The slight decrease in phenolics suggests the ring opening of some aromatics generating carboxylic compounds and this, along with the side chain oxidations, may explain the dramatic enhancement in the total carboxylics.

Similar changes in OH functionalities were also observed with the increase in reaction pressure (at constant temperature of 80°C and 2:1 H_2O_2 concentration) except that the slight decrease in the total phenolics and aliphatics are independent of pressure while for the total carboxylics (where the maximum enhancement was at least three times) decreased monotonically with pressure.

The changes in the OH structures of lignin were more dramatic when KIL oxidation was performed in CO_2-expanded liquid as shown in Figures 3 and 4. At all conditions, the total aliphatics increased almost twice and is independent of pressure and H_2O_2 concentration. This was unexpected and there are two possible explanations for this phenomenon. This dramatic increase in the total aliphatic OH may be due to the contribution of the aliphatic liquid solvent (alcohol) that reacted with lignin or due to unwashed liquid solvent adhering to the solid residue. The reaction of alcohol with lignin in acidic conditions is possible via the benzylic (α) carbon via a substitution reaction [41]. However, results from the KIL oxidation using the liquid solvent at atmospheric pressure showed the same magnitude of aliphatic increase with negligible changes in the carboxylic and phenolic moieties, hence, it is likely that the main cause is the latter.

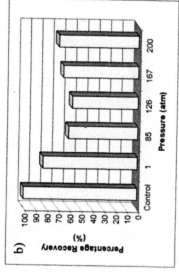

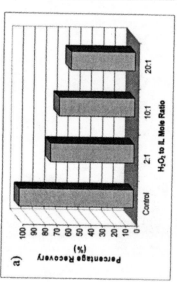

Figure 1. Percentage recovery of the reaction mixture after oxidation in CO_2-expanded ethylene glycol as: a) a function of H_2O_2 concentration at $80^\circ C$ and 126 atm; b) a function of pressure at $80^\circ C$ and $20{:}1$ H_2O_2 to IL mole ratio.

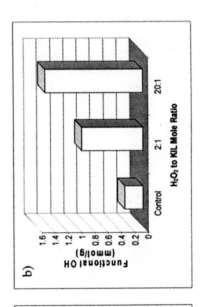

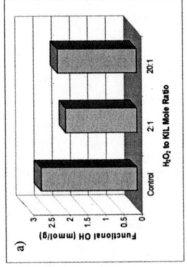

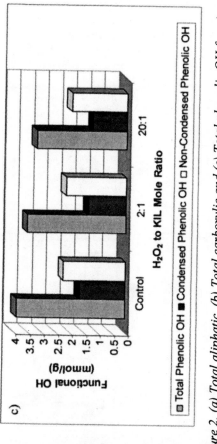

Figure 2. (a) Total aliphatic, (b) Total carboxylic and (c) Total phenolic OH functionality of KIL oxidized in neat CO₂ at 80°C and 85 atm as a function of H₂O₂ concentration.

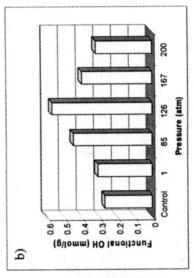

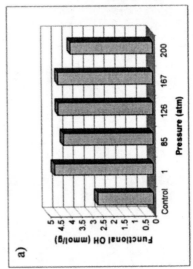

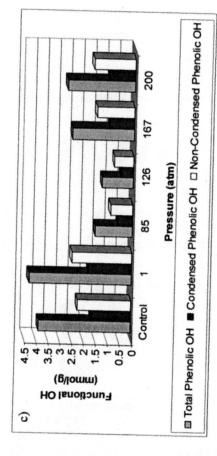

Figure 3. (a) Total aliphatic, (b) Total carboxylic and (c) Total phenolic OH functionalities of KIL oxidized in CO_2-expanded ethylene glycol at $80^\circ C$ and $20:1$ H_2O_2 to KIL mole ratio as a function of pressure. Note that at $P=1$, the oxidation was done in ethylene glycol at atmospheric pressure.

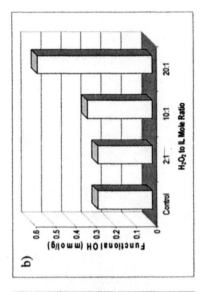

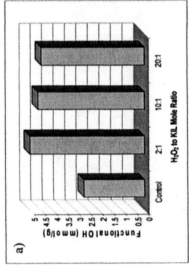

325

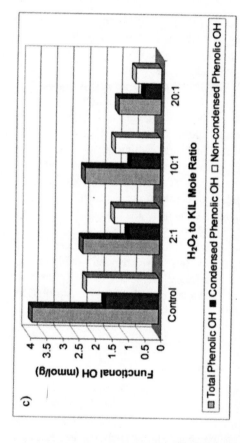

Figure 4. (a) Total aliphatic, (b) Total carboxylic and (c) Total phenolic OH functionality of KIL oxidized in CO₂-expanded liquid at 80°C and 85 atm as a function of H₂O₂ concentration.

The oxidative changes in the total carboxylics and total phenolics with pressure (see Figure 3) are noteworthy. The application of CO_2 pressure decreased markedly the total phenolics, suggesting an intensified aromatic ring opening reactions in the CO_2-expanded liquid oxidation due to the elimination of mass transfer limitations. The total carboxylics also increased, as would be expected, with the marked decrease in the phenolics. But it is not clear whether side chain oxidations also contributed to this enhancement. The increase in carboxylics goes to a maximum while the total phenolics correspondingly goes to a minimum at a pressure of 126 atm.

The existence of a minimum and maximum value suggests that there appears to be two competing effects occurring during the reaction. The increase in pressure, which enhances oxidation, augments the solubility of CO_2 in the liquid solvent. But in doing so, the concentration and/or the solubility of the KIL in the liquid solvent is diminished. Since no precipitation was observed during the qualitative solubility study, it is envisaged that the further dilution of the KIL solution with the increase in pressure, was the predominant factor that undermined the oxidation of the phenolic moieties. It is interesting to note that the correlation of total phenolics versus pressure is similar to that of the corresponding correlation of percentage recovery versus pressure (see Figure 1b and Figure 3c). On the other hand, the increase in H_2O_2 concentration consistently decreased the total phenolics and increased the total carboxylics, as expected, although the magnitude of oxidative changes at 2:1 and 20:1 H_2O_2 to KIL mole ratio were similar.

The oxidation of KIL was also done in water alone as well as in EG alone at the same conditions as those done in neat CO_2 and CO_2-expanded liquid (except that the pressure was atmospheric) to understand the effect of CO_2. In both instances, the changes in the lignin structures were not substantial except that in ethylene glycol (see Fig. 3 at P=0*) the aliphatic OH increased at least twice.

Molecular Weight Distribution

The molecular weight distributions of the oxidized KIL were measured by gel permeation chromatography to complement the findings from ^{31}P NMR spectroscopy. The average molecular weight as a function of pressure for samples oxidized in neat CO_2 and CO_2-expanded liquid is shown in Figure 5. The average molecular weight decreased as a result of hydrogen peroxide oxidation in neat CO_2, while it increased in CO_2-expanded liquid. The decrease buttresses the oxidative changes observed, such as the slight decreases in the aliphatic and phenolic structures as well as the remarkable increase in the carboxylic structures. The magnitude of the decrease in the molecular weight, though, indicates that the fragmentation did not result in the substantial formation of low molecular weight compounds. This is further evidenced in the

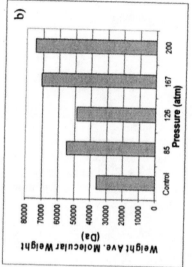

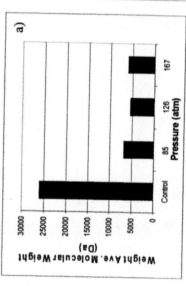

Figure 5. The average molecular weight of oxidized lignin as a function of pressure. a) Oxidation in neat CO_2; b) Oxidation in CO_2-expanded ethylene glycol.

measured high lignin yields (see Table 1). The rather surprising result is the increase in the molecular weight after oxidation in CO_2-expanded liquid, considering that more intensified oxidative changes were seen in CO_2-expanded media from ^{31}P NMR spectroscopy. This may have been due to the concentration of higher molecular weight lignin molecules after the separation of up to 40% fragmented lignin structures. It is interesting to note though, that the trend in the increase of the average molecular weight with pressure is consistent with the trend in the oxidative changes of the lignin structures and yield.

Oxidative Degradation Mechanism

An oxidative degradation mechanism under $scCO_2$ with H_2O_2 is suggested, taking into account the reactivity of a 5,5'-dimer (Scheme 1).

Primarily, it is assumed that the oxidative degradation is initiated by the hydroxyl radical reactions of hydrogen peroxide:

$$H_2O_2 \rightarrow HO-OH \rightarrow 2OH^{\cdot} \qquad (2)$$

The hydroxyl radicals are capable of abstracting a proton from the phenolic hydroxyl of the lignin, creating resonance-stabilized free radicals [i and ii], and this mechanism is analogous to the formation of diphenol monoradicals as shown using a 5,5'-dimer.

Moreover, there is evidence for the direct reduction of hydrogen peroxide [i] when the pH is less than 6 and its reduction is preceded by its decomposition (Eq. 3):

$$H_2O_2 \rightarrow 2H_2O + 2H^+ + 2e^- \qquad (3)$$

Alternatively, the possibility of generating the sufficiently stable superoxide ion, $O_2^{\cdot-}$, in the aprotic solvent system such as supercritical carbon dioxide, is quite high. The superoxide radical anion $O_2^{\cdot-}$ can be involved in a wide range of reactions, such as nucleophilic attack, reduction and oxidation reactions. Sawyer and co-workers [i-iii] have demonstrated that the superoxide radical anion $O_2^{\cdot-}$ can form from dissolved oxygen gas in aprotic solvents (Eq. 4):

$$O_2 + e^- \rightarrow O_2^{\cdot-} \qquad (4)$$

It is thus possible that in these oxidative degradation reactions, oxygen released from the decomposition of hydrogen peroxide results in the generation of superoxide ions $O_2^{\cdot-}$. This anion is a strong nucleophile known to disproportionate in water, producing oxygen, hydroperoxide and hydroxide ions [ii] (Eq. 5):

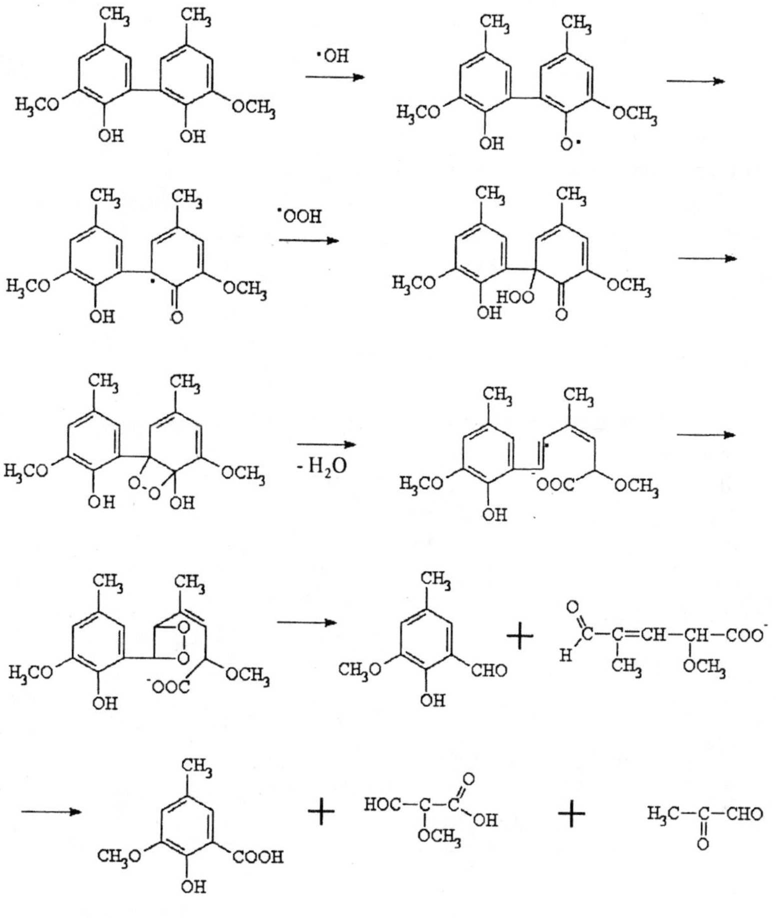

Scheme 1. The possible oxidative mechanism for 5,5'–dimer compound in the presence of supercritical carbon dioxide.

$$2O_2^- + H_2O \rightarrow O_2 + HOO^- + HO^{\cdot} \qquad (5)$$

Consequently, a mechanism that could account for the oxidation of the lignin in the presence of supercritical carbon dioxide is shown in Scheme 1 and is represented once again by the 5,5'–dimer. The preliminary results from the analysis of the separated low molecular weight portion of the reaction mixture is consistent with this mechanism. The functional structures are mostly aliphatics and carboxylics with minimal amount of phenolics.

Conclusions

This study has shown that it is possible to oxidize recalcitrant kraft lignin in scCO$_2$ in the absence of alkali. This highly valuable piece of scientific information when coupled with the multitude of attractive features and recent advances in supercritical technologies, offers some very exciting prospects. Namely, the development of new, environmentally benign, fragmentation technologies that do not require alkali. This will add value to lignin streams which are currently underutilized or in need of new markets.

References

1. Dorrestijn, E., Laarhoven, L. J. J., Arends, I. W. C. E., Mulder, P. *J. Anal. Appl. Pyrolysis*, **2000**, *54*, p 153.
2. Freudenberg, K. *Angew. Chem.*, **1939**, *52*, p 362.
3. *The Electrochemistry of Biomass and Derived Material;* Chum, H.L. and Baaizer, M.M. Eds. ACS Monograph: Washington DC, 1985.
4. Janson, J. and Fullerton, T. *Holzforschung*, **1987**, *41(6)*, p 359.
5. Bertaud, F., Croue, J.P. and Legube, B. *Ozone: Sci. Eng.*, **2001**, *23(2)*, p 139.
6. Pardini, V.L., Vargas, R.R., Viertler, H. and Utley, J.H.P. *Tetrahedron*, **1992**, *48(35)*, p 7221.
7. Crestini, C. and Argyropoulos, D. in, *Oxidative Delignification Chemistry: Fundamentals and Catalysis.* ACS Symposium Series no. 785, Argyropoulos, D. S., Ed.; Washington, ACS Books, 2001.
8. Sun, Y.J., Fenster, M., Yu, A., Berry, R. M., and Argyropoulos, D. S. *Canadian Journal of Chemistry* , **1999**, *77(5/6)*, p 667.
9. Drago, R.S., Corden, B.B. and Barnes, C.W. *J. Am. Chem. Soc.* , **1986**, *108*, p 2453.
10. Bozell, J.J., Hames, B.R. and Dimmel, D.R. *J. Org. Chem.* , 1995, 60, 2398.
11. Elder, T. and Bozell J.J. *Holzforschung*, **1996**, *50 (1)*, p 24.
12. Haikarainen, A., Kervinen, K., Sipila, J.,Repo, T., Brunow, G. and Leskala, M. - *Proceeding of International Symposium on Wood and Pulping Chemistry*, Nice France, 2001, p 231.
13. Hawthorne, S. B. *Anal. Chem.*, **1990**, *62*, p 633a.
14. McDaniel, L., Khorassani, M. and Taylor, L. *Journal of Supercritical Fluids*, **2001**, *19*, p 275.
15. Stahl, E., Schilz, W., Schutz, E., Willing, E. *Angew. Chem., Int. Ed. Engl.*, **1978**, *17*, p 731.
16. Hubert, P., Vitzthum, O.G. *Angew. Chem., Int. Ed. Engl.*, **1978**, *17*, p 710.
17. Amer, G.I. *US Patent* 4422966, Dec 27, 1983.
18. Froment, H. A. *U.S. Patent* 4 308 200, Dec 28, 1981.
19. McDonald, E.C., Howard, J., Bennett, B. *Fluid Phase Equilib.*, **1983**, *10*, p 337.
20. Avedesian, M.M. *US Patent* 4493797, Jan 1985.
21. Calimli, A.; Olcay, A. *Holzforschung*, **1978**, *32 (1)*, p 72.

22. Labrecque, R.; Kallaguine, S.; Grandmaison, J.L. *Ind Eng Chem Prod Res Dev* **1984**, *23*, 177.
23. Calimli, A.; Olcay, A. *Sep. Sci. Technol.* **1982**.
24. Demirbas A., *Energy Sources*, **2004**, *26(12)*, p 1095.
25. Koll, P.; Bronstrup, A.; Metzger, J.V. *Holzforschung*, **1979**, *33*, p 112.
26. Antal, M.J.; Allen, S.G.; Schulman, D.; Xu, X. *Ind Eng Chem Res*, **2000**, *39*, p 4040.
27. Yoshida, T.; Matsumura, Y. *Ind Eng Chem Res*, **2001**, *40*, p 5469.
28. Sasaki, M.; Kabyemala, B.M.; Malaluan, R.M.; Hirose, S.; Takeda, N.; Adschiri, T.; Arai, K. *J. Supercrit. Fluids*, **1998**, *13*, p 280.
29. Matsumura, Y.; Minowa, T.; Potic, B.; Kersten, S.R.A.; Prins, W.; van Swaaij, W.P.M.; van de Beld, B.; Elliott, D.C.; Neuenschwander, G.G. Kruse, A.; Antal, M.J. *Biomass and Bioenergy*, **2005**, *29*, p 269.
30. Saisu, M.; Takafumi, S.; Watanabe, M.; Adschiri, T.; Arai, K. *Energy & Fuels*, **2003**, *17*, p 922.
31. Li, L.; Kiran, E., *Ind Eng Chem Res*, **1988**, *27*, p 1301.
32. Modell, M. in *Fundamentals of Thermochemical Biomass Conversion*, R.P Overend, T.A. Milne and L.R. Mudge Eds., Elsevier Applied Science, 1985, p 95.
33. Lawson, J.R.; Klein, M.T. *Ind Eng Chem Fundam*, **1985**, *24*, p 203.
34. Wei, M.; Musie, G.; Busch, D.; Subramanian, B. *J.Am.Chem. Soc*, **2002**, p 124.
35. Sun, Y.; Fenster, M.; Yu, A.; Berry, R.M.; Argyropoulos, D.S. *Can J Chem*, **1999**, *77*, p 667.
36. Francis, R.C.; *TAPPI Press Anthol.* 1989, p 418.
37. Gellerstedt, G.; Pettersson, I. J. *Wood Chem Technol.*, **1982**, *2(3)*, p 231.
38. Argyropoulos, D., Sun Y. and Paluš, E. *Pulp and Paper Report*, PPR 1352.
39. Koda, K., Gaspar, A.R., Yu, L., Argyropoulos, D.S. *Holzforschung*, **2005**, *59*, p 612.
40. Ni, Y.; Hu, Q. *J. Appl. Polym. Sci.*, **1995**, *57(12)*, p 1441.
41. Allan, G.G. in *Lignins: Occurrence, formation, structure and reactions* Sarkanen K.V., Ludwig, C.H. Eds., Wiley-Interscience, New York,1971, p 511.
42. *Lignin Structure and Reactions*, Marton, J. Ed., American Chemical Society, 1966.
43. Argyropoulos, D.S. and Menachem S.B. *Advances in Biochemcial Engineering Biotechnology*, **1997**, *57*, p 127.
44. Schiffrin, D.J. *The Royal Society of Chemistry*, **1983**, *8*, p 126.
45. Merritt, M.V. and Sawyer, D.T. *J. Org. Chem.*, **1970**, *35(7)*, p 2157.
46. Sugimoto, H., Matsumoto, S. and Sawyer, D.T. *Environ. Sci. Technol.*, **1988**, *22(10)*, p 1182.
47. Sawyer, D.T. and Roberts, J.L. *U.S. Patent* 4:410, 1983.
48. AlNashef, I.M., Leonard, M.L., Kittle, M.C., Matthews, M.A. and Weidner, J.W. *Electrochemical Solid-State Letters*, **2001**, *4(11)*, D16.

Chapter 21

New Stereoselective Functionalization of Cellulose-Derived Pyrolysis Derivatives: Levoglucosenone and Its Dimer

Zbigniew J. Witczak

Department of Pharmaceutical Sciences, School of Pharmacy, Wilkes University, Wilkes-Barre, PA 18766

As a part of a study to develop methods to obtain high-value nonracemic chiral building blocks from waste cellulosic materials and biomass, a convenient modification method to pyrolyze these materials to produce levoglucosenone is presented in this chapter. Its functionalized dimer has been obtained by base catalyzed oligomerization of levoglucosenone. Both chiral bicyclic enones are convenient precursors for the synthesis of many attractive synthons and are generating steady interest due to their rigidity and stereoselective functionalization in many synthetic organic methodologies. Some of the examples of steroselective functionalization of these enones are presented in this chapter.

Historical Background

Practical methods for deriving economically useful chemicals from renewable biomass such as plant material, are highly desirable making possible the use biomass instead of petroleum as a source of chemicals and fuels. A major component of biomass is cellulose and the major products from the pyrolysis of cellulose are small saturated chiral molecules. Since biomass is chiral, these saturated molecules could also be chiral and could be useful chiral synthetic building blocks called chirons. Levoglucosenone *(1)*, is a classical representative of this class and with unique rigidity with several important functional groups including a ketone group double bond conjugated with ketone, a protected aldehyde and two protected hydroxyl groups by 1,6-anhydro ring. The enone an attractive chiral carbohydrate building block, is produced by the pyrolysis of cellulose composed materials. Despite the disadvantages of its low yield and the amount of solid cellulosic material necessary for pyrolysis, the efficiency and the economy of the pyrolysis process makes it an effective method. In addition, pyrolysis reduces the amount of waste cellulosic materials, which is clearly beneficial to the environment.

Although levoglucosenone has been known for more than 30 years *(2)*, it continues to have only limited applications in organic synthesis. This can be attributed to the rather conservative opinion regarding the process, purification and stability, etc.

This simple and small bicyclic enone molecule is an important and efficient chiral starting material for the synthesis of many analogs of complex natural products (Fig. 1). Despite the efforts of various laboratories *(1,3-6)* to promote the chemistry of levoglucosenone, the interest of chemical and pharmaceutical industries in this chemical remains low. We hope continuous promotion of this remarkable molecule will make levoglucosenone a commodity product, a status that should have been granted to this molecule long ago. Thus, the goal of this chapter is to put levoglucosenone and its functionalized analogs on the map as a valuable chiral building block to the synthesis of value added products.

Structural Studies and Physicochemical Properties

This small molecule (M.W. 126), with remarkable potential applications in synthetic organic chemistry, first attracted the attention of the chemists in the early 1970s. Since then, detailed structural studies have been published *(1-5)* and the revision of the previously published data for the MS and ^{1}H NMR spectra by Broido and coworkers *(6-7)* clearly established its structure as 1,6-anhydro-3,4-dideoxy-α-D-*glycero*-hex-3-enopyranose-2-ulose. These studies were based on the combined GC/MS using both electron impact (EI) and chemical ionization (CI) techniques and allowed for the determination of the correct molecular ion at m/z 126.

Among the first attempts to elucidate the structure of levoglucosenone through
^{1}H and ^{13}C NMR spectral analysis was with Broido data. The ^{13}C NMR chemical
shifts are shown in Scheme 1.

Relevant data on optical rotatory dispersion and circular dichroism were
reported by Ohnishi *(8)* whereas Domburg *(9,10)* reported conformational and
structural studies. Halpern and Hoppech conducted a detailed NMR study of the
levoglucosenone and its functionalized derivatives *(11)* and 1,4 adducts.

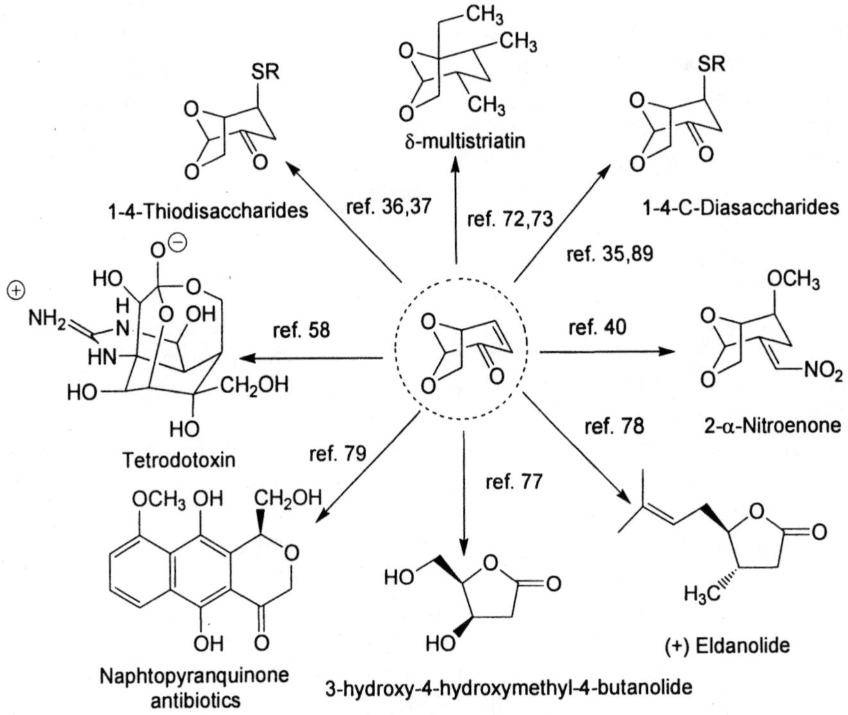

Figure 1.

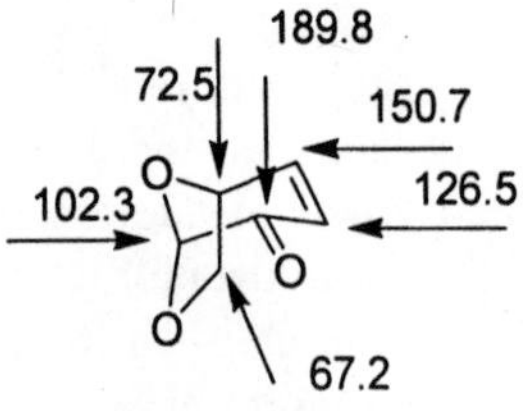

Scheme 1.

Yamada and Matsumoto *(12,13)* reported the photochemical α-cleavage of levoglucosenone and pointed out the general pathway of photolysis and its application to the synthesis of intermediates as convenient chiral building blocks.

Mechanism of the Formation of Levoglucosenone

A mechanism for acid catalysed thermal decomposition of cellulose *(15-26)* is a 1,2-hydride shift from the C-3 to the carbenium center at C-2 with the formation of a more stable hydroxycarbenium ion.

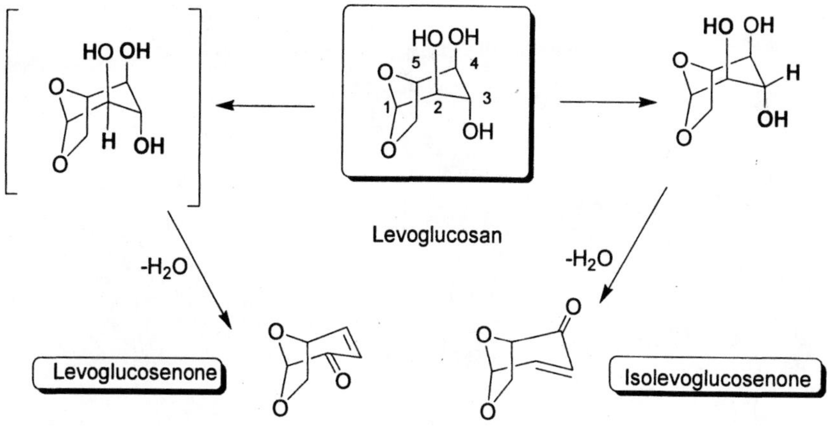

Scheme 2.

The intermediate levoglucosan is formed first *via* formation of a 1,6-anhydro ring. An alternative 1,2-hydride shift leading to a hydroxycarbenium ion at C-4 does not occur since the corresponding levoglucosenone isomer known as isolevoglucosenone was not found in the pyrolysate (Scheme 2).

The hypothetical mechanism of the formation of levoglucosenone via three alternative routes is depicted in Scheme 3 *(21)*.

The new modified method produces levoglucosenone in a relatively pure form (ca. 85%) and can be further purified by fractional distillation to give a pure 98% fraction. In this procedure, waste cellulosic materials such as paper and fibers from paper mill waste were pretreated with methanolic solution of phosphoric acid (1.5 wt %) and pyrolyzed by fast pyrolysis under reduced pressure (30-40mm Hg) in the Kugelrohr apparatus. The yellow distillate containing water and levoglucosenone was quickly neutralized with solid sodium bicarbonate (with very small portions) to a pH of 7.0, and extracted with methylene chloride. The extract was dried over anhydrous magnesium sulfate and evaporated to an oily syrup. The yield was ca. 8-10% based on the weight of cellulosic material and the purity of levoglucosenone was ca. 85%.

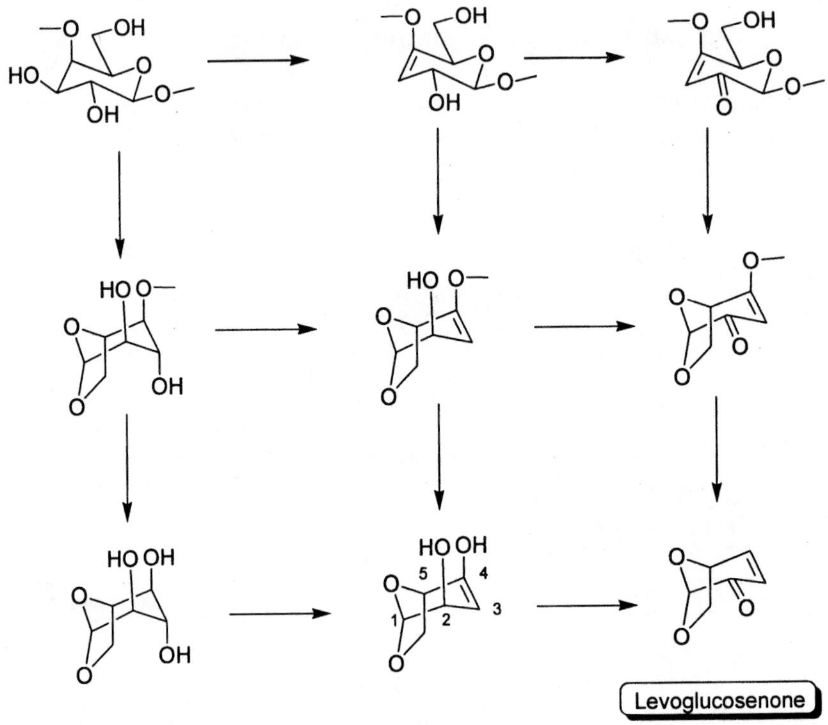

Scheme 3.

Synthesis of Levoglucosenone

The traditional method of cellulose pyrolysis for the production of levoglucosenone is still a viable and economically feasible procedure, however, the synthetic methods utilizing various and cheap starting carbohydrate precursors are highly competitive and cost-effective alternatives.

Koll and coworkers *(27)* reported the first practical synthesis of levoglucosenone as part of a study on the utilization of 1,6-anhydrosugars in the synthesis of convenient derivatives of 1,6-anhydrosugars (Scheme 4). The key step proceeds *via* rearrangement of one of the Cerny epoxides to the intermediate allyl alcohol which upon oxidation, produces levoglucosenone. The most efficient approach uses a cheap 1,6-anhydrogalactopyranose precursor through functionalization with thiocarbonyldiimidazole (TCDI) followed by desulfurization to the previously prepared allyl alcohol, which upon oxidation with manganese dioxide produces levoglucosenone in moderate 46% yield.

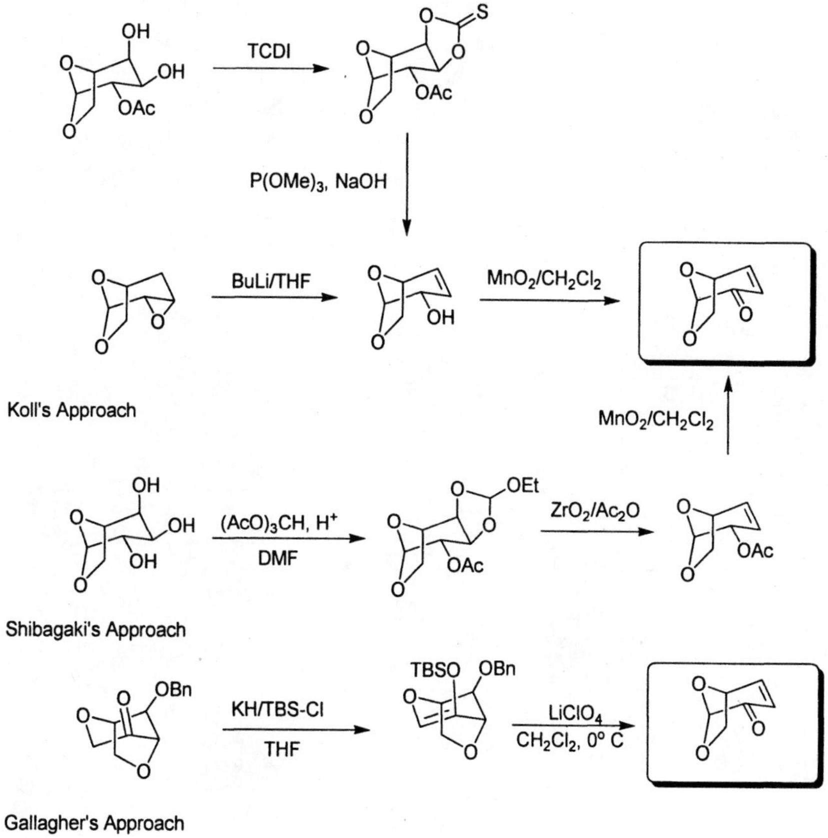

Koll's Approach

Shibagaki's Approach

Gallagher's Approach

Scheme 4.

Shibagaki and coworkers *(28)* reported an alternative practical and efficient approach to the synthesis of levoglucosenone utilizing a galactose derivative as a starting material. This route (Scheme 4) utilizes oxidative decarboxylation of 2,3-functionalized orthoesters with zirconium dioxide as a critically important key step.

In Gallagher's herbicidin synthesis, *(29)* an interesting and serendipitous discovery presented a possible alternative route to levoglucosenone. A bicyclic ketone (prepared from 1,5-anhydro-D-mannitol) underwent efficient silylation and gave the silyl enol ether, which when treated with Lewis acid promoters (TiCl$_4$, ZnBr$_2$, TMSOSO$_2$ CF$_3$ or LiClO$_4$) gave levoglucosenone. In the examination of the crude reaction mixtures by ^{1}H-NMR, it was evident that in all cases extensive rearrangement of silyl enol ether had always taken place. The researchers were able to isolate and characterize levoglucosenone as the major product.

A cleaner process was observed when the silyl enol ether was treated with LiClO$_4$ in the absence of an additional electrophilic component, and levoglucosenone was isolated in moderate 40% yield (Scheme 4).

Earlier efforts *(30)* pioneered the synthesisis of the (+)-enantiomer of levoglucosenone and its new 5-hydroxymethyl analog, starting from the known precursor, 5-hydroxymethyl-1,6-anhydro-α-*altro*-hexopyranose as depicted in Scheme 5.

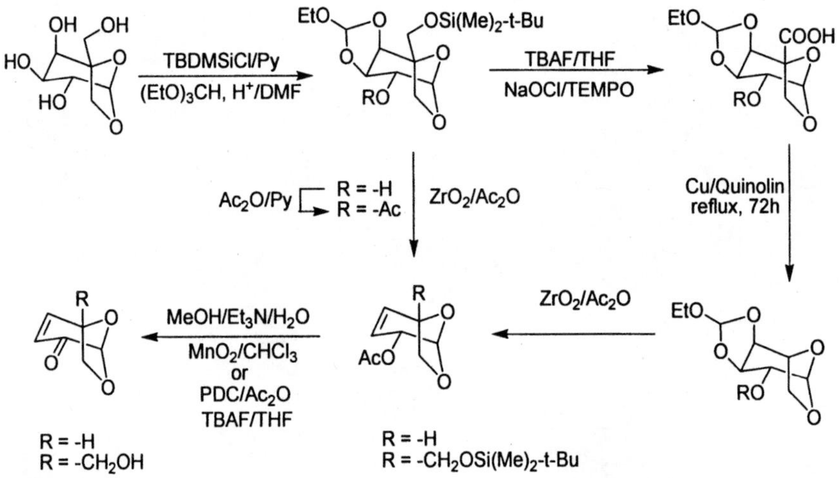

Scheme 5.

Functionalization of Levoglucosenone: Reactivity of the Conjugated System

The high chemical reactivity of the conjugated system in levoglucosenone is an excellent reason to further develop new synthetic approaches for the synthesis of a variety of natural products targets that require stereoselective coupling with the sugar unit. As levoglucosenone is by far the most prominent carbohydrate molecule used in conjugate addition reactions, some of its tandem reactions involving the initial conjugate addition will be discussed in separate sections.

Shafizadeh's laboratory *(31-34)* reported the first successful studies of thiol addition. In this research *(35-37)* and in other research *(38)* on the Michael addition reaction of reactive thiols to levoglucosenone and its isomeric isolevoglucosenone *(39)*, they clearly indicate the complete stereoselectivity of the addition and its general synthetic approach to 1,4- and 1,2-thiodisaccharides *(40)*.

These stereoselective, one-step synthetic approaches to (1,4)-3-deoxy-thiodisacharides *(36,37)* and (1,2)-3-deoxy-thiodisaccharides *(39,40)* are classical examples of exploiting the excellent functionality of both levo- and isolevoglucosenone enones.

Reactivity of the Keto Function

The reactivity of the carbonyl group in the conjugated system of levoglucosenone is well defined and it has been shown that the carbonyl group and the double bond can be reduced selectively *(19,33)*.

One of the important functionalizations of the keto function is the epoxidation using the Corey reagent (dimethylsulfoniumethylide in DMSO and THF) as reported by Gelas and Gelas *(41,42)*.

Many other laboratories *(43-104)*, through significant contributions to the utilization of the chemistry of levoglucosenone stimulated the field and increased attempts to make clear impact on the potential of levoglucosenone and functionalized analogs as value added derivatives.

One of the classical examples of the utilization of levoglucosenone and exploiting its universal functionalization approaches is the synthesis of natural product δ-multistriatin. The improved synthetic procedure for the synthesis of δ-multistriatin *(72)* was reported from our laboratory *(73)* utilizing previously developed Paton protocol for functionalization of levoglucosenone with nitromethane *(74,75)*. This approach of simultaneous double 1,4- and 1,2-addition reaction of nitromethane to the conjugated system and keto function at C-2 prompted us to develop new synthetic methods to functionalized nitroenones from levoglucosenone.

These specific approaches of additional modification of saturated levoglucosenone derivatives were achieved through addition of nitromethane to the carbonyl group at C-2. The subsequent mesylation of the geminal secondary hydroxyl group followed by *in situ* elimination under basic conditions produces highly valuable nitroenones *(40)* Scheme 6. Both nitroalkenes exist as E/Z (1:1) isomeric mixture as detected by UV and H-NMR.

The conjugated system of the C-2 nitroalkenes is highly reactive toward strong nucleophiles and it is an excellent Michael reaction acceptor of reactive 1-

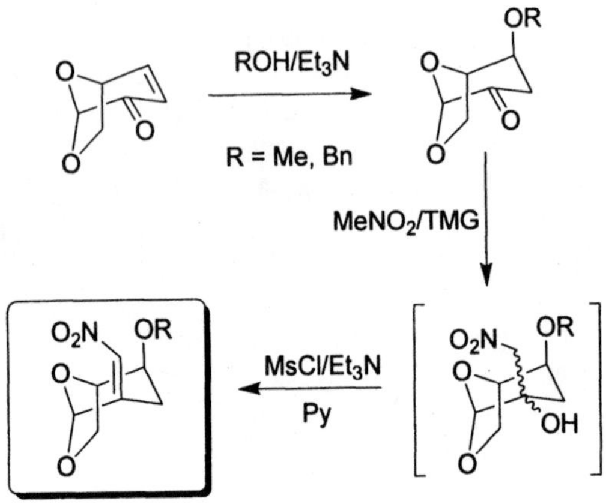

Scheme 6.

thiosugars nucleophiles as reported by us *(40)* in the synthesis of (1-2)-2,3,-dideoxy-2-C-nitromethyl-thiodisaccharides in 63-70% yield.

Dimerization of Levoglucosenone and Functionalization of its Dimer

Levoglucosenone's conjugated system undergoes base catalysed oligomerization *(46-52)* with the formation of variety of complex products including a dimer, and two cyclic isomeric trimers such as alkenic and nonalkenic. The structures of both alkenic and nonalkenic cyclic trimers were confirmed by x-ray crystallography as reported by Shafizadeh and coworkers *(46-47)*.

The hypothetically possible isomeric dimer as depicted in Scheme 7 was not isolated. The multicomponent mixture can be effectively separated by column chromatography into single pure products, and the dimer constitutes the predominant derivative isolated from the mixture (Scheme 7). The complex stereochemistry of the molecule has been examined by H-NMR is in agreement with original data as reported by Shafizadeh et al *(24)*.

The original oligomerization method reported by Shafizadeh et.al. *(46-52)* was improved and modified in this laboratory to produce levogloucosenone dimer in relatively moderate c.a. 38% yield. This functionalized molecule possesses a quite unique chemical character of double functional groups which provide excellent reactivity toward strong nucleophiles. The particularly reactive character of the dimer molecule offers possibilities for further functionalization into valuable amino or thio derivatives. Such compounds could have potential biological activity.

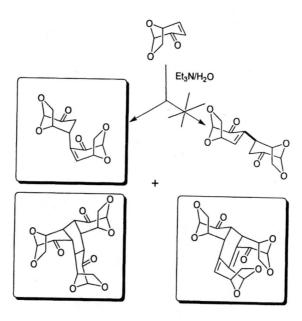

Scheme 7.

The previously reported data on reactive nucleophile additions to the conjugated system of levoglucosenone *(36-37)* prompted the investigation of the reactivity of the enone system of the dimer. The conventional Michael addition of reactive thiols such as 1-thiosugars proceeded smoothly with the formation of β-(1-4)-2,3,-dideoxythioadducts (Scheme 8).

The stereoselectivity of the conjugate addition and the correct stereochemistry of the adducts were proved by measuring the coupling constants of $J_{3e,5} = 1.5$ Hz. This coupling indicates that substituent at C-4 is axial with *quasi -equatorial* relationship between H-3e and H-5. Additionally, lack of coupling between H-4 and H-5, and coupling constants $J_{3e,4} = 3.18$ Hz indicate the axial disposition of the new substituent at C-4. Accessible in a single step process from dimer and formed in a completely sterospecific manner, thiodisaccharide is a potentially useful template for the synthesis of aminosugars having three chiral centers in the D-*ribo* configuration.

The functionalization of a C-2 keto group of levoglucosenone, as reported *(40)* earlier (Scheme 6), also prompted further expansion of the methodology to the functionalization of saturated analog of levoglucosenone dimer. This saturated analog of dimer is routinely produced in our laboratory *via* conventional hydrogenation of the double bond.

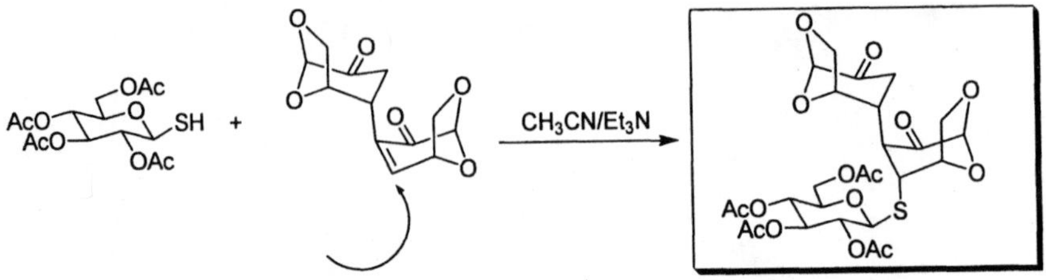

Scheme 8.

The base catalyzed addition of nitromethane to both C-2 keto functions and subsequent mesylation of the geminal secondary hydroxyls groups followed by *in situ* elimination under basic conditions produces valuable double functionalized nitroenone (Scheme 9).

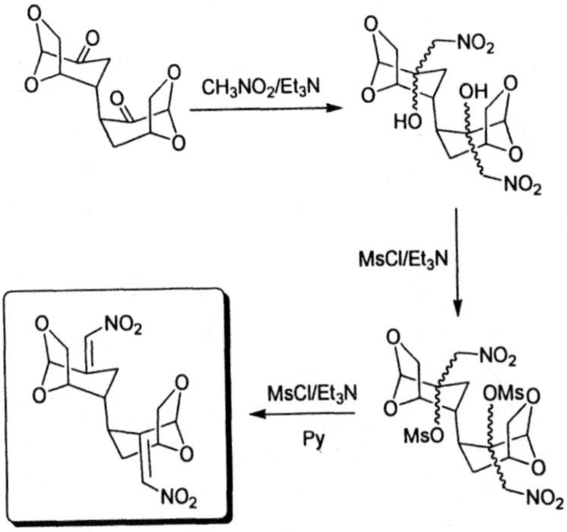

Scheme 9.

The reactivity of the double functionalized nitroenone of the levoglucosene dimer was tested in the reaction with 1-thiosugars *via* conventional Michael addition reaction catalyzed by triethylamine. The stereoselctive 1,2 addition proceeds *via* formation of an S-linkage from a less-hindered face of the molecule and at both nitroalkene functional groups of enone dimer. As expected, the shielding effect of both 1,6-anhydro rings prevents the formation of the 2-equatorial products, yielding only 2-axial products with new quaternary centers at C-2 of each rings. This provides a stable molecule, as no epimerization or β-elimination was observed during the reduction of the nitro group.

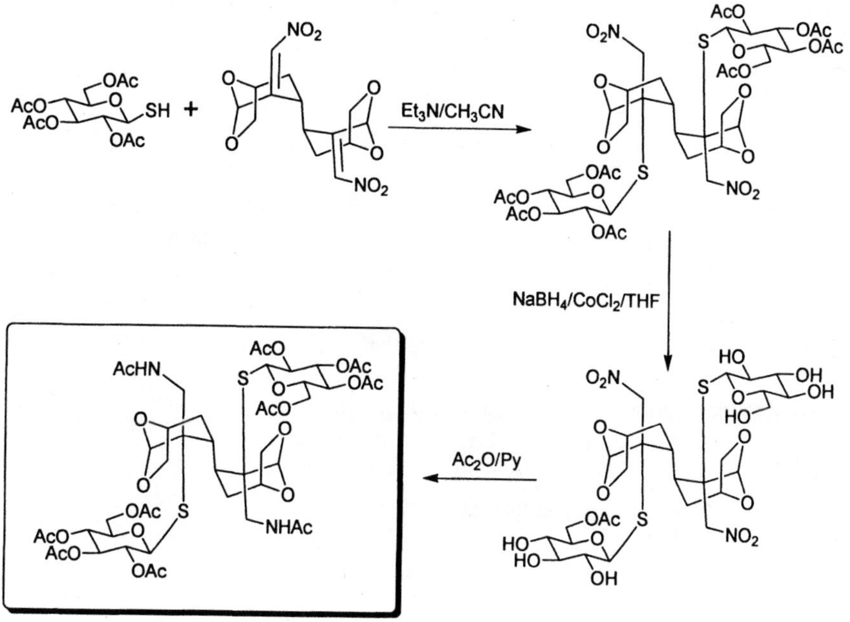

Scheme 10.

All the above observations clearly indicate the preferred stereochemistry of the double functionalized adducts. The most conventional way to prove the correct stereochemistry of the adducts is by measuring the coupling constants between H-3a and the -CH$_2$- of the nitromethyl group at C-2, i.e. J$_{CH}$ =2.8-3.2 Hz. The magnitude of these coupling constants strongly supports the proposed arrangements with equatorial substituents at C-2. Additionally, a strong NOE effect is observed between the H at C-1 and one of the hydrogens on the nitromethylene groups at C-2 of each ring which further proves the correct sterochemistry at C-2. The 1H NMR spectra of these adducts show a lack of coupling between H-4 and H-5, indicating that the pyranose ring of the adducts are in a $^1C_4$ conformations and are slightly distorted due to the presence of an equatorially oriented both nitromethylene groups. The reduction of the nitro group at C-2 in each ring of the thiodisaccharides was efficiently carried out with sodium borohydride/cobalt chloride complex, followed by conventional acetylation.

This geminal type of functionality occurs when the sugar moiety is in specific stereo orientation, and with acetamido functionality. Additionally, the basic functional group (-NHAc) may act as a binding site with receptors. Such functionalized thiodisaccharides should be valuable tools to probe any enzyme inhibitory activity of this new class of valuable derivatives.

New Perspectives

Although developments in the chemistry of levoglucosenone which employ modern reagents as tools in its organic synthesis presented in this review will definitely publicize levoglucosenone's vital potential, there is a great need for further attention.. This may encourage more research in different directions. Moreover, the chiral functionality of levoglucosenone and its functionalized new synthons including its dimer will create a variety of possibilities for interdisciplinary approaches not only in pure synthetic organic chemistry but in polymer and combinatorial chemistry. The latter approach is ideally suited in functionalizing this molecule by creating a number of useful scaffolds. The most useful scaffolds would have modified functional groups such as $-NH_2$, -COOH, -SH, at C-2, C-3, C-4, and C-6.

Conclusions

Through a number of new developments and synthetic methods devoted to the subject during the last fifteen years, one can easily conclude that this fascinating topic is growing and will continue to grow.

Additionally, the critically important environmental issue of utilizing waste cellulosic material and waste biomass products should be considered as an alternative feedstocks for green chemistry application to the production of many value-added products based on levoglucosenone and its functionalized derivatives.

Despite the low level of interest from the pharmaceutical industry, levoglucosenone chemistry will be one of the frontiers in carbohydrate chemistry, especially in the area of small molecules and complex oligosaccharides of medicinal interest. The variety of methods for the stereoselective functionalization of this classical building block provides a quite large number of attractive and highly stereoselective approaches to various classes of optically active derivatives of particular interest including sulfur and nitrogen heterocycles as well as rare carbohydrates.

This rich selection of potential approaches, combined with further developments of new procedures and modern reagents, creates an enormous opportunity for the field to be at the frontier for many years to come.

References

1. For reviews see; *Levoglucosenone and Levoglucosans Chemistry and Applications* Witczak, Z. J. Ed. ATL Press Science Publishers; Mt. Prospect, IL **1994**; Witczak, Z. J. in *Studies in Natural Products Chemistry,* Atta-Ur-Rahman, Ed. Vol. 14, Elsevier Science Publishers, Amsterdam, **1993**, pp 267-282; Miftakhov, M. S.; Valeev, F. A.; Gaisina, I. N. *Uspekhi Khimi,* **1994**, *63*, 922; B. Becker, *J. Carb. Chem.* **2000**, *19*, 253.; Trahanovsky, W. S.; Ochoda, J.M.; Wang, C.; Chang S. 217th Am. Chem Soc. Meeting, **1999**, Abstract FUEL 35.: Witczak, Z.J. in *Chemicals and materials from Renewable Resources,* J.J. Bozell, Ed. Vol.784 ACS Symposium Series, Oxford University Press, Washington D.C. **2001**, pp 81-97.

2. Tsuchiya, Y.; Swami, K. *J. Appl. Polym. Sci.* **1970**, *14*, p 2003.

3. Lam, L. K. M.; Fungi, D. P. C.; Tsuchiya, Y.; Swami, K. *J. Appl. Polym. Sci.* **1970**, *17*, p 391.

4. Woodley, F. A. *J. Appl. Polym. Sci.* **1971**, *15*, p 835.

5. Lipska, A. E.; McCasland, G. E. *J. Appl. Polym. Sci.* **1971**, *15*, p 419.

6. Halpern, Y.; Riffer, R.; Broido, A. *J. Org. Chem.* **1973**, *38*, p 204.

7. Broido, A.; Evett, M.; Hodges, C. C. *Carbohydr. Res.* **1975**, *44*, p 267.

8. Ohnishi, A.; Takagi, E.; Kato, K. *Bull. Chem. Soc. Jpn.,* **1975**, *48*, p 1956.

9. Domburg, G.; Berzina, I.; Kupce, E.; Kirshbaum, I. Z. *Khim. Drev.* **1980**, p 99.

10. Domburg, G.; Berzina, I.; Kirshbaum, I. Z.; Gavars, M. *Khim. Drev.* **1978**, p 105.

11. Halpern, Y.; Hoppech, J. P. *J. Org. Chem.* **1985**, *50*, p 1556.

12. Yamada, S.; Matsumoto, M. *Chem. Lett.* **1992**, p 2273.

13. Matsumoto, M. *J. Chem. Soc. Chem. Commun.* **1984**, p 658.

14. Shafizadeh, F.; Fu, Y. L. *Carbohydr.Res.* **1973**, *29*, p 113.

15. Shafizadeh, F.; Chin, P. P. S. *Carbohydr.Res.* **1976**, *46*, p 149.

16. Shafizadeh, F.; Chin, P. P. S. *Carbohydr.Res.* **1977**, *58*, p 79.

17. Koll, P.; Metzger, P. *Angew. Chem.* **1978**, *90*, p 802.

18. Shafizadeh, F.; Furneaux, R. H.; Stevenson, T. T. *Carbohydr. Res.* **1979**, *71*, p 169.

19. Ward D. D.; Shafizadeh F. *Carbohydr. Res.* **1981**, *93*, p 284.

20. Furneaux, R. H.; Gainsford, G. T.; Shafizadeh, F.; Stevenson, T. T. *Carbohydr. Res.* **1989**, *146*, p 113.

21. Shafizadeh, F. *Pure Appl. Chem.,* **1983**, *55*, 705-720; Shafizadeh, F. *J. Anal. Appl. Pyrolysis,* **1982**, *3*, p 283.

22. Smith, C. Z.; Chum, H. L.; Utley, J. H. P. *J. Chem. Res. Synopses* **1987**, *3*, p 88.

346

23. Shafizadeh, F.; Ward, D. D.; Pang, D. *Carbohydr. Res.* **1982**, *102*, p 217.
24. Shafizadeh, F.; Furneaux, R. H.; Pang, D.; Stevenson,T. T. *Carbohydr. Res.* **1982**, *100*, p 303.
25. Furneaux, R. H.; Mason, J. M.; Miller, I. J. *J. Chem. Soc. Perkin Trans. 1* **1984**, p 1923.
26. Morin, C. *Tetrahedron Lett.* **1993**, p 5095.
27. Koll, P.; Schultek, T.; Rennecke, R. W. *Chem. Ber.* **1976**, *109*, p 337.
28. Shibagaki, M.; Takahashi, K.; Kuno, H.; Honda, I.; Matsushita, H. *Chem. Lett.* **1990**, p 307.
29. Griffin, A.; Newcombe, N. J.; Gallagher, T.; In *Levoglucosenone and Levoglucosans Chemistry and Applications* Witczak, Z. J. Ed. ATL Press Science Publishers; Mt Prospect, IL 1994; p 23.
30. Witczak, Z. J.; Mielguj, R. *Synlett* **1996**, p 108.
31. Ward, D. D.; Shafizadeh, F. *Carbohydr. Res.* **1981**, *95*, p 155.
32. Shafizadeh, F.; Essig, M. G.; Ward, D. D. *Carbohydr.Res.* **1983**, *114*, p 71.
33. Essig, M. G.; Shafizadeh, F. *Carbohydr. Res.* **1984**, *127*, p 235.
34. Essig, M. G. *Carbohydr. Res.* **1986**, *156*, p 225.
35. Witczak, Z. J. *Pure Appl. Chem.* **1995**, *66*, p 2189.
36. Witczak, Z. J.; Chhabra, R.; Chen, H.; Xie, Q. *Carbohydr. Res.* **1997**, *301*, p 167.
37. Witczak, Z. J.; Sun, J.; Mielguj, R. *Bioorg. Med. Chem. Lett.* **1995**, *5*, p 2169.
38. Becker, T.; Thimm, B.; Thiem, J. *J. Carbohydr. Chem.* **1996**, *15*, p 1179.
39. Witczak, Z. J.; Chen, H.; Kaplon, P. *Tetrahedron: Asymmetry*, **2000**, *11*, p 519.
40. Witczak, Z. J.; Chhabra, R.; Boryczewski, D. *J. Carbohydr. Chem.* **2000**, *19*, p 543.
41. Gelas-Mialhe, Y.; Gelas, J.; Avenel, D.; Brahmi, R.; Gillie-Pandraud, H. *Heterocycles* **1986**, *24*, p 931.
42. Gelas-Mialhe, Y.; Gelas, J. *Carbohydr.Res.* **1990**, *199*, p 243.
43. Bahte, P.; Horton, D. *Carbohydr. Res.* **1983**, *122*, p 189.
44. Bahte. P.; Horton, D. *Carbohydr. Res.* **1985**, *139*, p 191.
45. Blake, A .J.; Forsyth, A. C.; Paton, M. R. *J. Chem. Soc. Chem. Commun.* **1988**, p 440.
46. Stevenson, T. T.; Furneaux, R. H.; Pang, D.; Shafizadeh, F.; Jensen, L. Stenkamp, R. E. *Carbohydr. Res.* **1983**, *112*, p 179.
47. Stevenson, T. T.; Stenkamp, R. E.; Jensen, L. H.; Shafizadeh, F.; Furneaux. R. H. *Carbohydr. Res.* **1983**, *104*, p 11.
48. Stevenson, T. T.; Essig, M. G.; Shafizadeh, F.; Jensen, L. H.; Stenkamp, R. E. *Carbohydr. Res.* **1983**, *118*, p 261.
49. Essig, M. G.; Shafizadeh, F.; Cochran, T. G.; Stenkamp R. E. *Carbohydr. Res.* **1984**, *129*, p 55.

50. Essig, M. G.; Stevenson, T. T.; Shafizadeh, F.; Stenkamp, R. E.; Jensen, L. H. *J. Org. Chem.* **1984**, *49*, p 3652.

51. Dawson, I. M.; Johnson, T.; Paton, M. R.; Rennie, R.A.C. *J. Chem. Soc., Chem. Commun.* **1988**, p 1339.

52. Blake, A. J.; Dawson, I. M.; Forsyth, A. C.; Gould, R.O.; Paton, M. R.; Taylor, D. *J. Chem. Soc. Perkin Trans. 1* **1993**, p 75.

53. Blake, A. J.; Cook, T. A.; Forsyth, A. C.; Gould, R. O.; Paton, R. M. *Tetrahedron* **1992**, *48*, p 8053.

54. Blake, A. J.; Gould, R. O.; Paton. R. M.; Taylor, P. G. *J. Chem. Res. Synopses* **1993**, p 289.

55. Isobe, M.; Fukami, N.; Nishikawa,T.; Goto, T. *Heterocycles* **1987**, *25*, p 521.

56. Ward, D. D.; Shafizadeh, F. *Carbohydr. Res.* **1981**, *93*, p 287.

57. Bamba, M.; Nishikawa, T.; Isobe, M. *Tetrahedron Lett.* **1996**, *37*, p 8199.

58. Isobe, M.; Fukuda, Y.; Nishikawa, T.; Chabert, P.; Kawai, T.; Goto, T. *Tetrahedron Lett.* **1990**, p 3327.

59. Chew, S.; Ferrier, R. J. *J. Chem. Soc. Chem. Commun.* **1984**, p 911.

60. Chew, S.; Ferrier, R. J.; Sinnwell, V. *Carbohydr. Res.* **1988**, *174*, p 161.

61. Ferrier, R.; Furneaux, R. *Aust. J. Chem.* **1980**, *33*, p 1025.

62. Ferrier, R.; Tyler, P. C. *J. Chem. Soc. Perkin Trans.1* **1980**, p 2767.

63. Mori, M.; Chuman, T.; Kato, K.; Mori, K. *Tetrahedron Lett.* **1982**, p 4593.

64. Isobe, M.; Nishikawa, T.; Yamamoto, N.; Tsukiyama,T.; Ino, A.; Okita, T. *J. Het. Chem.* **1992**, *29*, p 619.

65. Isobe, M.; Ichikawa, Y.; Goto, T. *Tetrahedron Lett.* **1981**, p 4287.

66. Isobe, M.; Fukami, N.; Goto, T. *Chem. Lett.* **1985**, p 71.

67. Isobe, M.; Nishikawa, T.; Pikul, S.; Goto, T. *Tetrahedron Lett.* **1987**, p 6485.

68. Matsumoto, K.; Ebata, T.; Koseki, K.; Okano, K.; Kawakami, H.; Matsushita, H. *Heterocycles* **1992**, *34*, p 1935.

69. Kettmann,V.; Bystricky, S.; Sticzay, T.; Koos. M. *Acta Crystallographica Sec. C. Crystal Structure Commun.* **1989**, *45*, p 1580.

70. Matsumoto, K.; Ebata, T.; Koseki, K.; Okano, K.; Kawakami, H.; Matsushita, H. *Carbohydr. Res.* **1993**, *246*, p 345.

71. Ebata, T.; Matsumoto, K.; Yoshikoshi, H.; Koseki, K.; Kawakami, H.; Mashushita, H. *Heterocycles,* **1990**, *31*, p 423.

72. Mori, M.; Chuman, T.; Kato, K. *Carbohydr. Res.* **1984**, *129*, p 73.

73. Witczak, Z. J.; Li, Y. *Tetrahedron Lett.* **1995**, 36, p 2595.

74. Forsyth, A. C.; Paton, R. M.; Watt, A. *Tetrahedron Lett.* **1989**, p 993.

75. Forsyth, A. C.; Gould, R. O.; Paton, R. M.; Sadler, I. H.; Watt, I. *J. Chem. Soc., Perkin Trans.1* **1993**, p 2737.

76. Matsumoto K.; Ebata T.; Koseki, K.; Kawakami, H.; Matsushita H. *Heterocycles* **1991**, *32*, p 2225.

77. Matsumoto, K.; Ebata, T.; Koseki, K.; Kawakami, H.; Matsushita, H. *Bull. Chem. Soc. Jpn.* **1991**, *64,* p 2309.
78. Koseki, K.; Ebata, T.; Kawakami, H.; Mashushita, H.; Hoshitake N.; Itoh, K. *Heterocycles* **1990**, *31*, p 1585.
79. Freskos, J. N.; Swenton J. *J. Chem. Soc. Chem. Commun.* **1984**, p 658.
80. Swenton, J. S.; Freskos, J. N.; Dalidowicz, P.; Kerns. M. L. *J. Org. Chem.* **1996**, *61*, p 459.
81. Taniguchi, T.; Nakamura, K.; Ogasawara, K. *Synlett* **1996**, p 971.
82. Witczak, Z. J.; Kaplon, P.; Kolodziej, M. *J. Carbohydr. Chem.* **2002**, *21*, p 1.
83. Tolstikov, G.A.; Valeev, F.A.; Gareev, A. A.; Khalilov, L. M.; Miftakhov, M.S. *Zh. Org. Khim.* **1991**, *27*, p 565.
84. Laikhter, A. L.; Niyazymbetov, M. E.; Evans, D. H.; Samet, A.V.; Semenov, V. V. *Tetrahedron Lett.* **1993**, *34*, p 4465.
85. Matsumoto, K, Ebata, T.; Koseki, K.; Okano, K.; Kawakami, H. Matsushita, H. *Carbohydr. Res.* **1993**, *246*, p 345.
86. Efremov, A. A.; Slaschinin, G. A.; Korniyets, E. D.; Sokolenko,V.A.; Kuznetsov, B. N. *Sibirskii Khim. Zhur.* **1992**, *6*, p 34.
87. Efremov, A. A.; Konstantinov, A. P.; Kuznetsov, B. N. *J. Anal. Chem.* **1994**, *49*, p 742.
88. Blattner, R.; Page, D. M. *J. Carbohydr. Chem.* **1994**, *13*, p 27.
89. Witczak, Z. J.; Chabra, R.; Chojnacki, J. *Tetrahedron Lett.* **1997**, *38*, p 2215.
90. Samet, A. V.; Laikhter, A. L.; Kislyi, V. P.; Ugrak, B. I.; Semenov, V. V. *Mendeleev Commun.* **1994**, p 134.
91. Niyazymbetov, M. E.; Laikhter, A. L.; Semenov, V. V.; Evans, D. H. *Tetrahedron Lett.* **1994**, *35*, p 3037.
92. Matsumoto, K.; Ebata, T.; Matsushita, H. *Carbohydr. Res.* **1995**, *279*, p 93.
93. Taniguchi, T.; Ohnishi, H.; Ogasawara, K. *Chem. Commun.* **1996**, 1477.
94. Samet, A. V.; Kislyi, V. P.; Chernyshova, N. B.; Reznikov, D. N.; Ugrak, B. I.; Semenov, V. V. *Rus. Chem. Bull.* **1996**, *45*, p 393.
95. Samet, A. V.; Yamskov, A. N.; Ugrak, B. I.; Vorontsova, L. G.; Kurella, M. G.; Semenov, V. V. *Rus. Chem. Bull.* **1996**, *45,* p 393.
96. Valeev, F. A.; Gaisina, I. N.; Miftakhov, M. S. *Rus. Chem. Bull.* **1996**, 45, p 2453.
97. Valeev, F. A.; Gaisina, I. N.; Sagitdinova, K. F.; Shitikova, O. V.; Miftakhov, M. S. *Zh. Org. Khim.* **1996**, *32*, p 1365.
98. Valeev, F. A.; Gorobets, E.V.; Tsypysheva, L.P.; Shingizova, G.Sh.; Kalimullina, L.Kh.; Safarov, M.G.; Shitikova, O. V.; Miftakhov, M. S. *Chemistry of Natural Compounds*, **2003**, *39,* p 563.
99. Samet, A. V.; Niyazymbetov, M. E.; Semenov, V. V.; Laikhter, A. L.; Evans, D. H. *J. Org. Chem.* **1996**, 61, p 8786.

100. Miftakhov, M. S.; Gaisina, I. N.; Valeev, F. A. *Rus. Chem. Bull.* **1996**, *45*, p 1942.

101. Nishikawa, T.; Araki, H.; Isobe, M. *Biosci. Biotechnol. Biochem.* **1998**, *62*, p 190.

102. Takeuchi, M.; Taniguchi, T.; Ogasawara, K. *Synthesis* **1999**, p 341.

103. Gomez, M.; Quincoces, J.; Peske, K.: Michalik, M. *J. Carbohydr. Chem.* **1999**, *18*, p 851.

104. Trahanovsky, W. S.; Revell, K. D.; Arvidson, K. B.; Wang, C.; Wang Y. 219th Am. Chem Soc. Meeting, **2000**, Abstract CARB 85.

105. Marquis, C.; Cardona, F.; Robina, I.; Wurth, G.; Vogel, P. *Heterocycles*, **2002,** *56*, p 181.

106. Zhu, Y-H.; P.Vogel. *Chemm. Commun.,* **1999**, p 1873.

107. Sarotti, A.M.; Spanevello, R. A.; Suarez, A.G. *Tetrahedron Lett.* **2004**, *45*, p 8203.

Chapter 22

Knots in Trees: A Rich Source of Bioactive Polyphenols

Bjarne Holmbom, Stefan Willfoer, Jarl Hemming, Suvi Pietarinen, Linda Nisula, Patrik Eklund, and Rainer Sjoeholm

Process Chemistry Centre, Aabo Akademi University, Turku/Aabo, Finland, FI 20500

Knots in trees, i.e. the branch bases inside stems, contain extraordinary high amounts of polyphenols, which are potent natural antioxidants and biocides. Studies of more than 50 tree species have shown that knots, in most of the studied species, contain remarkably higher amounts of polyphenols than the adjacent stemwood, for many species 20–100 times higher. Knots of softwood species typically contain 5–15% (w/w) of polyphenols, with lignans as the dominating group. Pine species contain a high percentage of stilbenes in their knots, while flavonoids are abundant in knots of certain hardwood species. Spruce knots from Northern Finland contain on average about 10% of lignans. The dominating spruce lignan 7-hydroxymatairesinol (HMR) is a strong antioxidant and has moreover been found to inhibit growth of certain tumors. Production of HMR and marketing it as dietary supplement in the USA has recently been started. In a large pulp mill using Norway spruce wood it is possible to sort out knots and extract up to 100 tons of HMR per year.

Knots in Trees

Knots are the branch bases inside the tree stems. Already in the 1930s, knots were reported to contain high contents of extractives (*1*). In studies by Wegelius (*2*) and Boutelje (*3*), the distribution of extractives in stemwood, branches and knots of spruce was determined. However, the knot extractives were not analyzed. The first studies that included analysis of knot extractives were on radiata pine (*Pinus radiata*) (*4*), and parana pine (*Araucaria angustifolia*) (*5*). Although the knots of parana pine were reported to contain much lignans, over 20% (w/w), the paper did not arouse any new interest in knot extractives. Ekman (*6*) investigated the distribution of lignans in Norway spruce trees and found high contents in the heartwood of branches (4–6%) and in roots (2–3%), but he did not analyze any knots.

In 1998, we analyzed a spruce knot, and found that it contained as much as 10% (w/w) of lignans. This discovery gave the starting signal for extensive research on knots; first on knots in Norway spruce (*Picea abies)* and Scots pine (*Pinus sylvestris)*, and later on knots in many more tree species. Up to now we have analyzed knots in more than 50 species.

Knots in trees are of little value in manufacturing of pulp. In fact, the knots are detrimental to pulping and pulp quality, and should preferably be separated before pulping. A process for separation of clean knot material for further extraction of polyphenols has recently been developed (*7*).

In this paper we summarize some of our research on polyphenols in knots. We have documented that knots are a very rich source of polyphenols, and especially of lignans, maybe even the richest in all nature. Many different polyphenols could be produced in industrial scale from knots. The main lignan in spruce knots, 7-hydroxymatairesinol (HMR), has been found to be a strong antioxidant, and furthermore to be anticarcinogenic. Recently, clearance was obtained from FDA to market HMR in dietary supplements in the USA. The industrial production of HMR that now has started includes separation of spruce knot material, extraction with ethanol, and precipitation of HMR as adduct with potassium acetate.

Polyphenols in Knots

We have during some years analyzed the polyphenols present in knots and compared the contents in knots with those in heartwood and sapwood samples from the same trees. We have analyzed fresh wood samples from at least 2–3 trees per species, including at least one living knot (having a living branch attached) and one dead knot from each tree. The wood samples have first been extracted with hexane to remove the lipophilic extractives, and thereafter the polyphenols have been extracted with acetone or ethanol containing 5% water (8). The polyphenols have been analyzed primarily by GC and GC-MS as trimethylsilyl derivatives. Most of the polyphenols have also been isolated in pure form for structural studies, primarily by NMR spectroscopy, and for various biotests. A large number of lignans have been identified in softwood species. Structures of some common softwood lignans are presented in Fig. 1. In addition to ordinary lignans we have also found substantial amounts of lignans with three or more phenylpropane units, called oligolignans (9).

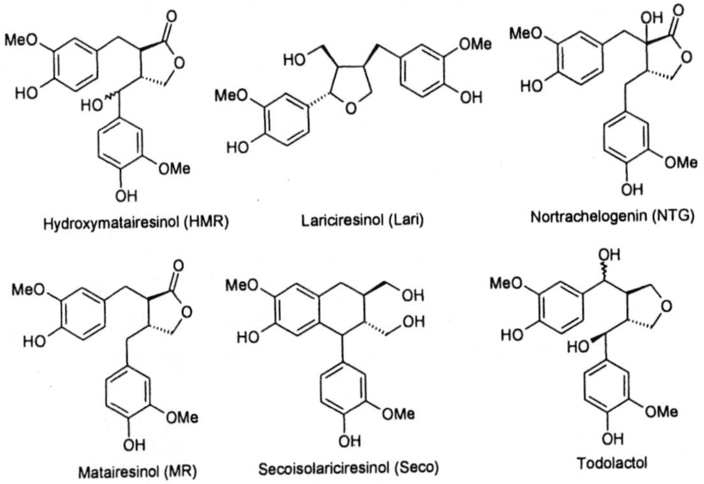

Hydroxymatairesinol (HMR)   Lariciresinol (Lari)   Nortrachelogenin (NTG)

Matairesinol (MR)   Secoisolariciresinol (Seco)   Todolactol

Figure 1. Structures of common lignans in softwood tree knots.

Large Amounts of Lignans in Knots of Spruce Species

Norway Spruce (Picea abies)

A study of more than 30 knots in seven Norway spruce trees (8) verified that knots contain much larger amounts of lignans than the ordinary stemwood, commonly over 100 times larger than in the adjacent stem heartwood. The studied knots contained 6–24% (w/w) of lignans. Later we have found even higher amounts, up to 29%, of lignans in spruce knots from Northern Finland. The lignan contents vary widely between trees, and even between knots in the same tree. Lignans are mainly found in the knots, i.e inside the stem, and their content drops to a level below 1% at 10–20 cm outwards in the branches. The typical distribution of lignans in Norway spruce trees is illustrated in Fig. 2.

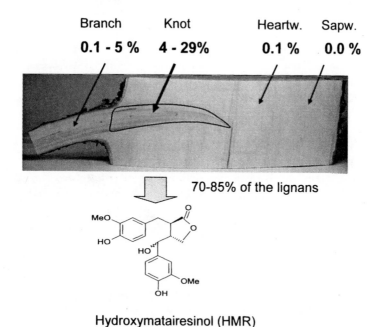

Hydroxymatairesinol (HMR)

Figure 2. Typical amounts of lignans (in % w/w) in Norway spruce trees, oligolignans excluded.

The lignans in spruce knots are composed mainly of HMR, comprising 70–85% of the lignans in the knots. HMR occurs in two stereoisomeric forms, in a ratio of about 3:1. It was recently unambiguously proven that the major isomer has the *7S,8R,8′R* configuration, while the minor isomer, eluting earlier in GC, has the *7R,8R,8′R* configuration (*10*).

The lignan contents were found to be at the same level from the middle of the tree stem out to the outer branch (*8*). However, the lignan contents were found to decrease in the radial direction from the center of the knot towards the outerwood (*11*). The contents were generally higher in the upper part of the knots (opposite wood) than in the lower parts, where dense compression wood is formed.

In addition to the ordinary dimeric lignans, Norway spruce knots also contain lignans with three and four phenylpropane units. Their amounts are 2–6% (w/w) of the spruce knots. The trimeric lignans, also called sesquilignans, are comprised of the major spruce lignans with an additional guaiacylglycerol unit bound to one of the phenolic hydroxyl units in the lignans (*9*).

Other Spruce Species

We have analyzed the polyphenols in knots and stemwood of other spruce species (*12*). White spruce (*Picea glauca*) contains almost the same amounts of lignans as Norway spruce (*Picea abies*) (Fig. 3). The lignan composition is also similar, with HMR as the main lignan. Black spruce (*Picea mariana*) contains less lignans and the relative amount of HMR is also lower. Sitka spruce (*Picea sitchensis*) contains only a small amount of lignans. Korean spruce (*Picea koraiensis)* contains much HMR, but Serbian Spruce (*Picea omorika)* contains only a small amount of HMR. Red spruce *(Picea pungens)* contains only a small amount of lignans and has secoisolariciresinol as the main lignan component. The todolactol isomers in paper (*12*) were incorrectly reported as isomers of liovil.

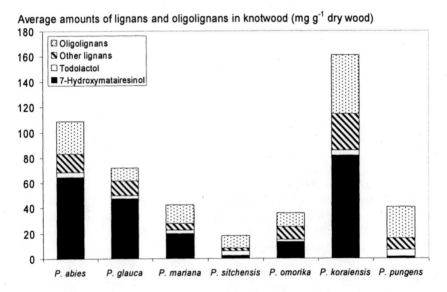

Figure 3. Lignans and oligolignans in knots of spruce species (12).

High Lignan Content in Knots of Fir Species

Fir (*Abies*) species are also used for pulp and paper production, especially in Canada, Northern USA, China and Central Europe. Like spruces, firs also have many knots since the branches usually remain on the tree stems even at mature age and in dense forest stands.

We have studied several fir species (*13*) and average data are presented in Fig. 4.

Fir knots contain typically 3–7% of lignans, and 3–5% of oligolignans. The oligolignans are composed mainly of guaiacylglycerol ethers and coumarates of the dominating lignans. Alpine fir (*Abies lasiocarpa*), growing in western Canada, contained only small amounts of lignans. In all firs, secoisolariciresinol was the most abundant lignan. Balsam fir (*Abies balsamea*) contained a substantial amount also of lariciresinol.

Typical for firs is that they also contain juvabiones, which are nonphenolic, cyclic sesquiterpenoid acids. Like lignans, also juvabiones are accumulated in the knots, with typical contents of 1–2%.

Average amounts of lignans, oligolignans, and juvabiones in knotwood (mg g^{-1}) dry wood

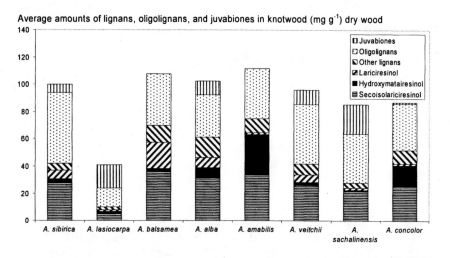

Figure 4. Lignans, oligolignans and juvabiones in knots of fir species (13).

Lignans Found in Knots of Pine and Other Softwood Species

Considerable amounts of lignans have also been found in knots of several pine species. Scots pine (*Pinus sylvestris*), the predominant pine in Europe, contained 0.4–2.9% of lignans in the knots (*14*) (Fig. 5). In addition to lignans, the knots contain 9-30% of resin acids and 1-7% of stilbenes.

The predominant lignan, comprising 80–95% of the knot lignans, is (–)-nortrachelogenin, also named (–)-wikstromol (*15*). No lignans could be detected in the ordinary stemwood. Pine knots also contain stilbenes of the pinosylvin type, as well as small amounts of oligolignans. However, the main extractives in pine knots are resin acids and other terpenoids making the recovery of pure lignans more difficult than for spruce and fir knots, which contain lignans as the dominating extractives.

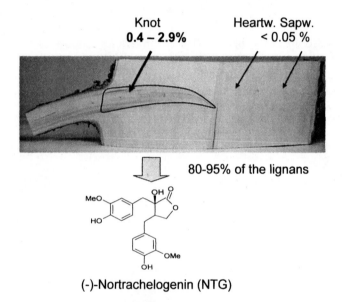

(-)-Nortrachelogenin (NTG)

Figure 5. Typical amounts of lignans in Scots pine (Pinus sylvestris) (14).

Several other pine species, as well as species of the families *Larix, Pseudotsuga, Tsuga* and *Thuja* have also been analyzed (*16, 17*). There are large differences between the pine species. Some pine species also contain flavonoids, in substantial amounts. Knots of *Larix* species (larches) contain mainly secoisolariciresinol and lariciresinol, similar to fir species. Knots of *Tsuga* species (hemlocks) are similar to spruce knots and contain HMR as the main lignan. Knots of *Pseudotsuga menziesii* (Douglas fir) contain both flavonoids and lignans.

Polyphenols in Knots of Hardwoods

A number of hardwood species have also been analyzed. Results have been published for a willow species (*Salix caprea*) (*18*), three aspen species (*19*) and two acacia species (*20*). Most hardwoods contain flavonoids in their knots. *Salix caprea* was found to contain the flavonoids dihydrokaempferol, catechin, gallocatechin, dihydromyrcetin and taxifolin. Knots contained more flavonoids than the stemwood of the same tree.

Knots in aspen trees contained 3–5% of flavonoids (Fig. 6). The dominating components were dihydrokaempferol and naringenin (Fig. 7).

Average amounts of flavonoids in knotwood (mg g-1 dry wood)

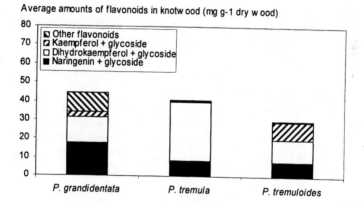

Figure 6. Flavonoids in knots of aspen species (19).

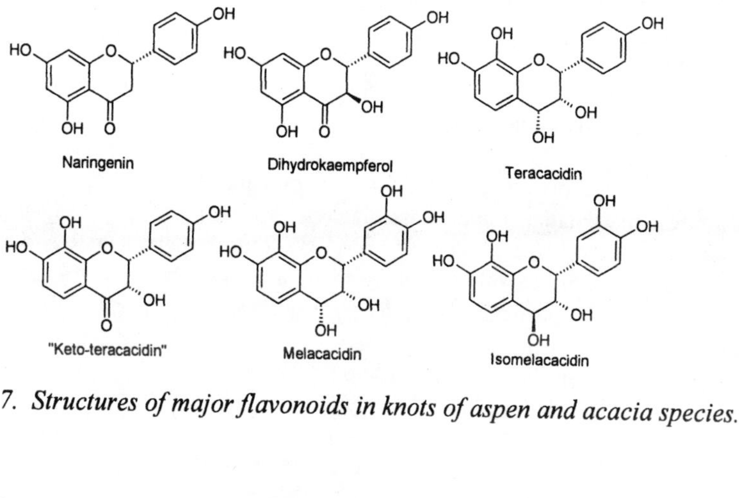

Figure 7. Structures of major flavonoids in knots of aspen and acacia species.

The main polyphenols in *Acacia mangium* were teracacidin and keto-teracacidin, whereas *Acacia crassicarpa* contained mainly melacacidin and iso-melacacidin (*18*). Also in these species, knots contained larger amounts of polyphenols than the stemwood.

Polyphenols are Bioactive Compounds

The lignans, flavonoids and stilbenes found in large amounts in knots most probably have a defense function in the trees, although the defense mechanisms have not been elucidated. Common to all these polyphenols is their antioxidative and radical-scavenging properties. Hydrophilic extracts of knots were documented to have a high antioxidative potency and/or a radical scavenging capacity (*16*). Pure lignans and the flavonoid taxifolin were also strong antioxidants. However, when tested on bacteria isolated from paper mill waters, only extracts of knots containing the stilbenes pinosylvin and its monomethyl ether, (mainly from pine species) showed antibacterial effects (*21*). Pinosylvin is also know to be a natural fungicide.

The large interest in the bioactivity of polyphenols is mainly due to their widespread occurrence in foods and their favorable effects on health. Flavonoids and stilbenes have been in the focus of health research for a long time, whereas lignans only more recently have attained broader interest (*22*). Lignans are of special interest because they are converted by bacteria in the large intestine to enterolactone and enterodiol. These lignan metabolites are considered to be associated with decreased risk for breast cancer and cardiovascular disease (*23*).

HMR in Spruce Lignan has Promising Bioactive Properties

The bioactivity of HMR has been studied extensively during more than ten years by the research group of Risto Santti and Sari Mäkelä at the University of Turku, Finland. The research has been carried out in cooperation with our and other groups.

It has been shown that in both rats and humans HMR can be metabolized to the known mammalian lignan enterolactone, which has antitumorigenic effects for hormone-dependent cancers (*23, 24, 25*). Furthermore, HMR is a powerful antioxidant, similar to other lignans (*16*), and may be able to decrease the oxidation of LDL-particles and thus support cardiovascular health. HMR and other lignans can also inhibit overactivity of phagocytes and lymphocytes (*26*). HMR has also been shown to be non-toxic (*27*).

Based on the research, HMR has been proposed as a chemopreventive agent against certain cancers, hormone-dependent diseases and cardiovascular diseases

(*28*). Recently, clearance was obtained from the FDA to market HMR in dietary supplements in the USA and in 2006, products containing HMR came on the market.

HMR is also of great interest because it can be used as a starting material for the production of other potentially valuable lignans. HMR is easily converted to matairesinol (Fig. 1) by catalytic hydrogenolysis (*29, 30*). HMR can also be reduced to secoisolariciresinol (*29*). Several other lignans can also be prepared from HMR, such as lariciresinol, cyclolariciresinol, enterolactone and enterodiol (*10, 29*). In alkaline conditions HMR is converted mainly to conidendric acid and in acidic conditions to conidendrin (*31*).

Industrial Production of HMR

When tree stems are chipped industrially, most of the wood knots will form thick chips. In the following chip screening knots will be enriched in the over-thick chip fraction. Today, this over-thick fraction is usually rechipped and used in the pulp production. However, knots are not desired in the wood furnish for mechanical pulping since knots are detrimental to the pulp quality (*32*). In chemical pulping the knots mostly remain uncooked and must be sorted out after the pulping (*33*). Therefore, knots should be taken out from the pulp wood furnish.

A convenient method for separation and purification of knotwood material from wood chips, named ChipSep, has been developed (*7*). In the method, over-thick chips are ground to splinters, the splinters are mildly dried and then separated by sedimentation in water. The knotwood material sediments while the normal stemwood material floats to the water surface.

Industrial production of HMR was started in 2005. Over-thick spruce chips from the UPM-Kymmene pulp and paper mill in Kajaani in Northern Finland is used as raw material. The chips are taken out in the wood handling before the thermomechanical pulping (Fig. 8). Knot material is then separated by the ChipSep process operated in continuous mode. The obtained knot material, with over 90% knot material, contains about 10% of lignans and has a HMR content of 6–8% (w/w). The lignans are extracted from the knot material by ethanol, and the major isomer of HMR is precipitated from the solution by addition of potassium acetate, in principle according to the method described already in 1957 (*34*). The HMR potassium acetate adduct has a purity of 90–95%.

In the UPM-Kymmene mill, processing about 1000 tons of spruce wood per day, it would be possible to separate knot material for an annual production of over 100 tons of HMR.

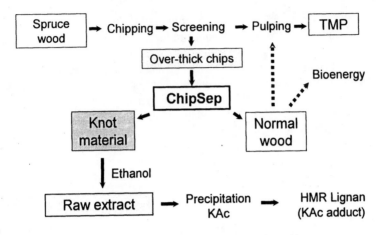

Figure 8. Outline of production of HMR lignan from Norway spruce.

Conclusions

The conclusion is that knots of certain spruce and fir species constitute an exceptionally rich source of lignans. In certain spruce knots, the amount of lignans can exceed even 20% (w/w) and the knots can contain hundreds of times more lignans than the adjacent stemwood in the same tree. Some spruce species, especially Norway spruce and white spruce, contain on the average 5–10% lignans in the knots, with 7-hydroxymatairesinol (HMR) comprising 70–85% of the lignans. Knots in firs commonly contain 3–7% of lignans. The most common lignans in firs are secoisolariciresinol followed by lariciresinol. Pine species contain both lignans, stilbenes and flavonoids in their knots, but the large amounts of resin acids and other terpenoids complicate their extraction and purification. Certain hardwood species contain substantial amounts of flavonoids in their knots.

The polyphenols occur in knots in free form and are easily extracted by ethanol, or other polar solvents. Not only HMR, but also other lignans, could be produced in large scale from knots at pulp and paper mills, either directly by extraction and purification, or by semisynthesis from HMR. Pure lignans, stilbenes and flavonoids can be extracted from knots and provide material for research on these potentially valuable bioactive compounds. Much research is presently being done in order to understand their biological and physiological properties, and consequently to create the base for applications in medicine and nutrition, or as natural antioxidants and antimicrobial agents in technical products.

Acknowledgements

Financial support has been received from the National Technology Agency (TEKES), the Academy of Finland, the Foundation for Research of Natural Resources in Finland, and the companies UPM-Kymmene, Raisio Chemicals (now Ciba Specialty Chemicals) and Hormos Nutraceutical. We are also grateful for the fruitful cooperation with Risto Santti, Sari Mäkelä, Niina Saarinen and Markku Ahotupa at University of Turku. This work is part of the activities at the Åbo Akademi Process Chemistry Centre within the Finnish Centre of Excellence Programme by the Academy of Finland.

References

1. Hägglund, E.; Larsson, S. *Svensk Papperstidn.* **1937**, *40*, pp 356-360.
2. Wegelius, T.H. *Acta For. Fenn.* **1940**, *48*, pp 1-191.
3. Boutelje, J.B. *Svensk Papperstidn.* **1966**, *69*, pp 1-10.
4. Hillis, W.E.; Inoue, T. *Phytochemistry* **1968**, *7*, pp 13-22.
5. Anderegg. R.J.; Rowe, J.W. *Holzforschung* **1974**, *28*, pp 171-175.
6. Ekman, R. *Acta Academiae Aboensis, Ser B.* **1979**, *39*(3), pp 1-6.
7. Eckerman, C.; Holmbom, B. *US Pat.* 6,739,533, **2004**.
8. Willför, S.; Hemming, J.; Reunanen, M.; Eckerman, C.; Holmbom, B. *Holzforschung* **2003**, *57*, pp 27-36.
9. Willför. S.; Reunanen, M.; Eklund, P.; Kronberg, L.; Sjöholm, R.; Pohjamo, S.; Fardim, P.; Holmbom, B. *Holzforschung* **2004**, *58*, pp 345-354.
10. Eklund, P.C.; Sillanpää, R.; Sjöholm, R. *J. Chem. Soc., Perkin Trans. 1* **2002**, *16*, pp 1906-1910.
11. Willför, S.M.; Sundberg, A.C.; Rehn, P.W.; Saranpää, P.T.; Holmnom, B.R. Distribution of lignans in knots and adjacent stemwood of *Picea abies. Holz Roh-Werkst.* **2005**, *63*, pp 353-357.
12. Willför, S.; Nisula, L.; Hemming, J.; Reunanen, M.; and Holmbom, B. *Holzforschung* **2004**, *58*, pp 335-344.
13. Willför, S.; Nisula, L.; Hemming, J.; Reunanen, M.; Holmbom, B. *Holzforschung* **2004**, *58*, pp 650-659.
14. Willför, S.; Hemming, J.; Reunanen, M.; Holmbom, B. *Holzforschung,* **2003**, *57*, pp 359-372.
15. Ekman, R.; Willför, S.; Sjöholm, R.; Reunanen, M.; Mäki, J.; Lehtilä, R.; Eckerman, C. *Holzforschung* **2002**, *56*, pp 253-256.
16. Willför, S.M.; Ahotupa, M.O.; Hemming, J.E; Reunanen, M.H.T.; Eklund, P.C.; Sjöholm, R.E.; Eckerman, C.S.E.; Pohjamo, S.P.; Holmbom, B.R. *J. Agr. Food Chem.* **2003**, *51*, pp 7600-7606.

362

17. Pietarinen, S.; Willför, S.; Ahotupa, M.; Hemming, J.; Holmbom, B. *J. Wood Sci.* **2006**, in press (DOI: 10.1007/s10086-005-0780-1).
18. Pohjamo, S.P.; Hemming, J.E.; Willför, S.M.; Reunanen, M.H.T.; Holmbom, B.R. *Phytochemistry* **2003**, *63*(2), pp 165-169.
19. Pietarinen, S.P.; Willför, S.M.; Ahoptupa, M.O.; Holmbom B.R. *J. Wood Chem. Technol.* **2006**, in press.
20. Pietarinen, S.; Willför, S.; Sjöholm, R.; Holmbom, B. *Holzforschung* **2005**, *59*, pp 94-101.
21. Lindberg, L.E.; Willför, S.M; Holmbom, B.R. *J. Ind. Microbiol. Biotechnol.* **2004**, *31*, pp 137-147.
22. Corsby, G.A. *Food Technol.* **2005**, *59*(5), pp 32-35.
23. Saarinen, N.M.; Wärri, A.; Mäkelä, S.I.; Eckerman, C.; Reunanen, M.; Ahotupa, M.; Salmi, S.M.; Franke, A.A.; Kangas, L.; Santti, R. *Nutr. Cancer* **2002**, *36*, pp 207-216.
24. Saarinen, N.M.; Penttinen, P.E.; Smeds, A.I.; Hurmerinta, T.T.; Mäkelä, S.I. *J. Steroid Biochem. Mol. Biol.* **2005**, *93*, pp 209-219.
25. Bylund, A.; Saarinen, N.; Zhang, J.-X.; Bergh, A.; Widmark, A.; Johansson, A; Lundin, E.; Adlercreutz, H.; Hallmans, G.; Stattin, P.; Mäkelä, S. *Exp. Biol. Med.* **2005**, *230*(3), pp 217-223.
26. Ahotupa, M.; Eriksson, J.; Kangas, L.; Unkila, M.; Komi, J.; Perala, M.; Korte, H. *US Pat. Appl. No. US* **2003**/0100514 A1.
27. Lina, B.; Korte, H.; Nyman, L.; Unkila, M. *Regul. Toxicol. Appl. Pharmacol.* **2005**, *41*(1), pp 28-38.
28. www.hmrlignan.com (August 22, **2005**).
29. Eklund, P.; Lindholm, A.; Mikkola, J.-P.; Smeds, A.; Lehtilä, R.; Sjöholm, R. *Org. Lett.* **2003**, *5*(4), pp 491-493.
30. Markus, H.; Mäki-Arvela, P.; Kumar, N.; Kul'kova, N.V.; Eklund, P.; Sjöholm, R.; Holmbom, B.; Salmi, T.; Murzin, D.Yu. *Catalysis Lett.* **2005**, *103*(1-2), pp 125-131.
31. Eklund, P.C.; Sundell, F.J,; Smeds, A.I., Sjöholm, R.E. *Org. Biomol. Chem.* **2004**, *2*, pp 1-8.
32 Sahlberg, U. *Tappi J.* **1995**, *78*(5), pp 162-168.
33. Allison, R.W.; Graham, K.L. *Appita* **1988**, *41*, pp 197-206.
34. Freudenberg, K.; Knof, L. *Chem. Ber.* **1957**, *90*, pp 2857-2869.

Chapter 23

Pyrolysis Mechanism of Woody Biomass Relating to Product Selectivity

Haruo Kawamoto and Shiro Saka

Graduate School of Energy Science, Kyoto University, Yoshida-honmachi, Sakyo-ku, Kyoto 606–8501, Japan

Thermochemical processes, especially gasification and fast pyrolysis are promising ways to convert woody biomass into fuels and useful chemicals. However, low product selectivity exists as a drawback to commercialization. In order to increase product selectivity, knowledge of the pyrolysis mechanism, a fundamental process of these thermochemical processes, provides useful information. In this chapter, recent data of the pyrolysis mechanism of woody biomass is presented. Particular emphasis is made on the key reactions which govern product selectivity between low-molecular-mass (MW) products and solid carbonized products. Data on the controlled pyrolysis of cellulose into low MW products and other useful chemicals such as levoglucosenone and furfurals, based on the proposed mechanism, is also discussed.

Pyrolysis is an underlying principle of thermochemical conversion processes of woody biomass into fuels and chemicals. Traditionally, carbonization and dry distillation processes are conducted to obtain charcoal and chemicals such as acetic acid and methanol. However, low product selectivity between gaseous, liquid and solid carbonized products presents a challenge for these applications in the production of fuels and chemicals in an economical manner.

Heating conditions are known to affect the product selectivity: high temperature pyrolysis (>600 °C) gives gaseous products more selectively (gasification), while quick heating conditions with quick quenching of the volatile products between 400-600 °C substantially increase the liquid product formation (fast pyrolysis). Even under these conditions, product selectivity is insufficient for effective utilization. For example, although gasification is a promising way to convert woody biomass to electricity and liquid fuels through more efficient gas turbine or gas engine and C1 catalysts, respectively, tar by-products produced in gasification processes make these processes difficult to manage. Tar causes several troubles including pipe fouling, engine damage and deactivation of the catalyst through coking.

Consequently, improving the product selectivity is a very important and challenging issue in the establishment of effective utilization systems based on pyrolysis. The pyrolysis mechanism on a molecular basis provides useful information in achieving this end goal. Information about the molecular mechanism of wood pyrolysis is very limited, although many studies have been conducted on the kinetics of the formation of volatile and carbonized products.

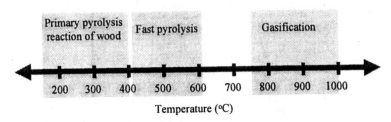

Figure 1. Temperature ranges of the primary pyrolysis of wood, fast pyrolysis and gasification.

In the laboratory, pyrolysis mechanisms of wood and its constituent polymers were studied on a molecular basis with the aim of improving product selectivity through controlled pyrolysis. Figure 1 shows the temperature range where the primary pyrolysis of wood occurs, as well as those of fast pyrolysis and gasification. The primary pyrolysis of wood, which occurs at substantially lower temperature than coal or petroleum pyrolysis, must play an important role in the fast pyrolysis and gasification processes. Therefore, in this study, there is an emphasis on the key reactions which govern the product selectivity in the temperature range of 150-400 °C. Mechanisms which are responsible for the characteristic low-temperature pyrolysis of wood are also targeted.

Cellulose Pyrolysis

Among the wood constituent polymers, cellulose pyrolysis has been studied very extensively. In 1918, Pictet reported levoglucosan (1,6-anhydro-β-D-glucopyranose) was a major product of cellulose pyrolysis, and also reported the ring-opening polymerization of levoglucosan occurring at 240 °C (2). Subsequently, many studies were conducted on cellulose pyrolysis related to the levoglucosan intermediate, which include the formation mechanism (ionic or homolytic) (3,4) and the effects of crystalline structure (5,6) and inorganic substances in cellulose (7,8) on the formation of levoglucosan and other products of pyrolysis (9,10). Thermal analysis including thermogravimetry (TG) and differential thermal analysis (DTA) provide important information about the volatilization of pure cellulose (10). This process occurs in a narrow temperature range around 350°C as an endothermic process, and this characteristic behavior is substantially modified in the presence of inorganic substances. The effects of inorganic substances have also been studied related to the fire retardants of wood and cellulosic materials (11). Many kinetic studies based on thermogravimetry have been conducted to explain the volatilization processes, and several kinetic models including the Broido-Shafizadeh model have been proposed (9). As for low MW products, extensive studies have been conducted with pyrolysis-gas chromatography (GC) techniques (12).

Pyrolysis Behavior of Levoglucosan

In this data, there is an emphasis on the behavior of levoglucosan since it is the major pyrolysis intermediate from cellulose (13). In cellulose pyrolysis, products were found to be formed in a stepwise manner: levoglucosan → polysaccharides →carbonized products. From these results, the ring-opening polymerization of levoglucosan has been proposed as a key reaction which governs the product selectivity between low MW and solid carbonized products. This proposal was further supported by the experimental results that no solid carbonized products were formed from levoglucosan during pyrolysis in sulfolane (tetramethylene sulfone, an aprotic polar solvent). These results are explained by the retardant effects on the levoglucosan polymerization in the homogeneous solution through inhibiting the accessibility.

Mechanism of Cellulose Pyrolysis

Based on the results of levoglucosan, pyrolysis conditions in sulfolane were applied to cellulose. Cellulose in sulfolane was found to be pyrolyzed selectively into soluble low MW products without any formation of solid carbonized products in the temperature range of 200-330 °C (*14*). These results strongly suggest that inhibition of the levoglucosan polymerization is quite effective for controlling the pyrolysis into low MW products.

Furthermore, residues in sulfolane were obtained only as colorless substances under all experimental conditions, although cellulose pyrolysis in nitrogen or dioctylphthalate (a poor solvent for levoglucosan) gave dark-colored carbonized substances under the same pyrolysis conditions except for the pyrolysis media (*14*). These results are interesting in relation to the pyrolysis mechanism of crystalline cellulose. Infra-red (IR) and microscopic analyses under cross-polar conditions indicated that the residue in sulfolane is crystalline cellulose itself, although the residues obtained in nitrogen or dioctylphthalate showed the strong peaks at $1700cm^{-1}$ appearing in relation to carbonization and the decreased birefringence showing the decreased crystallinity. Furthermore, from the X-ray diffraction analysis, changes in the crystallite size and crystallinity were found to be very small during pyrolysis in sulfolane and suddenly decrease and then disappear just prior to the complete disappearance of the residues (*15*). Similar results were also obtained in nitrogen and in dioctylphthalate in relation to the decrease of the X-ray diffraction signals of the samples during pyrolysis.

From these lines of evidence, a heterogeneous mechanism of cellulose pyrolysis is proposed as shown in Fig. 2 (*13-15*). Cellulose molecules in the ordered regions are considered to be stabilized through intermolecular interactions such as hydrogen bonding and hydrophobic interaction. Thus, the pyrolysis reaction is considered to start from the less stabilized molecules such as disordered or end molecules.

The reaction of monomeric levoglucosan in pyrolysis lead to low MW product formation more selectively due to lowering the boiling points of the products, while polysaccharides formed through repolymerization of levoglucosan stay in the heated zone to be pyrolyzed into carbonized substances more selectively. In sulfolane, the solvent is expected to penetrate the disordered region of cellulose during the swelling process, and then, inhibit the levoglucosan polymerization.

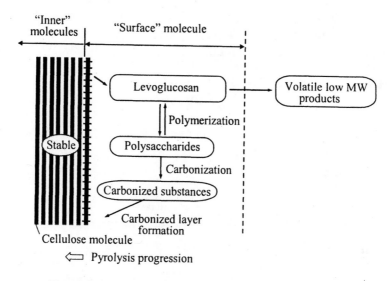

Figure 2. A proposed mechanism of cellulose pyrolysis.

Controlled Pyrolysis into Low MW Chemicals

Even in sulfolane, the obtained products are complex mixtures of low MW compounds including levoglucosan, levoglucosenone, furfural and 5-hydroxymethyl furfural (5-HMF). Use of a catalyst in sulfolane is effective to further control the reactions into more specific chemical formations. Cellulose pyrolysis in sulfolane containing sulfuric acid or polyphosphoric acid (0.1-1.0 wt%) at 200-280 °C gave levoglucosenone and furfural in maximum yields of 42.2 and 26.9 mol%, respectively (*16*). Sulfuric acid, a strong acid [pKa: -5.2, 1.99 (secondary)], tended to give furfural, while polyphosphoric acid, a weak acid [pKa: 2.15, 7.20 (secondary), 12.38 (tertially)], gave levoglucosenone more selectively. From the detailed study with the intermediates, the pathways were indicated including the conversion from levoglucosenone to furfural and interconversion between levoglucosan and its furanose isomers to give an equilibrium mixture of 3:1 under such acidic conditions (Fig. 3). Furthermore, water was found to be necessary to convert levoglucosenone into furfural. This finding is useful for further control of the pyrolysis products from levoglucosenone and furfural. Even under similar pyrolysis conditions (0.1 wt% sulfuric acid / sulfolane / 200 °C / 6min), product selectivity significantly varies depending on the water content in the pyrolysis system (Table I). Furthermore,, nitrogen flow conditions at atmospheric or slightly reduced pressure, aimed at removing the byproduct water, substantially increases the levoglucosenone yield from 24.8 to 42.2 mol% with the decreased yield of furfural + 5-HMF from 11.8 to 3.1 mol%. Alternatively, steam distillation conditions significantly improve the yields of furfural + 5-HMF from 11.8 to 35.7 mol% with concomitant decrease of the levoglucosenone yield from 24.8 to 8.5 mol%.

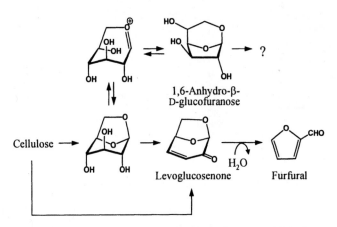

1,6-Anhydro-β-D-glucofuranose

Cellulose

Levoglucosenone

Furfural

H_2O

Figure 3. Pathways of cellulose pyrolysis in sulfolane with some acidic catalysts.

Table I. Effects of vacuum and steam distillation conditions on the product selectivity between levoglucosenone and furfurals (0.1 wt% sulfuric acid / sulfolane / 200 °C / 6 min).

Conditions	Yield (mol %)			
	Levoglucosenone	Furfural	5-HMF	Furfural + 5-HMF
In nitrogen Under nitrogen flow (100ml/min)	24.8 (100)	8.0 (100)	3.8 (100)	11.8 (100)
1atm	30.3 (122)[*1]	8.6 (108)	0.8 (21)	9.4 (80)
0.16atm	38.1 (154)	3.8 (48)	0.6 (16)	4.4 (37)
0.12atm	42.2 (170)	2.3 (29)	0.8 (21)	3.1 (26)
Steam distillation[*2]	8.5 (34)	26.9 (336)	8.8 (232)	35.7 (303)

*1: number in the parenthesis shows the relative yield (%).
*2: nitrogen (50ml/min) + steam (60ml/min).

These results indicate that pyrolysis in solvent with catalyst is a promising way for highly selective conversion of cellulosic biomass into various useful chemicals.

Lignin Pyrolysis

Lignin is a heterogeneous polymer consisting of the phenyl propanes linked through several ether-types (α-ether, β-ether etc.) and condensed (C-C)-types (β-aryl, biphenyl, β-β etc.) of linkages. Thus, understanding the pyrolysis behaviors of these substructures is quite important for understanding the overall behavior of lignin in wood.

Lignin pyrolysis has been studied with isolated lignin such as milled wood lignin (MWL) and Kraft lignin, as well as model compounds. Thermogravimetric analysis showed that the volatilization starts from 150°C and the major volatilization occurs between 300 and 400 °C with relatively large amount of solid carbonized residue (1). Volatile products from lignin have been studied with TG-IR (17) and pyrolysis-GC/MS techniques (18,19) and include aromatic monomers, methanol, formaldehyde, formic acid, CO and CO_2 (20). Kinetic models have also been studied to explain the volatilization process (20).

As for model compound study, Domberg's group reported the decomposition behavior of several model dimers, including β-ether, biphenyl and pinoresinol types based on their thermogravimetric analysis (20). Many studies were conducted with very simple model compounds such as phenethyl phenyl ether with no aromatic substituent groups (20-22). However, there are very few papers which describe the pyrolytic reactions of lignin model dimers with guaiacyl or syringyl type of the aromatic ring (23).

Under these circumstances, two approaches have been chosen in the research; one is with lignin model dimers which gives the details of the pyrolytic reactions occurring in substructures along with their relative reactivities; the other one is with the milled wood lignin (MWL) isolated from wood which gives the pyrolysis behavior of lignin macromolecules, including condensation, depolymerization and carbonization.

Pyrolysis Pathways of Model Dimers

Figures 4 and 5 show the pyrolytic pathways of ether-types and condensed-types of model dimers with guaiacyl type of aromatic rings investigated in 1 minute pyrolysis in nitrogen at 400 °C (24). Since ether linkages [β-ether (48), α-ether (16-21), 5-O-4 (4)] and C-C linkages [β-aryl (10), biphenyl (9.5-11), β-β (2)](number in the parenthesis: number of the linkage in 100 phenyl propane

α-Ether type:
  α-Ether cleavage ➡ Depolymerization

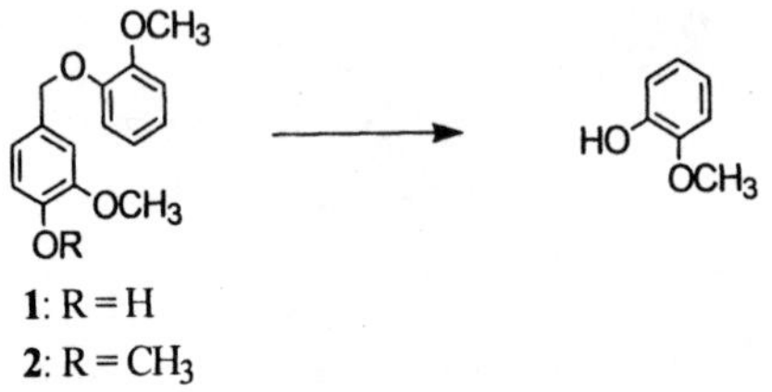

1: R = H
2: R = CH₃

β-Ether type:
  β-Ether cleavage (reaction **a**) ➡ Depolymerization
  C_γ-Elimination (reaction **b**)
      [Phenolic form (**3**): reaction **a**: favored]

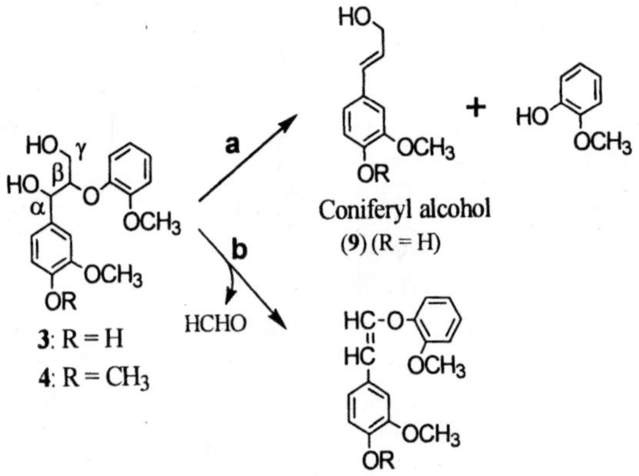

Coniferyl alcohol
(**9**) (R = H)

3: R = H
4: R = CH₃

HCHO

Figure 4. Pyrolysis pathways of ether-types of model dimers.

units) are reported in spruce MWL (*25*), the model dimers **1-8** give information about most of the major linkages in softwood lignin.

Reactivity and pyrolysis pathways were quite different depending on the model compound structure: phenolic compounds were much more reactive than the corresponding non-phenolic compounds; condensed types were more stable toward depolymerization; phenolic ether types were easily depolymerized. Alfa-ether linkages were easily cleaved at 400 °C in both phenolic and non-phenolic forms (**1** and **2**). For β-ether types (**3** and **4**), two types of reactions were observed: β-ether cleavage to form cinnamyl alcohols (**a**) and C_γ-elimination to form nol ethers (**b**). In the phenolic form (**3**), reaction **a** was substantially favored and the β-ether cleavage also occurred from the enol ether. The C_α-C_β cleavage (c) and C_γ-elimination (d) to form stilbenes were observed in β-aryl types (**5** and **6**), and reaction **d** was substantially favored in the phenolic form **5**.

β-Aryl type:
    C_α-C_β Cleavage (reaction c) ➡ Depolymerization
    C_γ-Elimination (reaction d)
        [Phenolic form (**5**): reaction **d**: favored]

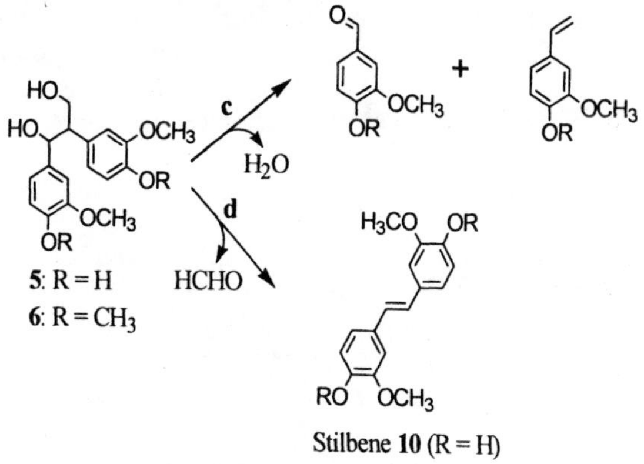

Stilbene **10** (R = H)

Aryl-aryl type: relatively stable

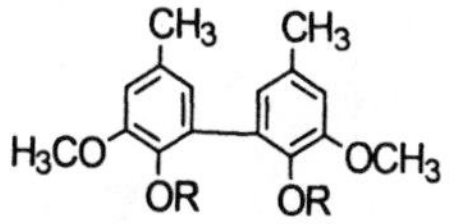

7: R = H
8: R = CH₃

Figure 5. Pyrolysis pathways of condensed types of model dimers.

Pyrolysis Behavior of Milled Wood Lignin and Roles of Substructures

The pyrolysis behavior of MWL isolated from Japanese cedar (*Cryptomeria japonica*) wood was investigated in nitrogen at 200-400 °C by analyzing the products fractionated into dioxane-soluble and insoluble fractions (*26*).

Figure 6 summarizes the temperatures at which condensation, depolymerization and carbonization become significant in the MWL pyrolysis. Temperatures at which the model dimers become reactive are also included.

Pyrolysis pathways of model dimers **1-6** at this temperature range were similar to those shown in Figures 4 and 5.

Condensation occurring in MWL was indicated by the formation of the dioxane-insoluble fraction, since the original MWL was soluble in dioxane. Substantial depolymerization starting at 300-350 °C was indicated from the GPC analysis of the dioxane-soluble fractions as well as the ultra-violet (UV) spectra of the fractions; formation of coniferyl alcohol (**9**) from the β-ether cleavage caused the UV λmax shift from 280 to 272nm with increased absorption around 300nm. At 400 °C, the signal at 1510cm^{-1} in the infra-red (IR) spectra of the dioxane-insoluble fractions, which is assigned as the aromatic skeleton in lignin, started to decrease substantially and completely disappeared during a longer heating time of 10 minutes. This change is considered to correspond to the polycyclic aromatic hydrocarbon (PAH) formation which is closely related to the carbonization.

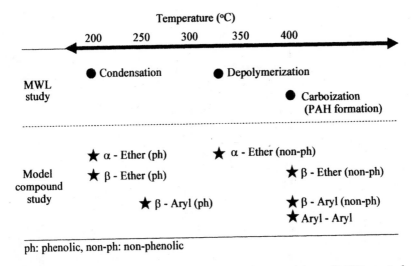

Figure 6. Summary of the temperatures at which conversion of MWL started or model dimers became reactive in 1 min pyrolysis.

Comparison of the results between MWL and model dimers indicates the important structures for each pyrolysis behavior of MWL. For condensation, phenolic structures are indicated to be key structures. Depolymerization of MWL occurs close to the temperature where the cleavage of the non-phenolic α-ether structure becomes reactive. From these relations, a chain depolymerization mechanism starting from the cleavage of the α-ether linkage in the endwise lignin ether chains is indicated. At the temperature where the carbonization of

MWL starts, all of the model dimers become reactive, and this indicates that the carbonization is related to the reaction of the aromatic ring itself.

Condensation and Depolymerization Mechanisms in Lignin Pyrolysis

Modification of the relative reactivities for condensation, depolymerization and carbonization is important to control the lignin pyrolysis. For example, activation of the depolymerization reaction by inhibiting the condensation and carbonization reactions may improve the low MW product formation from lignin. However, for controlling the key reactions, the detailed molecular mechanisms involved in these reactions should be known.

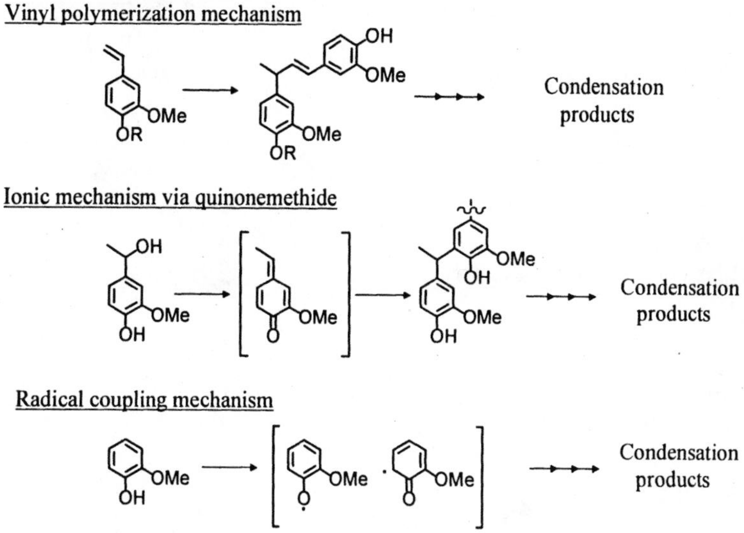

Figure 7. Condensation mechanisms at relatively low temperature.

Condensation mechanism was investigated at 250 °C with some simple model compounds including guaiacol, veratrol and their derivatives with C_α-hydroxyl and $C_\alpha=C_\beta$ structures (27). As a result, three types of the mechanisms were indicated as relatively low temperature condensation mechanisms, which include vinyl polymerization, ionic polymerization via quinonemethide intermediate and radical coupling between radicals derived from phenoxy radicals (Fig. 7).

Reactivities of the ether linkages were quite different depending on the structure; phenolic α- and β-ether types of model dimers were cleaved at 200-250 °C, while the non-phenolic α- and β-ether types were cleaved at 350 and

400 °C, respectively. Thus, different mechanisms are suspected to be involved in these ether cleavages.

Model dimers or trimers with C_α-O-phenyl groups having several substituent groups at the p-positions were useful to identify the mechanism of the α-ether cleavage [ionic vs. homolytic (radical) mechanisms] (28). If the cleavage proceeds in an ionic mechanism, the reactivity is expected to be positively related with the order of the Hammett substituent constant σ_p: -COMe (0.502) > -Cl (0.227) > -H (0) > -OMe (-0.268). Contrary to this, if the cleavage proceeds in a homolytic mechanism, the reactivity is expected to be positively related with the ΔBDE (bond dissociation energy) value (kcal/mol): -OMe (3.9) > -Cl (1.1) > -H (0) > -COMe (-0.6) [ΔBDE (kcal/mol) = 63.5 (anisol) –BDE (p-sustituted anisol)]. Reactivities positively related to the σ_p value indicated that phenolic α-ether linkage was cleaved in an ionic mechanism (Fig. 8), while non-phenolic α-ether cleavage was indicated to proceed in a homolytic mechanism from the positive relation between reactivity and ΔBDE value (Fig. 9). Much higher reactivity that was observed for the phenolic form is explainable by these different mechanisms.

Beta-ether cleavage in the α,β-diether types of model compounds was regulated by their reactivities for the α-ether cleavage. In non-phenolic form, the initial homolytic cleavage of the α-ether linkage to form benzyl radical is a rate-determining step for the subsequent β-ether cleavage (Fig. 8). From the benzyl radical intermediate, β-ether linkage was easily cleaved through β-fragmentation.

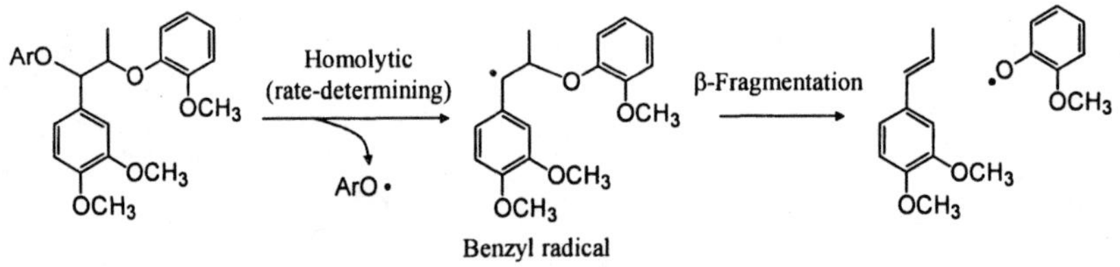

Figure 8. Cleavage mechanisms of non-phenolic α- and α,β-diether types of model compounds.

The phenolic α,β-diether types of model compounds showed somewhat complex substituent effects on their reactivities, which indicates the change in the mechanism of α-ether cleavage from ionic to radical. This strongly indicates radical species formation during pyrolysis. Additionally, with the α-ether groups, the β-ether cleavage occurred at substantially lower temperature of 250 °C than the similar model compounds without α-ether groups (350-400 °C). The pyrolytic cleavage mechanism of the β-ether linkage is still controversial because of of the lack of proofs supporting the proposed mechanism including 6-centered retro-ene (*21*), homolytic (via benzyl radical) (22) and oxirane mechanisms (*23*). The present results give a proof for a proposed mechanism of low temperature homolytic β-ether cleavage via quinonemethide intermediate as shown in Fig. 9.

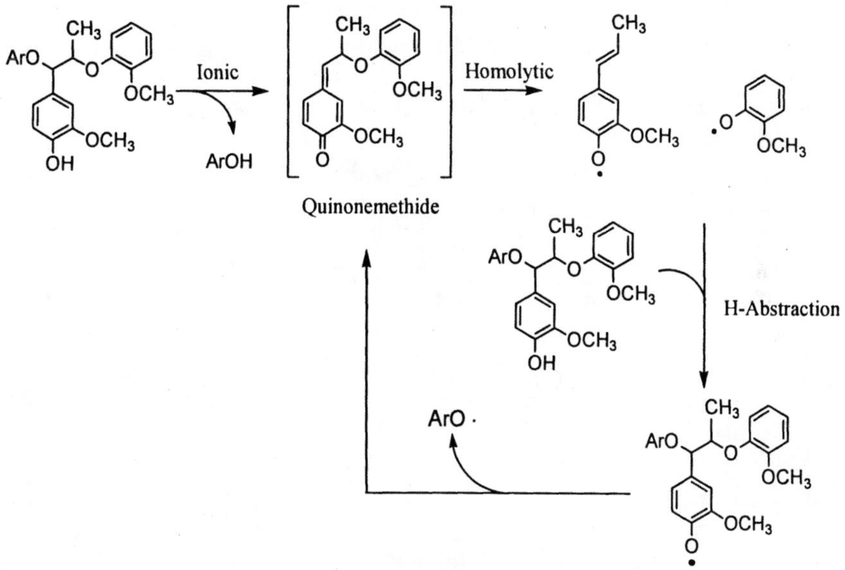

Figure 9. Cleavage mechanisms of the phenolic α- and α,β-diether types of model compounds.

Conclusions

Some of the pyrolysis mechanisms of wood and its constituent polymers have been determined and the controlled pyrolysis of cellulose has been demonstrated. These data provide useful information in developing selective conversion methods based on pyrolysis, including tar-free gasification and selective chemical production.

Acknowledgments

Some of these studies have been carried out under the COE program in the 21st Century of "Establishment of COE on Sustainable-Energy System" and by a Grant-in-Aid for Scientific Research (C)(2) (No.11660164 and No.16580132) from the Ministry of Education, Culture, Sports, Science and Technology, Japan.

References

1. Ramiah, M. V. J. Appl. Polymer Sci. **1970**, *14(5)*, pp 1323-37.
2. Pictet, A. J. Helvetica Chimica Acta **1918**, I, pp 226-230.
3. Kislitsyn, A.N.; Rodionova, Z.M.; Savinykh, V. I.; Guseva, A. V. Zhurnal Prikladnoi Khimii **1971**, *44(11)*, pp 2518-24.
4. Shafizadeh, F. J. Polymer Sci., Polymer Symposi **1971**, No. 36, pp 21-51.
5. Golova, O.P.; Pakhomov, A.M.; Andrievskaya, E.A.; Krylova, R.G. Doklady Akademii Nauk SSSR **1957**, *115*, pp 1122-5.
6. Basch, A.; Lewin, M. J. Polymer Sci., Polymer Chemistry Edition **1973**, *11(12)*, pp 3071-93.
7. Fung, D.P.C.; Tsuchiya, Y.; Sumi, K. Wood Science **1972**, *5(1)*, pp 38-43.
8. Shafizadeh, F.; Furneaux, R.H.; Cochran, T.G.; Scholl, J.P.; Sakai, Y. J. Appl. Polymer Sci. **1979**, *23(12)*, pp 3525-39.
9. Antal Jr., M. J. Adv. Solar Energy **1983**, *1*, pp 61-111.
10. Shafizadeh, F. J. Anal. Appl. Pyrolysis **1982**, *3*, pp 283-305.
11. Kandola, B. K.; Horrocks, A. R.; Price, D.; Coleman, G. V. J.M.S.-Rev. Macromol. Chem. Phys., **1996**, *C36(4)*, pp 721-794.
12. Powels, A. D.; Eijkel, G. B.; Boon, J. J. J. Anal. Appl. Pyrolysis **1989**, *14*, 237-280.
13. Kawamoto, H.; Murayama, M.; Saka, S. J. Wood Sci. **2003**, *49*, pp 469-473.
14. Kawamoto, H.; Hatanaka, W.; Saka, S. J. Anal. Appl. Pyrolysis **2003**, *70*, pp 303-313.
15. Kawamoto, H.; Saka, S. J. Anal. Appl. Pyrolysis, **2006**, *76*, pp 280-284.
16. Saito, S.; Kawamoto, H.; Saka, S. J. Anal. Appl. Pyrolysis, in press.
17. Fenner, R.A.; Lephardt, J.O. J. Agric. Food Chem. **1981**, *29(4)*, pp 846-9.
18. Obst, J.R. J. Wood Chem. Technol. **1983**, *3(4)*, pp 377-97.
19. Jakab, E.; Faix, O.; Till, F. J. Anal. Appl. Pyrolysis **1997**, *40-41*, pp 171-186.
20. Antal Jr., M. J. Adv. Solar Energy **1985**, *2*, pp 175-256.
21. Klein, M.T.; Virk, P.S. Report **1981**, MIT-EL-81-005.
22. Britt, P.F.; Buchanan, A.C.III; Thomas, K.B.; Lee, S.-K. J. Anal. Appl. Pyrolysis **1995**, *33*, pp 1-19.
23. Brezny, R.; Mihalov, V.; Kovacik, V. Holzforschung **1983**, *37(4)*, pp 199-204.
24. Kawamoto, H.; Horigoshi, S.; Saka, S. J. Wood Sci. submitted.
25. Adler, E. Wood Sci. Technol. **1977**, *11*, pp 169-218.
26. Kawamoto, H.; Nakamura, T.; Saka, S. In Science in Thermal and Chemical Biomass Conversion, Victoria, **2004**; in press.
27. Nakamura, T.; Kawamoto, H.; Saka, S. unpublished data.
28. Nakamura, T.; Kawamoto, H.; Saka, S. unpublished data.

Chapter 24

Antifungal Compounds Isolated from Tropical and Temperate Woods

Irawan W. Kusuma[1,2] and Sanro Tachibana[1]

[1]Department of Applied Bioscience, Faculty of Agriculture, Ehime
University, 3–5–7 Tarumi, Matsuyama, Ehime 790–8566, Japan
[2]Department of Forest Products Technology, Faculty of Forestry,
Mulawarman University, JI. Ki Hajar Dewantara, Kampus Gn., Kelua,
P.O. Box 1013, Samarinda 75123, Indonesia

Tropical woods, Amboyna (*Pterocarpus indicus*) and
Belangeran (*Shorea belangeran*), and a temperate wood, Sugi
(*Cryptomeria japonica*), were examined for antifungal
activities against wood-rotting fungi, *Pleurotus pulmonarius*
and *Pleurotus ostreatus*. Bioassay-guided fractionation of the
wood extracts resulted in the isolation of four antifungal
compounds: pterocarpol, liquiritigenin, isoliquiritigenin and β-
eudesmol from Amboyna wood; two antifungal compounds:
resveratrol and belangeranol from Belangeran wood; and two
antifungal compounds: (+)-2, 7(14),10-bisabolatriene-1-ol-4-
one and (-)-secoabietane dialdehyde from Sugi wood.
Furthermore, wide-ranging activities of the compounds against
six plant pathogenic fungi: *Fusarium oxysporum*, *Corynespora
cassiicola*, *Cochiliobolus miyabeanus*, *Trichoderma
harzianum*, *Aspergillus niger* and *Penicillum italicum* were
also investigated. (+)-2, 7(14),10-bisabolatriene-1-ol-4-one
was found to be active on antifungal assays against six plant
pathogenic fungi by revealing low minimum inhibition
concentrations for all fungi tested.

377

Fungi are one of the most harmful groups of plant pathogens, associated with damage to agricultural crops, forest trees, wood and wood-based products (*1*). Fungi also pose a threat to humans in that they produce mycotoxins, poisonous chemical compounds, as well as cause contamination and spoilage of foods and feeds (*2*). However, some fungi are also beneficial to humans, being used in the pharmaceutical and agricultural industries and in biotechnological processes such as biopulping, bioremediation of toxic substances and so on. Several chemicals, so called fungicides, are used to prevent and kill fungi. Although fungicides have been used with good results, they have potential negative effects on the environment, for instance toxicity to humans, accumulation in soil and so on.   Recent investigations showed that the application of synthetic fungicides, amphotericin B (*3*), griseofulvin and azole derivatives such as fluconazole (*4*) can lead to significant nephrotoxicity and have adverse effects on human health. However, Schaller *et al.* (*5*) isolated an antifungal compound, a phenanthrene-3,9-dione, whose activities are superior to those of amphotericin B and fluconazole from *Bobgunnia madagascariensis*. Given these problems, the development of more effective and less toxic naturally-occurring antifungal agents is an attractive alternative.   Various different natural substances, such as essential oils and plant extracts, have been tested for antifungal activities.   For example, Hinokitiol (β-thujaplicin) was found to have a broad antibacterial spectrum including activity against methicillin-resistant *Staphylococcus aureus* (MRSA) (*6*) and *Legionella sp.* (*7*). Actually, hinokitiol is used as a fungicide, food antiseptic and for other practical uses (*8*).   This paper describes not only the isolation and identification of antifungal compounds in Amboyna, Belangeran and Sugi woods, but also the wide-ranging activities of these compounds against several wood-rotting fungi and fungi pathogenic to plants.

Identification of Antifungal Compounds Isolated from Amboyna Wood

Amboyna, *Pterocarpus indicus* (Leguminosae), a large deciduous tree, is an excellent source of timber and material for furniture in southern Asia.   The red latex of amboyna is used as a remedy for tumors, especially of the mouth, while the leaves were reported to significantly inhibit the growth of Ehrlich ascites carcinoma cells in mice (*9*).

Isolation of Antifungal Compounds from Amboyna Wood

Amboyna woodmeal (1.2kg) was extracted twice with methanol for 8 hours at 65 °C and the methanol solution was evaporated with a rotary evaporator to give methanolic extracts (420g).   The methanolic extracts were suspended in water and extracted successively with *n*-hexane, diethyl ether, ethyl acetate and *n*-butanol, respectively to give each soluble. Antifungal assays of methanol extracts and each respective soluble against *Pleurotus pulmonarius* confirmed that the *n*-hexane and diethyl ether solubles were more active than the others and the residue, as shown in Table 1. Separation of active soluble fraction, a portion (8g) of the diethyl ether soluble (D) (62g) into 6 fractions (D1-D6) by column chromatography (CC) on silica gel with gradient elution of chloroform and methanol (CHCl₃-MeOH) gave fraction D3 (2.5g), which showed most active, causing 91% inhibition (91% activity) in the antifungal test. Repeated CC of the fraction D3 (2.4g) over silica gel resulted in the isolation of fractions D34 (1.52g), D36 (0.31g), D344 (403mg) and D346 (87mg) with 52%, 38%, 26% and 35% activity, respectively. Compound **1** (38.6mg) was isolated as a white crystal from fraction D344 after recrystallization from *n*-hexane and ethyl acetate (EtOA) in a yield of 0.0026% relative to Amboyna wood. Compounds **2** (13.3mg) and **3** (10.8mg) were isolated as each of yellowish crystal in their order of separation from D346 by silica gel CC with gradient elution of CHCl₃-MeOH and followed by recrystallization from ethanol and water in a yield of 0.009 and 0.0073% relative to Amboyna wood, respectively (*10,11*). Separation of another active soluble fraction, *n*-hexane soluble (26.5g) (C) into 5 fractions (C1-C5) by silica gel CC with gradient elution of CHCl₃-MeOH gave fraction C4 (6.4g) with 75% activity. Repeated CC of the fraction C4 (4.4g) over silica gel resulted in the isolation of the active fractions, C43 (2.8g), C432 (370mg) and C4322 (82mg) with 83, 85 and 60% activity, respectively. Compound **4** (24.7mg) was isolated as a white crystal from fraction C4322 by recrystallization from *n*-hexane and EtOAc in a yield of 0.003% relative to Amboyna wood. The antifungal activities of Compounds **1-4** against *P. pulmonarius* were 7, 29, 42 and 38% relative to Amboyna wood, respectively (*10,11*).

Table 1. Antifungal Activities against *Pleurotus pulmonarius* and *Pleurotus ostreatus* of each Soluble Fraction in Methanolic Extracts of Amboyna, Belangeran and Sugi Woods

Extracts/Soluble	Antifungal activity[a] (%, of wood)		
	P. pulmonarius	*P. ostreatus*	
	Amboyna	Belangeran	Sugi
Methanolic extracts	100	39	38
n-Hexane	35	0	36
Diethyl ether	46	36	22
Ethyl acetate	0	27	15
n-Butanol	0	0	0
Water layer (Residue)	0	0	0

Note: [a]: Values are the mean of triplicates.

Identification of Antifungal Compounds Isolated from Amboyna Wood and Their Biological Activities

Compound 1 was a white crystal, with a melting point (mp) of 101-103 °C. The electron ionization (EI) mass spectrum of Compound 1 showed a molecular ion peak at m/z 238 corresponding to the molecular formula $C_{15}H_{26}O_2$. Another ion peak at m/z 220 (M^+-H_2O) and a base ion peak at m/z 59 [$(CH_3)_2CHO^+$] suggested Compound 1 was a sesquiterpenoid having a eudesmane skeleton with a hydroxyl-isopropyl and a hydroxyl moiety, pterocarpol. Elucidation of Proton (^1H) nuclear magnetic resonance (NMR) spectrum of Compound 1 showed an angular methyl group at δ 0.71 ppm, which was characteristic of transfused eudesmane and a gem-dimethyl group at δ 1.16 and 1.17 ppm, respectively, attributed to a hydroxyl-isopropyl moiety as identified in the mass spectrum. Furthermore, signals for two exomethylene protons at δ 4.80 ppm and one allylic proton at δ 1.68-1.73 ppm were observed. The ^1H and Carbon 13 (^{13}C) NMR data of Compound 1 were consistent with those recorded for pterocarpol by Parthasarathy and Seshadri (12) and Nasini and Piozzi (13), respectively. Therefore, Compound 1 was identified as pterocarpol (Figure 1).

Compound 2 was a pale yellowish crystal, with an mp of 206-208 °C. The EI mass spectrum of Compound 2 showed a molecular ion peak at m/z 256 corresponding to the molecular formula $C_{15}H_{12}O_4$ and two typical fragment ions $[A_1+H]^+$ (m/z 137) and B_3^+ (m/z 120), characteristic of 7,4'-dihydroxyflavanone. Acetylation of Compound 2 gave a crystalline diacetate, $C_{19}H_{16}O_6$ (M^+=340), mp 181-183 °C. Elucidation of ^1H NMR spectrum of Compound 2 acetate confirmed signals for a para-disubstituted ring B, a 1,2,4-trisubstituted ring A, and an ABX system in ring C at δ 5.49 and 3.05 and 2.28 ppm attributed to one C-2 proton and two C-3 protons, respectively. The ^1H and ^{13}C NMR data of Compound 2 coincided with those of synthesized 7,4'-dihydroxyflavanone, liquiritigenin (14). Therefore, Compound 2 was identified as liquiritigenin (Figure 1).

Compound 3, $C_{15}H_{12}O_4$ (M^+=256), was a yellowish crystal, with an mp of 202-205 °C. Acetylation of Compound 3 gave a crystalline triacetate, $C_{21}H_{18}O_7$ (M^+=382), mp 120-121 °C. Elucidation of ^1H NMR spectrum of Compound 3 acetate confirmed signals for a para-substituted ring A, a 1,2,4-trisubstituted ring B, and a trans disubstituted double bond at δ 7.58 (H-β, d, J=15.6Hz) and 7.12 (H-α, d, J=15.6Hz) ppm. The ^1H and ^{13}C NMR data of Compound 3 coincided with those of synthesized 2',4',4-trihydroxychalcone, isoliquiritigenin (15). Therefore, Compound 3 was identified as isoliquiritigenin (Figure 1).

Compound 4 was a white crystal, with an mp of 80-81 °C. The EI mass spectrum of Compound 4 was similar to that of Compound 1 and showed a characteristic of eudesmane-type of sesquiterpenoid with a molecular ion peak at m/z 222 ($C_{15}H_{26}O$) and a base peak at m/z 59 [$(CH_3)_2CHO^+$]. The ^1H NMR spectrum of 4 showed signals for an angular methyl group at δ 0.70 ppm, which was characteristic of trans-fused eudesmane, a gem-dimethyl group at δ 1.21

ppm, two exomethylene protons at δ 4.72 and 4.55 ppm, one allylic proton at δ 1.77 ppm, respectively. The ^{1}H and ^{13}C NMR data of **4** coincided with those recorded for β-eudesmol (*12,16*). Therefore, Compound **4** was identified as β-eudesmol (Figure 1).

Pterostilbene, homopterocarpin, pterocarpin, pterocarpol, β-eudesmol, isoliquiritigenin et al. were isolated from the *genus* Pterocarpus trees by Arisanya *et al.* (*17*). β-Eudesmol can act as a hypertensive, antihepatotoxic or antiepileptic agent (*18*) and has antimicrobial activity (*19*). However, reports about the activity of natural β-eudesmol and pterocarpol against wood-rotting fungi are limited. Flavonoids have been reported to possess various biological activities, including antiinflammatory, antihepatotoxic, antiulcer actions (*20*), and antiallergic, antiviral and antitumor effects (*21*). Among naturally-occurring flavonoids, liquiritigenin and isoliquiritigenin are considered characteristic constituents of the Legminosae family (*22*). These compounds have been proven to act as monoamine oxidase inhibitors, inhibitors of cancer cells (*23*), antiinflammatory and xanthine oxidase inhibitors (*24*). To the best our knowledge, this is the first study to show that liquiritigenin and isoliquiritigenin possess antifungal activity (*25*).

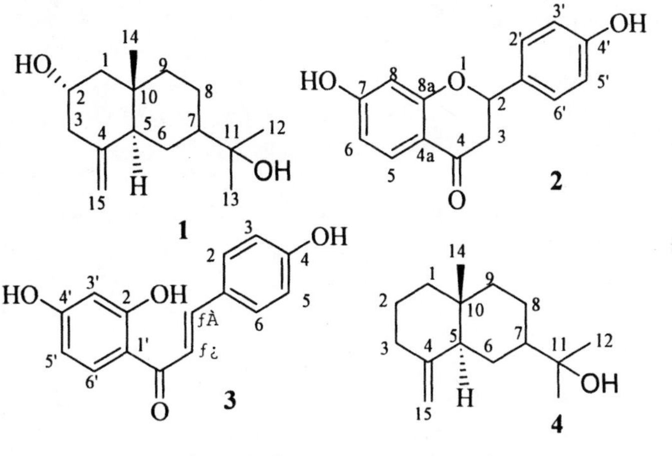

Figure 1. Chemical structure of pterocarpol (1), liquiritigenin (2), isoliquiritigenin (3) and β-eudesmol (4).

Identification of Antifungal Compounds Isolated from Belangeran Wood

Belangeran, *Shorea belangeran* (Dipterocarpaceae), a large deciduous tree, is an important timber native to tropical countries. *Shorea* is one of 16 genera in the family Dipterocarpaceae, the most recognized plant family in tropical rain forests. However, no reports about the photochemical of Belangeran have been found.

Isolation of Antifungal Compounds from Belangeran Wood

Belangeran woodmeal (2.3kg) was extracted twice with methanol for 8 hours at 65 °C and the methanol solution was concentrated *in vacuo* to give methanolic extracts (198g). The methanolic extracts were suspended in water and extracted successively with *n*-hexane, diethyl ether, ethyl acetate and *n*-butanol, respectively, to give each soluble. Antifungal assays of methanolic extracts and each respective soluble fraction against *Pleurotus ostreatus* confirmed that the diethyl ether and ethyl acetate solubles were more active than the others and the residue, as shown in Table 1. Separation of diethyl ether soluble (45g) (K) into 10 fractions (K1-K10) by silica gel CC with gradient elution of CHCl₃-MeOH gave two fractions K4 (0.2g) and K7 (6.3g) with 28% and 37% activity, respectively. Repeated CC of the fraction K4 (192mg) over silica gel resulted in the isolation of fractions K42 (149mg) and K422 (33mg) with 21 and 15% activity, respectively. Compound **5** (44.1mg) was isolated as a white crystal from fraction K422 by recrystallization from *n*-hexane and EtOAc in a yield of 0.002% relative to Belangeran wood. Separation of another active fraction K7 (6.2g) into 5 fractions (K7A-K7E) by silica gel CC with gradient elution of CHCl₃-MeOH gave fraction K7A (3.3g) with 33% activity. Repeated CC of the fraction K7A by silica gel CC with gradient elution of *n*-hexane-EtOAc in the isolation of fractions K7A8 (1.1g), K7A83 (345mg) and K7A83C (261mg) with 31, 29 and 21% activity. Compound **6** (58.4mg) was isolated as a pale yellow solid from fraction K7A83C by preparative HPLC (high pressure liquid chromatography) in AcCN/H₂O (7:3) in a yield of 0.005% relative to Belangeran wood. The antifungal activities of Compounds **5** and **6** against *P. ostreatus* were 13% and 18% relative to Belangeran wood, respectively (*26*).

Identification of Antifungal Compounds Isolated from Belangeran Wood and Their Biological Activities

Compound **5** was a white crystal, with an mp of 258-259 °C. The ultraviolet (UV) spectrum of Compound **5** showed maximum absorption at 322 [shoulder (sh)], 306 and 215 nm which is characteristic of stilbene. The EI mass spectrum of Compound **5** showed a molecular ion peak at m/z 228 corresponding to the molecular formula $C_{14}H_{12}O_3$. Acetylation of Compound **5** gave triacetate, $C_{20}H_{18}O_6$ (M⁺=354), mp 114-115 °C. The ¹H NMR spectrum of Compound **5** and its acetate indicated the occurrence of an A_2X spin system (3,5-dihydroxyphenyl group), an A_2B_2 spin system (4-hydroxyphenyl group), a *trans* disubstituted double bond [δ 6.96 (H-β, d, J=16.6Hz) and 6.83 (H-α, d, J=16.6Hz) ppm]. Elucidation of 1H and 13C NMR of Compound **5** and its acetate strongly suggested Compound **5** to be 3,5,4'- trihydroxystilbene, resveratrol. The ¹H and ¹³C NMR data of Compound **5** coincided with those of an authentic resveratrol purchased from Sigma Co. Ltd. Therefore, Compound **5** was identified as *trans*-resveratrol (Figure 2).

Compound **6** was a pale yellow solid, with an mp >250 °C. Fast atom bombardment (FAB) mass spectroscopy in the positive mode revealed a molecular ion peak at m/z 925 ([M+H]$^+$=925) corresponding to the molecular formula $C_{56}H_{45}O_{13}$ (m/z 925.2876 for $C_{56}H_{45}O_{13}$). cetylation of Compound **6** gave a deca-acetate, mp 235-238°C. The FAB mass spectrum of Compound **6** acetate revealed a molecular ion peak at m/z 1345 ([M+H]$^+$=1345) corresponding to the molecular formula $C_{76}H_{65}O_{23}$ (m/z 1345.3909 for $C_{76}H_{65}O_{23}$). The ^{1}H NMR spectrum of Compound **6** acetate exhibited signals for four sets of an A_2B_2 spin system (1,4-disubstituted phenyl group), two sets of a 1,3,5,6-tetrasubstituted phenyl group, and two sets of an A_2X spin system (1,3,5-trisubstituted phenyl group). On comparison of the signal pattern of several oligostilbenes (27,28), the signals for two sets of aliphatic hydrogen at the position of 7a and 8a and of 7d and 8d suggested the presence of two dihydrobenzofuran moieties bearing 4-oxyphenyl and 3,5-dioxyphenyl groups characteristic of oligostilbenes derived from resveratrol molecules. The four aliphatic hydrogens of Compound **6** acetate coupled to each other at δ 5.22 to 3.49 ppm and δ 5.29 to 3.01 ppm, respectively, suggested the presence of a tetrahydrofuran ring system (26,27). Based on the ^{1}H and ^{13}C NMR data, and the mass spectra of Compound **6** and its acetate, Compound **6** was a resveratrol tetramer having a tetrahydrofuran ring system (Figure 2).

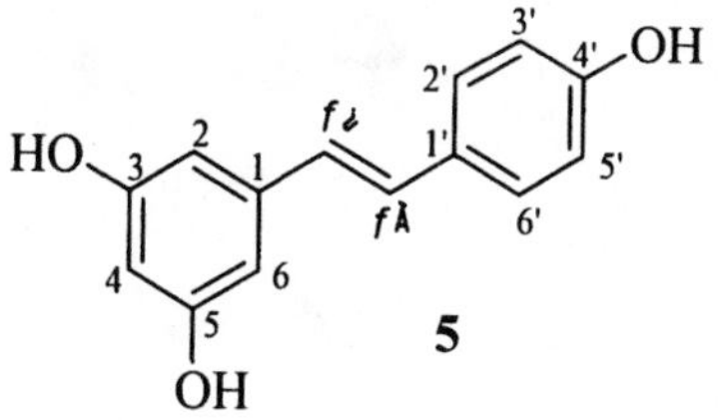

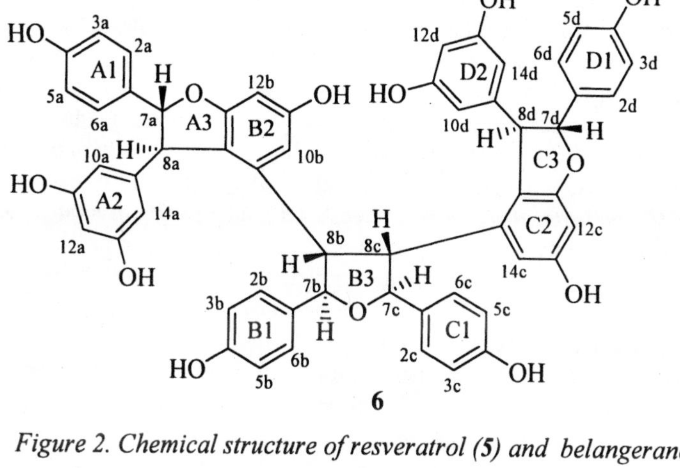

Figure 2. Chemical structure of resveratrol (5) and belangeranol (6).

The chemical structure and absolute configuration of Compound **6** was established by conducting X-ray crystal analyses of Compound **6** acetate (*26*). The crystal structure is shown in Figure 3. Several resveratrol tetramers have been isolated from the Dipterocarpaceae, Cyperaceae and Vitaceae species (*29*). However, Compound **6** has not been reported previously. Therefore, this novel resveratrol tetramer was tentatively named belangeranol. Recently, the stereoisomer of Compound **6** was isolated from grape fruit by Fujii *et al.* (*30*).

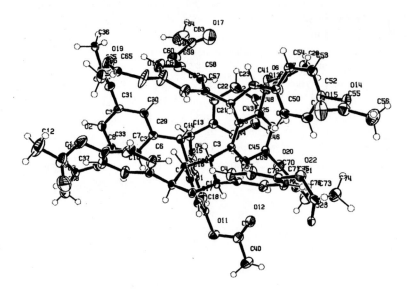

*Figure 3. X-ray structure of Compound **6** acetate.*

Resveratrol oligomers divided into two groups. One group contained at least one five-membered oxgene heterocyclic ring (*29*). The other group does not contain any oxygen heterocyclic ring (*29*). Stilbenes play important roles in plants, especially in heartwood protection as part of both constitutive and inducible defense mechanisms, and in dormancy and growth inhibition. Sotheeswaran and Pasupathy (*29*) reported that antifungal activity of resveratrol and some of the resveratrol oligomers have been studied and resveratrol have moderate antifungal activity, while the oligomers have broader activities. ε-Viniferin, a kind of oligostilbene, has antifungal activities of against some wood-rotting fungi, *Coriolus versicolor*, *Poria placenta* and *Gloeaphyllum trabeum* (*29*). Certain stilbenoids, besides being toxic to insects and other organisms, have mammalian antifeedant and nematicidal properties (*31*). Furthermore, these compounds have attracted much attention for their broad biological effects, which include antioxidant (*32*), antifungal (*33*), tyrosinase-inhibitory (*34*), anti-HIV-1 and cytotoxic effects (*35*), and antimicrobial inhibitors (*36*).

Identification of Antifungal Compounds Isolated from Sugi Wood

Sugi or Japanese cedar, *Cryptomeria japonica* (Taxodiaceae), one of the most economically important of all timber species for its excellent growth and wood quality, is widely grown in Japan. Essential oils of Sugi were reported to possess bioactivities, including larvicidal activity against *Aedes aegyptie* (*37*) and antifungal activity (*38*), while its wood and bark were found to exhibit activities against wood-rotting and pathogenic fungi (*39*).

Isolation of Antifungal Compounds from Sugi Wood

Sugi woodmeal (6kg) was extracted twice with methanol for 8 hours at 65 °C and the methanol solution was concentrated *in vacuo* to give methanolic extracts (225g). The methanolic extracts were treated in a similar manner as described above to give each soluble. Antifungal assays of methanolic extracts and each respective soluble fraction against *P. ostreatus* confirmed that the *n*-hexane soluble was more active than the others and the residue, as shown in Table 1. Separation of *n*-hexane soluble (S) (40g) into 13 fractions (S1-S13) by silica gel CC with gradient elution of *n*-hexane-EtOAc gave two fractions, S8 (0.3g) and S9 (1.4g) with 18 and 51% activity. Repeated CC of the fraction S8 (280mg) by silica gel CC with gradient elution of *n*-hexane-EtOAc resulted in the isolation of fraction S82 (228mg) and S823 (30mg) with 45 and 38% activity, respectively. Compound 7 (11.2mg) was isolated as a white solid from fraction S823 by preparative HPLC in MeOH/H$_2$O (4:1) in a yield of 0.0002% relative to Sugi wood. Separation of another active fraction S9 (1.3g) into 3 fractions by silica gel CC with gradient elution of *n*-hexane-Acetone gave fraction S92 (508mg) with 51% activity. Silica gel CC of the fraction S92 (472mg) with gradient elution of *n*-hexane-EtOAc resulted in the isolation of fraction S925 (254mg) with 40% activity. Compound 8 (55.7mg) was isolated as yellowish oil from fraction S925 by preparative HPLC in MeOH/H$_2$O (3:1) in a yield of 0.001% relative to Sugi wood. The antifungal activities of Compounds 7 and 8 against *P. ostreatus* were 35% and 38%, respectively, relative to Sugi wood (*26,40*).

Identification of Antifungal Compounds Isolated from Sugi Wood and Their Biological Activities

Compound 7, specific rotation ($[\alpha]_D^{19}$) –23.8°, was a white solid, with an mp of 189-191 °C. The EI mass spectrum of Compound 7 showed a molecular ion peak at m/z 316 corresponding to the molecular formula of C$_{20}$H$_{28}$O$_3$. The UV spectrum of Compound 7 showed the maximum absorption at 292, 233 and 203 nm, respectively, suggested the occurrence of an aromatic ring with a phenolic hydroxyl moiety. The ^{13}C NMR spectrum showed occurrences of two aldehyde groups at δ 205.7 and 191.3 ppm, six aromatic carbons at δ 110-158 ppm and three angular methyl carbons at δ 19-50 ppm, respectively. The ^{1}H NMR spectrum of Compound 7 indicated signals for two aromatic protons at δ 7.83 and 6.91 ppm and three angular methyl protons at δ 0.71, 1.01 and 1.51 ppm, signals for a methyne proton at δ 3.17 ppm and two methyl protons at δ 1.26 ppm, respectively, indicating the presence of an isopropyl moiety, which is a characteristic of abietane-type diterpenes. Based on an analysis of the spectral data (EI-MS, ^{1}H, ^{13}C and two-dimensional NMR) and specific rotation data, Compound 7 was identified as (-)-12-Hydroxy-6,7-secoabieta-8,11,13-triene-6,7-dial (Figure 4). The (+) form has been isolated from the bark of Taiwan red cypress, *Chamaecypris formosensis* (*41*), *Taiwania cryptomeriodes* (*42*) and *Cryptomeria japonica* (*43*). However, the (-) form, a diastereomer of 12-Hydroxy-6,7-secoabieta-8,11,13-triene-6,7-dial, has not been reported previously. Therefore, Compound 7 is a novel diastereomer of 12-Hydroxy-6,7-secoabieta-8,11,13-triene-6,7-dial.

Compound 8, $[\alpha]^{19}_{D}$ +160° was a yellowish oil. The UV spectrum of Compound 8 showed the maximum absorption at 232 nm. The EI mass spectrum of Compound 8 showed a molecular ion peak at m/z 234 corresponding to the molecular formula $C_{15}H_{22}O_2$. In the ^{13}C NMR spectrum, signal for 15 carbons including a ketone carbon at δ 198.4 ppm, a trisubstituted double bond at δ 132.5 ppm, a disubstituted vinyl methylene at δ 112.6 ppm, a carbon attached to hydroxyl group at δ 69.2 ppm, and three methyl groups attached to the C-12, C-13 and C-15 at δ 26.6, 17.8 and 15.3 ppm, respectively, were observed. The ^{1}H NMR spectrum of Compound 8 also showed signal for two methyl protons at δ 1.69 and 1.80 ppm, an olefinic proton at δ 5.10 ppm and two methylene protons at δ 2.17 ppm corresponding to the isoprenoid moiety. The two-dimensional NMR [proton-proton correlated spectroscopy (^{1}H-^{1}H COSY)] spectrum showed correlation between H-6 and H-5a and b, which were further coupled to H-1. These data, the mass spectrum and the specific rotation value enabled Compound 8 to be a bisabolane-type sesquiterpene, (+)-2,7(14),10-bisabolatrien-1-ol-4-one (Figure 4). Nagahama and Tazaki (*44*) first isolated the compound from Sugi.

Activities of (+)-2,7(14),10-bisabolatrien-1-ol-4-one as an antifeedant against a pest snail, an antirepellent against a pill-bug and an inhibitor of the germination of lettuce and rice seed, as well as activity of 12-Hydroxy-6,7-secoabieta-8,11,13-triene-6,7-dial as a termicidal agent have been reported (*43,45*). However, to the best of our knowledge, there has been no report about the antifungal activity of these two compounds. In this study, it was also found that fractions containing ferruginol, an antifungal diterpene alcohol previously isolated from *Cryptomeria japonica*, showed almost no activity against *P. ostreatus*. This might be correlated with the result of an antifungal study of *Cryptomeria japonica* conducted by Nakajima *et al.* (*39*), which found that ferruginol was only active against one species of the three fungi tested, *Lentinula edodes*. The results suggested that ferruginol possesses very specific antifungal activity.

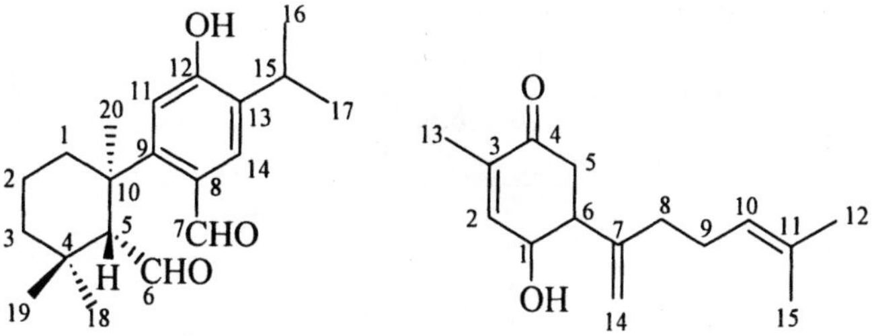

Figure 4. Chemical structure of Compounds 7 and 8.

Antifungal Activities of the Isolated Compounds from Amboyna and Sugi Woods Against Pathogenic Fungi

Liquiritigenin (2) and isoliquiritigenin (3) isolated from Amboyna wood, and Compound 8 isolated from Sugi wood were subjected to antifungal assays with several fungi pathogenic to plants, i.e. *Fusarium oxysporum, Corynespora cassiicola, Cochliobolus miyabeanus, Trichoderma harzianum, Penicillium italicum* and *Aspergillus niger*. The results are shown in Table 2. Several concentrations (liquiritigenin at 20-320 μM and isoliquiritigenin at 16-265 μM) were tested to determine the minimum inhibitory concentration (MIC) for each fungus. Liquiritigenin and isoliquiritigenin isolated from Amboyna wood had a significant effect on the mycelial growth of the test fungi. A low concentration of liquiritigenin or isoliquiritigenin was enough to inhibit *F. oxysporum* and *C. miyabeanus* growth, while the other fungi needed higher concentrations. A high concentration of the compound (320 μM of liquiritigenin and 265μM of isoliquiritigenin) was required to stop the growth of *A. niger*. New agents are needed to control pathogenic fungi in plants. The findings presented here show that liquiritigenin and isoliquiritigenin isolated from Amboyna wood had a significant inhibitory effect on the mycelial growth of fungal pathogens (*11*).

The minimum inhibitory concentrations of Compound 8 against fungi pathogenic to plants are summarized in Table 2. Based on the results of antifungal assays, it can be seen that Compound 8 was effective in inhibiting the growth of the six pathogenic fungi tested. Regarding the result that an increase in the amount of Compound 8 in the medium increased the inhibitory effect, it can be assumed that the compound possesses concentration-dependent activity. The results obtained here showed that Compound 8 isolated from Sugi wood, whose activities are superior to those of liquiritigenin (2) and isoliquiritigenin (3), had a significant inhibitory effect on the mycelial growth of fungal pathogens (*26,40*).

Table 2. Minimum Inhibitory Concentrations (MICs) of Liquiritigenin (2), Isoliquiritigenin (3) and Compound 8 against Six Pathogenic Fungi.

Compound	Minimum inhibitory concentration ($\mu mol/dish$)					
	FO	CC	CM	TH	AN	PI
Liquiritigenin (2)	20	40	20	40	320	40
Isoliquiritigenin (3)	16	33	16	33	265	33
Compound 8	0.15	0.07	0.03	0.15	0.03	0.03

Notes: FO; Fusarium oxysporum, CC; Corynespora cassiicola, CM; Cochliobolus miyabeanus, TH; Trichoderma harzianum, AN; Aspergillus niger, PI; Penicillium italicum.

Conclusions

Bioassay-guided fractionation of wood extracts resulted in the isolation of four antifungal compounds: pterocarpol, liquiritigenin, isoliquiritigenin and β-eudesmol from Amboyna (*Pterocarpus indicus*); two antifungal compounds: *trans*-resveratrol and a resveratrol tetramer having two dihydrobenzofuran rings and a tetrahydrofuran ring (tentatively named as belangaranol) from Belangeran (*Shorea belangeran*); and two antifungal compounds: (-)-12-Hydroxy-6,7-secoabieta-8,11,13-triene-6,7-dial and (+)-2,7(14),10-bisabolatrien-1-ol-4-one from Sugi (*Cryptomeria japonica*). (+)-2,7(14),10-Bisabolatrien-1-ol-4-one was found to be active on antifungal assays against six plant pathogenic fungi by revealing low minimum inhibitory concentrations against all fungi tested. The results of this study revealed potential antifungal agents from tropical and temperate woods, as part of the continuing effort to discover more effective and less harmful antifungal agents.

Acknowledgements

We thank Dr. Hidemitsu Uno of the Integrated Center for Science, Ehime University for obtaining the data of X-ray crystallography, and Dr. Satoshi Yamauchi of Faculty of Agriculture, Ehime University for measurement of optical rotation. We also thank Dr. Kazutaka Itoh and Mrs. Azuma Masataka, Ogawa Tomoko, Tapa Darma for their support and assistance with this research.

References

1. Punja, Z.K.; Ukhtede, R.S. *Trends in Biotechnology* **2003**, *21*, pp 400-407.
2. Katta, S.K.; Eskridge, K.M.; Bullerman, L.B. *J. Food Prot.* **1995**, *58*, pp 1014-1017.

389

3. Georgopapadakou, N.H.; Walsh, T.J. *Science* **1994**, *264*, pp 371-373.
4. Mackay-Wiggan, J.; Elewski, B.E.; Scher, R.K. *Dermatologic therapy*, **2002**, *15*, pp 78-88.
5. Schaller, F.; Rahalison, L.; Islam, N.; Potterat, O.; Hosrettmann, K. *Helv. Chim. Acta,* **2000**, *83*, pp 407-413.
6. Okabe, T.; Saito, K.; Fujii, T.; Iinuma, K. *Mokuzai Gakkaishi* **1994**, *40*, pp 1233-1238.
7. Morita, Y.; Matsumura, E.; Okabe, T.; Fukui, T.; Ohe, T.; Ishida, N.; Inamori, Y. *Biol. Pharm. Bull.* **2004**, *27*, pp 1666-1669.
8. Tachibana, S. *Mokuzai Gakkaishi* **1995**, *41*, pp 967-977.
9. Endo, H.; Miyazaki, Y. *Bul. Nat. Inst. Hyg. Sci.* **1972**, *90*, pp 69-71.
10. Kusuma, I.W.; Azuma, M.; Darma, T.; Itoh, K.; Tachibana, S. *Holzforschung* **2005**, *59*, pp 170-172.
11. Kusuma, I.W.; Azuma, M.; Darma, T.; Itoh, K.; Tachibana, S. *Pak. J. Biol. Sci.* **2004**, *7*, pp 1735-1740.
12. Parthasarathy, M.R.; Seshadri, T.R. *Curr. Sci.* **1965**, *34*, pp 115-116.
13. Nasini, G.; Piozzi, F. *Phytochemistry* **1980**, *20*, 514-516.
14. Kong, L.D.; Zhang, Y.; Pan, X.; Tan, R.X.; Cheng, C.H.K. *Cell. Mol. Life Sci.* **2000**, *57*, pp 500-505.
15. Yahara, S.; Ogata, T.; Saijo, R.; Konishi, R.; Yanahara, J.; Miyahara, K.; Nohara, T. *Chem. Pharm. Bull.* **1989**, *37*, pp 979-987.
16. Barton, D.H.R.; Beloeil, J.C.; Billion, A.; Boivin, J.; Lallemand, J.V.; Lelandais, P.; Mergui, S. *Helv. Chim. Acta,* **1987**, *70*, pp 2187-2199.
17. Arisanya, A.; Bevan, W.L.; Hirst, J. *J. Chem. Soc.* **1959**, pp 2679-2681.
18. Chiou, L.C.; Ling, J.Y.; Chang, C.C. *Neurosci. Lett.* **1997**, *231*, pp 171-174.
19. Bors, W.; Heller, C.; Michel, C.; Saran, M. *Methods Enzymol.* **1990**, *186*, pp 343-355.
20. Colerige, S.P.O.; Thomas, P.; Scurr, J.H.; Dormandy, J.A. *Br. Med. J.* **1980**, *296*, pp 1726-1727.
21. Hertog, M.G.L.; Hollman, P.C.H.; Katan, M.B.; Klomhout, D. *Nutr. Cancer* **1993**, *20*, pp 21-29.
22. Wollenweber, E.; Dietz, V.H. *Phytochemistry* **1981**, *20*, pp 869-932.
23. Kanazawa, M.; Satomi, Y.; Mizutani, Y.; Ukimura, O.; Kawauchi, A.; Sakai, T.; Baba, M.; Okayama, T.; Nishino, H.; Miki, T. *Eur. Urol.* **2003**, *43*, pp 580-586.
24. Kong, L.D.; Zhang, X.; Pan, X.; Tan, R.X.; Cheng, C.H.K. *Cell. Mol. Life Sci.* **2000**, *57*, pp 500-505.
25. Kusuma, I.W.; Ogawa, T.; Itoh, K.; Tachibana, S. *Pak. J. Biol. Sci.* **2005**, *8*, pp 136-140.
26. Kusuma, I.W. Ph. D. Thesis, Ehime University, Ehime, Japan, 2005.
27. Kurihara, H.; Kawabata, J.; Ichikawa, S.; Mishima, M.; Mizutani, J. *Phytochemistry* **1991**, *30*, pp 649-653.

28. Yan, K.X.; Terashima, K.; Takaya, Y.; Niwa, M. *Tetrahedron* **2002**, *58*, pp 6931-6935.
29. Sotheeswaran, S.; Pasupathy, V. *Phytochemistry* **1993**, *32*, 1083-1092.
30. Fujii, F.; He, Y.H.; Terashima, K.; Takaya, Y.; Niwa, M. *Heterocycles* 2005, *65*, pp 2461-2469.
31. Croteau, R.; Kutchan, T.M.: Lewis, N.G. In *Biochemistry and Molecular Biology of plants*; Buchanan, B.B.; Gruissem, W.; Jones, R.L. Eds.: American Society of Plant Physiologists, Rockville, MD, 2000, pp 1250-1318.
32. Faucooeau, B.; Teguo, P.W.; Huguet, F.; Barrier, L.; Decendit, A.; Merillon, J.M. *Life Science* **1997**, *61*, pp 2103-2110.
33. Pacher, T.; Seger, C.; Engelmeier, D.; Vajrodaya, S.; Hofer, O.; Greger, H. *J. Nat. Prod.* **2002**, *65*, pp 820-827.
34. Likhitwitayawuid, K.; Sritularak, B. *J. Nat. Prod.* **2001**, *64*, 1457-1459.
35. Dai, J.R.; Hallock, Y.F.; Cardellina J.H.; Boyd, M.R. *J. Nat. Prod.* **1998**, *61*, pp 351-353.
36. Docherty, J.J.; Fu, M.M.; Tsai, M. *J. Antimicrob. Chemotherapy* **2001**, *47*, pp 239-246.
37. Cheng, S.S.; Chang, H.T.; Chang, K.H.; Tsai, W.J. *Bioresource Technol.* **2003**, *89*, pp 99-102.
38. Cheng, S.S.; Lin, J.Y.; Chang, S.T. *J. Agric. Food Chem.* **2005**, *53*, pp 614-619.
39. Nakajima, K.; Yoshimoto, T.; Fukuzumi, T. *Mokuzai Gakkaishi* **1980**, *26*, 698-702.
40. Kusuma, I.W.; Tachibana, S. *Phytochemical Analysis* to be submitted.
41. Fang, J.M.; Jan, S.T.; Cheng, Y.S. *J. Chem. Research (S)* **1986**, pp 350-351.
42. Wang, S.Y.; Wu, J.H.; Shyur, L.-F.; Kuo, Y.H.; Chang, S.T. *Holzforschung* **2002**, *56*, pp 487-492.
43. Arihara, S.; Umeyama, A.; Bando, S.; Kobuke, S.; Imoto, S.; Ono, M.; Yoshikawa, K.; Amita, K.; Hashimoto, S. *Mokuzai Gakkaishi* **2004**, *50*, pp 413-421.
44. Nagahama, S.; Tazaki, M. *Mokuzai Gakkaishi* **1993**, *39*, pp 1077-1083.
45. Lin, W. H.; Fang, J.M.; Cheng, Y.S. *Phytochemistry* **1995**, *40*, pp 871-873.

Energy from Forest Biomass

Chapter 25

Pretreatments for Converting Wood into Paper and Chemicals

William R. Kenealy[1], Carl J. Houtman[1], Jose Laplaza[1], Thomas W. Jeffries[1], and Eric G. Horn[2]

[1]Forest Service, Forest Products Laboratory, U.S. Department of Agriculture, Madison, WI 53726
[2]Biopulping International Incorporated, 2912 Syene Road, Madison, WI 53713

Biorefining wood into paper and chemicals is not as easy as making a single traditional paper product. Paper is made from the cellulose-containing fractions of wood and processing may remove lignin and hemicellulose components. The yield and composition of the product depend upon the type of paper being produced. The paper process often alters the non-cellulose fractions, making them less suitable for conversion to other valuable products. This chapter will review methods of obtaining hemicellulosic sugars and investigate the potential for obtaining sugars by a wood pretreatment before it is made into pulp. As an example of hemicellulose removal the results of a new pretreatment using diethyl oxalate are also described. Hemicellulose removal by this system provides other benefits to the pulping process.

Biomass use is touted as a solution to the burgeoning petroleum importation of our nation and as a way to become more energy independent (1). When biomass is targeted for this purpose, carbohydrates are generally converted to products. Corn is used to make ethanol, and it is principally starch that is saccharified to glucose and fermented to ethanol. Sucrose from sugar cane is used to make ethanol in Brazil. Obtaining sugars from lignocellulose is much more difficult. Cellulose has very little solubility in water and a crystalline nature that must be overcome before it can be converted into glucose. The problem is compounded in lignocellulose when the cellulose has additional polymers such as hemicellulose and lignin that must be removed or destroyed before the cellulose is accessible.

The problems with biomass use are in part inherent in the structure of the lignocellulose and also the source of the material. Corn stover might be an easier material to convert to fermentable carbohydrates than wood, but collection, storage, and soil depletion are problems that must be solved for this feedstock. Wood is more easily transported and collected at processing sites; however, more energy is required for processing wood than for corn stover. The "Healthy Forests Initiative"(2) and Healthy Forest Restoration Act of 2003 identified small-diameter woody materials that exacerbate forest fires as a problem and its solution could provide materials for biomass use. However, these materials are often located in regions where there is little transportation infrastructure and little water present in the region for local processing.

The problems with lignocellulose use are daunting, but they can be addressed. Industry has long made paper products from wood and disposed of residual material by combusting the waste to gain energy for in-plant use. While the total conversion of wood into products (such as ethanol transportation fuel) may not be widely available, the pulp and paper industry does refine wood into paper products and energy. With initiatives like Agenda 2020's "Advancing the Forest Biorefinery" to develop "Integrated Forest Biorefineries" and obtain "Value Prior to Pulping" the industry is considering biorefining; the removal of the selected components before the paper product is made (3, 4). This is an opportunity for industry to expand its product portfolio.

Biorefineries are being considered for many different starting materials and products, not all of which consider paper as a product (5). Mutiple products can be made by fermentation or chemical methods from intermediates derived from wood. In the ideal case, for paper manufacturing all of material that does not end up in the paper product would be converted into valuable products and more useful forms of energy. In reality, there are trade-offs in energy efficiency, product yields, and potential waste streams that develop into a balancing act of trying to maintain yield and paper properties while obtaining source material for other products. Kraft pulping is often used as the target for biorefining in the pulp and paper industry. The hemicellulose is burned in the black liquor, but it contributes less to the fuel value than does the lignin. Recovery of the hemicellulosic sugars in front of the pulping operation will provide new product opportunities but may also change the operation of the recovery boilers. Research is required to evaluate the many approaches and processes that might be used.

Advances are being made in the conversion and processing of lignocellulose to useful products. When the cellulose is converted to ethanol or other products the hemicellulose must also be used since it is a significant fraction of lignocellulosic materials. If cellulose is to be a significant portion of the product, as in paper, the hemicellulose and lignin are the only materials that can provide new products. In both cases, the processing and ultimate fate of the hemicellulose is important for the economics and design of the process. In this chapter, we will consider a problem that vexes the proponents of biorefining in the pulp and paper industry: how hemicellulose is removed from wood, in a useable form, without harming the cellulose product.

Methods for Removing Hemicellulose

There are many processes that can remove hemicellulose from lignocellulose. Analytical and structural studies of hemicellulose often first isolate holocellulose obtained by oxidative means (6-8). Hemicelluloses can be removed from holocellulose and lignocellulosic materials by hot water, alkaline, and solvent extractive methods (6-15). The hemicellulose is obtained in a polymeric form facilitating the study of hemicellulose structure. Isolation of holocellulose and its conversion to chemicals and fuel would be very costly. Most studies concerning hemicellulose removal pertain to pretreatments of lignocellulose for its conversion to chemicals and fuels, and the isolation of polymeric hemicellulose is generally not a concern.

The carbohydrates present in plant cell walls depend upon the species, source, age, size, and treatment history of the material (6-13). Composition of plant cell mass is determined by isolation of polymeric forms and by determination of the sugar components in hydrolysates (10). A combination of methods shows that the glucan portion of plants is predominantly located in the cellulose, while xylan, galactan, arabinan, and mannan are located in the hemicellulose. Many extraction methods have been explored for the total conversion of lignocellulose into chemicals and fuel (16). In this latter regard, the goal of hemicellulose removal is to pretreat the lignocellulosic material so that the cellulose and hemicellulose can be converted into soluble sugars (17). Damage to the cellulose is of little concern when the purpose of treatment is to provide better access to the cellulose for enzymatic or acid digestion. Economics and prevention of process-generated metabolic inhibitors drive research into these pretreatments (17, 18). Metabolic inhibition is irrelevant when the carbohydrates are converted chemically to other products like furfural (19) and perhaps feed additives (*20*), although conversion and purification processes might be affected.

These pretreatments for the total conversion of lignocellulose to fuels and chemicals are good starting points for the investigation of processes to derive paper and other products from wood. Not all of these pretreatments will have applications in the biorefining of wood to a combination of paper, chemicals, and fuel. The effects of many of these pretreatments on the ability to make paper products have not been investigated. Another starting point is to examine pulping technologies that are proven to make paper products and discern if there are other products that can be obtained. While this chapter relates to hemicellulose-derived products, lignin can also be converted into products that have greater value than its fuel value.

There are three general chemical categories of hemicellulose extraction: alkaline, acidic, and solvent. Some of these methods have been used prior to making paper products; others have been developed for the purpose of facilitating total conversion of lignocellulose to fuels and chemicals. The less intensive extractions (in regards to temperature, acidity, and alkalinity) will liberate the water-soluble carbohydrates (*6, 10, 14*) in addition to what is liberated by the acid and base additions. Chemical methods can be used in conjunction with more aggressive treatments such as steam explosion pulping. Explosive decompression can be used with or without chemical impregnation of the chips to break apart wood fiber (*21*).

Alkaline Methods of Hemicellulose Removal

Efficient extraction and recovery of hemicellulose from wood by alkaline means requires finely ground wood or delignification of the wood (*6-9*). Grinding wood chips to extract hemicellulose prior to pulping would be cost prohibitive and damaging to the fiber, so less efficient extraction methods must be employed. The yield of extractable sugars from wood chips would be considerably less than that with ground wood or holocellulose. When the purpose of the extraction is to make paper and chemicals, the right treatment conditions can still release sugars.

Kraft pulping is the most used pulping process for making paper products. It is a high temperature alkaline process that removes both lignin and hemicellulose components from softwoods and hardwoods. The condition of the hemicellulosic materials, the complexity of the organics present, and the sulfur present in the black liquor make product development from this material very difficult. It is precisely this difficulty that has provided the incentive to explore the pretreatment of wood chips to recover the hemicellulosic fraction before pulping (*3, 4*). The objective is to make additional products along with the same kraft paper product. However, the yield might be affected if significant mannose is removed (*22, 23*). Removal of some hemicellulose without contaminating the fraction with lignin components might be possible with gentle alkaline treatments.

Cold soda pulping is a chemimechanical process that has been known for a long time (*24*). This process has been used to produce corrugating board and writing paper. On hardwoods such as aspen, the process reduces the content of hemicellulose in the product and increases both the glucan and klason lignin contents. Overall there is some loss in the net glucan, xylan, and acid-soluble lignin (*24*). The recovery of components from the treatment liquor has not been explored, and there is significant biological oxygen demand in the spent liquor. It is likely that hemicellulosic oligosaccharides, lignin components, and acetate contribute to the biological oxygen demand. A recent patent application reported cold caustic extraction of wood chips and hardwood pulp for the production of high purity xylose (*25*).

Ethylenediamine and monoethylamine in alkaline solutions and other amines (*26*) delignify pine and produce pulps with hemicellulose contents higher than kraft with roughly the same kappa and yield (*27*). Treatments such as ammonia fiber explosion and ammonia recycled percolation will degrade some of the hemicellulose and alter the crystallinity and accessibility of the cellulose to digestive enzymes (*17*). These treatments and lime or other alkaline treatments will alter lignin structure and also delignify the material. The potential for contamination of the carbohydrate stream with lignin digestion products may limit applications of an alkaline pretreatment of wood chips for the purpose of making paper and chemicals. There are a variety of alkaline treatments that are part of solvent extractions. These processes will be discussed in the solvent extraction section.

Acidic Methods of Hemicellulose Removal

Much is known about the dilute acid prehydrolysis for making cellulose accessible to further degradation (*16, 17, 28-45*). Within this knowledge base is information about temperatures and acid charges that do not directly degrade the cellulose. Early experiments on water and steam (acetic acid autohydrolysis) cooking of aspen showed little damage to the cellulose (*46*), and similar treatments of pine (*47*), eucalypt (*48*), and sweetgum (*49*) provided low hemicellulose sulphate pulps. The methods to produce high alpha cellulose pulps were reviewed and significant reduction in hemicellulose content was noted with prehydrolysis (*50*). These reports could be a starting point for investigating the use of these processes for converting wood chips to paper products and chemicals. The knowledge base is growing and new patents are continually being issued for both dilute acid (*51-53*) and strong acid hydrolysis (*54, 55*). However, little effect on cellulose degradation does not mean there will not be drastic effects on paper properties. Excessive temperatures will darken pulps, making them less useful for mechanical paper grades that require brightness. Small changes in the composition of the wood can dramatically affect the strength properties of the paper product. The historical data on acid prehydrolysis can be a very important starting point, but to employ this technology there must be empirical testing on the end products of interest. Both the paper product and the carbohydrate resource must be tested to prove the process is useful.

Acetic acid pulping and formic acid pulping are processes that can produce good paper products (*56-58*). Using performic acid in the Milox process produces a soluble carbohydrate stream from birch processing that can be separated from the lignin-containing stream. Acetic acid is an additional product recovered during the formic acid recycling (*58*). Since there is an appreciable degradation of the lignin during this process, problems are likely to occur with aromatic contamination of the hemicellulosic sugars; however, there is little sulfur in the product, so developing more refined uses would be less problematic than developing uses for kraft liquors.

Acetic acid pulping is an acidic method, but the use of large volumes of acetic acid and the precipitation of lignin components with ethyl acetate could also classify this as a solvent method (*56, 57, 59*). In some configurations, acetic acid pulping uses additional mineral acids that aid in the degradation of lignin components. The hemicellulose is increasingly degraded with higher temperatures, longer cook times, and greater loadings of acetic acid (*57*).

A gaseous source of acid has also been used. Sulfur dioxide can be introduced to the chips and sulfonic acid can be formed (*33, 36, 60-62*). This is an alternate way of introducing acid that does not involve impregnation of the chips. The extent of treatment can be varied by altering the time or the temperature, providing flexibility in the treatment (*63*). Sulfur dioxide treatment makes the sugar more accessible to digestion and has been used with wood chips in animal feed tests (*20*). With sulfur dioxide and steam explosion there is better recovery of the sugars when the chip size is increased (*64*). This is probably due to the increased damage to the lignocellulose components, which results from treating smaller chip sizes where the sugars are converted further from monosaccharides into derivatives. This is in contrast to alkaline extraction, where the extraction is much better with smaller material.

Solvent Methods of Hemicellulose Removal

Dimethyl sulfoxide (DMSO) is a solvent that has been used in the study of hemicellulose structure (*65, 66*). This solvent system extracts the hemicellulose and does not cleave the acetyl esters, allowing for the determination of the structure of the hemicellulose. DMSO is used on holocellulose and to extract hemicellulose from pulps. The use of this solvent for applications in pulping would be expensive and there would be potential hazard problems with handling large volumes of the solvent.

Many solvents have been reported that cause wood to swell (*67*). Swelling is apparently due to interactions with the cellulose component of the wood. Solvents that swell the wood might be useful in pretreatments that will allow access to various components of the wood, provided that the solvent can be displaced without shrinkage and the solvent can be recovered. Other solvents have been tested for the ability to make pulps; alcohols (*68-71*), phenols (*69, 72*), esters (*59*), and combinations of solvents with acidic (*73-77*) and alkaline additions (*26, 27, 68, 70*). The objective in most of these solvent pulping systems is to remove lignin from the lignocellulosic material, leaving the cellulose-containing fractions for paper making. In general, the hemicellulose is retained partly in the cellulose-containing fraction and partly in the degraded lignin fraction. Ethanol and methanol are the primary alcohols used in organosolv processes. These processes use aqueous alcohol solutions at elevated temperatures to degrade and remove lignin fractions. Many additions and modifications have been made to these solvent processes.

Ethanol and methanol are the most often used solvents. Phenol and butanol can delignify pine meal (*69*). Other alcohols have been tried, but the results have not been better than those obtained with ethanol and methanol (*27, 68*). There are many variations on using alcohols in the pulping process. The addition of compounds such as neutral alkali earth metal salts (*73, 75*) indicates that divalent cations are better than monovalent and trivalent cations when liberating softwood fibers. In these studies, lignin concentrations were reduced in the product but no characterization of the extract was made.

Hemicellulose Removal with Pulp and Paper Production

Biorefining for the pulp and paper industry is envisioned as the removal and use of material that does not contribute to the paper product. At present, there is little data in the published literature that indicates how much material can be removed and how to remove it without altering the paper product. The primary target for application of biorefining is in kraft pulping since it is the largest segment of the paper industry and has a low product yield of ~50%. This provides the potential for more carbohydrate to be removed for other uses. Low yield chemical pulps have the greatest potential, but there have been studies on the pretreatment of mechanical pulps that relate to the topic of biorefining.

Mechanical and thermomechanical pulps are high yield pulps. Generally the conversion of the wood chips to paper products is in the range of 90-95% by weight. There are losses in the process and these are generally from extractibles, water soluble carbohydrates and hemicellulose. It may not be possible to economically recover these small amounts of released materials with a stand alone pretreatment process. Waste disposal costs could be decreased, but without other advantages the recovery of hemicellulosic sugars from thermomechanical processes would be costly.

Mechanical pulps generally have low strength and the product specifications often require greater strength. To increase the strength, additives such as chemical pulps or starch are blended with the pulp. If the process that recovers the hemicellulosic sugars prior to pulping delivers a stronger product, the economics improve by saving on chemical pulp or other additions. There are also high cost factors in the production of mechanical pulps. One of the highest costs in the production of mechanical pulps is the electrical energy required for refining. If the energy of producing pulps can be lessened by a process of hemicellulosic sugar removal, the economics also improve.

Oxalic acid pulping is a process that pretreats wood chips for thermomechanical pulping, saves electrical refining energy, and delivers a product with increased strength (78, 79). We have recently determined that oxalic acid pulping also releases hemicellulosic sugars from the chips (80). The combination of electrical refiner energy savings and stronger pulps puts forth an economic argument for the use of oxalic acid pulping with a reasonable repayment for the investment (79). The addition of carbohydrate recovery and conversion to additional products should enhance the use of the process.

Oxalic acid pulping should be viewed as a dilute acid prehydrolysis. Since the process has been developed for thermomechanical pulping, the extent of carbohydrate extraction must be balanced against the electrical refiner energy savings and the paper properties, such as strength and brightness. In general, dilute acid pretreatments may be beneficial to thermomechanical pulps. The degree of hydrolysis must be controlled to be able to show a benefit. The hydrolysis can be affected by the amount of acid used, the time for which the wood is treated, and the temperature of treatment.

Mineral acids, oxalic acid, sulfur dioxide, diethyl oxalate, and acetic acid have all been used in dilute acid pretreatments. The oxalic acid and diethyl oxalate pretreatments are examples of processes that have been proven effective for thermomechanical pulps and are able to generate hemicellulosic sugar hydrolysates. Other methods may be applicable. However, most of the work on mineral acids is performed as a pretreatment to total carbohydrate conversion, and most of the work on hot water extraction (involving autohydrolysis via acetic acid) is directed towards kraft pulp since the temperatures used (160°C to 170°C) have detrimental effects on the brightness of thermomechanical pulps.

The sulfuric and acetic prehydrolyses are normally performed by suspending the wood chips in a given volume of water or acid and running the reaction. Oxalic acid is used in solution (0.3% wt/vol) and impregnated into the wood chips; the bulk of the solution is then drained away (78, 79). Diethyl oxalate (DEO) is injected into a digester containing preheated wood chips (80, 81). When DEO contacts water at temperatures around 140°C, there is a rapid hydrolysis of the ester, depositing oxalic acid in the water within the wood chip. Further heating hydrolyzes the carbohydrates. After the reaction, carbohydrates are extractable into water in a separate process step. As Figure 1 indicates, for both aspen and pine chips, increasing the amount of DEO added to the wood chips increases the amount of carbohydrates released from the chips. Similar carbohydrate extractions are also obtained by oxalic acid pulping (80).

The residual wood chips are enriched in cellulose and lignin relative to the control wood chips. The carbohydrates released represent the hemicellulose present, with the cellulose undegraded. If the duration of the treatment is extended, more carbohydrate is released. Figure 2 shows the amount of carbohydrate removed relative to the electrical refiner energy required to process the wood chips to thermomechanical pulp. The decrease of energy required to process wood could be related to the strength of the wood chips themselves. Wood strength has been correlated to the loss of arabinan or galactan components (82, 83), and these components are removed from the wood chips prior to any cellulose damage. The pulps from these studies have produced handsheets with strength properties (burst, tensile, and tear indexes) equal to or better than that of the control handsheets for chips with carbohydrate removal (total of arabinan, galactan, glucan, xylan, and mannan) up to 6% of the original wood chip weight. The electrical energy saved is generally between 30% and 50% for these pulps, and the handsheets show enhanced strength with brightness similar to the controls.

The relationship between carbohydrate extracted from the chips and electrical refiner energy savings may hold for other acid prehydrolysates. If so, the vast amount of literature on dilute acid hydrolysis might be mined for conditions that will produce a level of extractable carbohydrate under temperature conditions that will produce high quality bright paper.

Characterization of Hemicellulosic Extracts

The extraction of hemicellulosic sugars by an acidic prehydrolysis produces sugars that are mostly monosaccharides and small oligosaccharides. Acetyl esters normally present on the hemicellulose are partly cleaved, leaving some esters still attached to the sugars (80). This complicates a detailed analysis of the

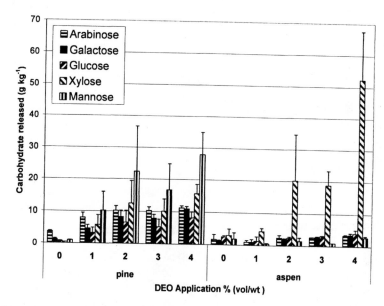

Figure 1. Carbohydrates released into aqueous solution from wood chips treated with DEO (0, 10, 20, 30 and 40 mL kg^{-1} wood) at 135 °C–140 °C for 30 min. Results represent the average and error bars the standard deviation for at least two treatments and extractions as described (80).

sugars present and necessitates a secondary hydrolysis in the analysis of the total carbohydrates present.

A second hydrolysis is even more important for the identification of the carbohydrate composition when the hemicellulose is extracted by an alkaline system. Here the hemicellulose comes out as a polymeric material and the acetyl esters as acetate. Polymeric oligosaccharides are easier to separate and concentrate from the aqueous solutions by neutralization and ultrafiltration. The acetate could be recovered as a salt, but if acetic acid is the desired product then acid addition will be required.

The separation of acetic acid and sugars can be accomplished with membrane filtration and has been proposed as a part of ethanol production from wood (*84, 85*). Reverse osmosis membranes can facilitate both separation and concentration of the carbohydrates. Systems have been configured for concentrating sugar solutions in industrial settings. One example is to recycle wash water from rinsing trucks that ship high fructose corn syrup (*86*). The sugar is recovered, the rinse water is reused, and the heat in the wash water is retained facilitating further cleaning.

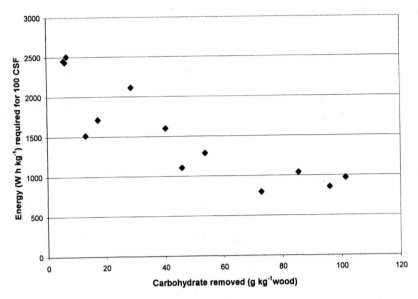

Figure 2. Energy required to process pine wood chips to 100 mL CSF. Chips were processed with DEO at 135 ℃–140 ℃ as described (81). Carbohydrate was calculated by the sum of identified sugars removed from wood chips and corrected for water of hydrolysis. Energy at 100 CSF was calculated from a power function fit of a plot of energy used in pressurized and atmospheric refiners for at least three pulps vs CSF.

Acidic pretreatments and hydrolysis are known to produce inhibitors of yeast that metabolize the sugars to ethanol (87). Little work has been performed to characterize inhibitors of metabolism produced by alkaline extraction methods. There are many post-extractive treatments for cleaning inhibitors from the extraction liquids (87, 88). Enzymatic (89), chemical (87, 88), and adsorbent methods (90, 91) can remove inhibitors and allow more extensive and faster conversion of the sugars to the product. The adsorbent methods may be particularly interesting because they offer the potential to recover other defined products from the adsorbent.

For the work on DEO pretreatments, we tested the raw extract for the metabolism of the sugars to determine if large concentrations of inhibitors were present. Ethanol production will require higher concentrations of sugars, which may result from different extraction procedures or reverse osmosis filtration. Figure 3 shows the result of a test of a supplemented DEO pine extract and metabolism by *Pichia stipitis*. The sugar concentrations were determined by

HPLC (*92*) and are reported in the figure as monosaccharides. There was some delay in the metabolism of the sugars, and then almost all the monosaccharides were consumed, with glucose, galactose, and mannose appearing to be used simultaneously, followed by xylose and then arabinose. A secondary hydrolysis was also performed, so the sugar derivatives could also be followed. The proportions of individual sugars that were present as monosaccharides in the extract were as follows: 89% arabinose, 38% galactose, 25% glucose, 56% xylose, and 30% mannose. By the end of the incubation, the following proportions of the total sugar present were consumed: 80% arabinose, 45% galactose, 66% glucose, 93% xylose, and 48% mannose. In all cases except arabinose, more sugars were consumed than could be accounted for by the monosaccharide content, indicating an ability of *P. stipitis* to metabolize sugar derivatives present in the extract.

A similar experiment was performed with *Saccharomyces cerevisiae*. Using the same extract and the conditions described above for *P. stipitis*, *S. cerevisiae* consumed the following percentages of the total sugar present: 5% arabinose, 46% galactose, 44% glucose, 6% xylose, and 35% mannose. *S. cerevisiae* does not normally use pentoses, and *P. stipitis* readily uses pentoses. These experiments also indicated that there might be some benefit in treating the extract with enzymes that might target mannose, galactose, and glucose derivatives so better conversion of the hexoses could be accomplished. The DEO and oxalic acid pretreatments have good potential in biorefining and thermomechanical pulping. Extending the reaction conditions by increasing heat, chemical loading, or time of pretreatment might also make these useful pretreatments for chemical pulps.

Summary

There are many methods of removing hemicellulose from lignocellulose. Acidic extraction results in short oligosaccharides or monomeric sugars. Alkaline and solvent methods provide more polymeric material. Polymeric material is easier to concentrate, using precipitation or ultrafiltration. If the extraction method does not deliver concentrated sugars, then reverse osmosis can be used. At the same time, reverse osmosis can separate the acetic acid from the sugars. Alkaline methods produce acetate, which will require acid addition and salt formation to derive acetic acid.

Acidic pretreatments are being used in commercial start-up ventures for the total conversion of lignocellulose to chemicals and fuels. The pretreatments are followed by further acid or enzymatic hydrolysis. The predominant product in these ventures is ethanol for use as a fuel. Acidic pretreatments are valuable in thermomechanical processes where electrical refiner energy savings and stronger

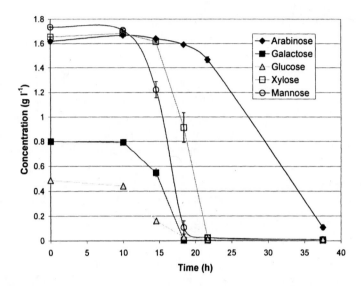

Figure 3. Carbohydrate usage (monosaccharide content) by Pichia stipitis (CBS6054) in 40-mL DEO kg^{-1} treated pine extract + yeast nitrogen base (pH 6.0). Data are average of two experiments, error bars represent standard deviation.

products are a result that is correlated with hemicellulose removal. Testing of other pretreatments might provide similar benefits. Pretreatments that are of value in thermomechanical pulping might also be of value in chemical pulping, and investigation of these techniques is underway.

References

1. Perlack, R.; Wright, L. L.; Turhollow, A. F.; Graham, R. L.; Stokes, B. J.; Erbach, D. C. *Biomass as feedstock for a bioenergy and bioproducts industry: The technical feasibility of a billion-ton annual supply*; Oak Ridge National Laboratories: Oak Ridge TN, April, 2005.
2. *Healthy forests; an initiative for wildfire prevention and stronger communities*; August 22, 2002.
3. Raymond, D.; Closset, G., *Solutions!* **2004**, (September), pp 49-53.
4. Thorp, B.; Raymond, D., *PaperAge* **2004**, *120(7)*, pp 16-18.
5. Kamm, B.; Kamm, M., *Appl. Microbiol. Biotechnol.* **2004**, *64*, pp137-145.

6. Baeza, J.; Freer, J., In *Wood and cellulosic chemistry*, ed.; Hon, D.; Shiraishi, N., Marcel Dekker, Inc.: New York, 2001.

7. Timell, T. E., In *Advances in carbohydrate chemistry*, ed.; Wolfrom, M. L.; Tipson, R. S., Academic Press: New York, 1964; 19, pp 247-302.

8. Timell, T. E., In *Advances in carbohydrate chemistry*, ed.; Wolfrom, M. L.; Tipson, R. S., Academic Press: New York, 1965; 20, pp 409-483.

9. Jacobs, A.; Dahlman, O., *Biomacromolecules* 2001, *2*, pp 894-905.

10. Pettersen, R. C., In *The chemistry of solid wood*, ed.; Rowell, R., American Chemical Society: Washington, DC, 1984; pp 57-126.

11. Aspinall, G. O., In *Biogenesis of plant cell wall polysacchardies*, ed.; Loewus, F., Academic press: New York, 1973; pp 95-115.

12. Aspinall, G. O., In *Advances in carbohydrate chemistry*, ed.; Wolfrom, M. L.; Tipson, R. S., Academic Press: New York, 1959; 14, pp 429-468.

13. Sun, X.-F.; Sun, R.; Fowler, P.; Baird, M. S., *Journal of Agricultural and Food Chemistry* 2005, *53*, pp 860-870.

14. Willfor, S.; Holmbom, B., *Wood Science and Technology* 2004, *38*, pp 173-179.

15. Adams, G. A., In *Methods in carbohydrate chemistry*, ed.; 1965; 5, pp 146-147.

16. Wyman, C.; Dale, B.; Elander, R. T.; Holtzapple, M.; Ladisch, M.; Lee, Y. , *Bioresource Technology* 2005, *96*, pp 1959-1966.

17. Mosier, N.; Wyman, C.; Dale, B.; Elander, R. T.; Lee, Y. Y.; Holtzapple, M.; Ladisch, M., *Bioresource Technology* 2005, *96*, pp 673-686.

18. Eggeman, T.; Elander, R. T., *Bioresource Technology* 2005, *96*, pp 2019-2025.

19. Rowell, R.; Hajny, G. J.; Young, R. A., In *Introduction to forest science*, ed.; Young, R. A., John Wiley & Sons, Inc.: 1982; 4, pp 451-469.

20. Hajny, G. J. *Biological untilization of wood for the production of chemicals and foodstuffs*; USDA Forest Service Forest Products Laboratory: Madison WI, 1981; 65.

21. Kokta, B. V.; Ahmed, A., In *Environmentally friendly technologies for the pulp and paper industry*, ed.; Young, R. A.; Akhtar, M., John Wiley & Sons, Inc.: New York, 1998; pp 191-214.

22. Vaaler, D.; Syverud, K.; Seem, B.; Moe, S. T., *Tappi Journal* 2005, *4(4)*, pp 23-27.

23. Copur, Y.; Makkonen, H.; Amidon, T. E., *Holzforschung* 2005, *59*, pp 477-480.

24. McGovern, J. N.; Springer, E. L. In *History of fpl cold soda cmp process*, TAPPI Pulping Conference, New Orleans, October 30-November 2, 1988; TAPPI Press: New Orleans, 1988; pp 641-648.

25. Svenson, D. R.; Li, J. United States 2005.

406

26. Abbot, J.; Bolker, H. I., *Svensk Papperstidning* **1984**, *87(12)*, pp R105-R109.
27. Green, J.; Sanyer, N., *Tappi Journal* **1982**, *65(5)*, pp 133-137.
28. Maloney, M. T.; Chapman, T. W., *Biotechnol. Prog.* **1986**, *2(4)*, pp 192-202.
29. Maloney, M. T.; Chapman, T. W.; Baker, A. J., *Biotechnology and Bioengineering* **1985**, *27*, pp 355-361.
30. Springer, E. L., *Industrial and Engineering Chemistry Product Research and Development* **1985**, *24*, pp 614-623.
31. Springer, E. L., *Tappi* **1966**, *49(3)*, pp 102-106.
32. Springer, E. L.; Harris, J. F., *Svensk Papperstidning* **1982**, *85(15)*, pp R152-R154.
33. Springer, E. L.; Libkie, K. A., *TAPPI* **1980**, *63(7)*, pp 119-120.
34. Springer, E. L.; Zoch, L. L., *Tappi* **1968**, *51(5)*, pp 214-218.
35. Conner, A. H.; Libkie, K. A.; Springer, E. L., *Wood Fiber Sci.* **1985**, *17(4)*, pp 540-548.
36. Harris, J. F.; Baker, A. J.; Conner, A. H.; Jeffries, T. W.; Minor, J. l.; Pettersen, R. C.; Scott, R. W.; Springer, E. L.; Wegner, T. H.; Zerbe, J. I. *Two-stage dilute acid hydrolysis of wood: An investigation of fundamentals*; USDA Forest Service: Madison WI, 1985.
37. Harris, J. F.; Scott, R. W.; Springer, E. L.; Wegner, T. H., In *Progress in biomass conversion*, ed.; Tillman, D. A., Academic Press: Orlando, FL, 1984; 5, pp 102-141.
38. Zerbe, J. I.; Baker, A. J., In *Energy from biomass*, ed.; Klass, D., Institute of Gas Technology: Chicago, 1987; pp 927-947.
39. Conner, A. H.; Lorenz, L. F., *Wood Fiber Sci.* **1986**, *18(2)*, pp 248-263.
40. Conner, A. H.; Wood, B. F.; Hill, C. G.; Harris, J. F., *Journal of Wood Chemistry and Technology* **1985**, *5(4)*, pp 461-489.
41. Carrasco, F.; Roy, C., *Wood Science and Technology* **1992**, *26*, pp 189-208.
42. Brennan, M. A.; Wyman, C., *Applied Biochemistry and Biotechnology* **2004**, *113-116*, pp 965-976.
43. Liu, C.; Wyman, C., *Bioresource Technology* **2005**, *96*, pp 1978-1985.
44. Lloyd, T. A.; Wyman, C., *Bioresource Technology* **2005**, *96*, pp 1967-1977.
45. Wyman, C.; Dale, B.; Elander, R. T.; Holtzapple, M.; Ladisch, M.; Lee, Y. *Bioresource Technology* **2005**, *96*, pp 2026-2032.
46. McGovern, J. N.; Brown, K. J.; Kraske, W. A., *TAPPI* **1949**, *32(10)*, pp 440-448.
47. Simmonds, F. A.; Kingsbury, R. M.; Martin, J. S.; Mitchell, R. L., *TAPPI* **1956**, *39(9)*, pp 641-647.
48. Meller, A., *TAPPI* **1950**, *33(5)*, 248-253.
49. Simmonds, F. A.; Kingsbury, R. M.; Martin, J. S., *TAPPI* **1955**, *38(3)*, pp 178-186.

50. Richter, G. A., *TAPPI* **1955**, *38(3)*, pp 129-150.
51. Torget, R. W.; Padukone, N.; Hatzis, C.; Wyman, C. US 6,022,419, 2000.
52. Nguyen, Q. A.; Tucker, M. P. US 6,423,145 B1, 2002.
53. Schmidt, A. J.; Orth, R. J.; Franz, J. A.; Alnajjar, M. US 6,692,578, 2004.
54. Lightner, G. E. US 6,258,175 B1, 2001.
55. Farone, W. A.; Cuzens, J. E. US 5,726,046, 1998.
56. Young, R. A., In *Environmentally freindly technologies for the pulp and paper industry*, ed.; Young, R. A.; Akhtar, M., John Wiley & Sons, Inc.: New York, NY, 1998; pp 133-156.
57. Young, R. A.; Davis, J. L., *Holzforschung* **1986**, *40*, pp 99-108.
58. Sundquist, J.; Poppius-Levin, K., In *Environmentally freindly technologies for the pulp and paper industry*, ed.; Young, R. A.; Akhtar, M., John Wiley & Sons, Inc.: New York, NY, 1998; pp 157-190.
59. Aziz, S.; Mcdonough, T. J., *TAPPI* **1987**, *70(3)*, pp 137-138.
60. Boussaid, A.; Cai, Y.; Robinson, J.; Gregg, D. J.; Nguyen, Q. A.; Saddler, J. N. *Biotechnol. Prog.* **2001**, *17*, pp 887-892.
61. Soderstrom, J.; Galbe, M.; Zacchi, G., *Biotechnol. Prog.* **2004**, *20*, pp 744-749.
62. Pan, X.; Zhang, X.; Gregg, D. J.; Saddler, J. N., *Applied Biochemistry and Biotechnology* **2004**, *113-116*, pp 1103-1114.
63. Overend, R. P.; Chornet, E., *Phil Trans R Soc Lond A* **1987**, *321*, pp 523-536.
64. Cullis, I. F.; Saddler, J. N.; Mansfield, S. D., *Biotechnology and Bioengineering* **2004**, *85(4)*, pp 413-421.
65. Hagglund, E.; Lindberg, B.; McPherson, J., *ACTA Chemica Scandinavica* **1956**, *10*, pp 1160-1164.
66. Bouveng, H. O.; Lindberg, B., *Methods in Carbohydrate Chemistry* **1965**, *5*, pp 147-150.
67. Mantanis, G. I.; Young, R. A.; Rowell, R., *Holzforschung* **1994**, *48*, pp 480-490.
68. Aziz, S.; Sarkanen, K., *Tappi Journal* **1989**, *72(3)*, pp 169-175.
69. April, G. C.; Kamal, M. M.; Reddy, J. A.; Bowers, G. H.; Hansen, S. M., *TAPPI* **1979**, *62(5)*, pp 83-85.
70. Dahlmann, G.; Schroeter, M. C., *Tappi Journal* **1990**, *73(4)*, pp 237-240.
71. Lonnberg, B.; Laxen, T.; Sjoholm, R., *Paperi ja Puu* **1987**, *69(9)*, pp 757-762.
72. Sakakibara, A.; Edashie, Y.; Sano, Y.; Takeyama, H., *Holzforschung* **1984**, *38*, pp 159-165.
73. Paszner, L.; Behera, N. C., *Holzforschung* **1985**, *39*, pp 51-61.
74. Paszner, L.; Cho, H. J., *TAPPI* **1989**, *72(2)*, pp 135-142.
75. Yawalata, D.; Paszner, L., *Holzforschung* **2004**, *58*, pp 7-13.

76. Goyal, G. C.; Lora, J. H.; Pye, E. K., *TAPPI* **1992**, *75(2)*, pp 110-116.
77. Sun, R.; Lu, Q.; Sun, X.-F., *Cell Chem. Technol.* **2002**, *36(3-4)*, pp 243-263.
78. Akhtar, M.; Swaney, R. E.; Horn, E. G.; Lentz, M. J.; Scott, G. M.; Black, C. C.; Houtman, C. J.; Kirk, T. K. 2002.
79. Swaney, R. E.; Akhtar, M.; Horn, E.; Lenz, M.; Klungness, J.; Sabourin, M. In *Oxalic acid pretreatment for mechanical pulping greatly improves paper strength while maintaining scattering power and reducing shives and triglycerides*, Tappi Fall Technical Conference: Engineering pulping &PCE & I, 2003; Tappi Press, Atlanta GA.: 2003.
80. Kenealy, W.; Horn, E.; Houtman, C. J.; Swaney, R. E.; Davis, M. W., *Holzforschung* **2006**, Submitted.
81. Kenealy, W.; Horn, E.; Houtman, C. J., *Holzforschung* **2006**, submitted.
82. Winandy, J. E.; Lebow, P. K., *Wood Fiber Sci.* **2001**, *33(2)*, pp 239-254.
83. Curling, S. F.; Clausen, C. A.; Winandy, J. E., *Forest Products Journal* **2002**, *52*, pp 34-39.
84. Ethanol research breakthrough:Wood feedstock. http://renewableenergyaccess.com/rea/news/story?id=22228
85. Suny researchers find way to make ethanol from wood. http://www.esf.edu/newspubs/news/2005/01.18.ethanol.htm
86. Reverse osmosis. http://www.gewater.com/library/tp/691_Sweet_Solutions.jsp
87. Klinke, H. B.; Thomsen, A. B.; Ahring, B. K., *Appl. Microbiol. Biotechnol.* **2004**, *66*, pp 10-26.
88. Palmqvist, E.; Hahn-Hagerdal, B., *Bioresource Technology* **2000**, *74*, pp 17-24.
89. Khiyami, M. A.; Pometto III, A. L.; Brown, R. C., *Journal of Agricultural and Food Chemistry* **2005**, *53*, pp 2969-2977.
90. Canilha, L.; Silva, J. B.; Solenzal, A. I., *Process Biochem.* **2004**, *39*, pp 1909-1912.
91. Horvath, I. S.; Sjode, A.; Nilvebrant, N.-O.; Zagorodni, A.; Jonsson, L. J., *Applied Biochemistry and Biotechnology* **2004**, *113-116*, pp 525-238.
92. Davis, M. W., *Journal of Wood Chemistry and Technology* **1998**, *18(2)*, pp 235-252.

Chapter 26

Lignin Biodegradation by Selective White Rot Fungus and Its Potential Use in Wood Biomass Conversion

T. Watanabe[1], Y. Ohashi[1], T. Tanabe[1], Y. Honda[1], and K. Messner[2]

[1]Laboratory of Biomass Conversion, Research Institute for Sustainable Humanosphere (RISH), Kyoto University, Gokasho, Uji, Kyoto 611–0011, Japan
[2]Laboratory of Industrial Microbiology, Institute for Bioengineering, University of Technology, Vienna, Austria

A selective white rot fungus, *Ceriporiopsis subvermispora* is shown to, degrade lignin without extensive damage to cellulose. This selective ligninolysis reaction is catalyzed by low molecular mass compounds at a site far from the enzymes. At an incipient stage of the wood decay, the fungus catalyzed *in situ* lipid peroxidation and secreted alkylitaconic acids and ceriporic acids. The extracellular metabolites are thought to attenuate the iron redox reactions, thereby inhibiting the production of a cellulolytic active oxygen species such as hydroxyl radicals. The selective lignolysis by this fungus can be applied to pretreatments of wood for ethanol fermentation, methane fermentation and feedstuff production.

Introduction

The continued use of fossil fuels is a serious environmental problem, causing the emission of carbon dioxide. In order to avoid serious global warming by the emission of carbon dioxide from fossil fuels, there is a growing demand for the production of energy and chemicals from renewable resources. In order to convert wood biomass by enzymatic saccharification and fermentation, the degradation of the lignin network is necessary because the cell wall polysaccharides are covered with lignin in the lignified plant cell walls. One potential approach to degrade lignin prior to saccharification and fermentation is to use the ligninolytic systems of white rot fungi. Among the numerous fungi so far isolated, a white rot fungus, *Ceriporiopsis subvermispora* is characterized as being highly potent toward biopulping, degrading lignin without extensive damage of cellulose[1-3]. Previous studies revealed that the selective ligninolysis by this fungus is catalyzed by low molecular mass compounds at a site far from the extracellular enzymes and fungal hyphae[2-5]. This fungus has been shown to decompose non-phenolic lignin model compounds[6]. In this chapter, the lipid-related extracellular metabolites of *C. subvermispora* and the conversion of wood biomass into ethanol, methane and other materials is reviewed.

Control of Active Oxygen Species and Radicals by Extracellular Lipid-Related Metabolites

The biodegradation of lignin is an extracellular free radical event that proceeds in concert with the activation of molecular oxygen and redox cycling of transition metals. When wood is colonized by wood-degrading fungi, their extracellular enzymes are not able to diffuse into the intact wood cell walls because the enzymes are too large to penetrate their pores. Hydroxyl radicals, ($\cdot$OH), a radical species highly destructive to cellulose and lignin model compounds[7,8], is proposed as a principal low molecular mass oxidant that erodes wood cell walls to enhance the accessibility of the extracellular enzymes of wood rot fungi to the components of wood cell walls. In brown rot, hydroxyl radicals disrupt cellulose and hemicelluloses in wood cell walls, with concomitant modification of lignin substructures like demethoxylation and hydroxylation of aromatic rings[8-11]. The production of hydroxyl radicals are also reported for non-selective white rot fungi[12-15]. Hydroxyl radicals are produced by the reaction of Fe^{2+} with H_2O_2 (Fenton reaction; $Fe^{2+} + H_2O_2 \rightarrow Fe^{3+} + OH^- + \cdot OH$), although other transition metals like Cu^+ are able to participate in the production of $\cdot$OH. In the Fenton system, catalysts for the reductive half cycle ($Fe^{3+} \rightarrow Fe^{2+}$) accelerate the hydroxyl radical formation. Wood rot fungi have versatile enzymatic and non-enzymatic systems to accelerate the reductive half cycle.[8,11,13-19]

In contrast to brown rot and non-selective white rot fungi, selective lignin-degrading fungi like *Ceriporiopsis subvermispora* are able to decompose lignin in wood cell walls without the intensive damage of cellulose. Wood decay by the biopulping fungus proceeds without the penetration of extracellular enzymes into the wood cell wall regions[2-5]. Lipid peroxidation has been proposed as a major pathway for the ligninolysis of this fungus at an incipient stage of wood decay[20-23]. Lignin biodegradation proceeds by free radical process in the presence of molecular oxygen and transition metals. Reductive radicals such as semiquinone radical reduce molecular oxygen to produce superoxide, which in turn reduce Fe^{3+} or disproportionate into H_2O_2. Fe^{3+} is directly reduced by lignin-derived phenols such as guaiacol and catechol[24]. Thus, if some inhibition systems for the iron redox reactions were not involved in the wood decaying systems, production of the cellulolytic oxidant, hydroxyl radical is inevitable (Fig. 1). This strongly suggests that the selective white rot fungus possesses unknown extracellular systems that attenuate the production of hydroxyl radicals.

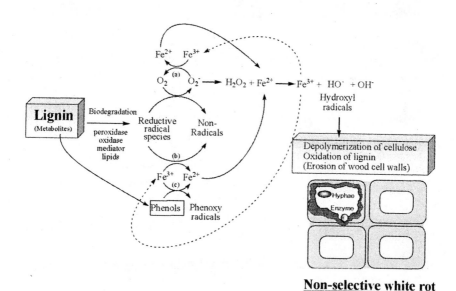

Non-selective white rot

Figure 1. A proposed pathway for the generation of hydroxyl radicals (•OH) during lignin biodegradation by non-selective white rot fungi. Phenols and reductive radical species reduce Fe^{3+} to promote the Fenton reaction. Wood cell walls are eroded by the hydroxyl radicals to assist penetration of extracellular enzymes.

412

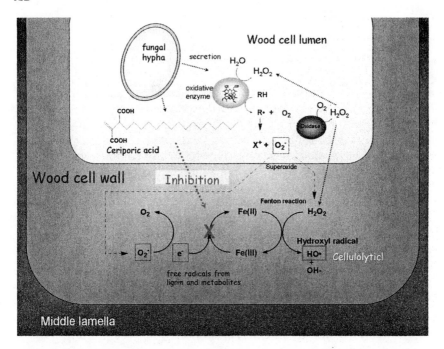

Figure 2. A proposed pathway for the inhibition of hydroxyl radicals (•OH) by the selective white rot fungus, Ceriporiopsis subvermisopora. The fungus secretes ceriporic acids. The metabolites inhibit production of a cellulolytic active oxygen species, hydroxyl radical by suppressing iron redox reactions.

First, a series of novel itaconic acid derivatives having a long alkyl side chain at position C-3 of its core (ceriporic acids) were isolated from the cultures of *C. subvermisrpora*[25-27]. Gutiérrez at al. reported the same compounds by GCMS analysis of crude extracts from eucalypt wood decayed by *C. subvermisrpora, Phlebia radiata, Pleurotus pulmonarius,* and *Bjerkandera adusta* in 2002[28]. Dodecanyl-, tridecanyl-, tetradecanyl-, pentadecanyl-, octadecenyl- and octadecanylitaconic acids, were also detected in very minor amounts or traces by the GCMS analysis of eucalypt wood cultures[29].

It was reported that the alkylitaconic acid strongly suppressed the Fenton reactions even in the presence of the Fe^{3+} reductants such as cysteine and hydroquinone[30] (Fig. 2). The inhibition of •OH production by the diffusible fungal metabolite accounts for the extracellular system of the fungus that attenuates the formation of •OH in the presence of iron, molecular oxygen and free radicals produced during lignin biodegradation. Recently, it was reported

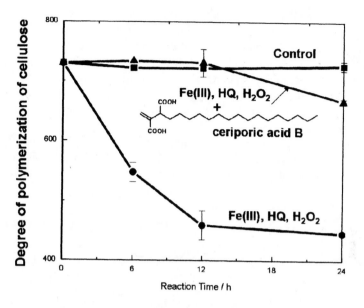

Figure 3.Ceriporic acid B, an extracellular metabolite of C. subvermispora suppresses the depolymerization of cellulose by the Fenton reaction.[31)]

that 1-nonadecene-2,3-dicarboxylic acid (ceriporic acid B), an extracellular metabolite of *C. subvermispora*, strongly inhibited •OH production and the depolymerization of cellulose by the Fenton reaction in the presence of iron ions, cellulose, H_2O_2 and a reductant for Fe^{3+}, hydroquinone (HQ), at the physiological pH of the fungus[31)] (Fig. 3). As expected by the *in vitro* experiments, this metabolite may play a key role in the mechanism of selective white rot because a high level of production of alkylitaconic acids has been observed only for *C. subvermispora* in SSF cultures of four white-rot fungi, *C. subvermisrpora, Phlebia radiata, Pleurotus pulmonarius, and Bjerkandera adusta* [28)].

In the incipient stage of wood decay, this fungus produced saturated (SFAs) and unsaturated fatty acids (USFAs), including linoleic acid (18:2n-6),and oxidized them with manganese peroxidase (MnP) to produce hydroperoxides and TBARS[21)]. The lipid peroxidation with MnP has been proposed as a ligninolytic system of this fungus because diffusible Mn^{3+} chelates can react with lipid and lipid hydroperoxides to generate free radicals[22)]. Analysis of the catalytic mechanisms of MnP for lipid peroxidation of linoleic acid revealed that the reaction starts from hydrogen abstraction from the enolic form of the fatty acids and proceeds by acyl radical chain reaction, accompanied by the production of aldehydes including glyoxal[23]. When free radicals were produced from lipid hydroperoxide models, TBHP or CHP with copper complexes, non-phenolic synthetic lignin was depolymerized[32], and both softwood and hardwood were delignified, leading to fiber separation, as observed in selective white rot[33]. The copper system was applied to biomimetic bleaching of pulp[34,35], and it is potentially applicable to pretreatments of wood for enzymatic saccharification and fermentation.

The reactions were carried out by mixing 0.5 mM $FeCl_3$, 0.25 mM HQ, 100 mM H_2O_2 and 0.1g cellulose in the presence and absence of 2.5 mM ceriporic acid B. Control contained cellulose and H_2O_2.

Conversion of Wood Biomass Using Selective White Rot Fungi

The development of conversion systems from lignocellulosics into biofuels and chemicals has received much attention due to immense potentials for the utilization of renewable bioresources. For example, ethanol production from lignocellulosics has been examined by saccharification with acids or cellulolytic enzymes and subsequent ethanol fermentation with yeast or gene-engineered bacteria. Since lignin makes the access of cellulolytic enzymes to cellulose difficult, it is necessary to decompose the network of lignin prior to enzymatic hydrolysis. Thus, effective pretreatments are needed for enzymatic saccharification and ethanol production from lignocellulosics. Biological pretreatment with lignin-degrading fungi is one possible approach (Fig. 4). It was reported that ethanol was produced by simultaneous saccharification and fermentation (SSF) from beech wood chips after bioorganosolve pretreatments by ethanolysis and white rot fungi, *C. subvermispora*, *Dichomitus squalens*, *Pleurotus ostreatus*, and *Coriolus versicolor*[36]. Beech wood chips were pretreated with the white rot fungi for 2-8 weeks without addition of any nutrients. The wood chips were then subjected to ethanolysis to separate them into pulp and soluble fractions (SFs). From the pulp fraction (PF), ethanol was produced by SSF using *Saccharomyces cerevisiae* AM12 and a commercial cellulase preparation, Meicelase, from *Trichoderma viride*. Among the four strains, *C. subvermispora* gave the highest yield on SSF. The yield of ethanol obtained after pretreatment with *C. subvermispora* for 8 weeks was 0.294 g g^{-1} of ethanolysis pulp (74% of theoretical) and 0.176 g g^{-1} of beech wood chips (62% of theoretical). The yield was 1.6 times higher than that obtained without the fungal treatments. The biological pretreatments saved 15% of the electricity needed for the ethanolysis.

Treatments with white rot fungi were also applied to the production of feed for ruminants from Japanese cedar wood[37]. Japanese cedar wood chips were

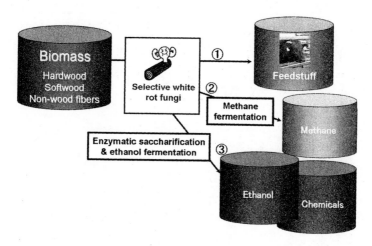

Figure 4. Production of ethanol, methane, feedstuff and other useful chemicals from biomass using pretreatments with selective white rot fungi.

treated with white rot fungi, *Pleurotus ostreatus*, *L. edodes*, *Pholiota nameko*, *Dichomitus squalens* and *C. subvermispora* (Fig. 5). In order to determine the digestibility of the wood, *in vitro* organic matter digestibility (IVOMD) and *in vitro* gas production (IVGP) were measured. The solubilization of the wood by the rumen microorganisms can be evaluated by IVOMD, and anaerobic fermentation via volatile fatty acid (VFA) production can be estimated by IVGP. Because increase in IVOMD is not proportional to the amount of organic matter (OM) available as an energy source for ruminants, measurement of IVGP is important to evaluate the digestibility. *In vitro* organic matter digestibility (IVOMD) in Japanese cedar wood without fungal treatments was between 0.047 and 0.068, while it was elevated to 0.446 by culturing with *C. subvermispora* for 20 weeks. The *in vitro* gas production (IVGP) in Japanese cedar wood cultured with *C. subvermispora* for 20 weeks increased to 107 ml/g organic matter (OM), while IVGP for *P. ostreatus*, *P nameko* or *D. squalens* was 37 ml/g OM, or lower. These results demonstrate that *C. subvermispora* has the highest potential to convert Japanese cedar wood into a feed for ruminants.

The pretreatments with white rot fungi were applied to methane fermentation from Japanese cedar wood (Fig. 6)[38]. Methane fermentation is advantageous for on-site energy supply. Methane gas can be converted to electricity using fuel cells or turbine systems, or combusted directly. The development of a bioconversion system from wood to methane should accelerate the establishment of bioenergy-based societies that use wood and forestry wastes to make electricity, heat and fuels. Fungal pretreatments were studied with

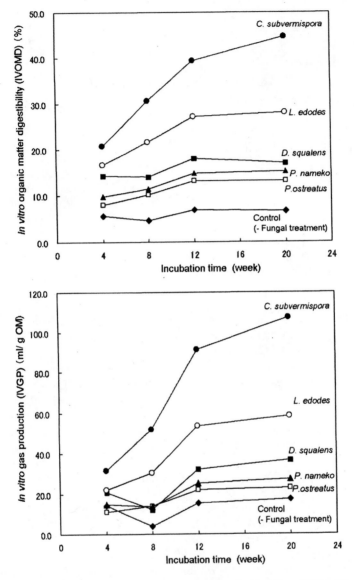

Figure 5. The change in in vitro organic matter digestibility (IVOMD) and in vitro gas production (IVGP) (ml/g organic matter) of Japanese red cedarwood cultured with basidiomycetes for different culture lengths. (Reproduced with permission from reference 37. Copyright 2005 Elsevier.)

several strains of *C. subvermispora* and *Pleurocybella porrigens* for methane fermentation of Japanese cedar wood. The methane fermentation of Japanese cedar wood was carried out after pretreatment with four strains of white rot fungi, *Ceriporiopsis subvermispora* ATCC 90467, CZ-3, CBS 347.63 and *Pleurocybella porrigens* K-2855. These fungi were cultivated on wood chip media with and without wheat bran for 4-8 weeks. The pretreated wood chips were fermented anaerobically with sludge from a sewage treatment plant. Pretreatments with *C. subvermispora* ATCC 90467, CZ-3 and CBS 347.63 in the presence of wheat bran for 8 weeks decreased 74-76% of beta aryl ether linkages in the lignin to accelerate production of methane. After fungal treatments with *C. subvermispora* ATCC 90467 and subsequent 30-day methane fermentation, the methane yield reached 35% and 25% of the theoretical yield based on the holocellulose contents of the decayed and original wood, respectively. In contrast, treatment with the three strains of *C. subvermispora* without wheat bran cleaved 15-26% of the linkage and produced 6-9% of methane. There were no significant accelerating effects in wood chips treated with *P. porrigens* which has a lower ability to decompose the lignin. Thus, it was found that *C. subvermispora* with a high ability to decompose aryl ether bonds of lignin, promoted methane fermentation of softwood in the presence of wheat bran.

Studies on the lignin biodegradation by selective white rot fungi will provide new insights into the development of environmentally-friendly processes for the production of chemicals, fuels and eco-materials from wood and non-wood lignocellulosics.

Conclusions

A selective white rot fungus, *C. subvermispora,* is able to delignify wood without penetration of extracellular enzymes into wood cell wall regions. The fungus has been applied to biopulping, and the production of feedstuff, ethanol and methane. In the incipient stage of wood decay, this fungus produces saturated and unsaturated fatty acids including linoleic acid. Subsequently oxidizing them with MnP generating free radicals. During the wood decay process, this fungus produces a series of alkylitaconic acids called ceriporic acids. *In vitro* experiments demonstrated that ceriporic acid B intensively inhibited •OH production by the Fenton reaction by suppressing the iron redox cycle. The fungal metabolite suppressed cellulose degradation by the Fenton reaction even in the presence of reductants for Fe^{3+} at the physiological pH of the fungal decay.

418

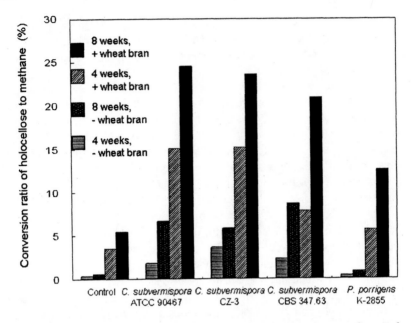

Figure 6. Conversion ratio of holocellulose of Japanese cedar wood to methane and after fungal pretreatments for 8 weeks and subsequent methane fermentation for 30 days. Values are expressed as percentage based on the weight of holocellulose in the original wood. (Reproduced with permission from reference 34. Copyright 2002 Elsevier.)

References

1. Akhtar, M.; Attridge, M. C.; Myers, G. C.; Kirk, T. K.; Blanchette, R. A. Biomechanical pulping of loblolly pine with different strains of the white-rot fungus *Ceriporiopsis subvermispora*. *Tappi J.* **1992**, 75, pp 105-109.
2. Messner, K. Biopulping, in Forest Products Biotechnology, Bruce A. B. and Palfreyman J. W. Eds.; Taylor & Francis: London, **1998**; pp 63-82.
3. Messner, K.; Srebotnik, E. Biopulping: An overview of developments in an environmentally safe paper-making technology. *FEMS Microbiol. Rev.* **1994**, 13, pp 351-364.
4. Srebotnik, E.; Messner, K. A simple method that uses differential staining and light microscopy to assess the selectivity of wood delignification by white rot fungi. *Appl. Environ. Microbiol.* **1994**, 60, pp 1383-1386.

5. Blanchettee, R. A.; Krueger, E. W.; Haight, J. E.; Akhtar, M.; Akin D. E. Cell wall alterations in loblolly pine wood decayed by the white-rot fungus, *Ceriporiopsis subvermispora*. *J. Biotechnol.* **1997**, 53, pp 203-213.

6. Srebotnik, E.; Jensen, K. A. Jr.; Hammel, K. E. Fungal degradation of recalcitrant nonphenolic lignin structures without lignin peroxidase. *Proc. Natl. Acad. Sci. U. S. A.* **1994,** 91, pp 12794-12797.

7. Gierer, J. Formation and involvement of superoxide (O_2^{-}/$HO_2^{\cdot}$) and hydroxyl (OH$^{\cdot}$) radicals in TCF bleaching processes: A review. *Holzforschung*, **1997**, 51, pp 34-46.

8. Henriksson, G.; Zhang, L.,; Li, J.; Ljungquist, P.; Reitberger, T.; Pettersson, G.; Johansson, G. Is cellobiose dehydrogenase from *Phanerochaete chrysosporium* a lignin degrading enzyme?. *Biochim. Biophys. Acta*, **2000**, 1480, pp 83-91.

9. Kirk, T. K. Effects of a brown-rot fungus, *Lenzites trabea*, on lignin in spruce wood. *Holzforschung*, **1975**, 29, pp 99-107.

10. Kirk, T. K.; Adler, E. Methoxyl-deficient structural elements in lignin of sweetgum decayed by a brown-rot fungus. *Acta Chem Scand*, **1979**, 24, pp 3379-3390.

11. Hyde, S. M.; Wood, P. M. A mechanism for production of hydroxyl radicals by the brown-rot fungus *Coniophora puteana*: Fe(III) reduction by cellobiose dehydrogenase and Fe(II) oxidation at a distance from the hyphae. Microbiology, **1997**, 143, pp 259-266.

12. Backa, S.; Gierer, J.; Reitberger, T.; Nilsson, T. Hydroxyl radical activity in cultures and wood colonised by white-rot fungi. *Holzforschung*, **1993**, 47, pp 181-187.

13. Tanaka, H.; Itakura, S.; Enoki, A. Hydroxyl radical generation by an extracellular low-molecular-weight substance and phenol oxidase activity during wood degradation by the white-rot basidiomycete *Trametes versicolor*. *J. Biotechnol.*, **1999**, 75, pp 57-70.

14. Barr, D. P., Shah, M. M.; Grover, T. A.; Aust, S. D. Production of hydroxyl radical by lignin peroxidase. from Phanerochaete chrysosporium., *Arch Biochem Biophys.* **1992**, 298, pp 480-485.

15. Guillén, F.; Gómez-Toribio, V.; Martínez, M. J.; Martínez, A. T. Production of hydroxyl radical by the synergistic action of fungal laccase and aryl alcohol oxidase. *Arch Biochem Biophys.* **2000**, 383, pp 142-7.

16. Hammel, K. E.; Kapich, A. N.; Jensen A.; Ryan, Z. C. Reactive oxygen species as agents of wood decay by fungi. *Enzyme and Microbial Technology* **2002**, 30, pp 445-453.

17. Kremer, S. M.; Wood, P. M. Evidence that cellobiose oxidase from *Phanerochaete chrysosporium* is primary an Fe(III) reductase. *Eur. J. Biochem.* **1992**, 205, pp 133-138.

18. Guillén, F. ; Martinez, M. J. ; Munoz, C. ; Martinez, A. T. Quinone redox cycling in the ligninolytic fungus Pleurotus eryngii leading to extracellular production of superoxide anion radical. *Arch. Biochem. Biophys.* **339**, pp 190-199.

19. Jensen, K. A. Jr.; Houtman C. J.; Ryan Z. C.; Hammel, K. E. Pathways for extracellular Fenton chemistry in the brown rot basidiomycete *Gloeophyllum trabeum. Appl. Environ. Microbiol.* **2001**, 67, pp 2705-2711.

20. Jensen, K. A.; Bao, J. R. W.; Kawai, S.; Srebotnik, E.; Hammel, K. E. Manganese-dependent cleavage of non-phenolic lignin structures by *Ceriporiopsis subvermispora* in the absence of lignin peroxidase. *Appl. Environ. Microbiol.* **1996**, 62, pp 3679-3686.

21. Enoki, M.; Watanabe,T.; Nakagame, S.; Koller, K.; Messner, K.; Honda, Y.; Kuwahara, M. Extracellular lipid peroxidation of selective white-rot fungus, *Ceriporiopsis subvermispora. FEMS Microbiol. Lett.,* **1999**, 180, pp 205-211.

22. Watanabe, T.; Katayama, S.; Enoki, M.; Honda, Y.; Kuwahara, M. Formation of acyl radical in lipid peroxidation of linoleic acid by manganese-dependent peroxidase from *Ceriporiopsis subvermispora* and *Bjerkandera adusta. Eur. J. Biochem.* **2000**, 267, pp 4222-4231.

23. Watanabe, T.; Shirai, N.; Okada, N.; Honda, Y.; Kuwahara, M. Production and chemiluminescent free radical reactions of glyoxal in lipid peroxidation of linoleic acid by the ligninolytic enzyme, manganese peroxidase. *Eur. J. Biochem.* **2001**, 268, pp 6114-6122.

24. Pracht, J.; Boenigk, J.; Isenbeck-Schröter, M.; Keppler, F.; Schöler, H. F. Abiotic Fe(III) induced mineralization of phenolic substances. *Chemosphere* **2001**, 44, pp 613-619.

25. Enoki, M.; Honda, Y.; Watanabe, T.; Kuwahara, M. A novel dicarboxylic acid produced by white-rot fungus *Ceriporiopsis subvermispora.* In Proceedings of the 44th Lignin Symposium; October 7-8 **1999**, Gifu, Japan; pp 69-72.

26. Enoki, M.; Honda, Y.; Kuwahara, M.; Watanabe, T. Chemical synthesis, iron redox interactions and charge transfer complex (CTC) formation of ceriporic acids from the selective lignin-degrading basidiomycete, *Ceriporiopsis subvermispora. Chem. Phys. Lipid* **2002**, 120, pp 9-20.

27. Amirta, R.; Fujimori, K.; Shirai, N.; Honda, Y.; Watanabe, T. Ceriporic acid C, a hexadecenylitaconate produced by a lignin-degrading fungus, *Ceriporiopsis subvermispora. Chem. Phys. Lipids* **2003**, 126, pp 121-131.

28. Gutiérrez, A.; del Rio, J. C.; Martinez-Inigo, M.; Martinéz, M. J.; Martinéz, A. T. Production of new unsaturated lipids during wood decay by ligninolytic basidiomycetes. *Appl. Environ. Microbiol.* **2002**, 68, pp 1344-1350.

29. del Rio, J. C.; Gutiérrez, A.; Martinéz, M. J.; Martinéz, A.T. Identification of a novel series of alkylitaconic acids in wood cultures of *Ceriporiopsis subvermispora* by gas chromatography/mass spectrometry. *Rapid Commun. Mass Spectrom.* **2002**, 16, pp 62-68.

30. Watanabe, T.; Teranishi, H.; Honda, Y.; Kuwahara, M. A selective lignin-degrading fungus, *Ceriporiopsis subvermispora* produces alkylitaconates that inhibit the production of a cellulolytic active oxygen species, hydroxyl radical in the presence of iron and H_2O_2. *Biochem. Biophys. Res. Commun.* **2002**, 297, pp 918-923.

31. Rahmawati, N.; Ohashi, Y.; Watanabe, T.; Honda, Y.; Watanabe, T. Ceriporic acid B, an extracellular metabolite of *Ceriporiopsis subvermispora* suppresses the depolymerization of cellulose by the Fenton reaction. *Biomacromolecules* **2005**, 6, pp 2851-2856.

32. Watanabe, T.; Koller, K.; Messner, K. Copper-dependent depolymerization of lignin in the presence of fungal metabolite, pyridine. *J. Biotechnol.* **1998**, 62, pp 221-230.

33. Messner, K.; Fackler, K.; Lamaipis, P.; Gindl, W.; Srebotnik, E.; Watanabe, T. Overview of White-Rot Research: Where we are today. *ACS Symposium Series* 845 Wood deterioration and preservation, American Chemical Society: Washington, DC, **2003**; pp 73-96.

34. Fackler, K.; Srebotnik, E.; Watanabe, T.; Lamaipis, P.; Humar, M.; Tavzes, C.; Sentjurc, M.; Pohleven, F.; Messner, K. Biomimetic pulp bleaching with copper complexes and hydroperoxides, Biotechnology in the Pulp and Paper Industry, Progress in Biotechnology Vol 21, L. Viikari and R. Lantto, Eds.; Elsevier, Amsterdam, **2002**; pp 223-230.

35. Rahmawati, N.; Ohashi, Y.; Honda, Y.; Kuwahara, M.; Fackler, K.; Messner, K.; Watanabe, T. Pulp bleaching by hydrogen peroxide activated with copper 2,2'-dipyridylamine and 4-aminopyridine complexes. *Chem. Eng. J.* **2005**, 112, pp 167-171.

36. Itoh H.; Wada, M.; Honda, Y.; Kuwahara, M.;. Watanabe, T. Bioorganosolve pretreatments for simultaneous saccharification and fermentation of beech wood by ethanolysis and white rot fungi. *J. Biotechnol.* **2003**, 103, pp 273-280.

37. Okano, K.; Kitagawa, M.; Sasaki, Y.; Watanabe, T. Conversion of Japanese red cedar (*Cryptomeria japonica*) into feed for ruminants by white-rot basidiomycetes. *Animal Feed Sci. Technol.* **2005**, 120, pp 235–243.

38. Amirta R.; Tanabe, T.; Watanabe, T.; Honda, Y.; Kuwahara, M. Watanabe, T. Methane fermentation of Japanese cedar wood pretreated with a white rot fungus, *Ceriporiopsis subvermispora. J. Biotechnol.*, **2006**, in press.

Chapter 27

Bioethanol Production from Lignocellulosics Using Supercritical Water

Shiro Saka and Hisashi Miyafuji

Graduate School of Energy Science, Kyoto University, Yoshida-honmachi, Sakyo-ku, Kyoto 606–8501, Japan

Bioethanol production from lignocellulosics with supercritical water (>374°C, >22.1MPa) technology was studied. Lignocellulosics after supercritical water treatment was fractionated into supercritical water insoluble residue and a supercritical water-soluble portion, which was further fractionated after 12 hours into the water-soluble portion, precipitates, methanol-soluble portion.. In the water-soluble portion, not only fermentable monomeric sugars but also oligomers were recovered. The oligomers and precipitates as polysaccharides could be easily hydrolyzed to glucose with cellulose, but not be fermented to ethanol with *Saccharomyces cerevisiae* due to the inhibitory contaminants. Thus, wood charcoal and overliming treatments were studied. As a result, wood charcoal was found to remove these inhibitors without removing the fermentable sugars. Furthermore, all fermentable sugars in the water-soluble portion could be converted to ethanol in the binary fermentation system of *Saccharomyces cerevisiae* and *Pachysolen tannophilus*. Consequently, supercritical water treatment is thought to be a promising pretreatment for ethanol production from lignocellulosics.

Introduction

Energy and environmental issues such as the exhaustion of fossil resources and global warming are of major concern. Due to its environmental friendliness, increased focus has been on ethanol production from biomass as an alternative to fossil fuels. Among various biomass resources, lignocellulosics such as wood can be used as one of the raw materials for producing ethanol. Various studies have been performed on the pretreatment of wood such as hydrolysis of lignocellulosics by acid catalysis (*1*, *2*) and enzymatic saccharification (*3*) to obtain the fermentable sugars for ethanol production.

Recently, supercritical fluid technology has been applied in the conversion of lignocellulosics to fuels and chemicals (*4-24*). In this chapter, recent progress in bioethanol production from lignocellulosics by supercritical water technology is presented.

Supercritical Water Treatment of Lignocellulosics

Supercritical Water

Supercritical water (>374°C, >22.1MPa) behaves very differently than water under normal conditions, exemplified by the following (*25*): The ionic product of water (Kw) is about three to four order magnitude larger in its supercritical state than that in a normal state. Therefore, supercritical water can act as an acid catalyst. Furthermore, the dielectric constant of supercritical water is in a range of 10 and 20, compared with about 80 in ordinary water. Thus, the hydrophobic substances can be solvated with the supercritical water. For the treatment of lignocellulosics with supercritical water, three different types of systems such as batch-type (*11, 17, 18*), semiflow-type (*19, 20*) and flow-type (*11, 21-24*) system have been developed by several researchers. More detailed information on these systems is described elsewhere (*11, 26*).

Separation of Treated Lignocellulosics

Figure 1 shows the separation scheme of the lignocellulosics treated in supercritical water. The supercritical water-soluble portion is separated by filtration from the supercritical water-insoluble portion, immediately after the treatment. While standing for 12 hours, the supercritical water-soluble portion results in the precipitates and oily substances due to the change of water in dielectric constant from supercritical state to ordinary one. They are filtrated to

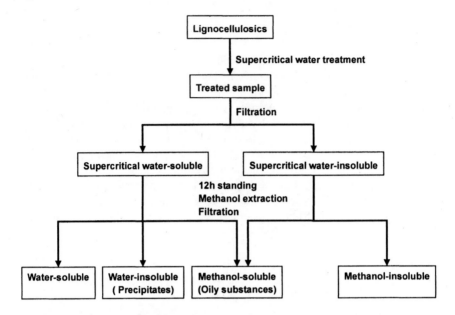

Figure 1. Separation of lignocellulosics treated in supercritical water.

separate from the water-soluble portion. The precipitates and oily substances are extracted with methanol, to separate into precipitates and methanol-soluble portion by filtration. Consequently, the water-soluble portion, water-insoluble portion (precipitates), methanol-soluble portion (oily substances) and methanol-insoluble portion are the results from the treated sample. Therefore, the former two and latter two portions mainly consist of carbohydrate-derived and lignin-derived products, respectively (5).

Decomposition of Cellulose and Hemicelluloses

Cellulose was decomposed by supercritical water and separated to water-soluble portion. This portion contained cello-oligosaccharides (oligomers) and glucose as hydrolyzed products, while levoglucosan, 5-hydroxymethyl furfural (5-HMF) and furfural were found as dehydrated products. In addition, fragmented products such as methylglyoxal, glycolaldehyde, dihydroxyacetone and erythrose were found. Furthermore, low-molecular weight organic acids, such as pyruvic acid, lactic acid, glycolic acid and acetic acid, were identified by HPLC or GPC analysis (11, 17-24).

It was also clarified by matrix assisted laser desorption ionization-time of flight mass spectrometric (MALDI-TOFMS) analysis that oligomers composed of cello-oligosaccharides up to cello-dodecaose (Degree of polymerization;

DP=12), and that reducing terminal glucose unit of some oligosaccharides were decomposed to levoglucosan, erythrose and glycolaldehyde. On the other hand, the precipitates were revealed to be glucan with DP between 13 and 100, which is insoluble in ordinary water but soluble in supercritical water (*11*).

Table I shows the chemical composition of cellulose treated in supercritical water. The total yield of hydrolyzed products, which are polysaccharides

Table I. Chemical Composition of Cellulose Treated in Supercritical Water at 380°C and 40MPa Using Flow-type System (*27*).

Product name	Yield (%)		
	0.12s	*0.24s*	*0.48s*
Hydrolyzed products			
Polysaccharides	31.2	12.1	n.d.
Oligosaccharides	41.1	37.6	7.5
Glucose	2.8	8.4	8.9
Fructose	0.5	3.4	10.0
(Subtotal)	(75.6)	(61.5)	(26.4)
Dehydrated products			
Levoglucosan	0.2	2.4	3.7
5-HMF	0.1	1.9	7.3
Furfural	n.d.	1.0	3.4
(Subtotal)	(0.3)	(5.3)	(14.4)
Fragmented products			
Erythrose	0.7	0.9	1.9
Methylglyoxal	0.6	3.0	6.7
Glycolaldehyde	2.5	6.7	14.6
Dihydroxyacetone	0.2	1.6	2.4
(Subtotal)	(4.0)	(12.2)	(25.6)
Organic acids			
Pyruvic acid	n.d.	0.5	0.7
Glycolic acid	1.3	2.9	3.9
Lactic acid	0.7	3.2	4.0
Formic acid	0.3	1.6	2.5
Acetic acid	0.1	0.7	1.0
(Subtotal)	(2.4)	(8.9)	(12.1)
Others*	17.7	12.1	21.5
Total	100.0	100.0	100.0

* Others consist of unidentified products, gasified products and water derived from dehydration of saccharides; n.d, not detected.

(precipitates), oligosaccharides, glucose and fructose, is 75.6% for 0.12s in treatment time, and decreased when the treatment time is prolonged. On the other hand, the yields of dehydrated products, fragmented products and organic acids are increased. Thus, the shorter treatment time is favorable to obtain the hydrolyzed products which can be used for ethanol production. Hemicelluloses were also decomposed by supercritical water into water-soluble portion, in

Table II. Lignin-derived Monomeric Products in the Methanol-soluble Portion from Japanese Cedar and Japanese Beech.

Product	Structure*
Guaiacol	G
Methylguaiacol	G–C
Ethylguaiacol	G–C–C
Vinylguaiacol	G–C=C
Syringol	S
Vanillin	G–CHO
Eugenol	G–C–C=C
Propylguaiacol	G–C–C–C
Methylsyringol	S–C
Isoeugenol (*cis, trans*)	G–C=C–C
Homovanillin	G–C–CHO
Acetoguaiacone	G–CO–C
Propioguaiacone	G–CO–C–C
Ethylsyringol	S–C–C
Guaiacylacetone	G–C–CO–C
Vinylsyringol	S–C=C
Propylsyringol	S–C–C–C
2-Methoxy-4-(1-hydroxypropyl) phenol	G–C–C–C–OH
Allylsyringol	S–C–C=C
Syringaldehyde	S–CHO
Propenylsyringol (*cis, trans*)	S–C=C–C
Homovanillic acid	G–C–COOH
Sinapylalcohol (*cis, trans*)	S–C=C–C–OH
2-Methoxy-4-(prop-1-en-3-one) phenol	G–CO–C=C
Sinapylaldehyde (*cis, trans*)	S–C=C–CHO
Acetosyringone	S–CO–C
Coniferylaldehyde (*trans*)	G–C=C–CHO
Propiosyringone	S–CO–C–C
Syringylacetone	S–C–CO–C
Ferulic acid	G–C=C–COOH

* G and S represent guaiacyl and syringyl nuclei, respectively.

which the hydrolyzed products could be expected such as glucose and mannose from glucomannan, and xylose from xylan. These monomeric sugars can also be utilized for ethanol production (5,19).

Decomposition of Lignin

The methanol-soluble portion was obtained only from the sample containing lignin, and its yield was close to the lignin content of the sample when the appropriate condition was selected. Therefore, the methanol-soluble portion mainly consisted of lignin-derived products. The molecular weight of the methanol-soluble portion was found to be less than 2,000 (5).

Table II shows lignin-derived monomeric products identified in the methanol-soluble portion from Japanese cedar and Japanese beech (14). The products composed of phenylpropane units (C6-C3) were derived through the cleavage of ether linkages of lignin. In addition to C6-C3 units, C6-C2 and C6-C1 units of products were also identified.

Ethanol Production

Ethanol Production from Cellulose

The oligosaccharides in the water-soluble portion or polysaccharides as precipitates obtained by supercritical water treatment of cellulose can be the substrates for ethanol production through their conversion to glucose which can be fermented with microorganisms. To obtain glucose from the water-soluble portion and the precipitates, enzymatic hydrolysis with cellulase and β-glucosidase were carried out. Both substrates were found to be hydrolyzed to glucose more easily than to untreated cellulose as shown in Fig. 2. Due to the lower DP and non-crystalline structure of the oligosaccharides and precipitates, higher accessibility of the enzymes must be attained (28, 29). These results clearly show that supercritical water treatment of cellulose can enhance enzymatic activity. However, the glucose concentration obtained after enzymatic hydrolysis was found to be lower than prospected. By the model experiments on the enzymatic hydrolysis of cellohexaose, it was shown that various compounds as in Table I, obtained by supercritical water treatment of cellulose, have inhibitory effects on enzymatic hydrolysis. Organic acids showed much higher inhibition to reduce the yield of glucose in enzymatic hydrolysis, compared to the fragmented and dehydrated products. These results could account for the lower glucose yield after the enzymatic hydrolysis of the water-soluble portion and precipitates. To increase glucose yield, overliming or wood charcoal treatment of the water-soluble portion was found to be effective as shown in Fig. 2. This is probably due to the conversion of the inhibitors to the non-inhibiting compounds seen in overliming treatment. In wood charcoal treatment, however, inhibitors can be adsorbed.

These results clearly show that supercritical water treatment coupled with overliming or wood charcoal treatment is effective to achieve ethanol from cellulose.

Ethanol Production from Lignocellulosics

In the water-soluble portion from Japanese cedar, various fermentable sugars such as glucose, fructose, mannose, galactose and xylose can be recovered from cellulose and hemicelluloses. However, these sugars could not be converted to ethanol by the fermentation with *Saccharomyces cerevisiae*.

From HPLC analysis on the water-soluble portion, vanillin, acetoguaiacone, guaiacol and coniferylaldehyde, which are lignin-derived products, were quantified as in Table III (described as "Untreated"). These compounds are

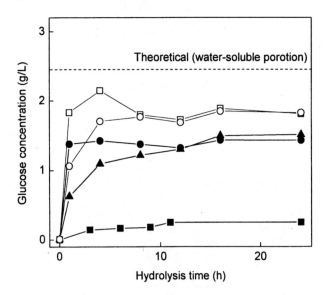

Figure 2. Changes in glucose concentration during enzymatic hydrolysis of the precipitates (▲), water-soluble portion (●) and that treated by overliming (○) or wood charcoal (□). Just comparison, untreated cellulose (■) is also included.

mainly contained in the methanol-soluble portion as in Table II, but contaminated in the water-soluble portion. In addition, furan compounds such as furfural and 5-HMF were also found as in Table III. It is known that various furan and lignin-derived products can inhibit the fermentation of sugars to ethanol (30). Therefore, the poor fermentability seen was due to the presence of these compounds. Thus, to achieve high ethanol production, some detoxification treatments are required.

Table III. Concentrations of Furans, Lignin-derived Products and Sugars in the Water-soluble Portion and that Treated with Wood Charcoal Prepared at 900°C.

	Concentration in the water-soluble portion (mg/L)	
	Untreated	Treated with wood charcoal
5-HMF	378	n.d.
Furfural	240	n.d.
Vanillin	818	n.d.
Acetoguaiacone	3	n.d.
Guaiacol	33	n.d.
Coniferylaldehyde	94	n.d.
Glucose	1.04×10^3	1.04×10^3
Other sugars*	0.92×10^3	0.94×10^3

*The total of fructose, mannose, galactose and xylose.   n.d.; Not detected.

Improvement of Fermentability with Wood Charcoal

Various detoxification methods to improve the fermentability of the hydrolysates obtained from lignocellulosics have been studied such as extraction with organic solvents (31), overliming (32-34), evaporation (33), steam stripping (35, 36), sulfite treatment (33, 37), ion-exchange (38), enzyme treatment (39, 40), zeolite treatment (41) and activated carbon treatment (35, 42). In our research, wood charcoal with high adsorption ability has been applied for a detoxification of the water-soluble portion. As in Table III, furan and lignin-derived products were adsorbed completely and not detected anymore after wood charcoal treatment. The adsorption ability of these compounds was higher in the wood charcoal prepared at higher temperature (43). On the other hand, sugars were revealed to remain constant in its concentration (Table III). The

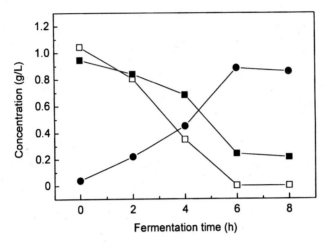

Figure 3. Concentration changes of sugars and ethanol during the fermentation of the water-soluble portion treated with the wood charcoal.
(□) glucose, (■) fructose + mannose + galactose + xylose, (●) ethanol

wood charcoals can, therefore, selectively remove the furan and lignin-derived products without removing the fermentable sugars. This absorption behavior of the wood charcoal is preferable to achieve the high fermentability of sugars to ethanol.

The water-soluble portion after wood charcoal treatment can be fermented with *Saccharomyces cerevisiae* to produce ethanol, as shown in Fig. 3. All consumed sugars in the water-soluble portion were found to be converted to ethanol. This result indicates that the water-soluble portion treated with the wood charcoal could be fermented effectively. However, xylose which cannot be fermented to ethanol by *Saccharomyces cerevisiae* still remained in the water-soluble portion after fermentation.

To achieve effective ethanol production from lignocellulosics, the remaining xylose should be fermented. Therefore, a binary fermentation system to mix *Saccharomyces cerevisiae* with *Pachysolen tannophilus*, with which xylose can be fermented to ethanol, was applied to the water-soluble portion after wood charcoal treatment. As shown in Fig. 4, after the ethanol production with *Saccharomyces cerevisiae*, the remaining xylose was found to be fermented to ethanol by *Pachysolen tannophilus*. In this binary system, therefore, all fermentable sugars can be utilized by both yeasts to produce ethanol.

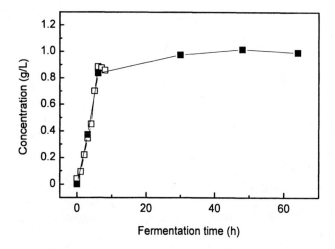

Figure 4. Changes of ethanol concentration in binary fermentation system of the water-soluble portion after wood charcoal treatment. (□) Saccharomyces cerevisiae, (■) Saccharomyces cerevisiae and Pachysolen tannophilus

Conclusions

Compared to other pretreatments of lignocellulosics for ethanol production, supercritical water treatment is a non-catalytic and rapid reaction. Although the various degradation products of lignocellulosics displayed some inhibitory effects on enzymatic hydrolysis and subsequent fermentation with yeasts, the inhibition can be reduced by the use of wood charcoal and/or overliming. Therefore, the supercritical water treatment followed by enzymatic hydrolysis is thought to be a promising process for bioethanol production from lignocellulosics.

Acknowledgements

This research has been carried out under the program of the Research for the Future (RFTF) of The Japan Society of the Promotion of Science (JSPS-RFTF97P01002), the COE program in the 21st Century of "Establishment of COE on Sustainable-Energy System" and by a Grant-in-Aid for Scientific Research (B)(2) (No.12460144) and a Grant-in-Aid for Young Scientists (B) (No.17780139) from the Ministry of Education, Culture, Sports, Science and Technology, Japan.

432

References

1. Torget, R. W.; Kim, J. S.; Lee, Y. Y. *Ind. Eng. Chem. Res.* **2000**, *39*, pp 2817-2825.
2. Iranmahboob, J.; Nadim, F.; Monemi, S. *Biomass Bioenergy* **2002**, *22*, pp 401-404.
3. Ortega, N.; Busto, M. D.; Perez-Mateos, M. *Int. Biodeter. Biodeg.* **2001**, *47*, pp 7-14.
4. Ueno, T.; Saka, S. *Cellulose* **1999**, *6*, pp 177-191.
5. Saka, S.; Konishi, R. In *Progress in Thermochemical Biomass Conversion*; Bridgwater, A.V., Ed.; Blackwell Sci.: Oxford, 2001; pp 1338-1348.
6. Ishikawa, Y.; Saka, S. *Cellulose* **2001**, *8*, pp 189-195.
7. Ehara, K.; Saka, S.; Kawamoto, H. *J. Wood Sci.* **2002**, *48*, pp 320-325.
8. Tsujino, J.; Kawamoto, H.; Saka, S. *Wood Sci. Technol.* **2003**, *37*, pp 299-307.
9. Minami, E.; Saka, S. *J. Wood Sci.* **2003**, *49*, pp 73-78.
10. Minami, E.; Kawamoto, H.; Saka, S. *J. Wood Sci.* **2003**, *49*, pp 158-165.
11. Ehara, K.; Saka, S. *Cellulose* **2002**, *9*, pp 301-311.
12. Takada, D.; Ehara, K.; Saka, S. *J. Wood Sci.* **2004**, *50*, pp 253-259.
13. Ehara, K.; Saka, S. *J. Wood Sci.* **2005**, *51*, pp 148-153.
14. Ehara, K.; Takada, D.; Saka, S. *J. Wood Sci.* **2005**, *51*, pp 256-261.
15. Minami, E.; Saka, S. *J. Wood Sci.* **2005**, *51*, pp 395-400.
16. Yoshida, K.; Kusaki, J.; Ehara, K.; Saka, S. *Appl. Biochem. Biotechnol.* **2005**, *123*, pp 795-806.
17. Sakaki, T.; Shibata, M.; Miki, T.; Hirosue, H.; Hayashi, N. *Bioresour. Technol.* **1996**, *58*, pp 197-202.
18. Sakaki, T.; Shibata, M.; Miki, T.; Hirosue, H.; Hayashi, N. *Energy Fuels* **1996**, *10*, pp 684-688.
19. Ando, H.; Sakaki, T.; Kokusho, T.; Shibata, M.; Uemura, Y.; Hatate, Y. *Ind. Eng. Chem. Res.* **2000**, *39*, pp 3688-3639.
20. Sakaki, T.; Shibata, M.; Sumi, T.; Yasuda, S. *Ind. Eng. Chem. Res.* **2002**, *41*, pp 661-665.
21. Sasaki, M.; Kabyemela, B.; Malaluan, R.; Hirose, S.; Takeda, N.; Adschiri, T.; Arai, K. *J. Supercrit. Fluids* **1998**, *13*, pp 261-268.
22. Sasaki, M.; Fang, Z.; Fukushima, Y.; Adschiri, T.; Arai, K. *Ind. Eng. Chem. Res.* **2000**, *39*, pp 2883-2890.
23. Sasaki, M.; Adschiri, T.; Arai, K. *AIChE J.* **2004**, *50*, pp 192-202.
24. Matsunaga, M.; Matsui, H. *Mokuzai Gakkaishi* (in Japanese) **2004**, *50*, pp 325-332.
25. Holzapfel, W.B. *J. Chem. Phys.* **1969**, *50*, pp 4424-4428.
26. Kusdiana, D.; Minami, E.; Ehara, K; Saka, S. In *12th European Conference and Technology Exhibition on Biomass for Energy, Industry and Climate Protection*: Amsterdam, 2002; pp 789-792.

27. Ehara, K.; Saka, S. In *Lignocellulose Biodegradation*; Saha, B.C.; Hayashi, K., Ed.; ACS Symposium Series 889; ACS: Washington, DC, 2005; pp 69-83.
28. Sasaki, M.; Iwasaki, K.; Hamatani, T.; Adschiri, T.; Arai, K. *Kobunshi Ronbunshu* (in Japanese) 2001, *58*, pp 527-532.
29. Nakata, T.; Miyafuji, H.; Saka, S. *Appl. Biochem. Biotechnol.* 2006, (in press).
30. Palmqvist, E.; Hahn-Hägerdal, B. *Bioresource Technol.* 2000, *74*, pp 25-33.
31. Wilson, J.J.; Deschatelets, L.; Nishikawa, N.K. *Appl. Microbiol. Biotechnol.* 1989, *31*, pp 592-596.
32. Palmqvist, E.; Hahn-Hägerdal, B.; Galbe, M.; Zacchi, G. *Enzyme Microb. Technol.* 1996, *19*, pp 470-476.
33. Larsson, S.; Reinmann, A.; Nilvebrant, N-O.; Jönsson, L.J. *Appl. Biochem. Biotechnol.* 1999, *77-79*, pp 91-103.
34. Martinez, A.; Rodriguez, M.E.; Wells, M.L.; York, S.W.; Preston, J.F.; Ingram L.O. *Biotechnol. Prog.* 2001, *17*, pp 287-293.
35. Maddox, I.S.; Murray A.E. *Biotechnol. Lett.* 1983, *5*, pp 175-178.
36. Yu, S.; Wayman, M.; Parekh, S.K. *Biotechnol. Bioeng.* 1987, *29*, pp 1144-1150.
37. Parajó, J.C.; Dominguez, H.; Domínguez, J.M. *Enzyme Microb. Technol.* 1997, *21*, pp 18-24.
38. Nilvebrant, N-O.; Reinmann, A.; Larsson, S.; Jönsson, L.J. *Appl. Biochem. Biotechnol.* 2001, *91-93*, pp 35-49.
39. Jönsson, L.J.; Palmqvist, E.; Nilvebrant, N-O.; Hahn-Hägerdal, B. *Appl. Microbiol. Biotechnol.* 1998, *49*, pp 691-697.
40. Palmqvist, E.; Hahn-Hägerdal, B.; Szengyel, Z.; Zacchi, G.; Reczey, K. *Enzyme Microb. Technol.* 1997, *20*, pp 286-293.
41. Eken-Saraçoğlu, N.; Arslan, Y. *Biotechnol. Lett.* 2000, *22*, 855-858.
42. Gong, C.S.; Chen, C.S.; Chen, L.F. *Appl. Biochem. Biotechnol.* 1993, *39-40*, pp 83-88.
43. Miyafuji, H.; Nakata, T.; Ehara, K.; Saka, S. *Appl. Biochem. Biotechnol.* 2005, *124*, pp 963-971.

Chapter 28

Integrating Black Liquor Gasification and Pulping: A Review of Current Technology

Mathias Lindström, Hasan Jameel, Ved Naithani, Adrianna Kirkman, and Jean Renard

Department of Forest Biomaterials Science and Engineering, North Carolina State University, Raleigh, NC 27695–8005

Gasification of black liquor could increase the flexibility and improve the profit potential of the paper industry. Its implementation would enable the application of modified pulping technologies, while creating a synthetic product gas that could be utilized in the production of value added products or electrical power. Black liquor gasification produces output streams that can be used with great benefit in modified pulping operations. Split sulfidity and polysulfide modifications to the kraft process lead to yield increases of 1-3% points with improved product quality. Modified sulfite pulping technologies resulted in yield increases of 5-18% points with much higher brightness and significant capital and operating cost savings.

Introduction

Biomass can be converted to power and/or fuels using a variety of technologies based around direct combustion, gasification and pyrolysis. In direct combustion, the amount of oxygen provided has to be sufficient for the efficient conversion of the carbon to carbon dioxide. In gasification the amount of oxygen is limited in order to produce a medium to low calorific gas, and in pyrolysis the biomass is heated to a very high temperature with limited oxygen to produce a mixture of gases and liquids with medium heating value.

One source of biomass that has typically been burned in a boiler to produce steam is black liquor, a by-product from the pulping process. Black liquor gasification (BLG) has some inherent advantages compared to the traditional combustion process. The efficiency of combustion is dependent on the mixing that occurs between the combusted material and oxygen, and gases burn more efficiently than either liquids or solids due to the improved contact between the oxygen and the fuel. In addition, the synthesis gas or syngas generated from gasification can be burned in a gas turbine. This is advantageous as the gas turbine can convert energy to electricity much more efficiently than a steam turbine, as used in a conventional chemical recovery system. The syngas can also be converted to other fuels, chemicals and materials via a wide range of proven chemical processes.

Presently, in a typical chemical pulp mill the black liquor is concentrated to greater than 65% dissolved solids and burned in a recovery boiler. The pulping chemicals are recovered in the smelt and the heat energy is converted to steam, which is used in a steam turbine generator to produce electricity. The typical thermal efficiency of a recovery boiler is generally 65-70%, and the thermal efficiency of the Rankine cycle for the conversion of steam to electricity varies from 30-38%, depending on the temperature and pressures of the different streams in the cycle. These values result in an overall system thermal efficiency of about 23% (*1*). On the other hand, if the black liquor is gasified, the syngas can after cleanup be combusted in a combined cycle for production of electricity. Combined cycle power generation entails the sequential utilization of a gas turbine followed by a steam turbine. The fuel gas is first burned in a gas turbine to produce electricity. The hot exhaust gas from the turbine is then passed through a heat exchanger to produce steam which is then used in a power-producing steam turbine. Implementing a gasifier with combined cycle cogeneration of power will increase the electricity production of the mill. A conventional steam cycle produces about 120-180 kWh/ton of steam, but a gasifier along with combined cycle power generation has the potential to generate 600-1000 kWh/ton of steam (*2*). Such a production of power would turn a pulp mill into a net exporter of electricity, and this potential is main motivation for the implementation of black liquor gasification.

In addition to the increased energy efficiency, gasification of black liquor has several other benefits relative to the traditional combustion recovery process. BLG process operation is inherently very stable and also flexible with regard to feed stock and load requirements. It is possible to process most any biomass material and stable operation can be maintained despite upset feed stock flows, even complete interruptions. BLG has the potential to revolutionize the chemical recovery cycle and, through the separation of sodium and sulfur, enable the utilization of modified pulping technologies. These pulping technologies will increase yield or reduce wood demand, improve product quality, decrease chemical usage but more importantly simplify the chemical recovery process. A simplification of the chemical recovery process will decrease the operating and capital cost for recovery. BLG would also decrease the malodor associated with the kraft process. Besides power generation, the resulting syngas can be used to generate bio-derived liquid fuels, bio-derived chemicals for the synthetic chemical and pharmaceutical industries, as well as H_2 for use in fuel cells.

Despite these benefits and opportunities, high capital cost and risk associated with new process implementation are impeding the implementation of BLG technologies in the industry. The synergy between BLG as increased energy generator and enabler of advanced pulping processes should increase the financial attractiveness of the realization of these new process concepts. Based on these observations, this chapter will address the following topics:

- Review of BLG technologies and their status
- Review of modified pulping technologies enabled by BLG
  - Split sulfidity pulping
  - Polysulfide pulping
  - Alkaline sulfite with anthraquinone (AS-AQ)
  - Mini-sulfide sulfite pulping with anthraquinone (MSS-AQ)
- Effects on process economics of BLG implementation

The effects on overall process economics of enhanced power production and modified pulping technologies will then be discussed, and how the combined implementation of BLG with these technologies can improve the financial attractiveness of the BLG technology.

Black Liquor Gasification Processes

Figure 1 describes the typical process elements included in the gasification of black liquor. The black liquor is initially introduced into a process vessel, the black liquor gasifier, which can either be pressurized or operate under

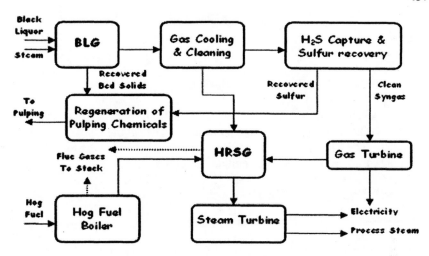

Figure 1. Simplified representation of BLGCC power/recovery systems.

atmospheric pressure. In general terms, the process involves the conversion of hydrocarbons and oxygen to hydrogen and carbon monoxide while forming separate solid and gaseous product streams.

The inorganic material, including all sodium salts, leaves as a bed solid or smelt depending on the gasifier operating temperature. The bed solids or smelt is then slaked and recausticized to form a caustic solution. The volatiles, including most of the reduced sulfur species, leave as a syngas of medium BTU value. The major components of the syngas are H_2S, CO_2, CO, H_2O, and H_2. To prepare the syngas for other applications and to regenerate the pulping liquor, all sulfur must be separated from the syngas and then dissolved into the caustic solution prepared from the bed solids. The clean product gas is burned in a gas turbine and the hot flue gases are combined and used to generate steam in heat recovery steam generators (HRSGs). This steam is then used in a steam turbine and other process applications.

A review of the development of alternative recovery technologies to the Tomlinson recovery boiler has been given (*3*). The following discussion will focus on the gasification processes currently in commercial operation. Black liquor gasification technologies can be classified by the operating temperature (*4*). High temperature gasifiers operate at about 1000°C and low temperature gasifiers operate at less than 700°C. In the high temperature gasifier, the inorganic material form a smelt and leave in the molten form, while in the low temperature system, they leave as solids. The fuel value of the syngas produced is also dependent on the gasifying technology. Typically, gasification produces a fuel gas with heating values of 3-4 MJ/Nm^3 using air and 89 MJ/Nm^3 using oxygen (5).

Low Temperature Gasifier/Steam Reformer

The development of low temperature fluidized bed gasifiers is being pursued by ThermoChem Recovery International (TRI) in the USA and by ABB in Sweden. The TRI system uses steam reforming to generate the product gases. As opposed to exothermic incineration or combustion technologies, steam reforming is an endothermic process. The steam reforming vessel operates at atmospheric pressure and at a medium temperature. The organics are exposed to steam in a fluidized bed in the absence of air or oxygen with the following reaction:

$$H_2O + C + Heat = H_2 + CO$$

The carbon monoxide produced in this first reaction then reacts with steam to produce more hydrogen and carbon dioxide.

$$CO + H_2O = H_2 + CO_2$$

The result is a synthesis gas made up of about 65% hydrogen. The TRI Steam Reformer technology, as shown in Figure 2, consists of a fluidized bed reactor that is indirectly heated by multiple resonance tubes of one or more pulse combustion modules. Black liquor is directly fed to the reactor, which is fluidized with superheated steam. The black liquor uniformly coats the bed solids, producing a char and volatile pyrolysis products which are steam cracked and reformed to produce a medium BTU gas. The residual char retained in the bed is more slowly gasified by reaction with steam. The sulfur and sodium are separated in that the sulfur becomes part of the gas stream and the sodium stays in solid form. Bed temperatures are maintained at 605-610 °C, thereby avoiding liquid smelt formation and the associated smelt-water explosion hazards.

Product gases are routed through a cyclone to remove the bulk of the entrained particulate matter and are subsequently quenched and scrubbed in a Venturi scrubber. A portion of the medium-BTU product gases can be supplied to the pulse combustion modules, and the combustion of these gases provides the heat necessary for the indirect gasification process. Low temperature gasification leads to complete separation of the sulfur and sodium in kraft black liquor to the gas and solid phase, respectively. Bed solids are continuously removed and mixed with water to form a carbonate solution. The inorganic chemical in the bed solids as well as the sulfur from the gas stream are recovered and used as cooking liquors for the mill. The product gas residence time in the

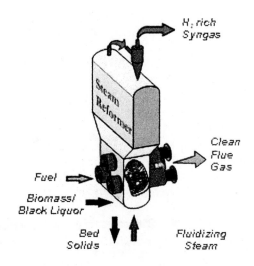

Figure 2. Schematic of MTCI Steam Reformer.

fluid bed is about 15 seconds because of the deep bed (20 ft) used, while the solids residence time is about 50 hours. These conditions promote extensive tar cracking and carbon conversion. In summary, the steam reforming reactor vessel has three inputs: fluidizing steam, black liquor, and heat; and has three outputs: bed solids, hydrogen rich product gas, and flue gas (*6*).

High Temperature Gasifier

High temperature gasification stems from work initiated by SKF in the 1970s. The original patent for the technology was issued in 1987 and it has since been developed through a sequence of demonstration projects. The gasifier, as developed by Chemrec, is a refractory-lined entrained flow reactor. In high temperature gasification (900-1000 °C), concentrated black liquor is atomized, fed to the reactor and decomposed under reducing conditions using air or oxygen as the oxidant. The initial chemical reactions involve char gasification and combustion and are influenced by physical factors like droplet size, heating rate, swelling, and the sodium and sulfur release phenomena. The resulting products, smelt droplets and a combustible gas, are then brought into direct contact with a cooling liquid in a quench dissolver. The two phases are separated as the smelt droplets dissolve in the cooling liquid forming green liquor. The exiting product gas is subsequently scrubbed and cooled for use in other unit operations. The split of sodium and sulfur between the smelt and gas phase is dependent on the process conditions. Typically, most of the sulfur leaves with the product gas and essentially all of the sodium with the smelt (7-9).

Current Status of BLG Technologies

The TRI steam reformer has been installed in two locations in North America, at the Norampac Mill in Trenton, Ontario and at the Georgia Pacific Mill in Big Island, Virginia. The Trenton mill produces 500 tpd of corrugating medium using a sodium carbonate based pulping process. Prior to the start-up of the low-temperature black liquor gasifier in September 2003, the mill had no chemical recovery system. For over forty years the mill's spent liquor was sold to local counties for use as a binder and dust suppressant on gravel roads. This practice was discontinued in 2002. The capacity of the spent liquor gasification system is 115 tpd of black liquor solids, and the syngas is burned in an auxiliary boiler (6).

Georgia-Pacific's mill at Big Island, Virginia, produces 900 tpd of linerboard from OCC and 600 tpd of corrugating medium from mixed hardwoods semi-chemical pulp. Like the Trenton mill, the Big Island mill uses a sodium carbonate process. In the past, the semi-chemical liquor was burned in two smelters providing chemical recovery but no energy recovery. Instead of replacing the smelters with a traditional recovery boiler Georgia-Pacific decided to install a low temperature black liquor gasification process. One difference between the two systems is that unlike Trenton, Big Island burns the generated product syngas in the pulsed combustors, so the product gas exiting the reformer vessel is cleaned prior to combustion (6).

The evolution of the high temperature gasifier has taken the technology from an air-blown process near atmospheric pressure to a high pressure (near 30 atm.) oxygen-blown process. Benefits realized through high pressure oxygen-blown operation are higher efficiencies, higher black liquor throughput and improved compatibility with down stream unit operations such as combined cycle power generation.

An air-blown pilot plant at Hofors, Sweden, was developed to verify the possibility of gasifying black liquor using an entrained-flow reactor operating at 900-1000 °C. The project showed that green liquor of acceptable quality could be generated,and the plant was dismantled in 1990. The Frövi, Sweden plant was designed as a capacity booster for the AssiDomän facility and was operated from 1991 to 1996, demonstrating the potential for black liquor gasification at a commercial scale. During its operation several technical problems were encountered and addressed. The identification of a suitable material for the refractory lining remained a problem. A subsequent commercial project was initiated in 1996 at the Weyerhauser plant in New Bern, North Carolina. The black liquor gasifier was more or less a scale-up of the Frövi plant, designed for a capacity of 300 tons of dissolved solids/day. In 1999 the process maintained greater than 85% availability. However, over the course of the project the plant experienced several technical problems, mainly related to the refractory lining, and it was shut down after cracks in the reactor vessel were discovered in 2000. After detailed studies and re-engineering, the gasifier operation at New Bern was resumed in the summer of 2003 (7). During the rebuild, it was retrofitted with spinel refractory materials developed at Oakridge National Labs in cooperation with other partners. The refractory material is in its second year of operation. The gasifier can burn up to 730,000 lb/day of solids or about 20% of the mill production (10). The syngas generated in the gasifier is currently burned in a boiler.

A pressurized air-blown demonstration project was established at the Stora Enso plant at Skoghall, Sweden, in 1994. The project showed the capability of a pressurized system to generate acceptable quality green liquor while maintaining high carbon conversion ratios. The process was converted to an oxygen-blown operation in 1997 resulting in a capacity increase of more than 60%. A second pressurized demonstration plant was completed in Piteå, Sweden, in 2005. The purpose of the project is to demonstrate high pressure operation (near 30 atm.) with associated gas cooling and sulphur handling unit operations required for a full-scale BLG process. Funding has been obtained for a scale-up project of the Piteå facility. The plant is designed for a capacity of 275-550 tDS/day and encompasses all the required unit operations, including the power island, for a BLG process with combined-cycle power generation (7).

BLG – The Cornerstone of the Biorefinery

The integrated forest biorefinery is a concept which, if implemented, has the potential to dramatically change the pulp and paper industry. The conversion of existing pulp mills into biorefineries is a natural progression when trying to realize the full potentials in the by-product streams from pulp and paper making. It is also a promising option for increasing the return on investment in an energy- and capital-intensive industry. The gasification of black liquor and biomass presents the best option for the generation of high value products from what today is essentially process waste (11). It is an inherently stable yet highly flexible technology that can be designed and sized according to the needs and requirements of individual mills. It allows for the recovery of any spent pulping liquor, and enables the generation of a wide array of liquors that can be used to optimize pulping chemistry, pulp yield and properties. It can be coupled with various other unit operations to generate power or feedstocks for liquid fuels and bio-chemical processes, and even hydrogen for utilization in fuel cells. It is environmentally superior to current recovery boiler technologies and presents a carbon-neutral source for power generation and synthetic products. BLG is an enabler of the biorefinery and the technologies it encompasses (12).

The underlying fundamental for implementation of any new technology is the impact it will have on the overall process economics. Some deciding factors that will influence the implementation of BLG involve the cost benefits associated with power generation and other high-value products that can be derived from the syngas. Another area of importance is the potential cost savings that can be realized through process modifications and optimization. The effect on wood, chemical and fuel demand from changes in the pulping process can have a significant effect on the variable operating cost, capital investment and maintenance costs. Therefore, research exploring the impact of BLG on pulping technologies will be of great importance for the eventual implementation of this technology.

Pulping Technologies Enabled by BLG

The implementation of black liquor gasification into the Kraft recovery cycle would present several opportunities and potential benefits regarding pulp mill operation and process economics. Using black liquor gasification, the recovered entities of sodium and sulfur can be split into two separate fractions with varying degrees of separation dependent on the operating conditions and the technology used. The separation of these chemicals creates some opportunities in the pulping process which can be employed to increase the pulp yield, extend delignification and improve product quality. The following modified pulping technologies can be used in combination with black liquor gasification to realize these potential benefits:

- Split Sulfidity Pulping
- Polysulfide Pulping
- Alkaline Sulfite Pulping
  - Alkaline Sulfite- AQ (AS-AQ)
  - Mini-Sulfite Sulfide AQ Process (MSS-AQ)

Split Sulfidity Pulping

Modified kraft pulping processes have gained widespread acceptance, because they can be used to either extend delignification or to enhance the yield and pulp properties at a given kappa number. The basic principles of modified extended delignification consist of a level alkali concentration throughout the cook, a high initial sulfide concentration, low concentrations of lignin and Na^+ in the final stage of the cook, and lower temperature in the initial and final stages of the cook (*13*). BLG would enable a mill to generate a high sulfidity liquor which can be used to provide a high sulfide concentration during the initial phase of the cook. Figure 3 shows the basic concept design for generating liquors of different sulfide concentrations.

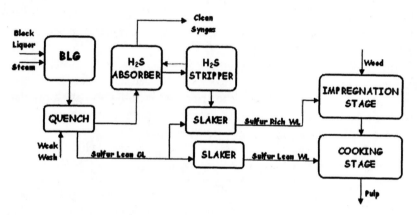

Figure 3. Schematic of unit operations in split sulfidity pulping.

In split sulfidity pulping, it would be necessary to generate two streams of white liquor – one that is sulfide rich and another that is sulfide lean. Sulfur profiling would be the lowest capital cost process to implement to modify the pulping process especially for mills with a modified continuous or batch pulping process.

The concept of sulfur profiling, or split sulfidity pulping, employing a sulfur rich stream in the rapid initial phase, followed by a sulfur lean stream in the bulk and residual phase, has been investigated as a method for extending delignification or increasing yield (*14-18*). Compared to conventional kraft cooks of similar H-factor, split sulfidity pulping has been shown to enhance selectivity of the pulping reactions, resulting in increases in both lignin removal and pulp viscosity. Moreover, split sulfidity pulping has been shown to increase pulp yield and strength properties (*19-21*).

The effects of multiple stage cooking using sulfur profiling, has also been studied. The process showed a significant improvement in selectivity (22,23). Increased sulfide sorption resulted in both higher lignin-free yields and increased viscosities. At 30% overall sulfidity, the lignin-free yield was 0.6 to

0.9% higher and viscosity 8.89 to 10.4 mPa higher than conventional kraft. At increasing overall sulfidities, the yield advantage was reduced. Screened yield increased only slightly with higher sulfidity levels during impregnation. Similar findings were reported in subsequent work (24). Pulping work conducted at STFI found that sorption of sulfide increases with increasing hydrosulfide concentration, time, temperature and concentration of positive ions, but decreases with an increasing concentration of hydroxide ions (25). The potential for modifying softwood kraft pulping, by sulfur profiling has also investigated, where all of the sulfide was added to the beginning of the cook, a high hydrosulfide concentration could be maintained both in the initial phase and near the transition point from the initial to the bulk delignification phase (26).

The work described above is difficult to implement in a mill that utilizes conventional recovery technologies. However, BLG generates separate streams of sulfur and sodium, which will allow for independent sulfur and alkali profiling. Thus, the alkali profile can be adjusted independent of the sulfur concentration at any point in the cook. These opportunities were investigated at NC State University, exploring split sulfidity pulping of southern pine with different initial alkali concentrations. Based on a modified continuous cooking (MCC) laboratory procedure, different approaches were devised to explore split sulfidity and different initial alkali profiles (27,28). Two levels of initial alkali were investigated where a fraction of the available hydroxide was charged in the initial stage. The low initial alkali procedure used 11% of the alkali; and the corresponding value for the high initial alkali procedure was 33%.

The effects of split sulfidity and different levels of initial alkali on delignification and total pulp yield are presented in Figure 4. As shown, split sulfidity pulping produced lower kappa pulps at similar H factors relative to the MCC procedure. The high initial alkali cooks generated pulps of lower kappa number compared to those of low initial alkali. The split sulfidity procedures produced pulp yields 1-2% greater than the MCC procedure, and the difference is more pronounced at higher kappa. Since the high initial alkali approach produced higher yields and lower kappa numbers than the low initial alkali approach, this would be the preferred option. At similar kappa numbers the split sulfidity pulps had viscosities 5 to 10 cps greater than those of the MCC pulps. The high initial alkali pulps produced higher tensile and burst index values relative the MCC pulps at a similar tear index. The MCC pulps were slightly easier to refine relative to the split sulfidity pulps.

The co-absorption of H_2S and CO_2 during the scrubbing in sulfur recovery, results in the production of NaHCO3. During recausticization all sodium exiting the gasifier will be converted to NaOH. The conversion of $NaHCO_3$ to NaOH requires twice as much lime compared to the conversion of $NaCO_3$ to NaOH. Thus, there is a two-fold increase in the amount of lime required to produce an

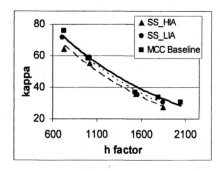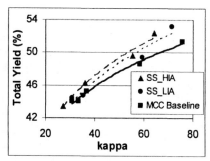

Figure 4. Delignification and yield results for split sulfidity pulping.

equivalent amount of NaOH, and as a result, BLG will increase the overall causticization load.

The potential for in-situ causticization within the gasifier could dramatically affect the load on the recaust cycle and lime kiln. In current recovery operations, the sodium carbonate obtained from the slaking of the boiler smelt is converted to sodium hydroxide using calcium oxide. The byproduct calcium carbonate is then calcined in large rotating kilns to regenerate the calcium oxide. A 1000 ton per day pulp mill will use about 100,000 barrels of fuel oil per year to fire its lime kiln. Through novel chemistries it may be possible to carry out the causticization reactions directly within a black liquor gasifier. This could potentially eliminate the need for the lime cycle and the associated fuel costs (*29*).

Another alternative to in-situ causticization, avoiding the increase in causticization requirements, would be to pretreat wood with green liquor. Previous work has demonstrated the feasibility of using green liquor in the impregnation stage, without increasing overall chemical usage (*15,21*). It has also been shown that the amount of sulfur adsorbed during the pretreatment decreases with higher [OH⁻] (*30*). By impregnating chips with high sulfidity, low pH liquor, a mill may enhance yield and further decrease the causticizing load. Figure 5 outlines the unit operations for a possible green liquor pretreatment process in conjunction with BLG.

Comparing green liquor pretreatment and kraft pulping, the greatest relative cost-benefit from a decrease in causticization using green liquor pretreatment would be achieved in a situation where the level of TTA was the same in both processes. This requires that similar pulp kappa numbers must be attainable through both processes at the same TTA charge. Green liquor pretreatment pulping has been investigated (*28*). Pulping results show that green liquor pretreatment would return pulps of higher kappa at the same level of total titratable alkali, as shown in Figure 6. However, as shown in the figure it would

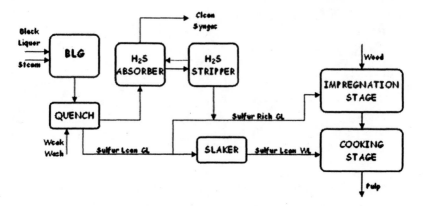

Figure 5. Outline of process using green liquor pretreatment.

be possible to achieve similar kappa number at the same level of system TTA by pulping to a higher H factor. Another option for decreasing the kappa number would be to increase the system TTA. Experiments performed with a 10% increase in TTA, labeled Hi-TTA, also resulted in a higher kappa number than the MCC baseline. In this study, if the TTA is increased by 20% the active alkali is the same in both processes, and the no causticizing benefits exist. The resulting pulp yield did not show any improvement with green liquor pretreatment, but the green liquor pretreated pulps had higher viscosity. These results did not show the yield benefit reported elsewhere, and may be the result from differences in pulping procedures (*21*).

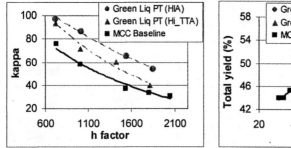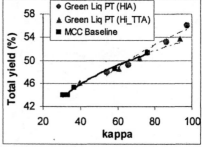

Figure 6. Obtained results for pulping using green liquor pretreatment.

Polysulfide Pulping with Anthraquinone

The effect on pulping chemistry of polysulfide (PS), often in conjunction with anthraquinone (AQ) as additives to the Kraft process, has been explored for some time (*31-36*). Its effectiveness has been established, and it is typically reported that each percent of PS added increases the pulp yield by one percent (*37,38*). However, efficiently generating high concentrations of PS within the Kraft chemical recovery cycle is difficult. There are currently three primary competing processes available for PS generation, Chiyoda, MOXY™ and Paprilox® (*39*). These processes, in general terms, produce pulping liquors with PS concentrations of five to eight grams per liter and PS selectivities ranging from 60 to 90 percent (*40-43*). This results in a PS limit of about 1% PS charge on oven dry wood for a mill operating at 25% sulfidity. However, a chemical recovery system based around BLG would allow for different pathways to generate PS liquors which would enable for higher charges of polysulfide. In addition, the separation of sodium and sulfur would allow for alkali profiling in conjunction with PS utilization. Figure 7 shows a schematic of PS process unit operations with BLG.

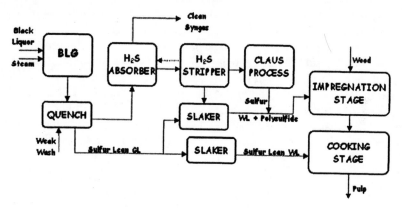

Figure 7. Outline of process using polysulfide.

Research efforts in the area of PS have generally been in one of two major areas: work on PS pulping associated with PS utilization in Kraft process operations and/or associated PS generation technologies, (*40-47*) and work investigating optimum parameters for PS pulping (*48-50*).

A smaller area of work has been based around the potential implementation of BLG and the opportunities created by the unrestricted management of sulfur and sodium as separate entities. The splitting of sulfur and sodium enables the application of polysulfide, sodium sulfide and sodium hydroxide independently of each other. Two processes have been described based on this concept. The ZAP process (Zero effective alkali in pretreatment), entails a two-stage pulping procedure, where the sulfur containing cooking chemicals (Na_2S and PS) were charged along with AQ to the wood in the pretreatment stage (*51*). In the subsequent cooking stage NaOH was added and the temperature increased. The obtained results using PS without AQ, indicate a potential yield benefit of 1% relative to conventional PS pulping at kappa 30. With the addition of AQ the yield benefit was increased to 1.5-2%. Additional results indicate even greater yield benefits at kappa 90. A different procedure called hyperalkaline polysulfide pulping has been suggested (*52*). The process utilizes two pretreatment stages followed by a cooking stage. In the first stage, alkali is charge to the wood at elevated concentrations, neutralizing the acids formed during the temperature elevation. PS is then charged in the second stage, followed by the cooking stage. The process resulted in a higher delignification rate, increased pulp viscosities and yield improvements of 1.5% as compared modified pulping without PS. It is worth noting that the bleachability and measured tear strength of the hyperalkaline PS pulps were similar to those of the Kraft reference pulp.

In addition to the capability for independent profiling of NaOH and PS, the implementation of BLG would allow for the conversion of all the sulfur in the cooking liquor to PS. This would enable higher PS charges than those available through conventional technologies. The total amount of sulfur that is available is dependent on the sulfidity of the pulping liquor. Table 1 illustrates this balance by displaying two examples of the partitioning of the total sulfur available in the system at 25 and 40% sulfidity. As seen in the table, PS charges slightly exceeding 2 % on wood is possible at 19.5% AA with 25% sulfidity. To enable higher PS charges the sulfidity must be increased, and as shown in the table, the corresponding value at 40% sulfidity is between 3 and 4% PS on wood.

Table 1. Demonstration of sulfur utilization as Na_2S (kraft) or Na_2S/PS (kraft-PS) and the system sulfur availability for PS generation

Cook Procedure	25% Sulfidity			40% Sulfidity		
	Tot. avail. Sulfur (S) (kg/ton)	S req. for PS (kg/ton)	S avail. as Na2S (kg/ton)	Tot. avail. Sulfur (S) (kg/ton)	S req. for PS (kg/ton)	S avail. as Na2S (kg/ton)
MCC	25.2	0	25.2	40.3	0	40.3
1% PS	25.2	10.0	15.2	40.3	10.0	30.3
2% PS	25.2	20.0	5.2	40.3	20.0	20.26
3% PS	25.2	30.0	- 4.8	40.3	30.0	10.3
4% PS	25.2	40.0	- 14.8	40.3	40.0	0.3

The effects of pulping southern pine with higher PS charges and alkali profiling was evaluated at NC State University. Table 2 shows a summary of the yield increases that were measured at various PS charges and sulfidities. At 40% sulfidity, the impact of alkali profiling in the initial stage was also evaluated. Three different levels of alkali were investigated. In the low initial alkali (LIA) cook, 56% of the alkali was charged to the impregnation stage. The corresponding values for the medium initial alkali (MIA) and the high initial alkali (HIA) were 65% and 75%, respectively. The flexibility to optimize the alkali profile and PS use would only be possible in combination with BLG (*53*).

Table 2. Total yield improvement from PS procedures compared to 25% S MCC pulp

Cook ID	Estimated Yield at 30 kappa	Yield Improvement (% pts.)	Average Viscosity (cps)
MCC 25% S	45.2	n.a.	39.1
1% PS 25% S	47.7	2.4	38.6
2% PS 25% S	48.1	2.9	38.0
Cook ID	Estimated Yield at 30 kappa	Yield Improvement (% pts.)	Average Viscosity (cps)
1% PS 40% S LIA	46.7	1.5	49.5
2% PS 40% S LIA	47.3	2.1	51.6
3% PS 40% S LIA	47.4	2.2	55.7
3% PS 40% S MIA	48.9	3.7	47.7
3% PS 40% S HIA	45.8	0.6	46.7

At 25% sulfidity, the kappa numbers with PSAQ were comparable to the MCC reference. The yield benefit was about 2% for a 1% PS charge, and about 3% for a 2% PS charge. At 40% sulfidity, the kappa number decreased with increasing PS charge and increasing levels of initial alkali. The level of initial alkali had a significant effect on the yield. There is indication of an optimum condition for initial alkali charge, where too great or too low of an initial hydroxide concentration negatively affects the pulp yield. The work exploring MIA and HIA indicates that there exists a maximum in yield benefit as a function of initial alkali concentration. In the ZAP process using PSAQ, a yield benefit of about 6% at 30 kappa was reported, where the initial alkali charge in the PS pretreatment stage was zero (*50*). They also showed that a minimum yield condition exists at a hydroxide concentration of about 0.3 mol/l. At hydroxide concentrations lower or greater than 0.3 mol/l, higher yields could be achieved. The results shown in Table 2 indicate that there also exists a maximum yield benefit at higher levels of initial alkali. The LIA procedure corresponds to an initial hydroxide concentration of about 0.6 mol/l, which is greater than the concentration reported for the yield minimum in the ZAP process. The maximum yield benefit was found to be an initial hydroxide concentration of 0.9 mol/l. The effect of initial alkali on pulp yield should be further investigated to optimize the benefits of PS pulping.

Alkaline Sulfite Pulping

The use of anthraquinone as an accelerator to alkaline sulfite pulping led to the development of the AS-AQ process (*54*). The Mini-Sulfide Sulfite anthraquinone (MSS-AQ) pulping process was investigated in Sweden in the mid 1980's (*55,56*). Both the AS-AQ and the MSS-AQ processes have some unique features outlined below (*55-60*):

- Pulp yield for linerboard is 10% pt higher than kraft at equivalent Kappa number and strength properties.
- The brightness of the AS-AQ/MSS-AQ linerboard pulp is considerably higher than the brightness of the corresponding kraft pulp: 40 % ISO and 18% ISO, respectively. This would be a considerable advantage for high quality printing on cardboard boxes for advertising.
- The need for the caustic room, lime kiln and associated energy usage is eliminated. This would be a very large capital savings, especially for greenfield mills and mills with major rebuilds.
- The yield advantage of AS-AQ/MSS-AQ over Kraft decreases rapidly as the kappa number decreases. However, because the AS-AQ/MSS-AQ pulps respond very well to oxygen delignification, it is possible to stop pulping at the defiberizing point (kappa 50) and to continue the delignification with oxygen and alkali while keeping most of the yield increase achieved at Kappa 50.
- The AS-AQ/MSS-AQ pulps are considerably brighter and also easier to bleach than the corresponding kraft pulps. ECF bleaching can be accomplished with lower ClO_2 usage. High brightness can be achieved with TCF sequences while producing a bleached pulp with acceptable pulp strength. If TCF bleaching becomes cost effective, the capital cost for the chlorine dioxide generators would also be eliminated.
- Lower TRS emissions results in a low odor mill.

The difference between the AS-AQ and MSS-AQ processes is the amount of sulfur that exists as sodium sulfide. In the AS-AQ process, all the sulfur exists as Na_2SO_3 with no Na_2S, while the MSS-AQ process uses a mixture of Na_2S and Na_2SO_3. The decision to utilize AS-AQ or MSS-AQ will depend on the amount of Na_2S that can be generated in pulping liquor during recovery operations, and on the potential improvements in delignification rate and pulp yield in MSS-AQ as compared to AS-AQ.

In the MSS-AQ process the total charge of alkali is about 22% Na_2O on OD wood, which is higher than for the kraft process (19% Na_2O). The delignification rate is slower than for the kraft process, and higher pulping temperatures (+ 18 °F) and anthraquinone addition (0.15% on OD wood) are required to obtain acceptable pulping rates. Preliminary optimization studies done at STFI (*55,56*) indicate that a sulfide ratio between 0.05 to 0.1 results in the lowest kappa number. The optimum also shifts to a lower value with increasing AQ charge. If the amount of sodium sulfide is increased above the optimum, the kappa number increases again.

Despite these significant advantages, the development of both the AS-AQ and MSS-AQ processes has not been pursued more aggressively because of the lack of an attractive chemical recovery process. Black liquor gasification combined with appropriate gas cleanup/absorber technology would present precisely such an alternative. The benefits of these two processes were compared for the production of linerboard.

AS-AQ Pulping Work

Two AS-AQ processes have been evaluated with southern pine: the traditional strongly alkaline, SA, and also the moderately alkaline, MA. It would be more attractive to operate at lower alkali charges since this would minimize the amount of NaOH that would be necessary in the process. When compared to the Kraft reference pulp, the moderately alkaline AS-AQ procedure produced pulps with a 10% yield benefit and a higher ISO brightness. The pulp refined somewhat more slowly, had a higher apparent sheet density, a lower tensile index, and a burst index at a similar tear index (*61*). Similarly, the strongly alkaline AS-AQ procedure returned a 5% yield benefit and a pulp of comparable ISO brightness.

MSS-AQ Pulping Work

When compared to the Kraft-AQ reference pulp, MSS-AQ pulping produced pulps with a 15-18% yield benefit and ISO brightness that was significantly higher. The yield and brightness of the different pulps are shown in Figure 8. The pulp refined more easily and at similar levels of tear index the MSS-AQ pulps had a slightly lower tensile and burst index (*61*).

Table 3 shows the chemical balance for the different processes in kg of chemicals required as Na_2O per oven dry metric ton pulp (ODtP). The integration of alkaline sulfite AQ into a mill would require the addition of

452

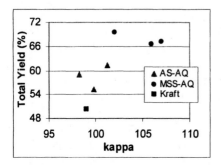 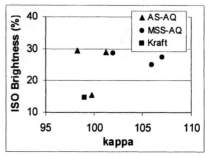

Figure 8. Total pulp yield and ISO Brightness versus kappa for Kraft, AS-AQ and MSS-AQ.

sodium hydroxide to adjust the pH. The AS-AQ process operated at strongly alkaline conditions would require the conversion of 163 kg of NaOH. This would require a recausticization system to be operated at the mill, and therefore make this process alternative less attractive. In the mildly alkaline AS-AQ process, the use of alkali can be decreased to 35 kg NaOH/ODtP.

Table 3. Chemical Requirements for Selected Options for production of 1 ODtP (all chemicals as Na$_2$O) converted to kg/ton

	Yield (%)	Wood (OD kg)	TTA (%)	NaOH (kg)	Na$_2$S (kg)	Na$_2$CO$_3$ (kg)	Na$_2$SO$_3$ (kg)
Kraft	50.0	2000	17.6	226	74	53	0
AS-AQ SA	55.4	1805	21.0	163	0	53	163
AS-AQ MA	60.0	1667	21.0	35	0	35	280
MSS-AQ1	66.6	1502	22.0	33	17	33	248
MSS-AQ2	67.3	1486	22.0	0	32	32	261
MSS-AQ3	69.6	1437	22.0	0	15	31	268

Some of the additional alkali required could be added as makeup chemical, but the 35 kg NaOH/ODtP is still greater than that required as makeup in most well operated mills. The MSS-AQ pulping process can be operated with no caustic, which would eliminate both the causticization process and the lime kiln. When the sodium sulfide charge is 5% of the total chemical, 17 kg of sodium sulfide will have to be generated by absorption of the H$_2$S from the syngas. Alkaline sulfite methods generate pulp yields significantly greater than the Kraft-AQ procedure at similar kappa. The moderately alkaline AS-AQ procedure resulted in an additional 10% yield benefit. MSS-AQ pulping resulted in yield increases ranging from about 15 to 20% as compared to the Kraft baseline. MSS-AQ and the moderately alkaline AS-AQ procedures generate pulps of significantly higher unbleached ISO brightness, ranging from about 25 to 30%, as compared to the Kraft and strongly alkaline AS-AQ procedures with ISO brightness around 15%. The higher pulp brightness indicates that there is a large potential for cost savings in bleaching operations if bleachable grade pulp was produced.

MSS-AQ with RTI Absorption Technology

Research Triangle Institute (RTI) has developed durable zinc oxide based regenerable desulfurization sorbents. The primary application for the technology has been removal of reduced sulfur compounds from hot coal derived synthesis gases and recovery of concentrated SO_2 streams. The product gases containing the sulfur compounds will be scrubbed to remove particulates and then sent to the RTI absorber. The sulfur containing compounds are absorbed in the ZnO bed. The clean gas can then be fed to the turbine generators for high efficiency power generation. When the absorber bed is regenerated using air, the sulfur is desorbed as sulfur dioxide which is the active chemical for the alkaline sulfite pulping processes. Tests were conducted at 260 to 530°C and at pressures from 240 to 2000 kPa. In fixed bed sulfidation reactor tests, inlet H_2S levels were decreased from percent levels to ppm levels. The sorbent capacity at breakthrough was as high as 17% S. The sorbent was regenerated with 3.5% oxygen at 566°C (*61*).

Black liquor gasification, RTI Absorption technology and MSS-AQ will have to be integrated into a chemical pulp mill to take advantage of the yield benefits, energy savings, capital savings and the increased power generation. The process flow-sheet with the above technologies integrated is shown in Figure 9.

The sulfur dioxide from the RTI absorber will be absorbed in the green liquor (sodium carbonate) to form the sodium sulfite liquor. An appropriate amount of the sulfur gases will be absorbed as hydrogen sulfide to produce the required amount of sodium sulfide. The RTI absorber serves the dual purpose of cleaning up the product gas and also regeneration of the chemicals. As shown in the process schematic, the causticization and lime kiln operations can be eliminated from the system as they are not needed in MSS-AQ pulping with BLG. This will result in significant capital and operating cost savings (*61*).

Economics of BLG Implementation

The overall economics associated with operating a BLG system will be impacted by the pulping process. The return on investment will also have a

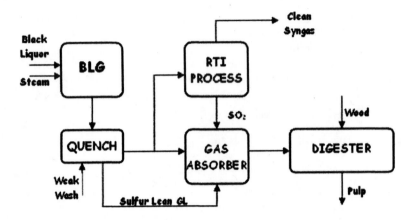

Figure 9. Schematic of the alkaline sulfite pulping processes with the RTI absorber

significant effect on how rapidly the technology will be implemented. The return on investment is dependent on the market price of oil, natural gas, and electricity, as well as the value of products that can be produced from the BLG syngas, and process savings achieved through BLG implementation. Work on forecasting the economic effects of BLG implementation, using analyses of existing and potential technologies will therefore be important. Areas of interests for these studies involve the exploration of markets for and economics of potential products, and the comparative analyses of different technologies and process options for their manufacture.

The BLG generation of a syngas high in hydrogen and carbon monoxide will allow for the production of a wide range of high value products currently unattainable with the Tomlinson recovery boiler. With different options for down-stream technologies following the BLG the capability exists to generate steam, power, or various bio-fuels and source materials for bio-chemicals processes such as (*12*):

- DME/Methanol
- Mixed alcohols
- Fischer-Tropsch liquids
- Syngas fermentation

Work has been done exploring the potential effects of modified pulping processes on the variable operating costs of pulp mill operations. In order to create a wider knowledge base for future decision-making regarding what products to manufacture and what process to use, more research is needed (*11,62-65*).

The Energy Group at the Princeton Environmental Institute has been a significant contributor to the work exploring the cost benefits of BLG operation and associated technologies. Work has been presented showing the potential for different alternative products within the framework of the biorefinery concept, focusing on the combined cycle generation of electrical power (*64-68*). A comparison of the reported estimated capital investment required for installation of the different process options, including operations and maintenance costs, is displayed in Table 4. The values in the table are given as percent changes for each process option compared to the Tomlinson base case (BASE) indexed at 100.

Table 4. Estimated Cost Comparisons of Tomlinson and BLGCC Power/Recovery Systems Relative the Tomlinson BASE (index = 100), (68)

	Tomlinson		*BLGCC System*		
	BASE	*HERB*	*Low-Temp Mill-Scale*	*High-Temp Mill-Scale*	*High-Temp Utility-Scale*
Direct Costs	100	130	172	162	202
Non-Direct Costs	100	140	267	151	186
TOTAL INSTALLED CAPITAL COST	100	132	192	159	199
Annual Operating and Maintenance Costs	100	154	154	154	162

The different process options include Tomlinson alternatives with either a conventional or a high efficiency recovery boiler, and mill scale LT and HT BLG, as well as a utility scale HT BLG. The values were based on "N[th] plant" level of technology maturity and reliability and have an estimated accuracy of ±30%. The capital requirement for the installation of BLG technologies is higher than for the Tomlinson recovery boiler. However, recovery operations built around a BLGCC system will convert the pulp mill to a significant net producer of electricity, compared to a mill using a Tomlinson boiler which is a net importer of electricity. The kraft reference mill used in the study required a net import of 36 MW, compared to a net production and export of power to the grid of 22 MW and 126 MW, for the low and high temperature BLG processes, respectively. To meet the same process steam demand, additional fuel or biomass would have to be processed. In detailed work comparing the kraft reference mill to the LT and HT BLG processes, it was shown that the conversion efficiency of biomass to power was about 15% for the non-pressurized LT BLG, nearly 50% greater than for the reference kraft mill. For the two HT BLG processes the same values were about 19% for the smaller utility design (85% increase) and about 17.5% for the mill-scale design (70% increase). The work also indicates superior financial performance of the BLG technologies relative to the kraft base comparison (*66-68*).

The internal rate of return (IRR) and net present value (NPV) were calculated for the different process options. Assuming a 2.5¢/kWh premium and a tax credit of 1.8¢/kWh for the production of renewable energy, the LT BLG process yielded an IRR of 20.9%/yr and a NPV of $73 million compared to the kraft base case. The corresponding values for the HT BLG mill and utility scale processes were 34.8%/yr and $138.5 million and 35.1%/yr and $216 million, respectively. Also indicated in the work is the distinct window of opportunity for the installation of the technology that exists in the near future, and that a lag in market penetration could be assessed as a loss of $9 billion over a 25 year period measured in terms of energy costs and emission reductions (*66-68*).

Summary and Conclusions

Biomass can be converted to power and/or fuels using gasification. The gasification of black liquor produces output streams that can be used with great benefit in modified pulping operations with the additional production of power and other high value products. BLG with combined cycle power production can generate more electricity than combustion in a boiler given the same fuel, turning an integrated pulp and paper mill into a net exporter of electricity. Other options for syngas utilization are the generation of bio-derived liquid fuels, bio-derived chemicals for the synthetic and pharmaceutical industries, and H_2 for use in fuel cells.

As discussed in this chapter, the implementation of a BLG system will have a dramatic effect on current pulp mill operations. In addition to the high value products made possible through the concepts related to the biorefinery, BLG will enable different modified pulping technologies through the separation of sulfur and sodium. A comparison of the different technologies enabled by BLG and their effects on the different pulping processes is given in Table 5.

As shown, modifications to the kraft process, using split sulfidity, green liquor pretreatment and polysulfide pulping, can generate increased rates of delignification, improvements in pulp yield, viscosity and strength properties, but not all at the same time. Further optimization of these technologies could have a significant effect on the kraft process. Alkaline sulfite pulping can generate significant yield benefits and significantly higher brightness. A unique opportunity exists in MSS-AQ pulping where it would be possible to eliminate the causticizing and lime kiln operations, resulting in very significant cost savings.

Table 5. General Comparison of BLG Enabled Pulping Technologies

Pulping technology	Effect on kappa	Effect on Yield	Effect on viscosity	Effect on properties
Split Sulfidity	Lower kappa at same chemical charge	1-2% higher yield	Comparable viscosity	Improved strength properties
Green Liquor pretreatment	Higher kappa at same chemical charge	No/small benefit	Increased viscosity	Improved strength properties
Polysulfide-AQ	Similar to lower kappa at same chemical charge	1-2% yield increase per % PS charged	Increased viscosity (ZAP slight decrease)	Similar properties, but lower tear strength (Hyperalkaline-PS has comparable tear)
AS-AQ	Higher kappa at same total chemical charge	5-10% yield increase	Comparable viscosity	Somewhat lower strength, much greater brightness, easy to bleach
MSS-AQ	Higher kappa at same total chemical charge	15-18% yield increase	Comparable viscosity	Somewhat lower strength, much greater brightness, easy to bleach

The overall economics associated with BLG implementation will have a significant effect on how rapidly the technology will be implemented. Analyses of the costs and benefits associated the installation of BLGCC technologies compared to conventional kraft operations using recovery boilers indicate increases of 20.9% in the internal rate of return and $73 million in the net present value for a mill-scale low temperature BLG process. The corresponding values for a mill-scale high temperature BLG process are 34.8% (IRR) and $138.5 million (NPV). Above and beyond these cost benefits, additional benefits could be realized from modified pulping operations and the production of other high value products.

References

1. Nilsson, L.J.; Larson, E.D.; Gilbreath, K.R.; Gupta, A. *Energy Efficiency and the Pulp and Paper Industry;* ACEEE: Washington D.C./Berkeley, CA, 1995.
2. Jahne, F. *"Commercial Success of Gasification Technology",* Tappi J. **1999,** *82(10):* 49.
3. Whitty, K.; Verrill, C.L. *"A Historical Look at The Development of Alternative Black Liquor Recovery Technologies and The Evolution of Black Liquor Gasifier Designs",* Proc. *2004 TAPPI Int. Chem. Rec. Conf.;* TAPPI Press: Atlanta, GA, 2004; pp 13-33.

458

4. Stigsson, L., *"ChemrecTM [Kvaerner Chemrec AB's] Black-Liquor Gasification"*, Proc. *1998 TAPPI Int. Chem. Rec. Conf.;* TAPPI Press: Atlanta, GA, 1998; pp 663-674.

5. Grace, T.M.; Timmer, W.M. *"A Comparison of Alternative Black Liquor Recovery Technologies"*, Proc. *1995 TAPPI Int. Chem. Rec. Conf.;* TAPPI Press: Atlanta, GA, 1995.

6. Rowbottom, B.; Newport, D.; Connor, E. *"Black Liquor Gasification at Norampac"*, Proc. *2005 TAPPI Engineering Pulping & Environmental Conference;* TAPPI Press: Atlanta, GA, 2005.

7. Chemrec ; URL http://www.chemrec.se; 2005.

8. Whitty, K.; Ekbom, T.; Stigsson, L. *"Chemrec Gasification of Black Liquor";* published on website URL http://www.chemrec.se/forsta.htm; 2000.

9. Lindblom, M. *"An Overview of Chemrec Process Concepts";* Presentation at the Colloquium on Black Liquor Combustion and Gasification; Park City, Utah, May, 2003.

10. Erickson, D.; Brown, C. *"Operating Experience with Gasification Pilot Project";* Tappi J. **1999,** *82(9).*

11. Farmer, M. *"Adaptable Biorefinery: Some Basic Economic Concepts to Guide Research Selection";* Proc. *2005 TAPPI Engineering, Pulping & Environmental Conference;* TAPPI Press: Atlanta, GA, 2005.

12. Burciaga, D. *"Black Liquor Gasification: The Foundation of The Forest Bio-Refinery "New Value Streams"",* Proc. *2005 TAPPI Engineering, Pulping & Environmental Conference;* TAPPI Press: Atlanta, GA, 2005.

13. Johansson, B.; Mjoberg, J.; Sandstrom, P.; Teder, A. *"Modified Continuous Kraft Pulping - Now a Reality";* Svensk Papperstidning--Nordisk Cellulosa. **1984,** *87(10):* pp 30-35.

14. Herschmiller, D.W. *"Kraft Cooking with Split Sufidity- A Way to Break the Yield Barrier";* Proc. *Breaking the yield barrier symposium;* TAPPI Press: Atlanta, GA, 1998; Vol. 1 p 59.

15. Andrews, E.K. Ph.D. thesis, North Carolina State University, Raleigh, NC, 1982.

16. Olm, L.; Tisdat, G. *"Kinetics of the Initial Stage of Kraft Pulping";* Svensk Papperstidning--Nordisk Cellulosa. **1979,** *82(15):* pp 458-464.

17. LeMon, S.; Teder, A. *"Kinetics of Delignification in Kraft Pulping (1). Bulk Delignification of Pine";* Svensk Papperstidning--Nordisk Cellulosa, **1973,** *76(11):* p 407.

18. Olm, L.; Tormund, D.; Jensen, A. *"Kraft Pulping with Sulfide Pretreatment-Part-1 Delignification and Carbohydrate Degradation";* Nord. Pulp Pap. Res. J. **2000,** *15,* 62.

19. Jiang, J.E; Herschmiller, D.W. "Sulfide Profiling for Increased Kraft Pulping Selectivity"; Proc. *1996 TAPPI Pulping Conference*; TAPPI Press: Atlanta, GA, 1996, pp 311-318.

459

20. Lownertz, P.P.H.; Herschmiller, D.W. *"Kraft Cooking with Split White Liquors and High Initial Sulfide Concentration: Impact on Pulping and Recovery";* Proc. *1994 TAPPI Pulping Conference;* TAPPI Press: Atlanta, GA, 1994, pp 1217-1224.

21. Andrews, E.K.; Chang, H-m.; Kirkman, A.G.; Eckert, R.C. *"Extending Delignification in Kraft and Kraft/Oxygen Pulping of Softwood by Treatment with Sodium Sulfur Liquors";* Proc. *Japan Tappi Symposium on Wood pulp chemistry;* TAPPI Press: Atlanta, GA, 1982.

22. Mao, B.; Hartler, N. *"Improved Modified Kraft Cooking. (1). Pretreatment with a Sodium Sulfide Solution";* Paperi ja Puu. **1992**, *74(6):* pp 491-494.

23. Mao, B.; Hartler, N. *"Improved Modified Kraft Cooking. (2). Modified Cooking Using High Initial Sulfide Concentration";* Nordic Pulp and Paper Research J., **1992**, *7(4):* pp 168-173.

24. Mao, B.; Hartler, N. *"Improved Modified Kraft Cooking,4: Modified Cooking with Improved Sulfide and Lignin Profiles";* Paperi ja Puu. **1995**, *77(6/7):* pp 419-422.

25. Olm, L.; Backstrom, M.; Tormund, D. "Pretreatment of Softwood with Sulfide Containing Liquor Prior to Kraft Cook*";* Proc. *1994 TAPPI Pulping Conference;* TAPPI Press: Atlanta, GA, 1994; p 29.

26. Jiang, J.E.; Greenwood, B.F.; Phillips, J.R.; Stromberg, C.B. *"Improved Kraft Pulping by Controlled Sulfide Additions";* Proc. *7th ISWPC Conference;* CICCST: Beijing, China, 1993.

27. Andritz Inc.; Personal communication; Oct, 2003.

28. Lindstrom, M.; Naithani, V.; Kirkman, A.; Jameel, H.; *"Effects on Pulp Yield and Properties Using Modified Pulping Procedures Involving Sulfur Profiling and Green Liquor Pretreatment";* Presented at 2004 Tappi Fall Technical Conference, Atlanta, GA, 2004.

29. Van Heiningen, A.; Schwiderke, E.; Chen, X. *"Kinetics of the Direct Causticizing Reaction Between Black Liquor and Titanates During Low Temperature Gasification";* Proc. *2005 TAPPI Engineering, Pulping & Environmental Conference;* TAPPI Press: Atlanta, GA, 2005.

30. Lopez, I.; Chang, H-m.; Jameel H.; Wizani, W. *"Effect of Sodium Sulfide Pretreatment on Kraft Pulping";* Proc. *1999 TAPPI Pulping Conference;* TAPPI Press: Atlanta, GA, 1999; p 135.

31. Li, Z.; Li, J.; Kubes, G.J. *"Kinetics of Delignification and Cellulose Degradation During Kraft Pulping with Polysulphide and Anthraquinone";* JPPS. **2002**, *28(7):* pp 234-239.

32. Griffin, C.W.; Kumar, K.R.; Gratzl, J.; Jameel, H. *"Effects of Adding Anthraquinone and Polysulfide to the Modified Continuous Cooking (MCC) Process";* Proc. *1995 TAPPI Pulping Conference*; TAPPI Press: Atlanta, GA, 1995; pp 19-30.

33. Jiang, J.E. *"Extended Delignification of Southern Pine [Pinus spp.] with Anthraquinone and Polysulfide";* TAPPI J. **1995**, *78(2):* 126-132.

460

34. Jiang, J.E. *"Extended Modified Cooking of Southern Pine [Pinus] with Polysulfide: Effects on Pulp Yield and Physical Properties"*; TAPPI J. **1994,** *77(2):* pp 120-124.
35. Landmark, P.A.; Kleppe, P.J.; Johnsen, K. *"Pulp Yield Increasing Process in Polysulfide Kraft Cooks"*; Tappi J. **1965,** *58(8):* p 56.
36. Vennemark, E. *"Some Ideas on Polysulfide Cooking"*; Svensk Papperstidning. **1964,** *67(5):* p 157.
37. Kleppe, P.J.; Minja, R.J.A. *"The Possibilities to Apply Polysulfide-AQ in Kraft Mills"*; Proc. *Breaking the yield barrier symposium*; TAPPI Press: Atlanta, GA, 1998, Vol. 1, p 113.
38. Sanyer, N.; Laundrie, J.F. *"Factor Affecting Yield Increase and Fiber Quality in Polysulfide Pulping of Loblolly Pine, Other Softwoods, and Red Oak"*; Tappi J. **1964,** *47(10):* p 640.
39. Luthe, C.; Berry, R. *"Polysulphide Pulping of Western Softwoods: Yield Benefits and Effects on Pulp Properties"*; Pulp. Pap. Can. **2005,** *106(3):* pp 27-33.
40. Munro, F.; Uloth, V.; Tench, L.; MacLeod, M.; Dorris, G. *"Mill-Scale Implementation of Paprican's Process for Polysulphide Liquor Production in Kraft Mill Causticizers - Part 2: Results of Pulp Mill Production Trials"*; Pulp. Pap. Can. **2002,** *103(1):* pp 57-61.
41. Tench, L.; Uloth, V.; Dorris, G.; Hornsey, D.; Munro, F. *"Mill Scale Implementation of Paprican's Process for Polysulfide Liquor Production in Kraft Mill Causticizers. Part-1 Batch Trials and Optimization"*; TAPPI J. **1999,** *82(10):* p 120.
42. Yamaguchi, A. *"Operating Experiences with the MOXY Process and Quinoid Compounds"*; In Anthraquinone Pulping: Anthology of Published Papers 1977-1996; Goyal, G.C., Ed.; TAPPI Press: Atlanta, GA, 1997; pp 287-291.
43. Nishijima, H.; et al., *"Review of PS/AQ Pulping to Date in Japanese Kraft Mills and the Impact on Productivity"*; Proc. *1995 TAPPI Pulping Conference;* TAPPI Press: Atlanta, GA, 1995; pp 31-40.
44. MacLeod, M.; Radiotis, T.; Uloth, V.; Munro, F.; Tench, L. *"Basket Cases IV: Higher Yield with Paprilox™ Polysulfide-AQ Pulping of Hardwoods"*; TAPPI J. **2002,** *1(10):* pp 3-8.
45. Olm, L.; Tormund, D.; Bernor Gidert E. *"Possibilities to Increase the Pulp Yield in a Kraft Cook of [the] ITC-Type"*; Proc. *Breaking the yield barrier symposium;* TAPPI Press: Atlanta, GA, 1998; Vol. 1, pp 69-78.
46. Hakanen, A.; Teder, A. *"Modified Kraft Cooking with Polysulfide: Yield, Viscosity, and Physical Properties"*; TAPPI J. **1997,** *80(7):* pp 86, 93, 100, 189-196.
47. Jiang, J.E.; Crofut, K.R.; Jones, D.B. *"Polysulfide Pulping of Southern Hardwood Employing the MOXYRG and Green-Liquor Crystallization Processes"*; Proc. *1994 TAPPI Pulping Conference;* TAPPI Press: Atlanta, GA, 1994; pp 799-806.

48. Gustafsson, R.; Ek, M.; Teder, A. *"Polysulphide Pretreatment of Softwood for Increased Delignification and Higher Pulp Viscosity";* JPPS. **2004,** 30(5): pp 129-135.

49. Mao, B. F.; Hartler, N.; *"Improved Modified Kraft Cooking,3: Modified Vapor-Phase Polysulfide Cooking";* TAPPI J. **1994,** 77(11): pp 149-153.

50. Lindstrom, M.; Teder, A. *"Effect of Polysulfide Pretreatment When Kraft Pulping to Very Low Kappa Number";* Nordic Pulp Paper Res. J. **1995,** 10(1): pp 8-11.

51. Olm, L.; Tormund, D. *"ZAP Cooking- Increase Yield in PS and PS-AQ Cooking with Zero Effective Alkali in the Pretreatment Stage";* Nordic Pulp Paper Res. J. **2004,** 19(1):p 6.

52. Brannvall, E.; Gustafsson, R.; Teder, A.; *"Properties of Hyperalkaline Polysulphide Pulps";* Nordic Pulp Paper Res. J. **2003,** 18(4): pp 436-440.

53. Lindstrom, M.; Naithani, V.; Kirkman, A.; Jameel, H. *"The Effect of Integrating Polysulfide Pulping and Black Liquor Gasification on Pulp Yield and Properties";* Proc. *2005 TAPPI Engineering, Pulping & Environmental Conference;* TAPPI Press: Atlanta, GA, 2005.

54. Nomura, Y.; Wakai, M.; and Sato, H. *"Process for Producing Pulp in the Presence of a Quinine Compound";* Japanese Patent 112,903, 1967; Canadian Patent 1079906, 1980.

55. Dahlbom, J.; Olm, L.; Teder, A. *"Characteristics of MSS-AQ [Minisulfide-Sulfite-Anthraquinone] Pulping - A New Pulping Process"* TAPPI J., **1990,** 73(3): pp 257-261.

56. Olm, L.; Wiken, J.E.; Olander, K. *"Alkaline Sulfite Pulp: Competitor to Sulfate Pulp";* Svensk Papperstidning--Nordisk Cellulosa. **1986,** 3(4): pp 75-77, 85.

57. Borgards, A.; Patt, R.; Kordsachia, O.; Odermatt, J.; Hunter, W. D. *"Comparison of ASAM and Kraft Pulping and ECF/TCF Bleaching of Southern Pine [Pinus spp.]"* Proc. *1993 TAPPI Pulping Conference;* TAPPI Press: Atlanta, GA, 1993; Vol. 2, pp 629-636.

58. Patt, R.; Schubert, H.L.; Kordsachia, O.; Oltmann, E.; Krull, M. *"Chlorine-Free Bleaching of Sulfite and ASAM Pulps on the Laboratory and Pilot-Plant Scale";* Papier. **1991,** 45(10A): V8-V16.

59. Ingruber, O.V.; Stradal M.; Histed, J.A.; *"Alkaline Sulfite-Anthraquinone Pulping of Eastern Canadian Woods";* Pulp and Paper Canada. **1982,** 83(12): 79-84, pp 87-88.

60. Virkola,N.E.; Pusa,R.; Kettunen, J.; *"Neutral sulfite AQ Pulping as an Alternative to Kraft Pulping";* Tappi J. **1981,** 64(5): pp 103-107.

61. Jameel, H.; Lindstrom, M.; Naithani, V.; Kirkman, A. *"Alkaline Sulfite Pulping and its Integration with the TRI Process";* Proc. *2005 TAPPI Engineering, Pulping & Environmental Conference;* TAPPI Press: Atlanta, GA, 2005.

62. Lindstrom, M.; Kirkman, A.; Jameel, H. et al., *"Economics of Integrating Black Liquor Gasification With Pulping: Part I – Effect Of Sulfur Profiling"*; Proc. *2002 TAPPI Engineering, Pulping & Recycling Conference;* TAPPI Press: Atlanta, GA, 2002.
63. Berglin, N.; Lindblom, M.; Ekbom, T. *"Preliminary Economics of Black Liquor Gasification with Motor Fuels Production";* Presentation at the Colloquium on Black Liquor Combustion and Gasification; Park City, Utah, May, 2003.
64. Larson, E.D.; Haiming, J.; *"Biomass Conversion to Fischer-Tropsch Liquids: Preliminary Energy Balances";* Proc. *4th Biomass Conference of the Americas;* Elsevier Science Ltd: Oxford, UK, 1999.
65. Williams, R.H.; Larson, E.D.; Katofsky, R.E.; Chen, J. *"Methanol and Hydrogen from Biomass for Transportation";* Energy for Sustainable Development. **1995,** *1(5):* p 18.
66. Larson, E.D.; Consonni, S.; Katofsky, R.E. *"A Cost-Benefit Assessment of Biomass Gasification Power Generation in the Pulp and Paper Industry – Final Report"*; prepared with support from the American Forest & Paper Association and the U.S. Department of Energy's Office of Biomass Programs (available from the authors) (Oct. 8, 2003).
67. Consonni, S.; Larson, E.D.; Katofsky, R.E. *"An Assessment of Black Liquor Gasification Combined Cycles Part A: Technological Issues And Performance Comparisons";* Proc. *ASME Turbo Expo 2004 Power for Land, Sea, and Air*; ASME: 2004.
68. Larson, E.D.; Katofsky, R.E.; Consonni, S. *"An Assessment of Black Liquor Gasification Combined Cycles Part B: Emissions, Costs and Macro-Benefits";* Proc. *ASME Turbo Expo 2004 Power for Land, Sea, and Air;* ASME: 2004.

Chapter 29

Production of Activated Carbon from Biochar Using Chemical and Physical Activation: Mechanism and Modeling

Ajay K. Dalai and Ramin Azargohar

Department of Chemical Engineering, University of Saskatchewan, Saskatoon, Saskatchewan, Canada

Biochar, a solid product of fast pyrolysis of biomass, was converted to activated carbon by physical (steam) and chemical (potassium hydroxide) activation. The effects of operating conditions on the BET surface area and the reaction yield of physically and chemically activated carbons were investigated. Two models for BET surface area and reaction yield of each activated carbon were developed. Using these models, the optimum operating conditions for production of activated carbons with large surface area and high yield were determined. The BET surface area and yield of products predicted by models and from experiments at optimum operating conditions showed good agreement. The effects of activating agent on the chemical structure of biochar, during chemical activation, were investigated by thermogravimetric method and infrared spectroscopy.

Introduction

Although the most familiar forms of carbon are cubic diamond and hexagonal graphite [1] and recently discovered fullerenes [2, 3], most carbon materials have less ordered structures. According to their crystallographic structure, they are categorized into graphitic and non-graphitic carbons [4]. The latter is classified as graphitizable (cokes) and non-graphitizable carbons (chars) [5]. Chars do not pass through a fluid phase during pyrolysis (carbonization) [4] and are the main precursors of activated carbon. These carbons are porous materials with highly developed internal surface area and porosity. They are used as catalyst [6, 7] and catalyst support [4, 7, 8] as well as in many adsorption processes for removal of impurities from liquids and gases. Activated carbons can be produced from most carbon-containing organic materials [9, 10], but commercial processes which make activated carbon use precursors, which originate from either degraded and coalified plant matter (e.g. peat, lignite and all ranks of coal) or botanical origin (e.g. wood, coconut shells and nut shells) [11, 12]. These materials have a high content of carbon and are inexpensive [5]. Coal is commonly used for producing high yields of activated carbon [13, 14]. Materials from botanical origin or in other words, lignocellulosic materials, have low inorganic and relatively high volatile content. The first characteristic results in producing activated carbon with low ash and the second characteristic helps to control the production process [9].

The physical and chemical activation methods are commonly applied in the production of activated carbon [5, 15]. In physical activation, char is produced during the first step, by carbonization (pyrolysis) of the precursor. This step removes non-carbon species [5] and produces char with a high percentage of carbon. Because of the blockage of the pores by tars [10], the internal surface area of char is too low. The second step of physical activation is high temperature gasification (activation) using oxidizing agents such as steam or carbon dioxide, which produces activated carbon with high porosity [5, 9, 10]. Porosity development is due to the penetration of the oxidizing agent into the internal structure of char and the removal of carbon atoms by reaction which results in the opening and widening of inaccessible pores [9, 10]. The overall reaction between steam and carbon (including heterogeneous water-gas reaction, shift reaction and methanation) [16] and the reaction between carbon dioxide and carbon [9] are endothermic. Oxygen is not used as an oxidizing agent because of the exothermic reaction between carbon and oxygen, which makes it difficult to control the process temperature and prevents the development of high porosity due to external burning of carbon particles [5, 9].

The chemical activation of the precursor with a chemical (dehydrating) agent is another important industrial process for producing activated carbon. The most common activating agents are potassium hydroxide, phosphoric acid and zinc chloride. In comparison with the mechanism in physical activation, the mechanism in chemical activation is not well understood [17] It seems that the chemical agent dehydrates the sample, inhibits the tar formation and volatile compounds evolution, and therefore enhances the yield of the carbonization process [5, 18]. After impregnating the precursor with the chemical agent followed by a heat treatment of the mixture, the chemical agent is eliminated by washing with acid/base and water. The washing step allows the creation of a pore structure [9].

For quality control of the activated carbon produced by the activation process, characterization methods are used to specify the physical and chemical structure of this product. Porosity is the most important physical characteristic of activated carbon which is used to study the textural properties. IUPAC (International Union of pure and Applied Chemistry) classifies porosity into three groups: micropores (width less than 2 nm), mesopores (width between 2 to 50 nm) and macropores (width greater than 50 nm) [19]. The main technique for studying the porosity is measurement of adsorption isotherms. An adsorption-desorption isotherm indicates the measured quantity of gas that is adsorbed or desorbed on the surface of a solid at different equilibrium pressures and constant temperature [20]. The chemical structure of activated carbon has a great effect on its adsorptive capacities. Due to the presence of many heteroatoms chemically bonded to the surface of activated carbon, many surface groups can be formed [10]. Oxygen surface groups, such as carboxylic, lactone, phenol, carbonyl, have the predominant role in the surface adsorption behavior of activated carbon [7]. Many methods, such as Diffuse Reflectance Infrared Fourier Transform Spectroscopy (DRIFTS), X-ray photoelectron spectroscopy (XPS), temperature programmed desorption (TPD) and chemical titration methods, can be used to determine the surface chemistry of activated carbons [21].

Biochar was used as the precursor in this study. It was provided by Dynamotive Canada Inc. (Vancouver, BC, Canada). This char is produced by fast pyrolysis of biomass. Agricultural wastes such as bagasse and wheat straw, and forest residues such as sawdust and bark are the main sources of biomass. Large amount of agricultural wastes and forest residues can be recycled and converted to value added products [22]. Biomass is converted to a mixture of liquid organic compounds, gases and biochar by fast pyrolysis [23], which is the rapid thermal decomposition of organic compounds in an inert atmosphere. The liquid product of fast pyrolysis (biooil) is used as a clean burning fuel in boilers and turbines instead of fossil fuels such as diesel oil and natural gas. Gaseous product is recycled to the process to supply some of the heat required for pyrolysis. The biochar yield in this process is 20-30 wt% [9]. This char is used as a high heating-value solid fuel in kilns and boilers.

The porosity developed during activation process depends on the type of precursor, type of activation process and operating conditions. Biochar, due to its availability, low price and low amount of ash content (inorganic contents), is a suitable precursor for activated carbon. Ash (mineral) content does not contribute in the formation of porosity and therefore its presence reduces the development of surface area and pore volume per unit mass [14, 24]. Steam and carbon dioxide are the most commonly used gases for physical activation. Carbon dioxide develops and widens micropores, but steam develops micropores as well as mesopores, and produces a wider range of pore size [5, 25]. Zinc chloride, phosphoric acid and potassium hydroxide are commonly used for chemical activation. The use of $ZnCl_2$ is decreasing due to environmental and corrosion problems [26]. Phosphoric acid produces finer pores in comparison to $ZnCl_2$ [5] and needs lower activation temperature in comparison with two other agents [9]. Potassium hydroxide develops highly microporous activated carbons [9]. The purpose of this study is to convert biochar to a high value-added product such as activated carbon and to further optimize the operating conditions for the production of activated carbon with large surface area and high yield.

The Biochar used in this work was produced by fast pyrolysis of biomass in a bubbling fluidized bed reactor by Dynamotive Energy System Corporation. The as-received char was sieved, and particles between 150 – 600 μm (100-30 mesh) were collected for activation.

Physical Activation

Steam was used as the oxidizing agent for physical activation in order to produce activated carbon from biochar. Steam, because of the smaller dimensions of water molecule in comparison with that of CO_2, provides faster diffusion into a porous network, easier access into the micropores and a faster reaction rate [9]. The effect of three parameters, activation temperature (T), mass ratio of steam to char (S/C) and activation time (t), on the process was studied. The range of activation temperature, mass ratio of steam to char and activation time were 600 – 900 °C, 0.4 – 2 and 0.9 – 4 h, respectively. The experiments were designed with a statistical method called central composite design (CCD). This method enables us to optimize the effective parameters with a minimum number of experiments, as well as to analyze the interaction effect between parameters. This method includes three kinds of run; factorial runs (2^k), axial or star runs (2k) and center runs (n_c=six replicates) [27], where k is the number of parameters. Therefore, the number of runs required for studying the effect of three parameters is as follows:

$$N = 2^k + 2k + n_c = 2^3 + 2*3 + 6 = 20 \qquad (1)$$

This method was used for the physical activation of the char. With such details presented elsewhere [28].

Using this statistical method and design-expert software, two models for BET surface area (Y_1) and reaction yield (Y_2) were developed, and the optimum operating conditions were obtained by using these models. These data are required for scaling-up the laboratory results to pilot-plant or full-scale levels [29]. Fig. 1 shows the schematic diagram of the experimental setup. The details of the reactor setup are mentioned elsewhere [28]. For each run, 20 g of biochar was used. The temperature of the reactor was increased to the desired activation temperature, at a rate of 3 ^{0}C/min, under nitrogen flow followed by the steam injection. At the end of the run, the reactor was cooled to room temperature by flowing nitrogen.

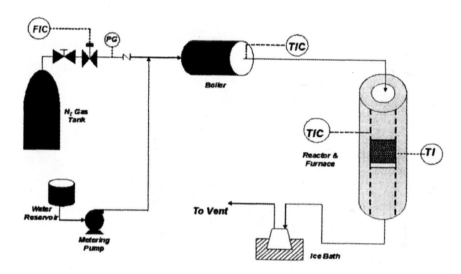

Figure 1. The schematic diagram of reactor set-up.

Results and Discussion

The ultimate and bulk ash analyses of biochar were done by Loring Laboratories, in Calgary, Canada. The ultimate analysis of biochar (moisture free) was: carbon-83.07, hydrogen-3.76, nitrogen-0.11, sulphur-0.01, oxygen-9.60, and ash-3.44 wt %. The bulk ash analysis of biochar was: SiO_2-53.48, Al_2O_3-7.73, TiO_2-0.10, Fe_2O_3-2.52, CaO-17.98, MgO-4.20, Na_2O-2.07, K_2O-6.93, P_2O_5-0.94, SO_3-1.06 and undetermined -2.99 wt %. The pH of biochar was 7.64 and its BET surface area was less than 10 m^2/g. Physically activated carbons produced in this study have an average pore diameter in the range of 13–26 ^{0}A, a maximum BET surface area of more than 950 m^2/g,, a reaction yield in the range of 17 – 79 wt% and maximum pore volume of more than 0.83 cc/g.

The pH of activated carbons was in the range of 9-11. The pH was measured according to ASTM D 3838-80.

After performing 20 experiments, two models were obtained for BET surface area (Y_1) and reaction Yield (Y_2), as shown in formulas (2) and (3), respectively. From a statistical point of view, three tests are required to evaluate the model; the test of significance of factors and interactions, the R-squared test and the lack-of-fit test. These tests showed that these models can navigate the experimental data.

$$(Y_1 - 25)^{2.5} = -3.600*10^6 + 5226.098*T - 2.040*10^7*(S/C) + 7.871*10^5*t \quad (2)$$
$$+35146.501*T*(S/C)$$

$$Y_2 = 94.85 - 0.01*T + 33.65*(S/C) + 0.31*t + 1.49*t^2 - 0.06*T*(S/C) - 0.01*T*t \quad (3)$$

Figure 2a is the three-dimensional plot of BET surface area model and shows the effects of temperature and mass ratio on the BET surface area of activated carbons prepared at a constant activation time of 2.46 hours.

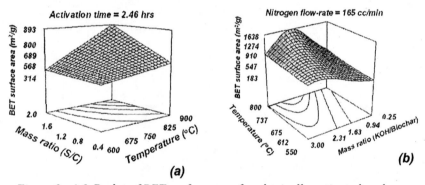

Figure 2. a) 3-D plot of BET surface area for physically activated carbons prepared at t=2.46 hrs b) 3-D plot of BET surface area for chemically activated carbons prepared at nitrogen flow rate of 165 cc/min.

As can be inferred from this figure, increasing the temperature increases BET surface area which is expected from an overall endothermic process such as steam activation. In addition, increasing the mass ratio, i.e., using more oxidizing agent, has a similar effect. The influence of time can be shown by the amounts of BET surface areas of three activated carbons prepared at the same temperature and mass ratio (T=750 °C and S/C=1.2), but at different activation times (0.6, 2.46, 4.32 hrs). For these samples, BET surface area is increased by increasing the time and these samples have surface areas equal to 554, 657 and 730 m²/g, respectively. This trend is in agreement with the equation shown above in formula 2 (2). In industry, it is more conventional to express these porosity characteristics per unit mass of the raw material (biochar) instead of product [9, 30]. Seven activated carbons were prepared at constant steam to feed mass ratio (S/C=1.2) and activation time (t=2.46 h) but at different temperatures, and their BET surface areas per unit mass of activated carbon and per unit mass of biochar are plotted against burn-off degree (percentage of feed mass loss due to activation) of these samples (see Fig. 3a).

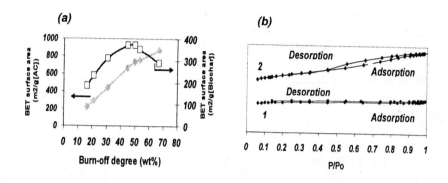

Figure 3. a) The effect of burn-off degree for activated carbons prepared at S/C=1.2 and t=2.46 hrs b) Typical isotherm plots for chemically activated carbon(1) and physically activated carbon(2).

It shows that the BET surface area (per unit mass of activated carbon) is increased by increasing the burn-off degree but for BET surface area per unit mass of biochar the curve exhibits a maximum in the range of 40-50 wt % of burn-off. Total pore volumes of these samples show similar trend, which was also observed by other researchers [25, 30, 31, 32]. It is due to widening of the porosity or the external ablation of the carbon particles at high degree of burn-off [5, 9]. The following constraints were used to find the optimum operating conditions for producing activated carbon with large BET surface area and high yield : (1) BET surface area $\geq$ 600 m^2/(g of activated carbon); (2) 50 $\leq$ Yield (wt %) $\leq$ 60; and (3) 0.91 $\leq$ activation time (hr) $\leq$ 1.5 .

The activation time should be as low as possible to obtain a good overall yield from a batch process. The optimum operating conditions were calculated based on the constraints, by Design Expert software, and are as follows:

$$T = 792 \ ^\circ C, \ \text{Mass ratio} = 1.06, \ t = 1.39 \ h$$

An experiment was performed at these operating conditions. The observed and predicted BET surface area and reaction yield were 664 and 643 m^2/g and 56.6 and 56.9 wt %, respectively. Therefore, the difference of BET surface area and product yield for experiment and model are less than 4 % and 0.6 %, respectively.

Chemical Activation

Potassium hydroxide was used as an activating agent for this process. In this study, the effects of three parameters on the BET surface area and reaction yield were investigated by CCD: activation temperature (T), mass ratio of KOH to biochar (R), and nitrogen flow-rate (F). The ranges of T, R, and F were 550-800 °C, 0.25-3.00, and 80-250 cc/min., respectively. For each run, a specified amount of potassium hydroxide was impregnated in biochar. The char was mixed with 100 ml of water having a desired concentration of KOH. This mixture was kept for 2 hours at room temperature to ensure the access of KOH to the interior of the biochar, and subsequently dried at 120 °C in an oven overnight. Then, 20 g of the prepared sample was placed in the fixed-bed reactor under a nitrogen flow and heated to 300 °C at 3 °C/min, and held at this temperature for 1h to prevent carbon loss through the direct attack of steam [33]. Then, the temperature was increased at 3 °C/min to reach the desired activation temperature. The chemical activation was carried out for 2 h at this temperature before cooling down. Then, the products were thoroughly washed with water, followed by treatment with 0.1 M HCl, and finally by distilled water to remove the soluble salts [9] and the potassium compounds [17]. The chlorine ions were eliminated with distillated water [34], when pH of the washed solution was between 6 and 7. Then the sample was dried at 110 °C for 12 h and characterized for its physical and chemical property measurements.

Results and Discussion

Chemically activated carbons produced in this study have an average pore diameter in the range of 13 – 15 °A, a maximum BET surface area of more than 1500 m^2/g,, a reaction yield in the range of 50 – 82 wt% and a maximum pore volume of more than 0.75 cc/g. The following models were obtained for BET surface area (Y_3) and reaction yield (Y_4):

$$(\frac{1}{\sqrt{Y_3}})=0.155-1.248*10^{-4}*T-0.0586*R-2.263*10^{-5}*F \tag{4}$$

$$+7.431*10^{-3}*R^2+3.794*10^{-5}*T*R$$

$$Y_4=73.05-3.5*A-10.25*B-5.21*B^2-3.17*A*D \tag{5}$$

Where A, B and D are as follows;

$$A = -1 + (T - 550)/125, B = -1 + (R - 0.25)/1.375, D = -1 + (F - 80)/85 \quad (6)$$

As shown in Fig. 2b, the three-dimensional plot of BET surface area model showing the effects of temperature and mass ratio on the BET surface area of activated carbon (prepared at constant nitrogen flow rate of 165 cc/min), it can be inferred that increasing the temperature increases the BET surface area. The BET surface area increases with an increase in the mass ratio up to a maximum, e.g., mass ratio of 1.93 at 800 °C, and then decreases. This phenomenon has been reported by some researchers for other precursors such as Spanish anthracite and Australian coal [17, 35]. It can be due to predominately pore widening at higher mass ratios [9] or severe gasification on the external surface of the carbon reducing the number of the pores [36]. Experimental data show that activated carbons prepared at a constant temperature of 675 °C and a mass ratio of 1.63, but at different nitrogen flow rates of 72, 165, and 258 cc/min, have BET surface areas equal to 582, 927, and 1210 m²/g, respectively. Therefore, the samples prepared at a higher nitrogen flow rate have a larger BET surface area. This could be due to faster removal of reaction products from the reactor at higher nitrogen flow rates [17].

The optimum operating conditions were defined according to the following constraints: (1) BET surface area ≥ 700 m²/g; (2) reaction yield ≥ 70 wt % .The solution is as follows:

$$T = 680^{\circ}C, KOH/biochar = 1.23, N_2 \text{ flow rate} = 240 \text{ cc/min}$$

The experiments were performed at these operating conditions and the BET surface area of product and reaction yield were measured. The observed and predicted BET surface areas were 836 and 783 m²/g, respectively. The observed reaction yield was 78.0 and the predicted one was 75.3 wt %. Therefore, the difference of BET surface area and product yield for experiment and model are less than 7 % and 4 %, respectively. Since mass ratio of KOH to biochar equals 1.23, more than 50 wt% of the feed is impregnating agent (KOH). The price of this chemical is several times higher than biochar price. Therefore, the recovery of the chemical agent and its related cost have strong effects on the economy of chemical activation [9] and an economical recovery method by acid washing should be developed for the production of low cost activated carbon by KOH chemical activation. Fig. 3b shows typical isotherm plots for physically (curve # 2) and chemically (curve # 1) activated carbons. The isotherm plots of physically and chemically activated carbons are type IV, which shows the formation of mesopores, [37] and type I, which is due to the development of microporous material.

The effect of activating agent (KOH) on the chemical structure of biochar was studied by some characterization methods such as TG/DTA (Thermogravimetric/Differential thermal analysis), DRIFTS and SEM. The TG/DTA study was carried out to investigate the effects of KOH on the reaction yield by a pyres-diamond TG/DTA (Perkin-Elmer instruments) under the flow of argon. TG/DTA analysis, after heating to 700 °C, for biochar and a mixture of biochar and KOH, with KOH/biochar ratio of 1.63, showed that weight loss of biochar was 8 wt% more than the mixture. It can be obviously inferred that the addition of the activating agent increases the carbon yield. KOH influences the pyrolitic decomposition, inhibits tar formation and therefore increases carbon yield [21]. One of the possible effects of activating agent in the chemical activation is cross-linking and aromatization of the precursor structure which reduces tar formation and increases carbon yield [26]. Therefore, DRIFTS was used to examine the effect of activating agent on the formation of aromatic structure during activation process. Spectra were obtained by a FTIR spectroscope (Spectrum GX, Perkin-Elmer) at a resolution of 4 cm^{-1}. Undiluted activated carbons, in the form of powder, were used for FTIR study in the scanning range of 450-4000 cm^{-1}.

The IR bands which can be associated with aromatic hydrocarbons are as follows: peak between 900 – 675 cm^{-1} (out-of-plane bending of the ring C-H bonds), peak between 1300-1000 cm^{-1} (in-plane bending bands), peaks between 1600-1585 cm^{-1} and 1500-1400 cm^{-1} (skeletal vibrations involving carbon to carbon stretching within the ring), peak between 3100-3000 cm^{-1} (aromatic C-H stretching bands), peak between 2000-1650 cm^{-1} (weak combination and overtone bands). Fig. 4 shows DRIFTS results for virgin biochar, activated carbon (prepared at 675 °C, KOH to biochar mass ratio of 1.63 and nitrogen flow rate of 165 cc/min), and processed biochar (without any activating agent and heated up to 675 °C). The effect of activating agent on the aromatization of the product can be seen from the spectra of biochars and activated carbon. This figure shows that there are low levels of aromatic compounds in biochar, which disappear after heat treatment without any activating agent. However, the aromatic structure in the product had developed due to use of the activating agent at the same heat treatment and gas flow-rate. Fig. 5 shows the effect of activation temperature on the aromatization of activated carbons prepared at mass ratio of 1.63 and gas flow rate of 165 cc/min. It can be inferred from this figure that an activation temperature of up to 675 °C increases the apparent aromatization of the structure of activated carbon, after which the aromaticity decreases.

The scanning electron micrographs were taken for biochar and activated carbon, prepared at temperature of 675 °C, KOH to biochar mass ratio of 1.63 and nitrogen flow rate of 165 cc/min (see Fig. 6). According to these micrographs, biochar does not have a porous structure, but activated carbon produced after acid washing step shows a highly porous structure.

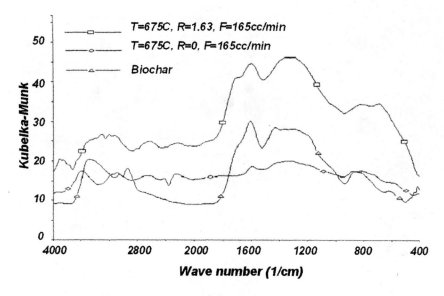

Figure 4. *Effect of KOH/biochar mass ratio on the surface chemistry.*

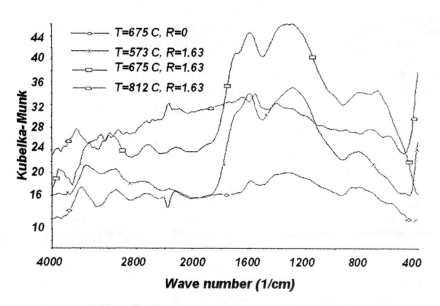

Figure 5. *Effect of activation temperature on the surface chemistry (All activated carbon prepared at F = 165 cc/min).*

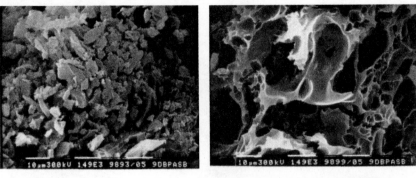

Biochar

Activated carbon prepared at
T=675 °C, R=1.63, F=165 cc/min

Figure 6. Scanning micrographs of biochar and chemically activated carbon.

Conclusions

Biochar as one of the products of the fast pyrolysis of biomass, was used as a precursor for the production of activated carbon. Physically and chemically activated carbons prepared from this material had BET surface area in the range of 300-950 and 180-1500 m^2/g, respectively. The BET surface area of physically activated carbon was increased by an increase in activation temperature, mass ratio of steam to char and activation time, whereas the BET surface area for chemically activated carbon was increased by an increase in temperature and nitrogen flow rate. An increase in mass ratio of KOH to biochar to a certain limit, increased surface area as well.

The optimum operating conditions for physical and chemical activation methods were calculated using models developed for BET surface area and reaction yield. The BET surface area and reaction yield of activated carbons prepared at optimum operating conditions were in good agreement with model predictions. Physically activated carbons had a relatively wide range of average pore diameter (13-26 °A). Chemically activated carbons, although had a narrow range of average pore diameter (13-15 °A), showed a highly microporous structure. Thermogravimetric study of chemical activation showed that activating agent increased activated carbon yield. DRIFTS confirmed the development of aromaticity by chemical activation which resulted in cross linking, reduction of tar formation and higher carbon yield.

References

1. McEnaney, B. In Carbon materials for advanced technologies; Burchell, T. D.Elsevier science Ltd: Oxford, UK, 1999.
2. Morgan, P. Carbon fibers and their composites CRC press: Boca Raton, FL, USA, 2005.
3. Rouquerol, F.; Rouquerol, J.; Sing, K. Adsorption by powders and porous solids: priciples, methodology and applications Academic press: San diego, CA, USA, 1999.
4. Byrne, J. F. and Marsh H. In Porosity in carbons: Characterization and applications; Patrick, Jr W. Edward Arnold: London, UK, 1995, Chapter 1.
5. Rodriguez-Reinoso, F. In Introduction to carbon technologies; Marsh, H.; Heintz, E. A.; Rodriguea-reinoso, F. Secretariado de publicaciones: Alicante, Spain, 1997, Chapter 2.
6. Dalai, A. K.; Majumdar A.; Chowdhury A.; Tollefson, E. L. Can *J Chem Eng* **1993**, *71*, pp 75-82.
7. Rodriguez-Reinoso, F. *Carbon* **1998**, *36*, pp 159-175.
8. Radovic, L. R. and Rodriguez-Reinoso, F. In: Chemistry and physics of carbon; Thrower, P. A.; Vol 25; Marcel Decker: Newyork, NY, USA, 1997.
9. Rodriguez-Reinoso, F.; in: Handbook of porous solids; Schuth, F.; Sing, K. S. W.; Weitkamp, J.; Vol 3, Wiley-VCH Verlag GmbH, Weinheim, Germany, 2002, Chapter 4.
10. Rodriguez-Reinoso, F.; Molina-Sabio, M. *Advances in colloid and interface science* **1998**, *76-77*, pp 271-294.
11. El-Hendawy, A. A.; Samra, S. E.; Girgis, B. S. *Colloid surf A: Physicochem eng aspects* **2001**, *180*, pp 209-221.
12. Lua A. C.; Yang, T. *J of coll and interface sci* **2004**, *274*, pp 594-601.
13. Munoz-Guillene, M. J.; Illian-Gomez, M. J.; Martinez, J. M.; Linares-Solano, A.; Salinas-Martinez de lecea, A. *Energy fuels* **1992**, *6*.
14. Linares-Solano, A.; Martin-Gullon, I.; Salinas-Martinez de lecea, C.; Serrano-Talaverra, B. *Fuel* **2000**, *79*, pp 635-643.
15. Kyotani, T. *Carbon* **2000**, *38,* pp 269-286.
16. Muhlen, H.; Van Heek, K. H. In Porosity in carbons: Characterization and applications; Patrick, W. Edward Arnold: London, UK, 1995, Chapter 5.
17. Lozane-Castello, D.; Lillo-Rodenas, M.; Cazorla-Amoros, D.; Linares-Solano, A. *Carbon* **2001**, *39*, pp 741-749.
18. Williams, P. T.; Reed, A. R. *J anal appl pyrolysis* **2004**, *71,* pp 971-986.
19. Sing, K. S. W.; verett, D. H.; Haul, R. A. W.; Moscou, L.; Pierotti, R. A.; Roquerol, J.; Siemieniewska, T. *Pure appl chem.* **1985**, *57*, p 603.
20. Lynch, J. Physico-chemical analysis of industrial catalysts: a practical guide to characterization; Editions Technip; Paris, France, 2003, chapter 1.

476

21. Figueiredo, J. L.; Pereira, M. F. R.; Freitas, M. M. A.; Orfao, J. J. M. *Carbon* **1999**, *37*, pp 1379-1389.
22. Zhang, T.; Walawender, W. P.; Fan, L. T.; Fan, M.; Daugaard, D.; Brown, R. C. *Chem Eng J* **2004**, *105*, pp 53-59.
23. Fan, M.; Marshall, W.; Daugaard, D.; Brown, R. C. *Bioresource technology* **2004**, *93*, pp 103-107.
24. Dalai, A. K.; Zaman, J.; Hall, E. S.; Tollefson, E. L. *Fuel* **1996**, *75*, pp 227-237.
25. Rodriguez-Reinoso, F.; Molina-Sabio, M.; Gonzalez, M. T. *Carbon* **1995**, *33*, pp 15-23.
26. Derbyshire, F.; Jagtoyen, M.; Thwaites, M. In Porosity in carbons: Characterization and applications; Patrick, Jr W. Edward Arnold: London, UK, 1995, Chapter 9.
27. Montgomery, D. C. Design and analysis of experiments; John Wiley & sons, USA, 1997, Chapter 13.
28. Azargohar, R.; Dalai, A. K. *Microporous and mesoporous materials* **2005**, *85*, pp 219-225.
29. Lazic, Z. R. Design of experiments in chemical engineering; Weily-VCH Verlag GmbH, Weinheim, Germany, 2004, Chapter 2.
30. Daud, W. M. A. W.; Ali, W. S. W. *Bioresource tech* **2004**, *93*, pp 63-69.
31. Rodriguez-Reinoso, F. In Fundamentals issues in control of carbon gasification reactivity; Lahaye, J.; Ehrburger, P. Kulver Academic, London, UK, 1990, p 533.
32. Rodriguez-Reinoso, F.; Molina-Sabio, M. *Carbon* **1992**, *30*, pp 1111-1118.
33. Otowa, T.; Nojima, Y.; Miyazaki, T. *Carbon* **1997**, *35*, pp 1315-1319.
34. Lillo-Rodenas, M.; Lozane-Castello, D.; Cazorla-Amoros, D.; Linares-Solano, A. *Carbon* **2001**, *39*, pp 751-759.
35. Teng, H.; Hsu, L. Y. *Ind Eng Chem Res* **1999**, *38,* pp 2947-2953.
36. Teng H.,; Ho J.; Hsu Y.-F. *Ind Eng Chem Res* **1996**, *35*, pp 4043-4049.
37. Shen, W.; Zheng, J.; Qin, Z.; Wang, J.; Lin, Y. *Colloids and surfaces A: Physiochemical and engineering aspects* **2003**, *229*, pp 55-61.

Novel Analytical Methods for Structural Elucidation of Forest Biomass

Chapter 30

Measurement of Cellulase Activity with Piezoelectric Resonators

Orlando J. Rojas, Changwoo Jeong, Xavier Turon, and Dimitris S. Argyropoulos

Department of Forest Biomaterials Science and Engineering, North Carolina State University, Raleigh, NC 27695–8005

The dynamics of cellulase binding and the activity on thin films of cellulose by using a piezoelectric sensing device (Quartz Crystal Microbalance with Dissipation monitoring, QCM-D) were examined. Upon exposure of the cellulose film to enzyme mixtures, a reduction in the sensor's frequency due to molecular binding is observed. Thereafter the frequency increases due to the loss of effective mass caused by the degradation of the film (enzymatic attack). In this study, it is demonstrated that the rate at which degradation of the film occurs can be monitored and quantified *in situ* and in *real time* as a function of enzyme concentration, temperature and pH of the incubating solution. Also, mass transport effects can be investigated by changing the flow conditions within the QCM-D reaction cell. The use of piezoelectric sensing to characterize and monitor the mechanisms and kinetics of enzyme activity on cellulosic substrates adds another dimension to our knowledge and to the methods available for such investigations.

Over the last two decades there has been remarkable progress in understanding the mechanisms by which enzymes interact with natural substrates such as lignocellulosic materials. Research has been carried out to manipulate and exploit enzymatic hydrolysis with the aim of producing fermentable sugars mainly driven by concerns over the escalating cost and reduced availability of fossil fuels. Such interest in cellulose conversion to energy and chemicals is not surprising since cellulose is the most abundant carbohydrate in the biosphere. Use of enzymes is also relevant to other industrial activities including wood pulping (*1*) and delignification (*2*); cellulose pulp bleaching (*3,4,5*); textile printing (*6*) and cotton pre-treatment (*7*); treatment of waste liquid effluents and detergency (*8*).

Most of the current work dealing with enzymes is focused on their adsorption and formation of enzyme-substrate complexes and on the development of enzyme binding domains and specificity (9). Active areas of research also include the three-dimensional structure of cellulolytic enzymes and the mechanisms and kinetics of cellulose binding and hydrolysis (10).

Lignocellulose degradation is a multienzymatic process due to the complex nature of lignified plant materials. Such degradation is an important step for the carbon cycle in the biosphere involving both hydrolytic and oxidative enzymes. Lignin removal is the key step for natural biodegradation of lignocellulosic materials, and is also required for converting wood into pulp for paper manufacture.

In this area cellulases are the most important enzymes due to their ability to cleave cellulose polymers into smaller units or simple sugars via hydrolysis reactions. Different oxidoreductases such as laccases and peroxidases which are involved in lignin degradation and other hydrolytic enzymes such as xylanases are widely used in industrial processes, especially in pulp and paper industries.

Cellulases

At present, cellulases are mainly used in food, beer and wine, animal feed, textile and laundry, pulp and paper, as well as in agricultural industries. The demand for these enzymes is growing rapidly, and this has driven extensive efforts to understand their mechanisms of activity.

Some microorganisms including fungi and bacteria produce various types of cellulase components. In several industrial applications genetic material is transferred to a host organism for commercial production. Because the original organism evolved to metabolize a mixed substrate containing cellulose, hemicellulose, lignin and extractives, the lower cost commercial enzymes are often a mixture of different specific isozymes.

Commercial cellulase mixes usually contain one or more exoglucanases such as cellobiohydrolase (CBH) which will proceed from either the reducing end or non reducing end of the cellulose chain and produce a shortened chain and cellobiose. The cellulase mixture may also contain several endoglucanases (EGI, EGII, etc.), which cleave randomly the internal β-1,4-glycosidic bonds on the cellulose chain along its length to produce free chain ends that will be acted upon by exoglucanases. Most cellulases also contain both β-1,4-glycosidase, which hydrolyze cellobiose units to glucose monomers, and various hemicellulases, which may have side chain cleaving capabilities. For complete hydrolysis of cellulosic material the synergistic combination of endoglucanase, exoglucanase and β-1,4-glycosidase is required.

Most cellulolytic enzymes are composed of two functionally distinct domains. The cellulose-binding domain is responsible for the close association of the enzyme with solid cellulose through strong binding to crystalline cellulose. The catalytic domain is responsible for catalyzing cellulose hydrolysis through an acid mechanism.

Cellulosic Substrates

Interactions of cellulases with cellulose play an important role in determining the efficiency of the enzymatic hydrolysis. Therefore, an understanding of enzyme interactions with the substrates is of great importance in processing wood and cellulosic fibers.

The rate of enzymatic hydrolysis and its yield are dependent on the adsorption of enzyme onto the substrate surface. Cellulose fibers are heterogeneous, porous substrates with both external and internal surfaces. The external surface area is determined by the shape and size of the cellulosic fibers, while the internal surface area depends on their capillary structure. The accessibility of cellulose to the cellulase is also controlled by the physicochemical properties of the substrate, the multiplicity of the cellulase complexes, and reaction parameters including those associated with mass transport and temperature. Furthermore, cellulose fibers contain both amorphous and crystalline regions. Crystalline regions are considered to be more difficult to degrade than amorphous regions and therefore significantly affect the rate and extent of enzyme hydrolysis.

Various kinetic models have been developed to describe the hydrolysis rate of cellulase. The models are based on parameters such as the amount of adsorbed enzyme on the cellulosic surface, the structural characteristics of the substrate including pore size distribution, crystallinity index and specific surface area, and cellulase-cellulose adsorption rates.

Enzyme Activity

There are two basic approaches to determine enzyme/cellulase activity. The first is based on the measurement of the individual activities of component enzymes and the second one is based on the activity of the total complex. Measurements of individual activities vary according to enzyme nature, namely, endoglucanases, exocellobiohydrolase and exoglucanases (11).

In the case of endoglucanases, activity is usually measured with any of the following techniques:

i) Viscometric method. It measures the endoglucanase activity according to changes in viscosity with time, as the enzyme degrades cellulose. It is basically simple but requires complex mathematical analysis in order to express the results in absolute units.

ii) The carboxymethylcellulose (CMC)-plate-clearing assay. Endoglucanase activity can be detected on agar plates; zymogram overlay gels, etc. by staining or precipitation of CMC present in the solid gel matrix.

iii) Measurement of reducing sugars. Reducing sugar released from CMC is often used jointly with measurements of changes of CMC viscosity in order to determine endoglucanase activity. Released sugars can be detected colorimetrically and the activity expressed in terms of moles of glucosidic bonds cleaved.

iv) Measurement of the chromophore and fluorescent groups released from modified CMC. This is based on chemically modified forms of cellulose that release soluble colored or fluorescent groups when their glucosidic bonds are hydrolyzed.

v) Chromatography. High performance liquid chromatography (HPLC) can provide valuable insight into the activities of components of cellulase complex, as transient cellooligosaccharide products can be detected, provided a means of stopping the hydrolysis at various stages is available. Separated hydrolysis products then could be detected photometrically.

Exoglucanases cleave cellobiose units from the non-reducing ends of cellulose or cellooligosaccharide molecules. Actually there is not a widely accepted method for measuring exoglucanases activity in cellulase complexes. When β-glucosidases are present then cellobiose is only a transitory product. A graphical method for determining exocellobiohydrolase activity is reported in the literature (9). Selective determination of exocello-biohydrolase activity in a mixture of cellulolytic enzymes using synthetic substrates is based on the selective cleavage of the agluconic bond and not on the holosidic or heterosidic bond (12-13).

Exoglucosidases cleave glucose units successively from the non-reducing end of the glucan. A single enzyme of this type, acting alone, could only attack a single terminal residue. Different approaches have been used to measure exoglucosidases activity. One of the methods is based on the different susceptibilities of cellotetraose and cellobiose to hydrolysis (*14,15,16,17*). A second method uses glucono-lactone, which is a potent inhibitor of β-glucosidase but has no effect on exo-β-1,4-glucosidase (*18*). A third method is based on the study of the rate of formation of D-glucose in the presence of various concentrations of added β-glucosidase (*10*). A fourth alternative uses as substrate a substituted cellodextrin supplemented with β-glucosidase. This substrate is susceptible to further hydrolysis by exoglucosidase from *Trichoderma viride* whereas endoglucanase, exocellobiohydrolase or β-glucosidase would not attack it (*19*).

Total Cellulase Complex Activity Measurement

"Total cellulase" assay methods measure the combined activities of the complex of cellulolytic enzymes towards an appropriate substrate, typically filter paper or powdered crystalline cellulose. Some other substrates, such as carboxymethylcellulose (CMC), trinitrophenyl-CMC (TNP-CMC) and cellulose azure, may be attacked by exo-acting cellulases as well as by endocellulases, and therefore provide some indication of their combined activities. However, hydrolysis of native crystalline cellulose is a much more stringent measure of the activity of a "complete" cellulase complex (9). Total cellulase activity can be measured by the residual cellulose by weight (*20*) or by determination of the sugars produced from insoluble cellulose. Such methods normally employ a colorimetric method to measure reducing sugars released. Substrates commonly used are filter paper or microcrystalline wood alpha-cellulose.

Overall, the mechanism of adsorption of most cellulase components onto cellulose is still unknown as well as the role of the different structural domains in cellulase action and their effects on substrates. Furthermore, with complex mixtures it can be difficult to understand how to dose enzymes to provide the correct level of activity. Often dosage rates are arrived at empirically based on a supposedly constant activity. In this investigation we propose the use of a piezoelectric technique, the Quartz Crystal Microbalance (QCM) to monitor, *in situ* and in *real time* the catalytic activity of cellulases on model substrates. This approach will allow unveiling the complex mechanism involved in enzymatic degradation of cellulosic materials.

Experimental

Materials and Methods

The enzyme used in this work was a commercially-available multi-component cellulase (Rocksoft[TM] ACE P150, Dyadic Int., FL). Sodium bicarbonate, sodium hydroxide, hydrochloric acid and acetic acid (all from Fisher Scientific), sodium acetate trihydrate (Fluka) and tris (hydroxymethyl) aminomethane (Acros) were used in the preparation of the buffer solutions of pH 4.5, 7 and 10. Water was obtained from a Milli-Q® Gradient system (resistivity >18MΩ cm).

Cellulose Thin Films and Related Methods

Cellulose thin films were used as models for cellulose. The process for preparing these surfaces (21) starts with either silicon wafers or QCM-D (Quartz Crystal Microbalance with Dissipation Monitoring) electrodes which are cleaned by standard chemical treatment and UV-ozone plasma. The QCM-D electrodes consisted of quartz crystals (Q-sense) coated with a conductive gold layer and a top 50-nm silica layer.

A cellulose solution was prepared by dissolving micro-crystalline cellulose (MCC) in 50 %wt water/N-Methylmorpholine-N-oxide (NMMO) at 115 °C. Dimethyl Sulfoxide (DMSO) was added to adjust the concentration of cellulose (0.05%) in the mixture. The cellulose solution was then spin coated (spin coater from Laurell Technologies model WS-400A-6NPP) on the substrates (silica wafer or QCM electrodes) at 5000 rpm for 40 seconds. The substrate was removed from the spin coater and placed in a milli-Q water bath to precipitate the cellulose. The cellulose-coated substrate was then washed thoroughly with milli-Q water, dried in a vacuum oven at 40 °C and stored at room temperature in a clean chamber for further use.

Cellulose Characterization

Ellipsometry, X-ray Photoelectron Spectroscopy (XPS), and Atomic Force Microscope (AFM) were used to characterize the cellulose films. X-ray Photoelectron Spectroscopy (XPS) was performed on bare silicon, polyvinyl amine (PVAm) and cellulose-coated wafers to quantify the respective chemical composition. This was accomplished by using a RIBER LAS-3000 XPS system with Mg Kα (1253.6 eV) as X-ray source. The thin cellulose films were characterized in terms of material distribution, surface roughness and topography using a Q-Scope 250 AFM model (Quesant Inst. Corp.), before and after enzyme treatment. The scans were performed in dried conditions using NSC16 standard wavemode Si_3N_4 tips with a force constant of 40 N/m (according to the manufacturer) and a typical resonant frequency of 170±20kHz. Finally, a Rudolph single-wavelength ellipsometer was used to determine the thickness of cellulose films on the surface of the silicon wafers (the refractive index of cellulose film was taken as 1.56).

Enzyme/Cellulose Surface Interactions

Quartz Crystal Microbalance

A Quartz Crystal Microbalance with Dissipation monitoring, QCM-D (Q-sense D-300, Sweden) was used to study enzyme binding and activity on cellulose thin films deposited on quartz/gold electrodes which are coated with a 50 nm SiO_2 layer. Temperature was controlled within $\pm$ 0.02 °C of the respective set point via a Peltier element.

QCM-D consists of a thin plate of a piezoelectric quartz crystal, sandwiched between a pair of electrodes. It measures simultaneously changes in resonance frequency, f, and dissipation, D (the frictional and viscoelastic energy losses in the system), due to adsorption on a crystal surface. f is measured before disconnecting the driving oscillator, and D is obtained by disconnecting the driving field and recording the damped oscillating signal as its vibration amplitude decays exponentially. Mechanical stress causes electric polarization in a piezoelectric material. The converse effect refers to the deformation of the same material by applying an electric field. Therefore, when an AC voltage is applied over the electrodes the crystal can be made to oscillate. Resonance is obtained when the thickness of the plate is an odd integer, n, of half wavelengths of the induced wave, n being an integer since the applied potential over the electrodes is always in anti-phase. If any material is adsorbed onto the crystal, it can be treated as an equivalent mass change of the crystal itself. The increase in mass, Δm, induces a proportional shift in frequency, f. There exists a linear relationship between m and f as described by Sauerbrey equation (22):

$$\Delta m = \frac{-\rho_q t_q \Delta f}{f_0 n} = \frac{-\rho_q v_q \Delta f}{2 f_0^2 n} = -\frac{C \Delta f}{n} \qquad [1]$$

where ρ_q and v_q are the specific density and the shear wave velocity in quartz respectively; t_q is the thickness of the quartz crystal, and f_0 the fundamental resonance frequency (when $n = 1$). For the crystal used in these measurements the constant C has a value of 17.8 ng cm^{-2} Hz^{-1}. The relation is valid when the following conditions are fulfilled: (i) the adsorbed mass is distributed evenly over the crystal. (ii) Δm is much smaller than the mass of the crystal itself (<1%), and (iii) the adsorbed mass is rigidly attached, with no slip or inelastic deformation in the added mass due to the oscillatory motion. The last condition is valid when the frequency decreases in proportion to the true mass of the adsorbate with no change in energy dissipation. Variations in the energy dissipation upon adsorption thus reflect the energy dissipation in the adlayer or at its interface. It is noted that the mass detected with the QCM-D device includes any change in the amount of solvent that oscillates with the surface.

This quantity may be significant for extended layers and thus one expects to obtain a higher value for the adsorbed amount from QCM-D studies than from, e.g., ellipsometry.

The Dissipation (D) Factor

A film that is "soft" (viscoelastic) will not fully couple to the oscillation of the crystal. The dissipation factor is proportional to the power dissipation in the oscillatory system and can give valuable information about the rigidity of the film:

$$D = \frac{E_{dissipated}}{2\pi E_{stored}}$$ [2]

Here $E_{dissipated}$ is the energy dissipated during one oscillation and E_{stored} is the energy stored in the oscillating system. Hence, the measured change in D is due to changes in contributions from, for example, slip and viscous losses. For QCM measurements in liquids, the major contribution to D comes from frictional (viscous) losses within the liquid contacting the crystal. According to Stockbridge (*23*), the shift in dissipation factor in a liquid environment is

$$\Delta D = \frac{1}{\rho_q t_q} \sqrt{\frac{\rho_l \eta_l}{2\pi f}}$$ [3]

where η_l and ρ_l are the viscosity and density of the fluid, respectively, and t and ρ are the thickness and the density of the quartz plate. When the adsorbed film slips on the electrode, frictional energy is created that increases the dissipation. Furthermore, if the film is viscous, energy is also dissipated due to the oscillatory motion induced in the film (internal friction in the film). Hence, a rigid adsorbed layer gives no change in dissipation while a loose layer gives a dissipation increase.

Batch and Continuous QCM Modes

Enzyme treatments were carried out using three different concentrations of the commercial enzyme in 50mM sodium acetate buffer (50, 16.7 and 5.6 ppm solids) in combination with different conditions of pH (4.5, 7 and 10) and temperature (28, 33 and 38 °C). Note that the maximum temperature allowable in our setup is 40 °C.

First, degassed (vacuum pump) buffer solution (no enzyme present) was injected into the QCM cell in which a cellulose-coated QCM electrode was installed. At least 18 hours were allowed for the cellulose to fully swell in the respective buffer solution. After this stage, enzyme solution was continuously introduced into the QCM cell with a syringe pump (Cole-Parmer 74900 series) at a flow rate of 0.2 ml/min (*continuous mode*). Alternative experiments were performed in *batch* conditions. In the *batch mode* the enzyme solution was introduced in the cell making sure that the buffer solution that was initially present was fully displaced. No further (continuous) enzyme injection was allowed. In both the *continuous* and *batch* modes the enzyme solution was injected only after the drift of the third overtone frequency (f_3) was lower than 2 Hz/hr. After the respective enzyme treatment the run was terminated when the change of frequency (f_3) was lower than 1 Hz/5min.

Results and Discussion

Substrate Characterization

The results from ellipsometry measurements on the PVAm/cellulose-coated substrates indicated a PVAm and cellulose layer thicknesses of ca. 1 and 12 nm, respectively. The AFM height obtained after an incision on the cellulose thin film with a scalpel gave similar thickness value (10-16 nm). Line profiles showed a uniform and homogeneous cellulose surface.

The chemical composition was confirmed by using XPS after curve fitting N and C peaks for the PVAm- and cellulose-coated silica surfaces. The ratio of O–C–O and C–O peaks was in agreement with the expected values calculated from the chemical composition of pure cellulose.

Enzymatic Degradation of Cellulose Films

Figure 1 depicts a typical plot of QCM frequency and dissipation signal for a cellulose film subject to enzyme treatment. The adsorption of the enzyme onto the cellulose surface is clearly shown in the form of a (transient) reduction in the QCM-D frequency. Enzyme binding occurs rather fast (less than 5 min in most cases). After binding, a reduction in mass of material on the surface is observed (increase in oscillation frequency). After some extended period of time the frequency signal reaches a plateau which indicates no further change in the film mass.

The replacement of the enzyme solution by buffer solution (rinsing) did not produce any noticeable change in the frequency response. This indicates that there is no further change in the adsorbed mass and also that the possible effect of variations in bulk density and viscosity is negligible. The QCM dissipation seen in *Fig. 1* (right) mirror the behavior observed for the frequency except for an initial dip after injecting enzyme solution (after 10 minutes operation) that we attribute to changes in temporal excess flow pressure on the crystal. Initially

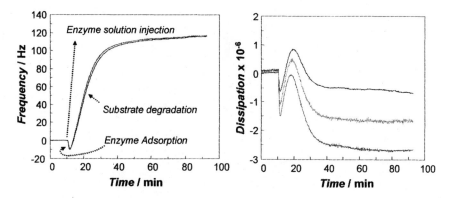

Figure 1. Change in frequency (left) and dissipation (right) by enzyme binding and degradation of cellulose thin films (38 °C, 50 ppm enzyme concentration and pH 4.5). The experiment was conducted in batch mode. The different plots represent the different overtones (or harmonics): $f_3/3$ (blue), $f_5/5$ (grey) and $f_7/7$ (red). The driving frequencies correspond to 15, 25 and 35 MHz for f_3, f_5 and f_7, respectively.

(<0 min) the dissipation response for the buffer solution at the various overtones (D_3, D_5 and D_7) are very similar. This reveals the existence of an initial thin, relatively rigid, film of cellulose. After about 10 min of enzyme solution injection (at *ca.* 20 min run time), the energy dissipation starts to decrease and more distinctive differences for the various overtones are observed. This indicates structural changes in the cellulose film upon enzyme attack.

Effect of Temperature

The frequency changes upon enzymatic degradation of cellulose films by cellulase under different temperature condition (38, 33 and 28 °C) and fixed concentration (50 ppm) and pH (4.5) in both batch and continuous modes were investigated. From these results plots of effective mass versus time were obtained according to Eqn. [1] (see Figure 2). The catalytic activity of cellulase is taken as the initial slope of the curve after completion of enzyme binding. That is, the slope after the curve's maximum (between 10 and 20 min incubation time for the cases shown in Fig. 2) is directly related to the (initial) maximum rate of degradation. By comparing the different initial slopes at different temperatures, it can be concluded that in both batch and continuous modes temperature plays a significant role in enzyme activity.

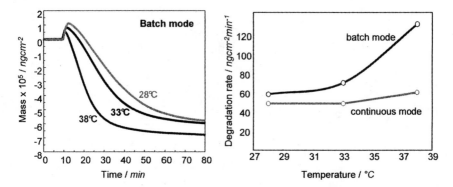

Figure 2. Degradation after incubation of cellulose thin films in cellulase solutions (50 ppm, pH 4.5) at three different temperatures: 38, 33 and 28°C, in batch mode (left). Results for the initial rate of degradation for batch and continuous operations are compared in the right figure.

Most of the features visible by AFM imaging on the untreated cellulose film disappear after its degradation by the enzyme. The roughness and average height values were reduced considerably after the treatment.

It is clearly demonstrated that as the system is operated closer to the suggested optimal temperature (50 °C according to the supplier) there is an increased rate of enzymatic hydrolysis. By comparing results from the two modes of operation it is also possible to conclude that the mass exchange plays a key role in defining the dynamics of the process. This issue will be discussed in later sections.

Effect of pH

The pH of the buffer solution affected significantly the initial rate of cellulose hydrolysis as can be observed in *Fig. 3* for mass plots at three different pHs. Enzyme inactivation was observed at pH 10 as judged by the lack of change in mass of films treated under this condition. More precisely, at pH 10 some minimal adsorption of the enzyme is observed on cellulose (small QCM frequency reduction) but no increase of frequency with time is observed (no material release or mass losses). Unlike the case of pH 10, the use of cellulase at neutral and acidic pH show (i) a distinctive adsorption (binding) step followed by (ii) cellulose degradation (increase in frequency or reduction of film mass).

The effect of pH on cellulose activity can be explained by hydrogen ion concentration in the system. Interactions between the ionizable groups of both the substrate and the active site of the enzyme can induce the optimal condition. However, the hydrolysis rate rapidly decreases with reaction time due to thermal instability of the cellulases (24).

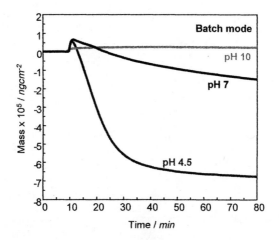

Figure 3. Enzyme binding and cellulose degradation after incubation of cellulose thin films in cellulase solutions (50 ppm, 38°C) at three different pHs: 10, 7 and 4.5, in batch mode.

for enzymatic degradation. The effect of the buffer solution pH is rather significant (*25*). It is clearly observed that the initial rate of hydrolysis is distinctively larger for acidic conditions. This observation is well documented in the literature where a low pH is recommended for higher hydrolysis rates (*26*).

Effect of Cellulase Concentration

The enzymatic activity increased with the concentration of enzyme in buffer solution (see *Fig. 4*). At the lowest enzyme concentration studied (5.6 ppm) both the binding and the rate of hydrolysis are very slow. This was confirmed by AFM images that showed that the topography of the surface of the cellulose film changed little after the enzyme treatment.

Dynamics and Energetics of Enzyme Activity

As explained before, kinetic parameters of enzymatic degradation can be obtained by using the QCM-D initial slope of frequency (or mass) vs. time curves. Table 1 shows the effect of the different conditions of temperature, pH and enzyme concentration on the initial reaction rates. The rates in Table 1 obtained from raw frequency data used for Figures 2, 3 and 4 allow estimation of the enzyme activity under different conditions of incubation.

All of the kinetic models developed so far assume a linear relationship between the initial rate of hydrolysis and the amount of enzyme-substrate complex formed by adsorption of cellulase.

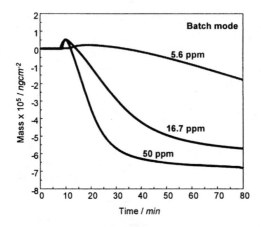

Figure 4. Degradation rate after incubation of cellulose thin films in cellulase solutions (pH 4.5, 38°C) at three different concentrations: 50, 16.7 and 5.6 ppm using batch operations.

It can be argued that the inactivation of the adsorbed cellulase is due to its diffusion into the cellulose fibrils (*ii*). In these experiments this effect can be ruled out as the existence of fibrils in thin films is not expected. It is more likely that the inactivation of enzyme with time is related with a strong inhibition by the by-products of hydrolysis, namely, cellobiose and glucose (*ii*) and also by the transformation of the cellulose into less digestible forms (*ii*). Since it is anticipated that the cellulose micro-structure plays a significant role (*ii*) additional experiments are underway to test the effect of the crystallinity of the surface on enzyme activity.

Table 1. Initial reaction rates from QCM slope in batch and continuous modes of operation.

	Initial slope (Hz/min) or reaction rate constant		
		Batch	
Continuous			
	28	3.29	2.74
Temperature, °C	33	3.74	2.75
(pH 4.5, 5.6 ppm)	38	7.38	3.40
	4.5	7.38	3.40
pH	7	0.94	0.27
(38 °C, 5.6 ppm)	10	0.02	0.01
	50	7.38	3.40
Concentration, ppm	16.7	3.21	2.78
(38 °C, pH 4.5)	5.6	0.25	0.89

It is worth mentioning the differences in the initial rate of hydrolysis as obtained by the two operational modes, i.e., batch and continuous modes. It was observed consistently that the reaction rates were higher for the batch operation which indicates that the controlling step in the dynamics of enzymatic reactions is not related to mass transport but rather to the cellulose-substrate binding. Furthermore, this observation was confirmed by comparison of degradation rates from QCM experiments (continuous mode) at different flow rates of enzyme solution (data not reported). Therefore, in the application of cellulase formulations special attention needs to be paid to the presence of binding factors that facilitate adsorption and interaction with the cellulose surface. Cellulose-binding domains (CBD) improve the process for the reasons explained above.

The activation energy, E_a, of the enzymatic hydrolysis of cellulose can be estimated from the reaction constant (k) at different temperatures by using an Arrhenius expression. Even though no specific model has been used to obtain the reaction constants, the initial rates of hydrolysis are directly related to the slopes of the QCM frequency vs. time curves. Any factor aimed to correct for the actual film mass is likely to be a linear function of frequency as seen in the Sauerbrey Eqn. [1]. Furthermore, any contributing factor will cancel out if one takes the ratio of two QCM slopes at two different temperatures and as a result it is possible to obtain E_a. By using this approach, the values of E_a were calculated and the obtained values were found to be similar to those obtained from other experimental techniques in cellulose/cellulase systems (*31*). Not only is the proposed QCM test valuable to describe the enzymatic degradation *in situ* and *real time* but it can also be used to obtain quantitative parameters in cellulase activity. There is work underway with a more extensive matrix of experimental conditions which will further explore issues related to kinetic models, rate constants and activation energies.

Due to the complexity of the system it is difficult to completely understand the enzymatic degradation of cellulose. The enzymatic activity in multi-component cellulases makes this even more difficult. The possibility of using *in situ* and *real time* detection with tools like the QCM-D justify more extensive efforts in this area.

Conclusions

A piezoelectric sensing technique was used to investigate the activity of cellulase based on the change in frequency of a cellulose-coated resonator during enzymatic degradation. The advantage of this approach is its simplicity and the ability to monitor enzyme activity on thin films of cellulose *in situ* and *real time*. The initial slope in mass (frequency) *versus* time curves is related to the initial reaction rates and is effective to estimate enzyme performance under different conditions of temperature, pH and concentration. We demonstrated that the pH of the enzyme solution plays a key role for an optimum enzyme activity. Mass transport effects are also relevant and they can be easily judged by changing the flow rates in the reaction cell. QCM-D is a valuable tool to study the mechanisms involved in enzyme transformations and provides a unique opportunity for development and screening of formulations.

Acknowledgements

We gratefully acknowledge financial support provided by the North Carolina Biotech Center grant NCBC #2005-CFG-8008.

References

1. Zhao J, Li XZ, Qu YB, Gao P. *Alkaline peroxide mechanical pulping of wheat straw with enzyme treatment.* Applied Biochemistry and Biotechnology, **2004**, *112 (1)*, pp 13-23.
2. Costa SA, Goncalves AR, Esposito E. *Ceriporiopsis subvermispora used in delignification of sugarcane bagasse prior to soda/anthraquinone pulping.* Applied Biochemistry and Biotechnology, **2005**, *121*, pp 695-706.
3. Sigoillot C, Camarero S, Vidal T, Record E, Asther M, Perez-Boada M, Martinez MJ, Sigoillot JC, Asther M, Colom JF, Martinez AT. *Comparison of different fungal enzymes for bleaching high-quality paper pulps.* Journal of Biotechnology, **2005**, *115 (4):* pp 333-343.
4. Roncero MB, Torres AL, Colom JF, Vidal T. *The effect of xylanase on lignocellulosic components during the bleaching of wood pulps.* Bioresource Technology, **2005**, *96(1):* pp 21-30.
5. Jimenez L, Ramos E, De la Torre MJ, Ferrer JL. *ECF and TCF bleaching methods as applied to abaca pulp.* Afinidad, **2005**, *62(515),* pp 14-21.
6. Kokol V, Heine E. *Effective textile printing using different enzyme systems.* Coloration Technology, **2005**, *121(4),* pp 209-215.
7. Pyc R, Sojka-Ledakowicz J, Bratkowska H. *Biosynthesis of enzymes by Aspergillus niger IBT-90 and an evaluation of their application in textile technologies.* Fibres & Textiles in Eastern Europe, **2003**, *11(4),* pp 71-77.
8. Stebbing DW, Zhang X, Soong G, Mansfield SD, Saddler JN. *Fungal enzyme treatment of newsprint mill white water: Impact on white water and paper properties.* Journal of Pulp and Paper Science, **2005**, *30(1),* pp 3-8.
9. Meyer M, Wohlfahrt G, Knablein J, Schomburg D. *Aspects of the mechanism of catalysis of glucose oxidase: A docking, molecular mechanics and quantum chemical study.* Journal of Computer-aided Molecular Design, **1998**, *12 (5),* pp 425-440.

10. Woodward J, Affholter KA, Noles K, Troy NT, Gaslightwala SF *Does cellobiohydrolaseii core protein from trichoderma-reesei disperse cellulose macrofibrils?* Enzyme and Microbial Technology, **1992**, 14 (8), pp 625-630.

11. Sharrock Kr. *Cellulase assay methods - a review.* Journal of Biochemical and Biophysical Methods, **1988**, *17 (2)*, pp 81-105.

12. Deshpande MV, Eriksson KE, Pettersson LG. *An assay for selective determination of exo-1,4,-beta-glucanases in a mixture of cellulolytic enzymes.* Analytical Biochemistry, **1984**, *138(2)*, pp 481-487.

13. Hrmova M, Petrakova E, Biely P. *Induction of cellulose degrading and xylan degrading enzyme systems in aspergillus terreus by homodisaccharides and heterodisaccharides composed of glucose and xylose.* Journal of General Microbiology, **1991**, *137*, pp 541-547 Part 3.

14. Reese ET. Estimation Of Exo-Beta-1-β-4-Glucanase *In Crude Cellulase Solutions.* Advances in Chemistry Series, **1969**, (95), pp 26-35.

15. Sternberg D, Vijayakumar P, Reese ET. *Beta-glucosidase microbial production and effect on enzymatic-hydrolysis of cellulose.* Canadian Journal of Microbiology, **1977**, *23 (2)*, pp 139-147.

16. Ait N, Creuzet N, Cattaneo J. *Properties of beta-glucosidase purified from clostridium-thermocellum.* Journal of General Microbiology, **1982**, *128*, pp 569-577.

17. Garsoux G, Lamotte J, Gerday C, Feller G. *Kinetic and structural optimization to catalysis at low temperatures in a psychrophilic cellulase from the Antarctic bacterium Pseudoalteromonas haloplanktis.* Biochemical Journal, **2004**, *384*, pp 247-253.

18. Reese ET, Maguire AH, Parrish FW. *Glucosidases and exo-glucanases.* Canadian Journal of Biochemistry, **1968**,*46(1)*, pp 25-34.

19. Rabinowitch ML, Martianov VA, Chumak GA, Klyosov AA. *Substituted cellodextrine, a new substrate for the selective assay of exo-1,4-beta-glucosidase in cellulase complexes.* Bioorganicheskaya Khimiya, **1982**, *8 (3)*, pp 396-403.

20. Halliwell G. *Microdetermination of cellulose in studies with cellulose.* Biochemical Journal, **1991**, *68*, pp 605-610.

21. Gunnars S, Wagberg L, Stuart M.A.C. *Model films of cellulose: I. Method development and initial results.* Cellulose Chemistry and Technology, **2002**, *9(3-4)*, p 239.

22. Rodahl M, Hook F, Krozer A, Brzezinski P, Kasemo B. *Quartz crystal microbalance setup for frequency and Q-factor measurements in gaseous and liquid environments.* Review of Scientific Instruments, **1995**, *66(7)*, pp 3924-3930.

23. Stockbridge C.D. *Vacuum Microbalance Techniques.* **1966**. Plenum Press: New York.

24. Caminal G, López-Santín J, Solà C. *Kinetic Modeling of the Enzymatic-Hydrolysis of Pretreated Cellulose.* Biotechnology and Bioengineering, **1985**, *27(9)*, pp 1282-1290.
25. Deng SP, Tabatabai MA *Cellulose activity of soils.* Soli Biol. Biochem **1994**, *26*, pp 1347-1354.
26. Criquet S. *Measurement and characterization of cellulase activity in sclerophyllous forest litter.* Journal of Microbiological Methods, **2002**, *50(2)*, pp 165-173.
27. Tanaka M, Ikesaka M, Matsuno R, Converse AO. *Effect of Pore-Size in Substrate and Diffusion of Enzyme on Hydrolysis of Cellulosic Materials with Cellulases.* Biotechnology and Bioengineering, **1988**, *32(5)*, pp 698-706.
28. Holtzapple MT, Caram HS, Humphrey AE. *Determining the Inhibition Constants in the Hch-1 Model of Cellulose Hydrolysis.* Biotechnology and Bioengineering, **1984**, *26(7)*, pp 753-757.
29. Lee YH, Fan LT. *Kinetic-Studies of Enzymatic-Hydrolysis of Insoluble Cellulose. 2. Analysis of Extended Hydrolysis Times.* Biotechnology and Bioengineering, **1983**, *25(4)*, pp 939-966.
30. Wald S, Wilke CR, Blanch HW. *Kinetics of the Enzymatic-Hydrolysis of Cellulose.* Biotechnology and Bioengineering, **1984**, *26(3)*, pp 221-230.
31. He D, Bao L, Long Y, Wei W, Yao S. *A new study of the enzymatic hydrolysis of carboxymethyl cellulose with a bulk acoustic wave sensor.* Talanta, **2000**, *50(6)*, pp 1267-1273.

Chapter 31

High-Throughput Material Characterization Techniques: Near Infrared and Laser-Induced Breakdown Spectroscopy

Nicole Labbé[1], Timothy G. Rials[1], Stephen S. Kelley[2], and Madhavi Z. Martin[3]

[1]Forest Products Center, University of Tennessee, 2506 Jacob Drive, Knoxville, TN 37996–4570
[2]Department of Forest Biomaterials Science and Engineering, North Carolina State University, Raleigh, NC 27695–8005
[3]Environmental Sciences Division, Oak Ridge National Laboratory, P.O. Box 2008, MS 6422, Oak Ridge, TN 37831–6422

High throughput technologies such as near infrared, mid infrared and laser induced breakdown spectroscopy are very powerful techniques that can be used to characterize heterogeneous materials. They have been developed in various fields to investigate properties and processes and to control the quality of manufactured products. This chapter gives a brief description of these technologies and some examples of their applications in the forest industry. Rapid methods to measure chemical composition and properties of materials and manufactured products are indispensable to control their quality and to optimize their processes. Analytical spectroscopy such as near infrared, mid infrared and laser induced breakdown spectroscopy are useful methods that can be developed as high throughput technologies and can be implemented as quality and process control tools in an industrial environment. These methods become outstandingly effective when coupled with statistic analysis to classify materials and predict their properties.

495

High Throughput Methods

Numerous research studies are addressing the development of high throughput analytical tools for the rapid assessment of complex materials mainly because standard methods are frequently time-consuming and expensive. Spectroscopic methods can be a remarkably efficient alternative to the traditional methods. Over the past thirty years, infrared spectroscopy, especially near infrared (NIR) spectroscopy, has rapidly becoming the problem solving tool of choice for heterogeneous systems. The recent development of equipment featuring improved electronic and optical components and advances in computers capable of effectively processing the information contained in spectroscopic data have ushered in the expansion of this technique in a number of different fields. Relevant studies have been reported in many different areas such as food industry (1,2,3) pulp and paper industry (4,5) panel manufacturing (6,7) polymer extrusion (8,9,10), plastic recycling (11), pharmaceuticals (12,13) fermentation processes (14,15), refinery (16,17) and oil extraction (18,19). Because of its advantages over other analytical techniques, such as its low sample preparation and its ability to collect spectra for solid and liquid samples, near infrared spectroscopy is one of the most recognized and preferred techniques in industry (20). Another spectroscopic-based technique that is increasingly used to assess product quality and to facilitate on-line process monitoring and control is Laser Induced Breakdown Spectroscopy (LIBS). This method has been used to assess elemental composition of a variety of materials (21,22), e.g., in the assessment of the inorganic composition of wood and wood based-products (23,24,25), in the identification of teeth affected by caries (26), and in the detection of airbone biological agents (27).

NIR spectroscopy and LIBS are complimentary methods. While NIR characterizes the organic constituents of a material, LIBS provides its inorganic composition. Due to better technology in computer and processors, these techniques are now very fast and require nearly no sample preparation making them very suitable for on-line applications. However, to predict any properties, these methods require the development of calibration models. The relationships between measured ("true") property and spectral data have to be strong to use these methods for quantitative application. The accuracy of the data (measured property) needed to construct the calibration models are essential for the success of spectroscopic analyses. The models can only be applied to materials similar to those used for the models.

Near Infrared Spectroscopy

Although near infrared energy was discovered by William Herschel in 1800, the first applications of NIR spectroscopy were only reported in the

1950s. In the late 1960s, work led by Karl Norris demonstrated the potential value of near infrared spectroscopy for quantitative work by making measurements of agricultural products. The NIR region covers the wavelength range 800 to 2500 nm. The absorption bands are based on overtone and combination of fundamentals vibrations of the investigated molecule. Near infrared spectra, which originate from characteristic molecular vibrations, can be used to get an insight into the molecular structure and composition of any materials. The technique is based on the fact that molecules are able to absorb infrared light, and this results in different kinds of molecular vibrations. The absorbed energy is characteristic for different molecular vibrations and depends on the chemical environment of the submolecular groups.

The acceptance of analytical methods in various fields, especially NIR spectroscopy, is mainly due to progress in computer and microprocessors. In the first works, to extract quantitative information from NIR spectra, a calibration curve was calculated based on the Beer-Lambert's Law. A coefficient, called the proportionality coefficient, was determined from the concentration and the optical density, measured at one specific wavelength. The first step was to adjust this curve by using a simple linear regression method (univariate calibration). With subsequent progress in computer technology, it was possible to establish more complex models. Methods using many measured available variables simultaneously were developed (multivariate calibration). These methods allow the selection of the significant absorption bands that were involved in the prediction. Among these methods, principal component regression (PCR) and partial least squares (PLS) are the most commonly used. Principal component analysis (PCA) was also introduced to facilitate NIR data analysis.

Nowadays, chemometric methods such as PCA, partial least square (PLS) and principal component regression (PCR) are commonly used to deal with large spectroscopic data sets (29,30). A great advantage of these multivariate analyses (MVA) over classical methods is that the whole spectral data range is taken into account instead of specific bands assigned to specific components. MVA will detect any significant correlations between bands (wavenumber and intensity). There are some very detailed books on MVA (28,29,30,31). The following section is not meant to cover the whole range of MVA, but instead, to provide some examples where these different methods are applied to extract useful information from large and complex data sets.

Principal Component Analysis

Principal component analysis (PCA) is a mathematical procedure for resolving sets of data into orthogonal components whose linear combinations approximate the original data to any desire of accuracy (32). Two important results from a principal component analysis are the scores and the loadings. The scores plot is used to detect sample patterns, groupings, similarities or differences. The loadings show which variables are responsible for the similarities or differences between the samples found in the scores plot. In the following application, PCA was used to classify and predict mechanical properties of composite wood products, demonstrating the viability of the NIR approach to characterize wood and wood based-products. Figures 1a, 1b show the NIR spectra and the results of PCA (score) of near infrared spectra collected on wood-composites samples. In this laboratory NIR is developed as a process sensor for wood-polymer composites. Wood-polymer composites are a rapidly emerging class of materials.

They have found uses in decking and architectural moldings, and other markets will undoubtedly emerge in the near future. Typical formulations for these extruded materials include wood flour (40-70%), a polyolefin matrix (24-29%), coupling agents (2-3%), and lubricants (2-3%). Currently, there are very few alternatives available for process monitoring, or even product quality control. Near infrared spectroscopy can be a solution to these problems since it can be used as a processor sensor for wood-polymer composite manufacturing. Wood composites were formulated using either polyethylene (PE) or polypropylene (PP) as the matrix polymer. Materials were also processed with, and without, a maleated polyolefin (MAPP or MAPE) compatibilizer. Figure 1a

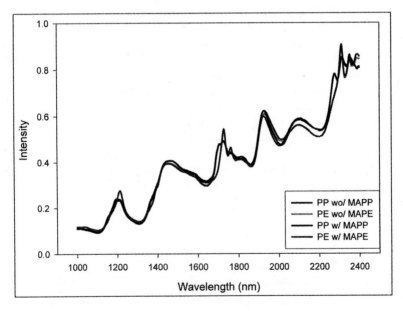

Figure 1a. Near infrared spectra of wood-composites.

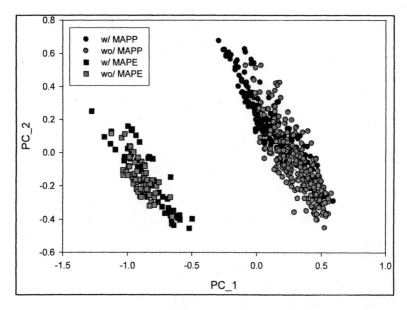

Figure 1b. Principal component analysis (score) of NIR spectra collected on wood-composites samples.

shows the near infrared spectra collected on the composites. It is important to specify that a visual inspection of the different composites is not successful in distinguishing the samples (by the type of polymer matrix, or by presence or absence of coupling agent). The spectra are very similar, however, differences between the polyethylene and the polypropylene matrix can be observed. This can be attributed to different –CH substitution patterns in the polymers. Principal component analysis was used to explore the spectra. In Figure 1b, the scores map plots Principal Component 2 against Principal Component 1, and provides some information on the composite variation. The clear separation is driven by the type of matrix polymer used in the formulation: polyethylene (to the right) and polypropylene (to the left). Interestingly, further clustering can be found from the principal component analysis of the spectral data of the polypropylene composite samples (Figure 2). Separation along the factor 1 axis can be attributed to the presence or absence of compatibilizer. This a promising observation since the effect of coupling agent can be detected, despite its presence in very small amounts.

This application highlights the power of principal component analysis of NIR spectra, since the contribution of the type of polymer (polyethylene versus polypropylene) and the presence or absence of coupling agent was not detectable from visual inspection of the wood polymer composites.

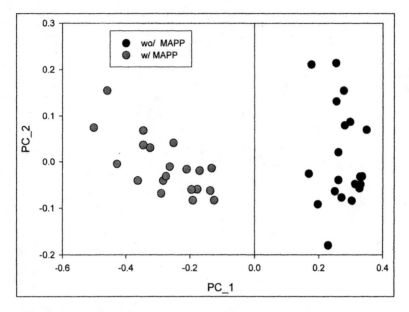

Figure 2. Principal component analysis of NIR spectra of wood-polypropylene composite with and without coupling agent.

Partial Least Square

Quantitative properties such as chemical composition and strength can also be obtained by NIR spectroscopy when coupled with the Partial Least Squares method. Unlike PCA, Partial Least Squares or Projection to Latent Structures (PLS) is a directed analysis. The method provides a model for the relationship between a set of predictor variables (spectral data) and a set of response variables (properties of interest). If the spectral data contain information about the properties of interest, a reliable calibration model can be constructed. A coefficient of correlation is calculated, indicating the strength of the model. The response has to be independently measured for each sample by standard methods. The model is then used to predict the properties of interest of unknown samples. Figure 3a and 3b show two models that were developed from NIR spectra of different softwoods samples (Loblolly Pine, Longleaf Pine, Shortleaf Pine, and Slash Pine). Figure 3a shows the model developed to predict the chemical composition (cellulose, hemicellulose, lignin and extractives content) of these wood species and Figure 3b shows the model built to predict the mechanical properties (modulus of rupture) of these samples. The coefficient correlation shows how well the models can perform to predict these properties.

In all these samples, NIR technique accurately predicted the chemical composition of the softwood samples.

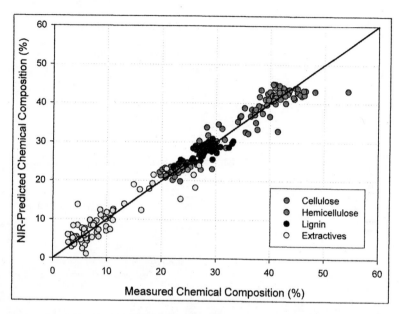

Figure 3a. Model to predict chemical composition of softwoods from near infrared spectra. Correlation between the measured and predicted chemical composition of four softwood species.

Most of the correlation coefficients between the "true" (measured) and "predicted" chemical composition were above 0.8 and the prediction errors were similar to the measurement errors of the wet chemical methods used to establish the "true" chemical composition. These models can now be apply to NIR spectra collected on unknown softwood and predict their chemical and physical properties in a matter of a second.

These examples highlight the value and potential of using NIR techniques to measure properties of complex materials such as wood and wood composites. By using MVA, very robust models were developed to predict the chemical, physical, and mechanical properties of woods and wood-based products. However, NIR spectroscopy has some limitations and drawbacks. It is not a direct method. It is necessary to develop a regression calibration model first that links spectral data to properties of interest and then the calibration model can be used to predict properties of a new set of samples. NIR measurements are very often not as accurate as those made by using standard methods and the samples must be similar to the calibration materials.

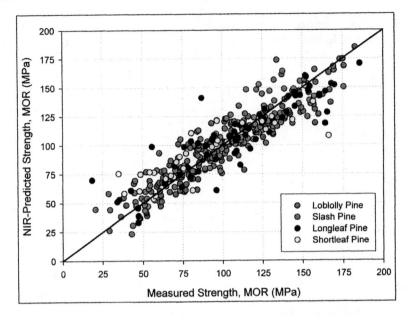

Figure 3b Model to predict mechanical proper ties (MOR) of softwoods from near infrared spectra. Correlation between the measured and predicted strength of four softwood species.

Two-Dimensional Correlation Analysis

One significant limitation of NIR spectroscopy is that due to significant overlap of the overtone and combination vibrations in the NIR spectral range, it is difficult to assign specific spectral features to specific sample components. MVA statistical tools can be applied to analyse these complex NIR spectra, extract useful information from large spectral data sets, highlight subtle features in the spectra and differentiate samples based on these subtle spectral differences. However, these MVA tools are much less effective for providing chemical insight or assigning NIR spectral features to different sample components or specific functional groups.

The principles of two-dimensional (2D) correlation infrared spectroscopy were published by Noda more than fifteen years ago (33), and recently, Noda and Ozaki et al have developed 2D correlation tools as a method for simplifying and visualizing complex spectra consisting of many overlapped bands (34,35,36). The method was based on the correlation analysis of perturbation-induced variations of spectral intensity and was especially useful in emphasizing spectral features not readily seen in the original spectra (37). However, in this original approach, the time-dependent behavior changes in the spectral-intensity were applied with a simple sinusoidal fashion to effectively employ the original data analysis algorithms. Initially, this approach was used for studying polymers (38,39) and liquid crystals (40,41) under a periodical stimulation.

In 1993, Noda presented a more generally applicable mathematical formalism for constructing 2D correlation spectra from any transient or time-resolved spectra having an arbitrary waveform, the generalized 2D correlation spectroscopy (34). The method was based on the Fourier transformation of the dynamic spectra matrix thus producing a synchronous spectrum matrix (real part) and an asynchronous spectrum matrix (imaginary part). Additional improvements in the calculation of the generalized 2D correlation spectra have been demonstrated by using the Hilbert transformation instead of the Fourier transformation (42). This general algorithm creates a 2D representation of the spectra measured under some variable perturbation, including not only time but also any other physical or chemical variables such as temperature, pressure, concentration, and sample composition. The two-dimensional representation spreads the spectral signal over the second dimension, thereby simplifying the visualization of complex spectra that contain many overlapped bands and enhancing spectral resolution. The type of spectral signals analyzed by the newly proposed 2D correlation method became virtually limitless, including Mid-IR, Near-IR, Raman, X-ray, Florescence, etc. (43,44) Using any one of these analytical tools and a perturbation, 2D homospectral correlation maps can be generated. Moreover, by applying two independent spectroscopic techniques to the same samples it is also possible to construct 2D correlation maps (2D heterospectral correlation spectrum) that highlight the in-phase and out-of-phase chemical features of the samples. The first 2D heterospectral analysis was reported by Noda and Ozaki using Raman and Mid-IR spectra (45). This approach provides the intriguing possibility of correlating various overtone and fundamental bands to establish unambiguous assignments in the NIR region (46).

Figure 4 shows the schematic contour map of the synchronous 2D NIR correlation spectrum in the range 882-2500 nm region for a set of wood samples. NIR spectra were collected on wood samples with cellulose content increasing (as the perturbation). The goal of the experiment was to use the 2D correlation approach to improve our understanding of the specific vibrations of wood components and assign vibrations to individual wood components. A one-dimensional reference spectrum (spectrum 1) is provided above and to the left of the contour map to show the basic features of the spectra of the system during the experiment. The synchronous spectrum is symmetric with respect to a diagonal line corresponding to the spectral coordinates wavelength 1 = wavelength 2. Correlation peaks appear at both diagonal and off-diagonal positions. Peaks located at diagonal positions, referred as auto-peaks, represent the overall extent of the dynamic variations in spectral intensity. The synchronous 2D NIR spectra reveal several auto-peaks at 2080, 1920, 1505 and 1250 nm. The appearances of these autopeaks mean that the intensities of these bands vary most significantly with increasing concentrations of cellulose. Synchronous cross peaks, appearing at off-diagonal positions, represent the simultaneous changes of spectral signals at two different wavelengths. Positive cross peaks indicate that the spectral intensities at corresponding wavelengths are either increasing or decreasing together as function of the perturbation while negative cross peaks show that one of the intensities is increasing while the other is decreasing. The band at 2080 nm is strongly negatively correlated to the peaks at 2295, 1688 nm and positively correlated to the bands at 1920, 1505, 1250 nm. This band is assigned to OH stretching vibration in wood. Because the lignin and hemicelluloses content is fixed in the spectral set, this band can be exclusively assigned to the cellulose hydroxyl vibrations. Once this assignment is done, it is easy from the 2D synchronous contour to assign the other bands that are changing with the perturbation. The band at 2080 nm is negatively correlated to the band at 2295 nm. The intensity of this peak is increasing while the other one is decreasing. Cellulose content is the perturbation and is increasing in the set so one can conclude that the band at 2295 nm is not a cellulose group and because the only other component that is changing in the set is extractives content one can assign this band to carbon/hydrogen vibrations of extractives. 2080 nm peak is positively correlated to 1920 nm band. This band is another band assigned to cellulose, specifically to interactions between carbohydrate hydroxyls and water. The band at 1920 nm, related to the intramolecular hydrogen-bond between water and OH from wood component, gives the same type of correlation than the band at 2080 nm, positive to 2080, 1505, 1250 nm and negative to 2295 and 1688 nm. The bands at 2295 and 1688 nm are negatively correlated to several bands off-diagonal, at 2080, 1920, 1505, 1370 and 1250 nm. The band at 2295 nm is assigned to the CH combination of extractives. The absorption band at 1688 nm represents mainly the contributions from the first overtone of CH stretching vibrations in extractives. Because the perturbation is cellulose content in the data set, one can assign numerous bands in the NIR region to cellulose by using the two-dimensional correlation method.

Laser Induced Breakdown Spectroscopy

Laser induced breakdown spectroscopy (LIBS) is another spectroscopic-based approach that can be used successfully to assess chemical composition of materials. This method gives the elemental composition (inorganic composition) of the material. Like NIR spectroscopy, LIBS has benefited from recent advances in component instrumentation both in terms of capabilities and application areas. In particular, the advent of the broadband (multispectrometer) detector allows LIBS to be sensitive to complex matter such as explosives,

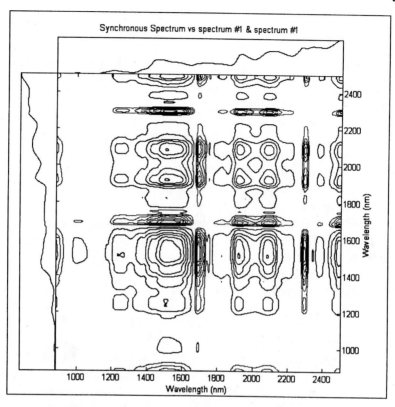

Figure 4. Synchronous contour map generated from the cellulose content-dependent spectral variation.

plastics, minerals, bacteria, etc. In fact, the broadband spectral response from 200-980 nm means that LIBS is now capable of detecting all chemical elements since all elements emit light somewhere in that spectral range. The ability of LIBS to provide rapid multielemental microanalysis of bulk samples (solid, liquid, gas, aerosol) in the parts-per-million (ppm) range with little or no sample preparation has been widely demonstrated (47). LIBS induces the vaporization of a small volume of sample material with sufficient energy for optical excitation of the elemental species in the resultant sample plume. The vaporized species then undergo de-excitation and optical emission on a microsecond time scale, and time-dependent ultraviolet-visible spectroscopy fingerprints the elements associated with the spectral peaks. The rapidity of sampling (typically 10 Hz laser repetition rate) and the ability to scan a sample surface by ablating a hole into a solid sample with repeated laser pulses, for depth profiling or focus the laser spark below the surface of a liquid sample permits more versatile analyses and provides sufficient statistics for bulk sampling. A schematic diagram of a classical experimental LIBS system is shown in Figure 5. A laser is used as the excitation source. The first step in LIBS is plasma initiation. Plasma can be formed when a laser beam of sufficient energy is focused onto a small volume of material ($\sim$ 14 x 10^{-9} cm^3), creating a power density inside the volume exceeding tens of gigawatts per square centimeter.

All species in the sample matrix are instantly released into their elemental constituents. A computer software data acquisition program is used to acquire the spectra, identify the peaks, calculate the full width at half maximum (FWHM) of the peaks of interest, and also to calculate the area under the peak which can be used in the semi-quantification of elements from a similar matrix. This software interfaces with the spectrometer and the ICCD detector to the computer which will monitor spectra from any sample that is tested using the LIBS technique. The intensity and signal-to-background ratio for the emitted wavelengths are used to determine the presence of the specific element. The emitted wavelengths are fingerprints for the different elements present in the material matrix. Recently, tremendous progress has been made in the quantitative analysis of the elemental composition of chemically treated wood and natural wood products (25), by using LIBS and MVA. The LIBS technique was performed on lumbers treated by different preservative chemicals (chromated copper arsenate CCA, ammoniacal copper zinc ACZA, alkaline copper quat ACQ) and principal component analysis (PCA) and partial least squares (PLS) regression were performed on the LIBS data. The results highlight the potential for this technique to identify and classify the type of preservative (Figure 6a), as well as to predict the concentration of inorganic preservatives present (Figure 6b) in the samples.

In addition to classifying the samples by the preservative type (ACZA, CCA, ACQ), the analysis shows the differences in the samples treated by the same chemical (Figure 6a). These differences come from the wood itself. and can be due to the way the substrate impacts the uptake and retention of the treatment chemicals. Alternatively, they may be due to the variations in the natural composition of the wood. In a manner similar to NIR spectroscopy, one is able to obtain quantitative information from LIBS data, however, a model must be developed by using samples with known elemental composition. Models were developed for all chemicals found in the preservatives (copper, zinc, chromium, and arsenic). For each chemical very strong correlation (higher than 0.9) was obtained, showing the efficiency of LIBS to predict the chemical composition of lumbers that have been treated by preservatives. This work showed that LIBS is capable of rapid, accurate, and simultaneous analysis of multiple inorganic constituents in materials. However, while sample preparation

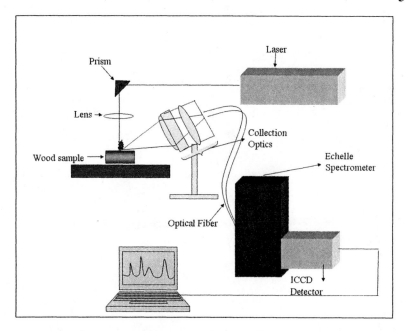

Figure 5. Laser Induced Breakdown Spectroscopy system.

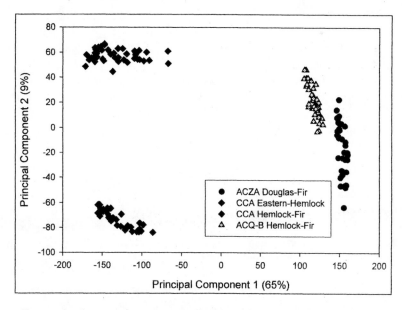

Figure 6a. Principal component analysis of LIBS spectra collected on preservative treated woods. (Reproduced with permission from reference 25. Copyright 2005 Elsevier.)

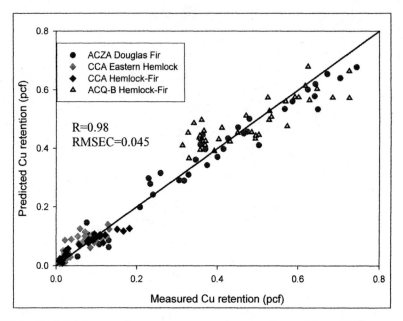

Figure 6b. Model to predict copper content in treated wood from LIBS spectra. Correlation between the measured and predicted copper content.

is very low, the technique is subject to the problem of enhanced "matrix effects". Emission lines of an element from a solid, a solution or a biological material may not preserve the same intensity ratios or even have preferential lines from different matrices. Therefore, it is important to control several parameters such as wavelength, energy/pulse, number of shots, repetition rate in order to obtain the best data for each matrix.

Conclusions

The purpose of this chapter is to provide an overview of two effective techniques that have been developed for several decades in the characterization and measurement of heterogeneous materials. Near infrared (NIR) spectroscopy and Laser Induced Breakdown Spectroscopy (LIBS), relative to other spectroscopic techniques, have a number of advantages that make them ideal tools for the characterization of heterogeneous and complex materials. In conjunction with statistical analysis they offer the unique ability to predict chemical, physical and mechanical properties.

References

1. Zachariassen, C.B.; Larsen, J.; van den Berg, F.; Balling Engelsen, S. Use of NIR spectroscopy and chemometrics for on-line process monitoring of ammonia in low methoxylated amidated pectin production. *Chemometrics and Intelligent Lab. Syst.* **2005**, 76, pp 149-161.
2. Delwiche, S.R.; Norris, K. Classification of hard red wheat by near-infrared diffuse reflectance spectroscopy. *Cereal Chem.* **1993**, 70, pp 29-35.
3. Downey, G.; Boussion, J. Authentication of coffee bean variety by near-infrared reflectance spectroscopy. *JNIRS* **1996**, 2, pp 85-92.
4. Michell, A.J. Pulpwood quality estimation by near-infrared spectroscopic measurements on eucalyptus woods. *Appita J.* **1995**, 48, pp 425-428.
5. Henriksen, H.C.; Naes, T.; Rodbotten, R.; Aastveit, A. Prediction of important sulfite pulp properties from near infrared spectra: a feasibility study and comparison of methods. *JNIRS* **2004**, 12, pp 279-286.
6. Dolezel-Howath, E.; Hutter, T.; Kessler, R.; Wimmer, R. Feedback and feedforward control of wet-processed hardboard production using spectroscopy and chemometric modeling. *Anal. Chim. Acta* **2005**, 544, pp 47-59.
7. Sjoblom, E.; Johnsson, B.; Sundstrom, H. Optimization of particleboard production using NIR spectroscopy and multivariate techniques. *Forest Prod. J* **2004**, 54, pp 71-75.
8. Watari, M.; Higashiyama, H.; Mitsiu, M.; Tomo, M.; Ozaki, Y. On-line monitoring of the density of linear low-density polyethylene in a real plant by near-infrared spectroscopy and chemometrics. *Appl. Spectrosc.* **2004**, 58, pp248-255.
9. Hansen, M.G.; Vedula, S. In-line fiber optic near-infrared spectroscopy: monitoring of reological properties in an extrusion process. Part I. *J. Appl. Polym. Sci.* **1998**, 68, pp 859-872.
10. Hansen, M.G.; Vedula, S. In-line fiber optic near-infrared spectroscopy: monitoring of reological properties in an extrusion process. Part II. *J. Appl. Polym. Sci.* **1998**, 68, pp 873-889.
11. Feldhoff, R.; Wienke, D.; Cammann, K. Fuchs, H. On-line consumer package identification by NIR spectroscopy combined with a FuzzyARTMAP classifier in an industrial environment. *Appl. Spectrosc.* **1997**, 51, pp 362-368.
12. Blanco, M.; Alcala, M.. Simultaneous quantification of five active principles in a pharmaceutical preparation: Development and validation of near infrared spectroscopic method. *Eur. J. Pharmaceutical Sci.* **2006**, 27, pp 280-286.
13. Kramer, K.; Ebel, S. Application of NIR reflectance spectroscopy for the identification of pharmaceutical excipients. *Anal. Chim. Acta* **2000**, 420, pp 155-161.

14. Sivakesava, S.; Irudayaraj, J.; Ali, D. Simultaneous determination of multiple components in lactic acid fermentation using FT-MIR, NIR, and FT-Raman spectroscopic techniques. *Process Biochemistry* **2001**, 37, pp 371-378.

15. Cimander, C.; Carlsson, M.; Mandenius, C.F. Sensor fusion for on-line monitoring of yoghurt fermentation. *J. Biotechnol.* **2002**, 99, pp 237-248.

16. Falla, F.S.; Larini, C.; Le Roux, G.A.C.; Quina, F.H.; Moro, L.F.L.; Nascimento, C.A.O. Characterization o crude petroleum by NIR. *J. Pet. Sci. Eng.* **2006**, 51, pp 127-137.

17. Kim, M.; Lee, Y.H.; Han, C. Real-time classification of petroleum products using near-infrared spectra. *Computers Chem. Eng.* **2002**, 24, pp 513-517.

18. Cozzolino, D.; Murray, I.; Chree, A.; Scaife, J.R. Multivariate determination of free fatty acids and moisture in fish oils by partial least-squares regression and near-infrared spectroscopy. *LWT* **2005**, 38, pp 821-828.

19. Marquez, A.J.; Diaz, A.M.; Reguera, M.I.P. Using optical NIR sensor for on-line virgin olive oils characterization. *Sensors and Actuators B* **2005**, 107, pp 64-68.

20. Blanco, M.; Villarroya, I. NIR spectroscopy: a rapid-response analytical tool. Trends in Analytical Chem. 2002, 21, pp 240-250

21. Amador-Hernandez, J.; Fernandez-Romero, J.M.; Luque de Castro, M.D. In-depth characterization of screen-printed electrodes by laser-induced breakdown spectrometry and pattern recognition. *Surf. Interface Anal.* **2001**, 31, pp 313-320

22. Jurado-Lopez, A.; Luque de Castro, M.D. Rank correlation of laser-induced breakdown spectroscopy data for the identification of alloys used in jewelry manufacture. *Spectrochim. Acta Part B* **2003**, 58, pp 1291-1299.

23. Moskal, T.M.; Hahn, D.W. On-line sorting of wood treated with chromated copper arsenate using laser-induced breakdown spectroscopy. *Appl. Spectrosc.* **2002**, 56, pp 1337-1344.

24. Solo-Gabriele, H.M.; Townsend, T.G.; Messick, B.; Calitu, V. Characteristics of chromated copper arsenate-treated wood ash. *J. Hazard. Mater.* **2002**, B89, pp 213-232.

25. Martin, M.Z.; Labbé, N.; Rials, T.G.; Wullschleger, S.D. Analysis of preservative-treated wood by multivariate analysis of laser-induced breakdown spectroscopy spectra. *Spectrochim. Acta Part B* **2005**, 60, pp 1179-1185.

26. Samsek, O.; Telle, H.H.; Beddows, D.CS. Laser-induced breakdown spectroscopy: a tool for real-time, in vitro and in vivo identification of carious teeth. *BMC Oral Health* **2001**, 1, pp 1-9.

27. Hybl, J.D.; Lithgow. G.A.; Buckley, S.G. Laser-induced breakdown spectroscopy detection and classification of biological aerosols. *Appl. Spectrosc.* **2003**, 57, pp 1207-1215.

28. Martens, H. and Naes, T. *Multivariate calibration.* John Wiley & Sons, New York, 1989

29. Adams, M.J. *Chemometrics in Analytical Spetroscopy,* Royal Soc. Chem. Cambridge, 1995

30. Kramer, R. *Chemometric techniques for quantitative analysis.* Marcel Dekker, New York, 1989

31. Beebe, K.R.; Pell, R.J.; Seasholtz, M.B. *Chemometrics- A practical guide.* Wiley, New York, 1998

32. Burns, D.A. and Ciurczak, E.W. *Handbook of near-infrared analysis.* Marcel Dekker, New York, 2001

33. Noda, I. Two-dimensional infrared spectroscopy of synthetic and biopolymers. *Bull. Am. Phys. Soc.* **1986**, 31, pp 520-528.

34. Noda, I. Two-dimensional infrared (2D-IR) spectroscopy: theories and applications. *Applied Spectroscopy* **1990**, 44, pp 550-561.

35. Noda, I. Generalized two-dimensional correlation method applicable to Infrared, Raman, and other types of spectroscopy. *Appl. Spectrosc.* **1993**, 47, pp 1329-1336.

36. Noda, I., Dowrey, A.E.; Marcott, C.; Story G.M.; Ozaki, Y. Generalized two-dimensional correlation spectroscopy. *Appl. Spectrosc.* **2000**, 54, pp 236-248.

37. Noda, I. Two-dimensional vibrational spectroscopy. *J. Am. Chem. Soc.* **1989**, 111, pp 8116-8117.

38. Palmer, R.A.; Manning, C.J.; Chao, J.L.; Noda, I.; Dowrey, A.E.; Marcott, C. Application of Step-Scan Interferometry to Two-Dimensional Fourier Transform Infrared (2D FT-IR) Correlation Spectroscopy. *Appl. Spectrosc.* **1991**, 45, pp 12-17.

39. Marcott C., Noda, I.; Dowrey, A.E. Enhancing the information content of vibrational spectra through sample perturbation. *Anal. Chim. Acta* **1991**, 250, pp 131-155.

40. Gregoriou, G.V.; Chao, J.L.; Toriumi, H.; Palmer, R.A. Time-resolved vibrational spectroscopy of an electric filed-induced transmition in a nematic liquid crystal by use of tep-scan 2D FT-IR. *Chem. Phys. Lett.* **1991**, 179, pp 491-496.

41. Nakano, T.; Yokoyama, T.; Toriumi, H. One- and Two-Dimensional Infrared Time-Resolved Spectroscopy Using a Step-Scan FT-IR Spectrometer: Application to the Study of Liquid Crystal Reorientation Dynamics *Appl. Spectrosc.* **1993**, 47, pp 1354-1366.

42. Noda, I. Determination of two-dimensional correlation spectra using the Hilbert transform. *Appl. Spectrosc.* **2000**, 54, pp 994-999.

43. Noda, I.; Dowrey, A.E.; Marcott, C.; Story, G.M.; Ozaki Y. Generalized two-dimensional correlation spectroscopy. *Appl. Spectrosc.* **2000**, 54, pp 236-248.

44. Ozaki, Y.; Noda, I. Two dimensional Correlation Spectroscopy, Eds.; American Institute of Physics, Melville, NY, 2000, AIP Conf. Proc. 503

45. Noda, I., Liu, Y., Ozaki, Y. Two-dimensional correlation spectroscopy study of N-methylacetamide in the pure liquid state. 2. Two-dimensional raman and infrared-raman heterospectral analysis. *J. Phys. Chem.* **1996**, 100, pp 8674-8680.

46. Noda, I.; Liu, Y.; Ozaki, Y.; Czarnecki, M.A. Two-dimensional Fourier Transform near-infrared correlation spectroscopy studies of temperature-dependent spectral variations of oleyl alcohol. *J. Phys. Chem.* **1995**, 99, pp 3068-3073.

47. Martin, M.Z.; Cheng, M.D.; Martin, R.C. Aerosol Measurement by Laser-Induced Plasma Technique: A Review. *Aerosol Sci. and Technol.* **1999**, 31(6): pp 409-421.

Chapter 32

Carbonyl and Carboxyl Profiles as Two Novel Parameters in Advanced Cellulose Analytics

Antje Potthast, Thomas Rosenau, and Paul Kosma

Department of Chemistry and Christian-Doppler-Laboratory, University of Natural Resources and Applied Life Sciences Vienna, Muthgasse 18, A–1190 Vienna, Austria

Profiles of oxidized functionalities in cellulosics (carbonyls and carboxyls in the μmol / g range) in relation to the molecular weight distribution (MWD) are obtained by a combination of group-selective fluorescence labeling and multi-detector gel permeation chromatography (GPC). Oxidative damage of cellulosic materials as well as oxidative changes by processing, chemical or physical treatments can be monitored with high accuracy much more comprehensively than hitherto possible.

Oxidized Groups in Cellulose

Cellulose is not the ideal homopolymer built up of β-1,4-glycosidically linked anhydroglucose units, but contains small amounts of various "irregular" structures, mainly oxidized groups. The reliable and accurate determination of such oxidized functionalities in cellulose represents a largely unsolved problem in cellulose chemistry (1), and is the topic of the present chapter.

Oxidized groups in cellulosics are introduced by pulping and bleaching processes, according to the respective conditions chosen (2,3). Especially bleaching – both chlorine-based and oxygen-based – is known to affect the integrity of the cellulose backbone by generating oxidized positions and causing subsequent chain cleavage. The creation of oxidized groups along the cellulose chain is a highly undesirable result, as these positions constitute "hot spots" along the carbohydrate chain, where a pronounced chemical instability is introduced and where subsequent cleavage will primarily occur. Thus, despite their small concentration, carbonyl and carboxyl contents are among the crucial factors to govern macroscopic properties and behavior of cellulosics.

Impact of Oxidized Functionalities on Cellulose Properties

The minute amounts of oxidized positions in cellulose is the main reason for generally decreased performance parameters in textiles, paper and other cellulosic materials, as shown by the following examples.

Decreased strength due to *alkaline degradation* can be traced back on a molecular level to the cleavage of the glycosidic bonds adjacent to a carbonyl group by β-alkoxy elimination. An increased number of carbonyl groups causes a decreased stability of the material under alkaline conditions, resulting in a more or less severe DP loss. During aging ("ripening") of sodium cellulose as used in the viscose process, carbonyl groups decrease drastically with new carboxyl groups being generated. Another important mechanism under alkaline conditions is the endwise peeling reaction starting from the aldehyde functionality at the reducing end of the cellulose chain. It is partly responsible for yield losses in alkaline pulping and bleaching.

An important feature of carbonyl groups is their ability to undergo n-π* electron transitions upon UV irradiation (240-320 nm). A number of secondary processes, such as reactions involving oxygen (*e.g.,* singlet oxygen) and generation of radical species, may start from the resulting excited state, finally causing *photo-yellowing* and photo-degradation of pulp, (4) an unfavorable result especially in the case of paper-grade materials and fibers. The role of carbonyl groups in light-induced yellowing is a controversial topic. While older studies showed a clear effect of carbonyls on light-induced yellowing (5,6,7), Chirat *et al.* (8,9) found an influence only in the case of heat-induced color reversion, but not upon UV irradiation. The comparability of results in this field is often limited, since radiation conditions, such as wavelength and power of the light source, moisture content and atmosphere, have a great impact on the outcome. Yellowing of cellulosic materials cannot only be caused by exposure to light, but can also be triggered by *thermal stress*. Large differences as to the underlying pathways and mechanisms exist, the dependence on the moisture content is pronounced. (10) Both carbonyl and carboxyl groups have a significant influence. However, the effects are complex, sometimes counteracting and sometimes synergistic. The overall outcome of thermal yellowing cannot be related to a single structural feature of the pulp, but depends on the interplay of conditions and on the pulp properties as a whole.

Higher temperatures or prolonged dry storage at room temperature cause a phenomenon referred to as ***hornification***, which is characterized by a pronounced decrease of pulp reactivity in subsequent derivatization reactions (11,12,13). This effect is accompanied by a decrease in pore volume, accessibility, specific surface area and reactivity in general, which is largely irreversible, as re-wetting of the dried samples does not restore the previous properties. Chromophore formation and yellowing phenomena are also consequences of hornification. The influence of water and the crucial role of oxidized groups have been demonstrated. (10,14,15)

An important subject, also with regard to economic issues, is the ***aging*** of cellulose (16) and paper, especially in the presence of air, air pollutants and moisture, which can sometimes lead to deterioration of the whole book stock in libraries. Also with respect to aging, both carboxyl and carbonyl groups play a major role. Acidic decay affected almost all modern papers from 1850 until 1990. Carbonyl and carboxyl groups formed by different oxidation reactions promote the autocatalytic decay of such cellulosic material (17).

Large amounts of carbonyl groups cause reduced filterability of xanthogenate in ***viscose production*** and decreased strength properties of the resulting fibers, whereas similar negative effects are unknown for carboxyl groups (18,19,20). Aging kinetic studies also revealed that viscose fibers exhibit a faster aging rate than other cellulosics (21).

In the ***Lyocell process***, carbonyls are the major precursor for chromophore generation, which finally result in discoloration of the spinning dope and decreased brightness properties of the resulting products (*22,23*). Carboxyl groups, in contrast, are mainly contributing to side reactions of the amine N-oxide cellulose solvent (24), and might even cause uncontrollable decomposition reactions of the dope (25).

General Aspects of the Analysis of Oxidized Groups in Cellulosics

The accurate determination of oxidized functionalities in cellulose has to contend with several inherent difficulties. First, the extremely low average contents of those groups, which range in the order of μmol / g, require very sensitive means of detection since conventional, direct instrumental techniques, such as IR, Raman, UV, fluorescence or NMR spectroscopy, fail to report such minor amounts. Hence, a chemical method must be applied, in which carbonyl or carboxyl structures are either "titrated" by a reagent, or converted by a selective reaction into structures, which can then be monitored by traditional spectroscopic techniques. The second major problem follows directly: the derivatization of oxidized groups will be a heterogeneous reaction with inherent problems of accessibility and reagent adsorption. But even if carried out homogeneously, the required solvents will be rather exotic, such as metal amine complexes, N,N-dimethylacetamide / LiCl or N-methylmorpholine-N-oxide, so that conventional carbonyl or carboxyl reactions cannot easily be transferred into those media. And third, no kinetic data are available, which additionally complicates the situation by the lack of means to report the completeness of a derivatization reaction.

To measure the content of carbonyls and carboxyls, some conventional methods are available, as discussed below. Apart from the drawbacks of specific methods, the available approaches suffer from general shortcomings: high limits of determination and detection, which are unacceptable for accurate methods, and rather large amounts of sample material needed, which is unacceptable in the case of valuable historic material. In addition, those methods provide sum parameters only, yielding an overall carbonyl or carboxyl content, respectively; possible differences between regions of the molecular weight distribution (MWD) remain undetected.

To address those analytical challenges appropriately, a novel analytical approach should be: (a) more sensitive: responding to even minute amounts of such functional groups (b) selective and quantitative toward one functionality and (c) able to monitor not just the sum of parameters, but analytical values in relation to the MWD. The two novel approaches in cellulose analytics described in the next sections, the CCOA method for carbonyl profiling and the FDAM method for carboxyl profiling, meet all these requirements to a large extent, providing a hitherto unmatched level of information on the oxidation state of cellulosic samples. Both method names are derived from the abbreviations of the semi-systematic chemical terms of the respective selective labeling reagents: CCOA for carbazole-carbonyl-oxyamine and FDAM for fluorenyl-diazomethane.

Carbonyls in Cellulosics

Occurrence and Generation of Carbonyls in Cellulosics

The main causes for the formation of carbonyl groups in cellulose are isolation and purification procedures. This applies in particular to cellulose from wood, which has undergone a number of process steps to be separated from lignin, hemicelluloses, and extractives. Cellulose from *Acetobacter xylinum* (26), which has not undergone such purification steps, contains an amount of carbonyl groups that corresponds to the number of reducing end groups, and with regard to introduced oxidized groups it can thus be considered rather "pure" material (27).

In acidic sulfite *pulping*, generally a larger number of carbonyl groups is introduced as compared to prehydrolysis kraft pulping. The alkaline conditions of kraft pulping favor dissolution of low-molecular weight material, which has a high relative carbonyl content. During sulfite pulping, hydrolysis of cellulose and hence the generation of more reducing end groups is promoted. The amounts of carbonyl groups introduced by ASAM (alkaline sulfite – anthraquinone – methanol) and Organocell pulping processes are comparable to the prehydrolysis kraft case, whereas for acetosolv a higher number was observed, *cf.* Table 1 and Figure 1. In general, acidic pulping processes introduce more carbonyl groups than alkaline ones.

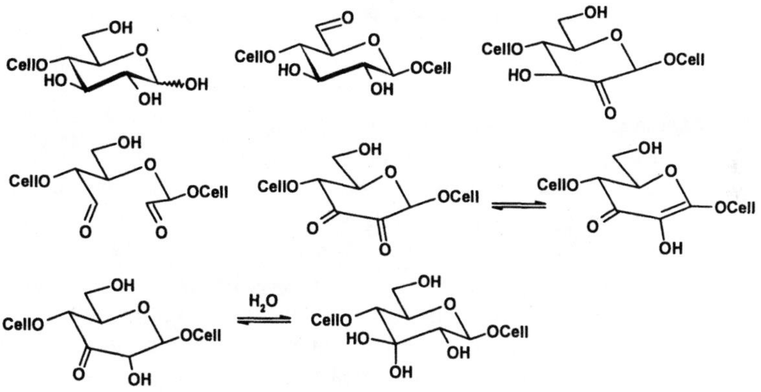

Figure 1. Examples of different carbonyl structures in cellulose.

Bleaching of pulp represents the main source of carbonyl groups in cellulosics. While pulping proceeds in the absence of strong oxidants, oxidizing agents are deliberately used in bleaching to achieve the desired brightness effect. Older procedures employ HOCl and Cl_2, whereas more recent pulp mills operate according to so-called ECF (elementally chlorine free) or TCF (totally chlorine free) bleaching sequences. Even though chlorine and chlorine dioxide are rather selective bleaching agents, their use is being debated due to environmental concerns.

Treatment of pulp with HOCl, performed in hypochlorite bleaching ("*H stage*") for many years, results in the formation of considerable amounts of carbonyl groups, if the treatment is carried out at a pH range between 2 and 8, whereas under alkaline conditions mainly carboxyl groups are formed. An H stage in all pH ranges is accompanied by a decrease in the molecular weight, which depends linearly on the oxidative damage done, *i.e.* on the carbonyls created, but not on the pH (28).

Table 1. Carbonyl groups in pulps after different pulping methods.

Pulping	Carbonyl content, μmol/g
Kraft	7-15
Prehydrolysis Kraft	2-10
Sulfite	7-30
ASAM	~ 5.2
Prehydrolysis ASAM (29)	5-17
Acetosolv	~ 16.2
Organocell	~ 2.1

TCF bleaching sequences utilize a combination of oxygen, hydrogen peroxide, and ozone as active species. These oxygen-based bleaching chemicals are environmentally more benign as compared to hypochlorite, elemental chlorine or chlorine dioxide, but unfortunately, their selectivity towards removal of lignin and residual chromophores is much lower, so that oxidative damage is done to the carbohydrates. In particular, the heavy metal management during pulping becomes an important issue, since a number of very reactive radical species such as hydroxyl and hydroperoxyl radicals, are formed in the presence of the bleaching agents and transition metal ions (*Fenton* and *Haber-Weiss* cycles). Gierer (30) and Gratzl (31) state the generation of different oxygen-derived radical species in all TCF sequences as the main cause of the limited selectivity and the introduction of oxidized groups.

Large amounts of carbonyl groups are introduced during an *ozone (Z) stage* (*32,33*). Besides oxidation by hydroxyl radicals, which can be generated in an ozone stage either by slow decomposition of ozone in water or be triggered in the presence of metal ions, also an ionic pathway by a 1,3-insertion mechanism of ozone has been proposed (34). Such processes are always accompanied by a DP loss (35). Investigations with methyl 4-*O*-ethyl-β-D-glucopyranoside as model compound showed the following order of oxidation sensitivity in an ozone treatment: C1>C6>C2,C3>C4,C5,(36) the β-anomer reacting faster than the α-counterpart.

During a *hydrogen peroxide (P) stage*, which is mainly appended at the end of a bleaching sequence to gain the desired final brightness, the situation is similar to an ozone stage if radicals are present and their subsequent reactions are allowed to proceed. However, in their absence, according to investigations on cotton samples, a P stage introduced predominantly keto groups, which neither cause a pronounced DP loss nor a decreased brightness and were thus thought to be located at C-3 (*37,38*).

High energy *radiation*, for instance γ-irradiation or β-irradiation ("electron beaming"), causes an increase in carbonyl groups, mainly through the action of generated radical species. Both procedures are accompanied by a DP loss, which can be used to adjust the molecular weight of the cellulose prior to utilization in the viscose process (39). The number of carbonyls introduced by high energy electron beams is directly correlated with the applied dosage (2). β-Irradiation is also considered as a means for cellulose activation, since the radiation can also affect highly ordered regions in cellulose, changing the morphology for subsequent reactions in a favorable way (39). Also, γ-irradiation showed a linear relationship between dose and carbonyl generation (40).

Conventional Determination of Carbonyls in Cellulosics

The quantification of carbonyl groups in cellulose is so far limited to the measurement of the total carbonyl content by different methods as briefly reviewed in the following text.

The so-called *copper number* is still the method of choice in the pulp and paper industry in process control, and also in research labs for its quick and easy handling. According to the underlying principle, the reducing power of cellulose is measured by reaction with an alkaline copper(II) complex under well-defined conditions. The formed copper(I) ions can be titrated after re-oxidation (41). The detailed reaction mechanism is still not entirely understood, neither are the types of oxidized structures recorded. If carried out under exactly similar conditions, the copper number determination produces a reliable and reproducible sum parameter, which cannot be directly linked to the quantity of a specific oxidized functionality, such as carbonyls. A correlation to the carbonyl group content as measured by the well-defined CCOA method was found (42), see below. Thus, for quick or high-throughput analysis the copper number remains a valuable parameter despite its obvious limitations.

In 1945, Geiger and Wissler (43) proposed a method for the quantification of keto and aldehyde groups in cellulose based on reacting carbonyls in cellulosics with a *hydrazine* reagent (*Girard's* reagent P, acetylhydrazide pyridinium chloride), and thus converting the cellulose into an anion exchanger. The anion exchange capacity was then measured by titration with a colored anion (*e.g.,* picrate), its concentration being measured colorimetrically after elution with sodium hydroxide. Besides the rather tedious and complicated procedure, the method severely underestimated the carbonyl quantities and was therefore recommended for semi-quantitative measurements only (44). Cyrot (45) proposed the reaction with *hydroxylamine* to the corresponding oxime as a means of carbonyl quantitation, since oximation gave higher conversion than the reaction with hydrazine. As analytical parameter, the degree of nitrogen fixation is measured, either by a *Kjeldahl* procedure, or by elemental analysis. According to Rehder *et al.* (46), running the oximation in a zinc acetate buffer eliminates side reactions with acidic groups and minimizes those with lactones. However, the oximation method fails in the case of small sample amounts or small quantities of carbonyls contained.

According to Lewin (47) the formation of cyanohydrins by reaction of carbonyl groups with cyanide can also be taken as gauge for oxidized functionalities. Excess cyanide was removed from the pulp and determined by titration with AgNO₃. This *cyanide method* frequently yields overestimated carbonyl contents due to adsorption phenomena, and is less often used today because of the toxicity of the reagents used.

The consumption of sodium borohydride (*reduction method*) upon reduction of cellulosic carbonyl groups was also used for their quantification. Excess NaBH₄ was measured by the amount of hydrogen formed in the reaction with acid (48). However, this method provided only rough estimates.

The reducing power of reducing end groups and other aldehydes was also used to convert TTC (2,3,5-triphenyltetrazolium chloride) into triphenyl-formazane, a red dye which was quantified spectrophotometrically (49). The same principle was applied by Strlic and Pihlar (50) in homogeneous solution in DMAc/LiCl yielding results more reproducible than the heterogeneous procedure in aqueous suspension. However, the TTC treatment causes sometimes severe cellulose degradation (51), probably by triggering β-elimination processes in the strong alkaline medium used for the oxidation, which limits accuracy and general applicability of the method.

Carboxyls in Cellulosics

Occurrence and Generation of Carboxyls in Cellulosics

Only second to carbonyl groups, carboxyls are a very important oxidized functionality in cellulose. Carboxyl groups in cellulose are artificially introduced by pulping and bleaching processes, whereas they are naturally contained in hemicelluloses in the form of glucuronic acids or hexenuronic acids in the side chains. The major position for oxidative attack in cellulosics that generates carboxyls is the reducing end group, which is converted into the corresponding gluconic acid. Oxidation of the primary hydroxyl groups at C-6 produces glucuronic moieties. By formation of acids at C-2 and/or C-3, the pyran ring structure of glucopyranose units is destroyed, which is possible with or without simultaneous cleaving of the cellulose chain (Figure 2). In acidic media the different acids are present in the form of their lactones, if an energetically favored 5- or 6-membered ring can be formed (46).

In unbleached pulps the bulk amount of carboxyl groups correlates with the content of residual lignin. Residual sulfonic acid groups, contributing to overall acidity, might be contained in unbleached sulfite pulps.

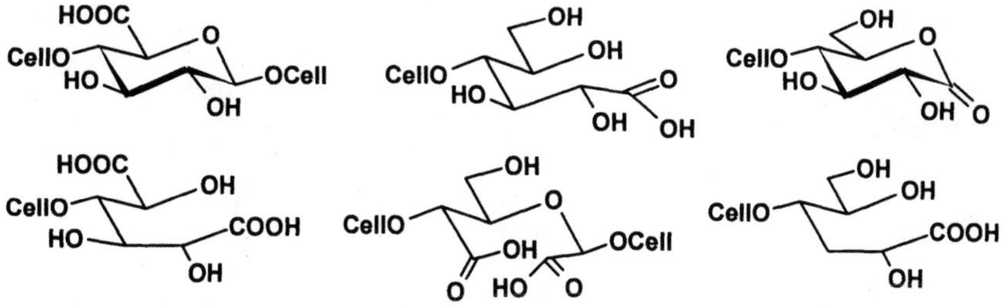

Figure 2. Examples of different carboxyl structures in cellulose.

In acidic pulping procedures, a relatively high amount of carboxyl groups is introduced, in particular as glucuronic and gluconic acids. In acidic sulfite pulping, the hydrogen sulfite ion is able to act as oxidant. It converts reducing end groups, especially in low-molecular weight material, into the corresponding acid, being in turn reduced to thiosulfate, a common byproduct in the sulfite process. Due to inductive effects of the uronic acid moieties, the adjacent β-1,4-linkages are activated in a way that the neighboring glycosidic bonds become more prone to hydrolysis. In contrast, the presence of gluconic acids results in a cellulose more resistant to acid degradation (46).

Table 2. Carboxyl group content of different cellulosic materials.

Pulping	Carboxyl content, μmol/g
beech prehydrolysis kraft	8-50
beech/spruce sulfite dissolving	20-30
hardwood Acetosolv dissolving	25-30
prehydrolysis ASAM spruce (52)	40-60
cotton linters	<10-20
paper grade pulps	20-300
viscose fiber	~ 40

After alkaline pulping, metasaccharinic and gluconic acid residues are the typical representatives of carboxyl groups in cellulose (53). Metasaccharinic acid is the final product of the stopping reaction after alkali-induced peeling and rearrangement reactions (*cf.* Figure 2). Under alkaline conditions, gluconic acid carboxyl groups are crucial with regard to cellulose stability as they hamper further degradation according to alkaline peeling pathways. This is utilized in pulping systems that employ low amounts of anthraquinone or polysulfide as redox auxiliaries to facilitate the oxidation of reducing end groups.

An exhaustive oxidation of C-6, e.g., by means of the TEMPO reagent, eventually results in water-soluble polyglucuronic acids. In contrast, oxidation of C-2 and C-3 by a combination of periodate and chlorous acid yielded insoluble material. (54)

Conventional Determination of Carboxyls in Cellulosics

All conventional methods for carboxyl quantification in cellulosics rely on the conversion of free acids into salts, and are thus, basically, mere variations of an acid–base titration. They depend on the cation exchange capacities of the material, and differ mainly in sample preparation, in the cation used, and in the direct or indirect measurement of the exchanged cations. Direct alkalimetric titration of the fully protonated carboxyl group with NaOH was used, as was ion exchange with chromophoric cations, and other *titrimetric methods* as reviewed recently (55). In addition to those, decarboxylation and quantification of the formed CO_2 has been reported for carboxyl quantification (56).

The most frequently used technique, the direct or reversible (57) *methylene blue method,* is based on the quantification of the colored organic base, measuring either its absorption to the polymer, or its depletion in the titer, respectively. Problems arise with accessibility, non-specific adsorption, and reproducibility, but still this method has been widely used due to the lack of alternatives. Presently, the reversible methylene blue method (58), the sodium bicarbonate procedure according to Wilson (59), and the zinc acetate method (60) are widely applied in pulp and paper laboratories. They all require large amounts of material, and naturally yield no information on the molecular weight-related distribution of the carboxyls. By far the major drawback is the low reproducibility: all these methods generate values of severely deficient comparability, both in inter-procedural and in inter-laboratory comparisons. This was again confirmed in a recent round-robin test (61).

Profiling of Carbonyls and Carboxyls in Cellulosics

Two main drawbacks of the existing methodology for the determination of oxidized groups in cellulosics can be extracted from the wealth of literature and are the following: the inaccuracy and poor reproducibility due to ill-defined reactions and low concentrations, and the availability of sum parameters only (overall contents). To overcome these obstacles, we recently introduced two methods into cellulose analytics, the CCOA method (62) for carbonyl profiling and the FDAM method (63) for carboxyl profiling, which were comprehensively discussed in the literature (64,65,66,67,68), these references contain also comprehensive information on the GPC system used, the eluant, calibration, validation of the method and detailed experimental procedures.

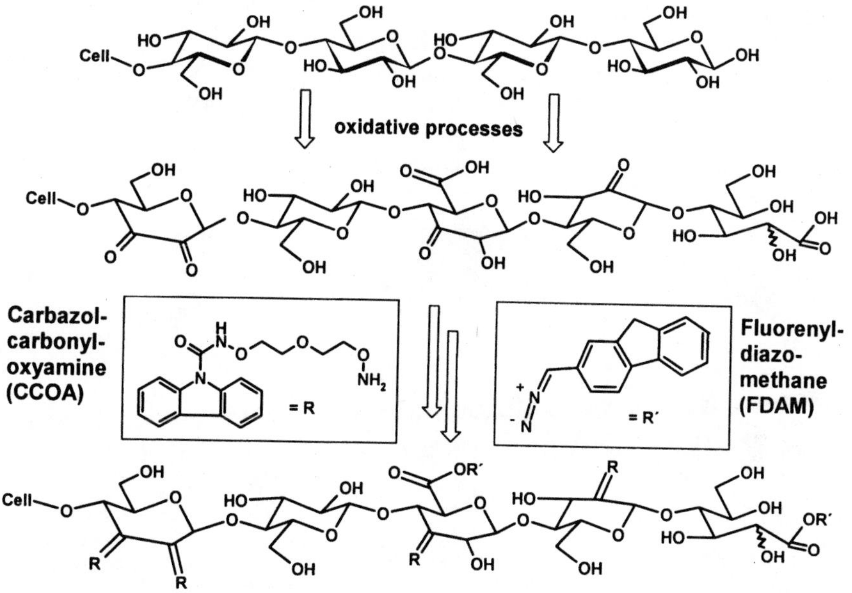

Figure 3. Selective fluorescence labeling with CCOA and FDAM.

Both methods use a group-selective reaction to attach a *fluorescence label* which is later used to report the presence of the oxidized group by a concentration-equivalent signal. As the concentration of the oxidized structures is extremely low, UV labels are still not sensitive enough in the case of low carbonyl or carboxyl contents, so that fluorescence markers were required. Comprehensive studies were performed to search for selectively and quantitatively reacting reagents, to improve the structure of the fluorescence label which must provide suitable spectroscopic properties compatible with a GPC-MALLS system, and to optimize the labeling conditions. By the selectivity of the labeling reaction and the highest possible degree of conversion comprehensively examined with model compounds (62,63), an optimum accuracy and reproducibility was assured, which is reflected by the validation parameters of the method (65). The requirements for sample amount are rather low, between 5 and 50 mg, so that the method can well be applied to samples from more valuable sources.

Both CCOA and FDAM labeling reactions are performed as a *pre-column derivatization step* prior to separation of the cellulose in a modified gel permeation chromatography (GPC) system. By combining group-selective fluorescence label with *size-exclusion chromatography*, quantification of oxidized groups relative to the MWD of the cellulosic material became possible for the first time, displayed in *carbonyl and carboxyl profiles*, respectively. For that purpose, the output of mass-proportional multi-angle laser light scattering (MALLS) and the concentration-dependent refractive index (RI) detectors is combined with the signal of a fluorescence detector, so that the extent of oxidation can be calculated for each slice of the MWD and displayed as degree-of-substitution plots for the respective functional group (Figure 4). These *DS plots* have a typical exponentially decaying curve shape.

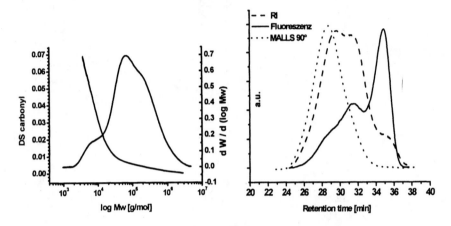

Figure 4. Calculation of MWD and carbonyl profile (right) from the detector output of a GPC with MALLS / RI / fluorescence detection (left).

While DS plots reflect the oxidation state of a given cellulosic material, dynamic oxidative changes are better visualized in *ΔDS plots*, which give the difference in DS_{CO} or DS_{COOH} between the starting material and the material after the respective treatment. This is illustrated by the example in Figure 5, which shows a comparison between a hypochlorite treatment of cellulose and an ozone bleaching step, both at different concentrations and dosages, with regard to the carbonyl content. For both treatments, an increase in carbonyl groups with higher treatment intensity was found.

However, hypochlorite-introduced carbonyls mainly in the low-molecular weight region of the pulp, whereas ozone caused a decrease in the lower Mw region, probably due to formation of acids and lactones, and an increase in the higher Mw part of the pulp (66). According to our investigations, chemical treatments generally affected the low and middle Mw part of the pulp more pronouncedly, whereas irradiation, either by UV light or by electron beams, also influenced high-molecular weight parts significantly (67). Such detailed monitoring of oxidative changes would have been simply impossible according to conventional methodology based on sum parameters. This can be visualized

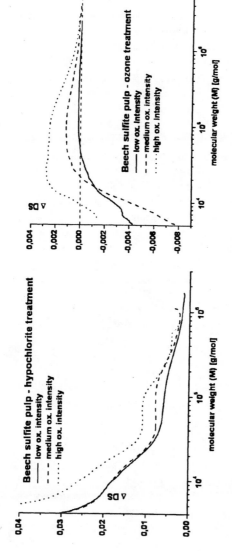

Figure 5. ΔDS plots of carbonyl groups after treatments of different intensities with hypochlorite (left) and ozone (right).

exemplarily by the right part of Figure 5. While conventional methods would report nearly no changes in the overall carbonyl content, the new methodology shows that the decrease in low-molecular weight carbonyls is largely compensated by an increase in high-molecular weight carbonyls to give that net outcome. This differentiation of the behavior of carbonyls or carboxyls generally allows new insights into the reactivity of cellulose and hemicellulose during different treatments in cellulose processing.

To even better visualize oxidative changes in one pulp or differences in the carbonyl/carboxyl content between different pulps, a numeric evaluation of the data is carried out. For that purpose, the amount of carbonyls or carboxyls in specific regions of the MWD is calculated from the DS curves. Using the four molecular weight ranges DP < 100, 100 < DP < 200, 200 < DP < 2000, and DP > 2000, proved to give the best results. Figure 6 shows the resulting pie diagrams for carbonyl and carboxyl content of two pulp samples.

Such detailed view of the functional groups in relation to the molecular weight is useful to study the response of cellulose towards any processing step. Of course the response to treatments involving oxidative steps will be most pronounced. In recent studies, for instance the behavior of celluloses during Lyocell or viscose processing was examined (66). It became evident, that different regions of the MWD reacted quite differently, and that there were significant differences according to the cellulosic substrates. Also bleaching stages and the effect of changes in the bleaching conditions on the MWD and oxidation state of pulps were monitored successfully.

It should be noted that so far the CCOA and FDAM approaches are only applicable to cellulosics soluble in DMAc/LiCl, and to pulps with quite low lignin contents. Studies to extend the applicability to other substrates are currently underway.

The carboxyl-selective FDAM method can be conveniently used to monitor hemicelluloses or changes in the hemicellulose-rich MWD region, using for instance the 4-O-methyl glucuronic acid residues in hardwood xylan side chains as the probes to be labeled. Thus, differences both in the state of oxidation and in the localization of hemicellulose become visible. In addition, surface labeling can be used to distinguish areas of higher and lower states of oxidative damage, for instance, in the case of historic paper samples.

Summarizing the advances brought about by the CCOA and FDAM methods, it can be stated that carbonyl and carboxyl group contents in cellulosics are detected for the first time very accurately, in a chemically unambiguous way, and as profiles (relative to the MWD). The methods have low levels of detection and quantification, and require low amounts of sample only. At the same time with the profiles, all data of the molecular weigh distribution are obtained (Mn, Mw, DPI, MWD). Measuring more than 500 different cellulosics samples in more than 3000 single runs proved the method to be quite effective and suitable for routine analysis.

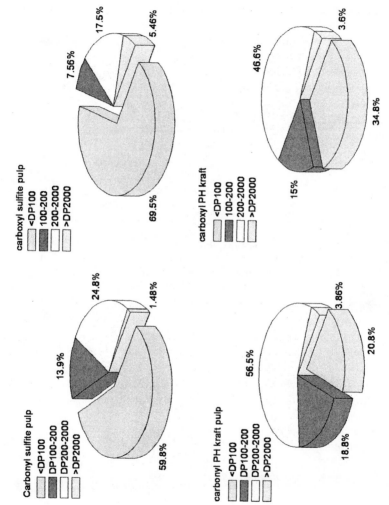

Figure 6. Distribution of carbonyl and carboxyl content in an acid sulfite pulp (upper) and a prehydrolysis kraft pulp (lower).

528

References

1. Klemm, D.; Philipp, B.; Heinze, T.; Heinze, U.; Wagenknecht, W. *Comprehensive Cellulose Chemistry*; vol. 2, Wiley-VCH: Weinheim, Germany, 1998, pp 302-314.
2. Schleicher, H.; Lang, H. *Das Papier* **1994**, *12*, pp 765-768. b) Sixta, H. Habilitation thesis, Technical University of Graz, 1995.
3. Sixta, H. Habilitation thesis, Technical University of Graz, Austria. 1995.
4. Gratzl, J. S. *Das Papier*, **1985**, *39 (10A)*, V14.
5. Dong Ping Pong, J., Master thesis, North Carolina State University, 1975.
6. Rapson, W. H.; Corbi, J. C. *J. Pulp. Paper Mag. Can.* **1964**, *65*, T-459.
7. Sjöström, E.; Eriksson, E. *TAPPI J.* **1968**, *51(1)*, p 16.
8. Chirat, C. ; de la Chapelle, V. *J. Pulp Paper Sci.* **1999**, *25*, p 201.
9. Chirat, C.; de la Chapelle, V.; Lachenal, D. *Proceedings of the 4th European Workshop on Lignocellulosics and Pulp*, Stresa, Italy, 1996, p 146.
10. See for instance: Beyer, M.; Bäurich, D.; Fischer, K. *Das Papier* **1995**, *10A*, pp 8-14.
11. For a review see: Kato, K. L.; Cameron, R. E. *Cellulose* **1999**, *6 (1)*, 23 and references cited therein.
12. Garcia, O.; Torres, A. L.; Colom, J. F.; Pastor, F. I. J.; Diaz, P.; Vidal T. *Cellulose* **2002**, *9 (2)*, pp 115-125.
13. Haggkvist, M.; Li, T. Q.; Odberg, L. *Cellulose* **1998**, *5 (1)*, 33-49.
14. Forsskahl, I.; Tylli, H.; Olkkonen, C. *J. Pulp Paper Sci* **2000**, *26(7)*, pp 254-249.
15. Beyer, M.; Lind, A.; Koch, H.; Fischer, K. *J. Pulp Pap. Sci.* **1999**, *25(2)*, pp 47-51.
16. Lewin, M. *Macromol. Symp.* **1997**, *118*, pp 715-724.
17. Blüher, A.; Vogelsanger, B. *Chimia* **2001**, *55*, p 981.
18. Meller, A. *Svensk Papperstidn.* **1957**, *60*, p 611.
19. Bartunek, R. *Das Papier* **1968**, *22(9)*, p 534.
20. Schleicher, H., Lang, H. *Das Papier*, **1994**, *12*, p 765.
21. Selli, E.; Beltrame, P. L.; Testa, G.; Bonfatti, A. M., Rossi, E.; Seves, A. *Angew. Makromol. Chem.* **1998**, *257*, p 63
22. Adorjan, I.; Potthast, A.; Rosenau, T.; Sixta, H.; Kosma, P. *Cellulose* **2005**, *12(1)*, pp 51-57.
23. Rosenau, T.; Potthast, A.; Milacher, W.; Adorjan, I.; Hofinger, A.; Kosma, P. *Cellulose* **2005**, *12(2)*, pp 197-208.
24. Rosenau, T.; Potthast, A.; Hofinger, A.; Sixta, H.; Kosma, P. *Holzforschung* **2002**, *56 (2)*, pp 199-208.

25. Rosenau, T.; Potthast, A.; Sixta, H.; Kosma, P. *Progr. Polym. Sci.* **2001**, *26 (9)*, pp 1763-1837.
26. Klemm, D.; Schumann, D.; Udhardt, U.; Marsch, S. *Prog. Polym. Sci.* **2001**, *26*, p 1561.
27. Christian-Doppler Laboratory Vienna, unpublished results.
28. Lewin, M.; Epstein, J. A. *J. Polm. Sci.* **1962**, *58*,p 1023.
29. Gause, E., PhD thesis, University of Hamburg, Germany, 1993.
30. Gierer, J. *Holzforschung* **1997**, *51*, p 34.
31. Gratzl, J. S. *Das Papier* **1992**, *10A*, V1.
32. Katai, A.; Schuerch, C. *J. Polym. Sci.* **1966**, *4*, p 2683.
33. Godsay, M. P.; Pearce, E. M. *Proceedings of the TAPPI Oxygen Delignification Conference*, 1984, p 55.
34. Pan, G.; Chen, C. L.; Chang, H. M.; Gratzl, J. S. *Proceedings, The Ekamn Days,* 1981, *2*, p 132.
35. Katai, A.; Schuerch, C. *J. Polym. Sci.* **1966**, *4*, p 2683.
36. Kishimoto, T.; Nakatsubo, F. *Holzforschung* **1996**, *50*, p 372.
37. Lewin, M.; Mark, H. F. *Macromol. Symp.* **1997**, *118*, p 715.
38. Lewin, M.: Ettinger, A. *Cell. Chem. Technol.* **1969**, *3*, p 9.
39. Fischer, K.; Goldberg, W.; Wilke, M. *Lenz. Ber.* **1985**, *59*, p 32.
40. Takacs, E.; Wojnarovits, L.; Földvary, C.; Borsa, J.; Sajo, I. *Radiat. Phys. Chem.* **2000**, *57*, 399.
41. TAPPI method T-430 om-94 (Braidy); Zellcheming IV/8/70 (Schwalbe-Sieber).
42. Röhrling, J., PhD thesis, University of Agricultural Sciences, Vienna, Austria, 2002.
43. Geiger, E.; Wissler, A. *Helv. Chim. Acta* **1945**, *28*, p 1648.
44. Wennerblom, A. *Svensk Paperstidn.* **1961**, *14*, p 519.
45. Cyrot, J. *Chimie Analytique* **1957**, *39*, p 449.
46. Rehder, W.; Philipp, B.; Lang, H. *Das Papier* **1965**, *19(9)*, p 502.
47. Lewin, L., In *Methods in Carbohydrate Chemistry;* Whistler, R.L.; BeMiller, J.N., Eds., vol. 6, Academic Press: New York, London, 1972, p 76.
48. Ströle, U. *Makromol. Chem.* **1956**, *20*,p 19.
49. Szabolcs, O. *Das Papier* **1961**, *15*, p 41.
50. Strlic, M.; Pihlar, B. *Fresenius'J. Anal. Chem.* **1997**, *357*, p 670.
51. Nagel, G., Potthast, A., Rosenau, Th., Kosma, P., Sixta, H. *Lenz.Ber.,* **2005**, *84*, pp 27-35.
52. Saake, B., PhD thesis, University of Hamburg, Germnany, 1992.
53. Alfredsson, B.; Samuelson, O.; Sandstig, B. *Svensk Papperstidn.* **1963**, *18*, p 703.
54. Chang, P. S.; Robyt, J. F. *J. Carbohydr. Chem.* **1996**, *15(7)*, p 819.
55. B. Phillipp, W. Rehder, H. Lang, Zur Carboxylgruppenbestimmung in Chemie-zellstoffen, *Das Papier* 19 (1965) pp 1-9.

530

56. [1] X.-S. Chai, Q.X. Hou, J.Y. Zhu, S.-L. Chen, S. F. Wang, L. Lucia, Carboxyl Groups in Wood Fibers. 1. Determination of Carboxyl Groups by Headspace Gas Chromatography, *Ind. Eng. Chem. Res.* 42 (2003) pp 5440 – 5444, X.-S. Chai, Q.X. Hou, J.Y. Zhu, Carboxyl Groups in Wood Fibers. 2. The Fate of Carboxyl Groups during Alkaline Delignification and Its Application for Fiber Yield Prediction in Alkaline Pulping, *Ind. Eng. Chem. Res.* 42 (2003) pp 5445 – 5449

57. Putnam, E. S. *Tappi J.* **1964**, *47*, pp 549–554.

58. Husemann, E.; Weber, O. H. *J. Prakt. Chem.* **1942**, *159*, pp 334–342.

59. Wilson, K. *Svensk Papperstidn.* **1948**, *51*, pp 45–49.

60. Doering, H. *Das Papier* **1956**, *10*, pp 140–141.

61. Carried out under the supervision of the German Zellcheming Committee for pulp and paper analytics.

62. Röhrling, J.; Potthast, A.; Rosenau, T.; Lange, T.; Borgards, A.; Sixta, H.; Kosma, P. *Synlett* **2001**, *5*, pp 682-684.

63. Bohrn, R.; Potthast, A.; Rosenau, T.; Sixta, H.; Kosma, P. *Synlett* **2005**, *20*, pp 3087-3090.

64. Röhrling, J.; Potthast, A.; Rosenau, T.; Lange, T.; Borgards, A.; Sixta, H.; Kosma, P. *Biomacromolecules* **2002**, *3*, pp 969-975.

65. Röhrling, J.; Potthast, A.; Rosenau, T.; Lange, T.; Ebner, G.; Sixta, H.; Kosma, P. *Biomacromolecules* **2002**, *3*, pp 959-968.

66. Potthast, A.; Röhrling, J.; Rosenau, T.; Borgards, A.; Sixta, H.; Kosma, P. *Biomacromolecules* **2003**, *4 (3)*, pp 743-749.

67. Potthast, A.; Schiehser, S.; Rosenau, T.; Sixta, H.; Kosma, P. *Holzforschung* **2004**, *58*, pp 597-602.

68. Bohrn, R.; Potthast; Schiehser, S.; Rosenau, T.; Sixta, H.; Kosma, P. *Biomacromol.* **2006**, in press.

Chapter 33

Chemiluminometry of Cellulosic Materials

Matija Strlič[1], Drago Kočar[1], and Jana Kolar[2]

[1]Faculty of Chemistry and Chemical Technology, University of Ljubljana, Slovenia
[2]National and University Library, Ljubljana, Slovenia

The emission of light as a result of chemical reaction during degradation of polymers has been studied since the early 1960s.[1,2] Many polymers have been studied in depth so far, and chemiluminometry has entered industrial labs as a routine investigation technique. It is particularly attractive due to the absence of sample preparation, its non-invasiveness and simplicity of instrumentation. The data can be rapidly obtained, often in the early stages of oxidation, and the technique is complementary with other approaches. However, due to the often-encountered multitude of simultaneous chemiluminescent reactions, the interpretation of data is rarely straightforward.

Introduction

Chemiluminescence is light emission as a consequence of relaxation of a species excited in an elementary process of a chemical reaction. This elementary process proceeds with a rate, which is related to the kinetics of the overall reaction. In the commercially available chemiluminometers, measurements proceed using photomultipliers and photon counters. The signal I, in s^{-1}, is thus proportional to

$$I = G \cdot \phi \cdot \frac{d[\mathsf{L}]}{dt},$$

where [L] is the concentration of the chemical species giving rise to chemiluminescence, t is time, G is the geometric parameter (including the size and specific surface of the sample, absorption of light in the sample, etc.) and ϕ is the quantum efficiency with a typical value of ~10^{-9}. Since photomultipliers usually in use have maximum sensitivity in the interval ~300-650 nm, the reaction should also comply with the condition: $\Delta H \geq 185 - 400\,kJ\,mol^{-1}$.

As a rule, although not exclusively, light emission is a consequence of an exothermal process and is frequently observed during atmospheric oxidation of materials. A comprehensive review of chemiluminescence measurements and data evaluation during polymer oxidation was edited by Zlatkevich.[3] Only recently, research on natural polymers, including cellulose, also intensified.[4]

The commercially available luminometers for polymers and non-volatiles are produced by Tohoku Electronic Industrial Co., Japan,[5] and by the Polymer Institute of the Slovak Academy of Sciences, Slovakia.[6] Lately, a separate sampling unit has been constructed enabling us to measure chemiluminescence in atmosphere containing a pre-set fraction of water vapor.[7]

Samples are heated in a compartment flushed with nitrogen, oxygen, or mixtures thereof. The temperature interval usually in use is 30-250 °C. Sample holders can accommodate samples in solid form (films, foils, powders etc.) or non-volatile liquids. The amount needed is a few milligrams. Thick samples or granules can lead to an unwanted temperature gradient across the sample. Sample preparation is usually not needed, however, sample size, specific surface and other morphological properties also influence the measurements.[8]

Typically, isothermal, dynamic, and perturbation experiments are performed. During the latter, an experimental parameter is changed abruptly, e.g. atmosphere composition, sample irradiation, etc. In the following subchapters, we will briefly review the state-of-the-art.

Chemiluminescence of Cellulose in Inert Atmosphere

At elevated temperatures, chemiluminescence of cellulose can readily be observed even in inert atmosphere. This indicates that even reactions other than oxidation can lead to light emission. In cellulose, this was shown to be a consequence of the transglycosidation reaction.[9] It was also shown that the decrease of degree of polymerisation of cellulose during isothermal heating in a nitrogen atmosphere can be correlated with an integrated chemiluminescence signal at 180, 190 and 200 °C. The rate of transglycosidation is affected by the presence of alkali-earth metal carbonates:[10] the higher the electronegativity of the alkali-earth metal, the higher the rate constant of chain scission and chemiluminescence intensity also increases correspondingly. The hypothesis that a complex between a metal ion and the glycosidic oxygen is formed has thus been put forward. Additionally, this indicates that even trace quantities of extraneous material can have a strong effect on the chemiluminescence signal. This should be taken into account when studying real paper samples. Also, if a sample is not flushed well with inert atmosphere prior to an experiment, oxygen remaining in the fibrous structure has a pronounced effect.[10]

Dynamic experiments in inert atmosphere, during which the temperature is increased at some constant rate, are of considerable importance. Several different processes lead to light emission during these experiments (Figure 1).[9] Without a pre-treatment, a monotonously increasing signal is indicative of thermolysis (Figure 1C). If the sample is pre-oxidised, a peak appears with the maximum situated at approximately 130-150 °C (Figure 1A). The peak area was shown to correlate with peroxide content in pulp, determined titrimetrically. The area can be calculated by deconvolution of the signal, supposing two independent phenomena: decomposition of peroxides and thermolysis.

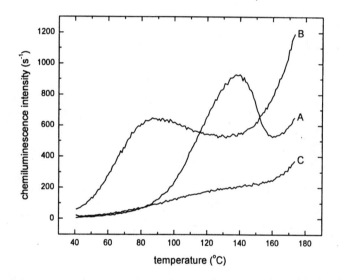

Figure 1. Dynamic chemiluminometric experiments in nitrogen atmosphere with sulphite bleached pulp: (A) pre-oxidised in oxygen atmosphere, 80 °C, 30 min; (B) irradiated under a 60-W incandescent light source at a 25-cm distance from sample; (C) without pre-treatment. Rate of temperature increase: 2.5 °C min⁻¹. (Reproduced with permission from reference 9. Copyright 2000 Elsevier.)

To explain the phenomenon leading to the peak with maximum at approximately 85 °C (curve B in Figure 1), it was proposed that recombination of charge-transfer complexes takes place after irradiation, supported by the low apparent activation energy of the process, 20.5 kJ mol[-1].[9] Other chemiluminescent phenomena at low temperatures were observed after plasma or laser treatments of paper.[11,12]

Chemiluminescence of Cellulose in Oxidative Atmosphere

Cellulose, being a heterochain polymer with a number of variously reactive hydroxyl groups, decomposes via the acid-catalysed hydrolytic mechanism.[13] The relative importance of this well-researched degradation pathway decreases with increasing pH of the macromolecular environment, while the importance of oxidative degradation increases.

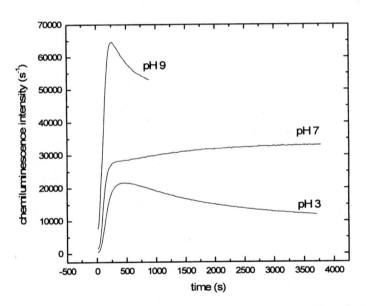

Figure 2. Isothermal chemiluminescence of cellulose samples (Whatman filter paper) impregnated with phosphate buffers of different pH as indicated. Conditions: 180 °C, O₂ atmosphere. (Reproduced with permission from reference 10. Copyright 2001 Elsevier.)

In samples of acidic character, oxidation and acid-catalysed hydrolysis can take place simultaneously and light emission is also observed, albeit less intensively than in alkaline ones (Figure 2). The relatively complex autoxidation degradation pathway is predominant in cellulose in a mildly alkaline environment[14,15] The reaction scheme in Figure 3 is thought to adequately describe the process of oxidation of organic polymers, including cellulose. The autoxidation scheme was originally used to explain autoxidation of simple hydrocarbons in solution, and its application to heterogeneous systems, such as atmospheric oxygen/cellulose, is not without risks. E.g., the addition of O_2 to $P^\bullet$ is a diffusion-controlled reaction,[16] with activation energy 0 kJ mol^{-1},[17] and at ambient conditions, other reactions of $P^\bullet$ are negligible. However, if diffusion of oxygen to reaction sites is impaired, e.g. due to slow diffusion in crystalline regions, the relative importance of other reactions of $P^\bullet$ may increase. Additionally, mobility of polymeric chains is lower in comparison to the mobility of low-molecular-weight compounds in solutions. Differences in mobility may easily lead to differences in rates of reactions; and instead of one rate of reaction we may well have to speak of a distribution of rates.

The shape of curves in Figure 3 is typical for polymers with a short length of autoxidation chains.[18] This conclusion is further supported by the generally low steady-state content of peroxides in cellulose samples during oxidative degradation.[19]

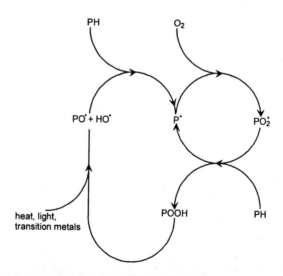

Figure 3. The Bolland-Gee autoxidation reaction scheme. Native cellulose polymer is denoted as PH.

Chemiluminometry was successfully used to study certain reactions taking place during cellulose autoxidation. Pre-oxidation of samples prior to a dynamic experiment in inert atmosphere gives rise to a chemiluminescence peak, the area

536

of which corresponds to the content of peroxides in the sample formed during the pre-oxidation treatment. This approach can be used for determination of the content of peroxides in samples of similar origin. Using pullulan samples of different molecular weights and thus different content of aldehyde end groups, the correlation between peroxide and aldehyde group content was obtained as shown in Figure 4. Chromatographic analyses of samples that were subjected to accelerated degradation at 80 °C, 65% RH, indicated a correlation between peroxide content and rate of chain scission.[20] Thus, the mechanisms of autoxidation initiation as discussed for simple carbohydrates[21,22] may be applied.

The build-up and the fate of peroxides during degradation is of primary importance in studies of oxidation. Using a chromatographic method, formation of hydroperoxides during accelerated ageing at 80 °C, 65% RH was followed and shown to be fairly rapid, while the content was found to be extremely low.[19] Using chemiluminometry, the formation or decomposition of peroxides can also be easily followed. At room temperature, peroxides formed during oxidation for 60 min in an oxygen atmosphere at 80 °C degrade quickly, with a steady state attained in 100 min, apparently following first-order kinetics with a calculated activation energy $E_a = 75$ kJ mol^{-1}.

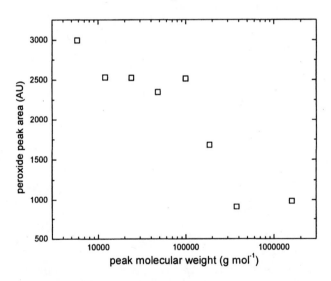

Figure 4. Peroxide peak areas for pre-oxidised (60 min, O₂, 100 °C) pullulan samples of different initial peak molecular weights. (Reproduced with permission from reference 20. Copyright 2003 Elsevier.)

Peroxides are also thought to be the species that give rise to elevated chemiluminescence during isothermal experiments in oxygen atmosphere (Figure 1). By switching the atmosphere between oxidative and inert (perturbation experiments), several additional features of the oxidation process are revealed (Figure 5).

An experiment in oxidative atmosphere starts with an initial peak. If after a period of heating in oxygen, the atmosphere is switched to nitrogen (at 3600 s in Figure 5), chemiluminescence decreases rapidly. Since the sample is heated continuously, peroxides, accumulated in oxidative atmosphere, decompose rapidly. If the decreasing chemiluminescence signal is treated as a first order process and the rate is calculated from the decaying chemiluminescence curves at different temperatures, we obtain the activation energy E_a = 76 kJ mol^{-1}, the same value as calculated for first-order decomposition of peroxides in dynamic experiments.

Another experiment reveals the same value, the so-called pseudo-stationary experiment. In semilogarithmic coordinates in Figure 6, the Arrhenius relationship between logarithmic values of chemiluminescence intensity and temperature is revealed. From such simple experiments, we can calculate the activation energy to be E_a = 78 kJ mol^{-1}. The three types of experiments yielding the same value of E_a, indicate that in oxidative atmosphere the emission of light is a consequence of peroxides decomposing in a first-order manner. According to the literature,[21,22] the primary decomposition products are a carboxylic acid,

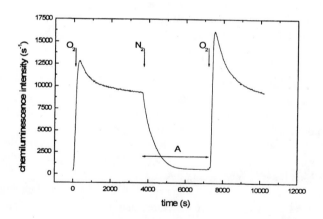

Figure 5. Isothermal chemiluminescence of Whatman filter paper at 180 °C. The atmosphere surrounding the samples was exchanged rapidly, as indicated. The period of heating in nitrogen ("A") was 3600 s.

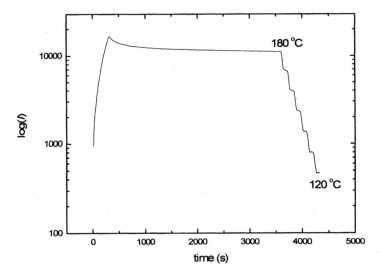

Figure 6. A pseudo-stationary chemiluminescence experiment with Whatman filter paper in oxygen atmosphere. After 3600 s, the temperature was decreased to 120 °C in a stepwise manner.

and a hydroxyl radical or hydrogen peroxide. Due to its higher mobility, the latter can decompose via Fenton-like mechanism, especially in an environment rich in transition metals. Hydroxyl radicals were already shown to accompany chain scission during oxidative degradation of cellulose. For their detection, however, techniques other than chemiluminometry were used.[23]

In the atmosphere-switch experiments, we can vary the time of sample heating in inert atmosphere after the switch from oxygen to nitrogen ("A" in Figure 5). During this period, transglycosidation will be the predominant mode of degradation. A variety of elimination and isomerisation reactions also take place, many of which lead to formation of carbonyls. If the atmosphere is switched to oxygen after a period of heating in nitrogen, the peak appearing after introduction of oxygen shows variation in height depending on the time of heating in nitrogen. The height of this peak correlates with the content of carbonyls which have formed during heating in nitrogen (Figure 7).

On the basis of chemiluminometric experiments complementing other experimental techniques, we can conclude that the unstable carbonyl groups initiate oxidation processes. If present in excess to the steady-state content at the chosen temperature, this will be demonstrated as an initial peak in isothermal experiments. If their content is low, chain scission processes will initially dominate until the steady state content is reached. We can confirm this by using pullulan samples of different molecular weights (and thus different contents of aldehyde end groups), as in Figure 8. The sharp initial increase of

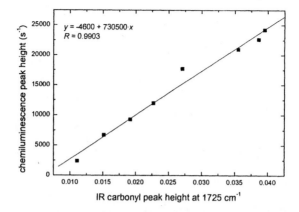

$$y = -4600 + 730500 \cdot x$$
$$R = 0.9903$$

Figure 7. Correlation of peak height after readmission of oxygen in atmosphere exchange experiments (cf. Fig. 5), with carbonyl peak height in FTIR spectra of samples after different periods of heating in nitrogen.

chemiluminescence in the low-molecular-weight sample indicates a high content of aldehydes. This peak is absent in the high-molecular-weight pullulan, however, the broad peak at long reaction times indicates that chain-scission leads to an increase and then to a decrease in the content of carbonyls. In the low-molecular-weight sample only the decrease is evident.

In dynamic experiments in oxygen atmosphere the rate of heating should allow sample equilibration, <10 °C min^{-1} is usually used. As an example, chemiluminescence intensity is plotted against inverse temperature in Figure 9 for two samples with different initial water content. From the curve slope, activation energy for the process leading to light emission can be calculated. The slopes of both curves are similar only at low temperatures, with E_a 51-56 kJ mol^{-1} indicating an effect of drying at elevated temperatures. The decreased emission intensity of the moist sample indicates a quenching role of water, to be discussed later.

Not only a single temperature scan, but several, e.g. in temperature cycling experiments, are also in use.[24] It has been shown that during the first temperature cycle, the material behaves differently as at each temperature it is still practically in its virgin state. However, the curves in Figure 9 are not linear and their shapes possibly reflect many phenomena, especially at higher temperatures. Various modelling approaches have been developed to take the curvature into account e.g. by Rychlý et al.,[24,25] who proposed a faster and a slower degradation process, an ionic and a radical one. Rates of both processes can be estimated using the modelling approach, a discussion of which is beyond the scope of this review.

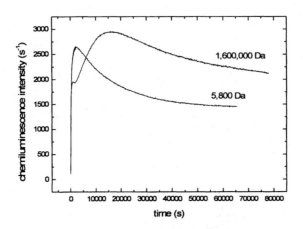

Figure 8. Isothermal chemiluminometric experiments with pullulan samples of two different molecular weights, as indicated. Conditions: 180 °C, O₂.

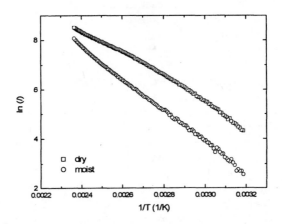

Figure 9. Dynamic experiment in oxygen atmosphere (sulphate pulp, 2.5 °C min⁻¹). Prior to the experiment, the two samples were kept either in dry conditions or at 100% RH, as indicated.

Chemiluminescence of Cellulose in Humid Oxidative Environments

In the first paper published on chemiluminometry of paper in 1979, Kelly et al.[26] cycled moist and dry air above a sample at 70 °C. They observed a repeatable behavior: on each switch to moist atmosphere a sharp CL peak appeared and the signal then dropped below the level of CL in dry atmosphere. The hypothesis was put forward that water-absorption-induced swelling leads to bond scission, and thus to enhanced oxidation. Contemporary chemiluminometers of higher sensitivity allowing better control of experimental parameters, allow us to study the effect of water during cellulose oxidation much more effectively.[27]

These studies are particularly important since oxidative degradation of cellulose at room temperature is often more interesting than at higher temperatures. In a hydrophilic macromolecular material, such as cellulose, water is expected to play an important role during degradation at T < 100 °C. During acid-catalysed hydrolysis, an increased amount of water in the material will lead to a higher rate of degradation. In conditions, during which oxidation prevails, the role of water is less clear, and studies of chemiluminescence provided a significant input.

The experiment in Figure 10 shows an isothermal experiment, during which the content of water vapor in oxygen was varied at 90 °C.[27] Typical for such experiments is an immediate increase of light emission when water is admitted, which is followed by a relatively fast decay. The increase can be correlated with relative humidity introduced after the drying period, which indicates a chemical role of water in the reactions leading to chemiluminescence. After the initial peak, the signal decreases to a level lower than in dry atmosphere, indicating the quenching effect of water discussed earlier. Higher mobility of cellulose chains could also lead to relaxation pathways other than light emission.

Immediately after readmission of dry oxygen, the detected light signal increases again, the increase depending on the prior humidity of the atmosphere surrounding the sample. In an earlier study,[10] the set of reactions

$$O_2^{\bullet-} + H_2O \rightarrow HO_2^{\bullet} + HO^-,$$
$$2HO_2^{\bullet} \rightarrow O_2^* + HOOH,$$
$$O_2^* \rightarrow O_2 + h\nu,$$

was proposed to explain the chemical role of water during oxidative degradation

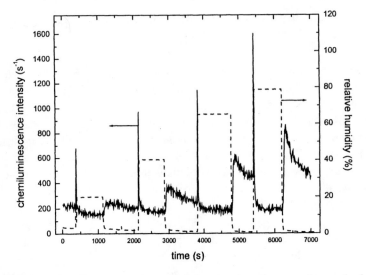

Figure 10. A chemiluminometric experiment in oxygen atmosphere of relative humidity as indicated, using four layers of paper made of cotton pulp at 90 °C. (Reproduced with permission from reference 27. Copyright 2005 Elsevier.)

of cellulose. The resulting singlet oxygen emission can be observed as dimol emission with a maximum at 634 nm. The topic was further investigated[27] using superoxide scavenger 4,5-dihydroxy-1,3-benzene disulphonic acid, disodium salt, which suppressed chemiluminescence. The presence of superoxide in quantities high enough to enter the above set of reactions is in agreement with the observed low reactivity of superoxide against sugars.[28]

Chemiluminescence of Cellulose Following Irradiation with Light

Irradiation of cellulose with visible light has been shown to lead to intense chemiluminescence after irradiation is discontinued (Figure 1B). It was shown that this phenomenon depends on irradiation time, sample type, and on the composition of the atmosphere during irradiation.[9] The phenomenon was studied earlier,[29] yet evidence on its dependence on oxygen present in the atmosphere surrounding the sample was presented by Strlič et al.[9] The reactive centers sensitive to light are supposed to be carbonyl or glycosidic oxygen while oxygen participates in the initiation process of light emission. The hypothesis of charge-transfer complex formation induced by light and subsequently destroyed by elevated temperature was put forward.[10] Both atmospheric humidity and heating in oxygen atmosphere lead to a fast decay of this luminescing species.

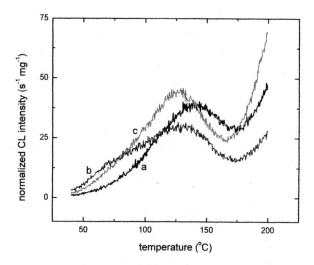

Figure 11. Dynamic chemiluminometric experiments in nitrogen atmosphere for a – untreated; b –Xenotest-irradiated sample for 10 min; and c – Xenotest-irradiated (10 min) and subsequently stored 580 min at 23 °C in darkness. (Reproduced with permission from reference 30. Copyright 2005 Elsevier.)

Changes in peroxide content during irradiation of cellulose in Xenotest are easily followed using dynamic experiments as in Figure 1B.[30] The non-irradiated sample exhibits a maximum at ca. 130-140 °C (Figure 11A), corresponding to peroxide content.[9] After 10 min of irradiation, peroxides are already partly decomposed (CL emission at 130 °C in curve B is considerably lower than in curve A). An enhanced radical activity thus induced leads to an elevated peroxide signal, and consequently degradation, upon subsequent storage in darkness (curve C).[30]

Conclusions

As an experimental technique, chemiluminometry complements the usual thermo- and photo-degradation studies. It represents a sensitive technique and can be used during very early stages of degradation.

However, the provided information is complex due to the many simultaneous processes leading to chemiluminescence. On the basis of modelling and through various experimental techniques, the information on the identity of luminescing species and on the rates of processes leading to luminescence can be extracted. In this way, new knowledge was gained on the role of aldehyde groups, peroxides, water and superoxide during oxidative degradation of cellulose.

544

Acknowledgements

The authors gratefully acknowledge the support of the European Community, 5th Framework Energy, Environment and Sustainable Development programme, contract No. EVK4-CT-2000-00038 (PAPYLUM). The work is the sole responsibility of the authors and does not represent the opinion of the Community. The Community is not responsible for any use that might be made of the data appearing herein. Further financial support from Ministry of Higher Education, Science and Sport of the Republic of Slovenia (programme no. P1-0153) is gratefully acknowledged.

References

1. Ashby, G.E. *J. Polym. Sci.*, **1961**, *50*, pp 99-106.
2. Schard, M.P. ; Russell, C.A. **1964**, *8*, pp 985-995.
3. Zlatkevich, L. (Ed.), *Luminescence Techniques in Solid-State Polymer Research*, Marcel Dekker, New York, 1989.
4. Strlič, M.; Kolar, J. (Eds,), *Ageing and Stabilisation of Paper*, National and University Library, Ljubljana, 2005.
5. http://www.tohokueic.com/cl_analyzers_index.html, accessed 25/10/2005.
6. http://www.lumipol.com, accessed 25/10/2005.
7. Constructed in the frame of the 5th Framework Programme Papylum project, financed by the EC (EVK4-CT-2000-00038), http://papylum.uni-lj.si, accessed 10/12/2005.
8. Kočar, D.; Pedersoli, Jr., J.L.; Strlič, M.; Kolar, J.; Rychlý, J.; Matisová-Rychlá, L. *Polym. Degrad. Stab.*, **2004**, *86*, pp 269-274.
9. Strlič, M.; Kolar, J.; Pihlar, B.; Rychlý, J.; Matisová-Rychlá, L. *Eur. Polym. J.*, **2000**, *36*, pp 2351-2358.
10. Strlič, M.; Kolar, J.; Pihlar, B.; Rychlý, J.; Matisová-Rychlá, L. *Polym. Degrad. Stab.*, **2001**, *72*, pp 157-162.
11. Rudolph, P.; Ligterink, F.J.; Pedersoli, Jr., J.L.; Van Bommel, M.; Bos, J.; Aziz, H.A.; Havermans, J.B.G.A.; Scholten, H.; Schipper, D.; Kautek, W. *Appl. Phys. A*, **2004**, *79*, pp 181-186.
12. Strlič, M.; Šelih, V.S.; Kolar, J.; Kočar, D.; Pihlar, B.; Ostrowski, R.; Marczak, J.; Strzelec, M.; Marinček, M.; Vuorinen, T.; Johansson, L.S *Appl. Phys. A*, **2005**, *81*, pp 943-951.
13. Nevell, T.P.;. Zeronian, S.H *Cellulose Chemistry and Its Applications*, Ellis Horwood, Chichester, 1985.
14. Kolar, J. *Restaurator*, **1997**, *18*, pp 163-176.
15. Kolar, J.; Strlič, M.; Novak, G.; Pihlar, B. J. *Pulp Pap. Sci.*, **1998**, *24*, pp 89-94.

16. Audouin, L.; Langlois, V.; Verdu, J.; de Bruin, J.C.M. *J. Mat. Sci.*, **1994**, *29*, pp 569-583.

17. Reich, L.; Stivala, S.S. *Autoxidation of Hydrocarbons and Polyolefins; Kinetics and Mechanisms*, M. Dekker, New York, 1969.

18. Rychlý, J.; Rychlá, L. *Chemiluminescence from polymers*, in: *Ageing and Stabilisation of Paper*, Strlič, M.; Kolar, J., (Eds.) , National and University Library, Ljubljana, 2005.

19. Kočar, D.; Strlič, M.; Kolar, J.; Pihlar, B. *Anal. Bioanal. Chem.*, **2002**, *374*, pp 1218-1222.

20. M. Strlič, P.J.; Kočar, D.; Kolar, J.; Rychlý, J.; Pihlar, B. *Carbohydr. Polym.*, **2003**, *54*, pp 221-228.

21. Thornalley, A. Stern, Carbohydr. Res., **1984**, *134*, pp 191-204.

22. Arts, S.J.H.F.; Mombarg, E.J.M.; van Bekkum, H.;   Sheldon, R.A. *Synthesis*, **1997**, pp 597-613.

23. J. Kolar, M. Strlič, B. Pihlar, Anal. Chim. Acta., **2001**, *431*, pp 313-319.

24. Rychlý, J.; Matisova-Rychlá, L.;Strlič, M. *Polym. Int.*, **2000**, *49*, pp 981-986.

25. Rychlý, J.; Matisová-Rychlá, L.; Strlič, M.; Kolar, J. *Polym. Degrad. Stab.*, **2002**, *78*, pp 357-367.

26. G.B. Kelly, J.C. Williams, G.D. Mendenhall, C.A. Ogle, in: G.B. Kelly, R.K. Eby (Eds.), *Durability of Macromolecular Meterials*, Advances in Chemistry Series, ACS, Washington, **1979**, *95*, pp 117-125.

27. Kočar, D.; Strlič, M.; Kolar, J.; Rychlý, J.; Pihlar, B.; Rychlá, L. *Polym. Degrad. Stab.*, **2005**, *88*, pp 407-414.

28. Schuchmann, M.N.; von Sonntag, C. *Naturforsch. B.*, **1978**, *33*, pp 329-331.

29. Mendenhall, G.D.; Agarwal, H.K. *J. Appl. Polym. Sci.*, **1987**, *33*, pp 1259-1274.

30. Malešič, J.; Kolar, J.; Strlič, M.; Kočar, D.; Fromageot, D.; Lemaire, J. Haillant, O. *Polym. Degrad. Stab.*, **2005**, *89*, pp 64-69.

Indexes

Author Index

Subject Index

582